Functional Polyurethanes—In Memory of Prof. József Karger-Kocsis

Functional Polyurethanes—In Memory of Prof. József Karger-Kocsis

Special Issue Editors

Sándor Kéki
József Karger-Kocsis

MDPI • Basel • Beijing • Wuhan • Barcelona • Belgrade

Special Issue Editors

Sándor Kéki
Department of
Applied Chemistry,
University of Debrecen
Hungary

József Karger-Kocsis
Department of Polymer
Engineering, Faculty of
Mechanical Engineering,
Budapest University of
Technology and Economics
Hungary

Editorial Office
MDPI
St. Alban-Anlage 66
4052 Basel, Switzerland

This is a reprint of articles from the Special Issue published online in the open access journal *Polymers* (ISSN 2073-4360) from 2018 to 2020 (available at: https://www.mdpi.com/journal/polymers/special_issues/functional_polyurethanes).

For citation purposes, cite each article independently as indicated on the article page online and as indicated below:

LastName, A.A.; LastName, B.B.; LastName, C.C. Article Title. *Journal Name* **Year**, *Article Number*, Page Range.

ISBN 978-3-03928-494-8 (Pbk)
ISBN 978-3-03928-495-5 (PDF)

Contents

About the Special Issue Editors

Sándor Kéki was graduated as a chemist (MSc) in 1989 and obtained his PhD in Chemistry in 1996 from the University of Debrecen (formerly Lajos Kossuth University). He also received the degree of Doctor of Science (DSc) from the Hungarian Academy of Sciences (HAS) in 2008. His research interests focus on the synthesis and characterization of smart polymers and low molecular weight "intelligent" compounds including e.g. shape memory polymers and solvatochromic light emitting materials. He is the coauthor of more than 170 peer-reviewed scientific papers and 3 book chapters. He is currently a full professor and the head of the Department of Applied Chemistry at the University of Debrecen.

József Karger-Kocsis served as a full professor at the Budapest University of Technology and Economics and was the Editor-in-Chief of the internationally respected polymer journal *Express Polymer Letters*. Professor Karger-Kocsis's research interest covered a wide range of scientific and practical areas, including the improvement of the mechanical properties of synthetic and natural polymers and their composites, development of new techniques and methodologies for the investigation of polymer composites, and studying the interaction between polymer matrices and reinforcing materials. Professor Karger-Kocsis also responded with great enthusiasm to the challenges of various environmental issues by focusing on the recycling of plastic and rubber wastes, as well as the application of renewable raw materials in the synthesis of polymers. He obtained excellent seminal results in the field of smart and functional materials. His outstanding scientific achievements are marked by his 4 books, 50 book chapters, approximately 500 scientific papers published in peer-reviewed journals, and more than 20,000 citations with a Hirsh-index of 72.

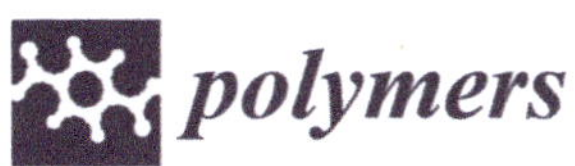 *polymers*

Editorial

Functional Polyurethanes—In Memory of Prof. József Karger-Kocsis

Sándor Kéki

Department of Applied Chemistry, Faculty of Science and Technology, University of Debrecen, H-4032 Debrecen, Egyetem tér 1., Hungary; keki.sandor@science.unideb.hu

Received: 4 February 2020; Accepted: 10 February 2020; Published: 13 February 2020

In the era of our "plastic age", polyurethanes (PUs) represent one of the most versatile polymers that are produced by the nucleophilic addition reaction between isocyanates and various polyols [1]. The broad range and excellent mechanical and chemical properties of PUs resulting from the huge number of possible variations in the types of isocyanates and polyols have made this fascinating group of polymers very useful and valuable materials that can now be found in almost all facets of our every day life [2]. Furthermore, depending on the chemical formulation, their chemical, mechanical, and biological properties can be designed and tailored for specific applications. The area of use can be further broadened by incorporation of additional polymer segments in and onto the polymer backbone to produce PUs with additional (e.g., shape-memory and/or self-healing) properties [3–6]. Thus, research interest in the preparation and characterization of various PUs is steadily and rapidly growing, as demonstrated by the number of research articles published per year in this field (Figure 1).

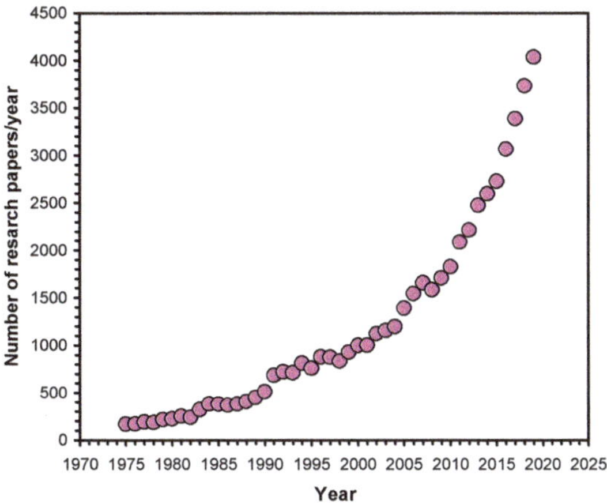

Figure 1. Variation of the number of research articles per year from 1975 to 2019. (Web of Science, 27. 01. 2020, search for topic: polyurethane, document type: article).

Indeed, as shown in Figure 1, the number of articles published annually increases exponentially, as indicated by an ascendant interest in PUs. Following this trend, this Special Issue focused on novel design strategies, and the synthesis and characterization of PUs with special and/or multifunctional properties. This Issue consists of 22 original research papers addressing various aspects of PUs, ranging from the synthesis to their applications, including some reports on high-level theoretical calculations supporting the experimental results.

Eight articles in this Issue aim at the synthesis of various PUs that are potentially applicable for biomedical purposes. Innovative research demonstrates an intriguing method for the synthesis of surface-grafted segmented poly(ester-urethane) by poly(ethylene glycol) (PEG) utilizing allophanate reaction with 1,6-hexanediisocynate and Michael addition to attach PEG of varying molecular weights onto the surface [7]. According to the in vitro tests, the resulting surface-grafted PUs showed elevated resistance to protein and platelet adsorptions and increased their number of potential uses in biomedical applications. Another report in this Issue focuses on the improvement of the hemocombatibility of a bioderadable poly(ether-ester-urethane) (PEEU) by blending it with a copolymer containing phosphorylcholine pendant groups [8]. The authors revealed that both the mechanical properties and the hemocompatibility of the blend met the requirements for potential blood-contacting applications. A subsequent paper from the same research group demonstrated that PEEUs synthesized from poly(ether-ester) and diurethane diisocyanate exhibit shape memory properties [9]. These materials are believed to be capable of fabricating medical devices with shape recovery near body temperature. Nagy et al. designed and prepared PUs with different crosslink densities containing biodegradable poly(ε-caprolactone)-diol, 1,6-hexamethylene diisocyanate (HDI), and sucrose as a chain-extender/crosslinker [10]. It was shown that the crosslink densities, and thus the mechanical properties, could be varied according to the amount of sucrose added to the reaction mixture. The PUs synthesized by the method proposed by the authors of this article are expected to be promising materials for making scaffolds for tissue engineering.

Injuries are almost unavoidable in our lives, and hence it is essential to protect wounds from infection and to use implants in case of serious injury. The report by A. Worsley et al. proposes an elegant approach to avoiding infection of wounds using polyhexamethylene biguanide (PHMB) embedded in PU nanofibrous membranes produced by electrospun. It was shown that after an initial burst, prolonged release of the PHMB from the polymer matrix with excellent antimicrobial effect could be achieved [11]. The stability of NanoShort titanium dental implants in PU foam sheets was evaluated in vitro in terms of foam densities and thicknesses, and high-level protection ability of PU foams was found [12].

New biocompatible PUs based on isorbide were prepared form poly(propylene glycol) and D-isosorbide aliphatic isocyanates such as isophorone diisocyanate. In vitro experiments with these PUs displayed low cytotoxicity towards HaCaT human skin cells [13]. Fluorescent labeling and magnetic response play an invaluable role in many areas of chemistry, biology, and materials science, including biomedicine, as shown by the authors of [14]. B. Li et al. [15] combined fluorescent and magnetic properties into biocompatible Janus particles [16] composed of fluorescent polyurethane and magnetic nano-Fe_3O_4. The authors were able to control both the core/shell and the Janus structure, thus making the rational design of composite nanoparticles for biomedical application possible.

Knowledge of the structure–property relationships is of great importance in designing new materials with tailored properties. The following seven papers are devoted to the presentations of structural modifications of PUs and some theoretical aspects of urethane forming reactions. An interesting paper is reported by Brzeska et al. This group prepared linear and branched polyurethanes containing polyhydroxybutyrate with tunable properties. The resulting PUs proved to be capable of controlling the water and oil sorptions through variations of the poly([R,S]-3-hydroxybutyrate) amount incorporated into the PUs [17]. Novel PUs containing fluorinated chain extenders were synthesized and investigated by J-W. Li et al. [18]. They demonstrated that introducing 1H,1H,10H,10H-perfluor-1,10-decanediol chain extender enhances the rigidity of the PU films by forming hydrogen bonds between the fluorinated moieties and the NH of the urethane bonds. In another paper, an intriguing approach to producing thermoplastic dynamic vulcanizates (TDVs) based on PUs is presented by A. Kohári et al. [19]. The authors dispersed rubbers such as NBR, XNBR, and ENR in "in situ"-generated thermoplastic PUs to obtain PU based TDVs. It was shown that these TDVs possess higher tensile strengths and a slightly lower elongation at break than commercial ones. Another way to improve the mechanical properties of PUs is to employ different kinds of modifiers. A. Strakowska et al. [20] proposes polyhedral oligomeric silsesquioxanes (POSSs) to enhance the properties of rigid PU foams. They report that POSSs, especially with amino and hydroxyl groups, were capable of reinforcing the rigid PU foams and both the morphology and the hydrophobicity of the foams could be varied with added POSS modifiers. The next theoretical study by D. Niedziela et al. [21] was motived by the experimental results obtained on the spatial distribution of solid foam, who, of course, have a practical interest in foam-forming industrial formulations. They set up a detailed mathematical model that successfully described the spatial inhomogeneity in the expanding PU-foams. The following paper by W. Cheikh et al. [22] focuses on the alcohol-isocyanate reaction, which is the base reaction of urethane formation. In this report, the authors propose a two-step mechanism for urethane formation, including two reaction routes: one is catalyzed by alcohol and the other by isocyanate. The two possible reaction paths were also confirmed by theoretical calculations employing a fourth generation Gaussian thermochemistry method combined with SDM (Solvent Model Density) model. In a subsequent publication from the same research group [23], a report on the ab initio study of the reaction leading to the formation of 4,4'-methylene diphenyl diisocyanate (4,4'MDI) from aniline and formaldehyde is also given.

In the next seven papers, some environmental aspects of PUs are emphasized. In recent years, due to the depletion of fossils in one hand and the desire to reduce the ecological footprint of plastics produced worldwide on the other [24], much effort has been devoted to manufacturing plastics from renewable resources. J.-W. Li et al. [25] prepared environmentally friendly PUs from castor-oil and carbon black. Carbon black served as a promoter in these PU formulations to facilitate microphase separation in order to enhance the mechanical properties. Evening primrose oil as a potential alternative to petrochemical polyols used in PU foams was demonstrated by J. Paciorek-Sadowska et al. [26]. The authors synthesized bio-based rigid PU foams with lower density, thermal conductivity, water sorption, and higher content of closed shell as compared to their petrochemical-based counterpart. W. Zhou et al. [27] revealed a nice example of how to combine an additive and a bio-based polyol to obtain flame-retardant rigid polyurethane foam. They synthesized phosphorous- and silicon-containing tung oil-based polyols by the reaction of the flame-retarding compounds with epoxidized tung oil, which were thereafter incorporated into PU foams. The resulting rigid polyurethane foams were proved to have enhanced thermal stability and flame retardancy. S. M. Choi et al. [28] developed a one-pot environmentally friendly approach to obtain a cellulose nanoparticles/waterborne PU nanocomposite with increased biodegradability and mechanical properties, while a contribution by B. Necasová et al. offers [29], in addition to the conventional ones (mechanical and chemical), and a new surface treatment method of woods called Multihollow Surface Dielectric Barrier Discharge (MHSDBD) plasma to fabricate wood/plastic composites (WPCs). The authors evaluated the surface-modification methods on PVC- and PE-based WPCs, and it was concluded that plasma treatment (MHSDBD) was the most effective one to produce WPCs with appropriate mechanical properties. The accumulation of heavy metal ions including, e.g., Cd^{2+}, Cu^{2+}, and Pb^{2+} ions, in industrial wastewaters, poses severe environmental problems [30]; therefore, efforts to remove and recover them are highly welcome. D. Xue et al. [31] proposed a new adsorbent with high adsorption efficiency, namely, dithiocarbamate-grafted PU/polyethyleneimine-polidopamine graphene-base PU composite for quick removal and recovery of Cd^{2+}, Cu^{2+}, and Pb^{2+} ions, from wastewaters. R. Yu et al. [32] used thermoplastic PUs (TPUs) to modify the properties of asphalt. They found that by adding TPUs to asphalt, not only the physical but also the chemical properties of the modified asphalt could be altered and the aging resistance could also be improved with respect to the base asphalt.

At the end of this section, I would like to recall that guest editing of this Issue was started by Prof. József Karger-Kocsis but unfortunately fate intervened and Prof. Karger-Kocsis passed away at the age of 68 on 13 December, 2018. I would like to salute the outstanding scientific achievements he made during his academic life. Accordingly, the editorial team has decided to dedicate this Special Issue to the Memory of Prof. József Karger-Kocsis.

Finally, I would like to thank all the authors and reviewers for their invaluable contributions to this Special Issue, and I hope that the scientific results presented here and in this Issue will stimulate further ideas and research interests in this prosperous field.

References

1. Caraculacu, A.A.; Soceri, S. Isocyanates in polyaddition processes. Structure and reaction mechanisms. *Prog. Polym. Sci.* **2001**, *26*, 799–851. [CrossRef]
2. Krol, P. Synthesis methods, chemical structures and phase structures of linear polyurethanes. Properties and applications of linear polyurethanes in polyurethane elastomers, copolymers and ionomers. *Prog. Mat. Sci.* **2007**, *52*, 915–1015. [CrossRef]
3. Hager, M.D.; Bode, S.; Weber, C.; Schubert, U.S. Shape memory polymers: Past, present and future developments. *Prog. Polym. Sci.* **2015**, *49*, 3–33. [CrossRef]
4. Ratna, D.; Karger-Kocsis, J. Recent advances in shape memory polymers and composites: A review. *J. Mat. Sci.* **2008**, *43*, 254–269. [CrossRef]
5. Karger-Kocsis, J.; Kéki, S. Biodegradable polyester-based shape memory polymers: Concepts of (supra)molecular architecturing. *Exp. Polym. Lett.* **2014**, *8*, 397–412. [CrossRef]
6. Syrett, J.A.; Becer, C.R.; Haddleton, D.M. Self-healing and self-mendable polymers. *Polym. Chem.* **2010**, *1*, 978–987. [CrossRef]
7. Liu, L.; Gao, Y.; Zhao, J.; Yuan, L.; Li, C.; Liu, Z.; Hou, Z. A Mild Method for Surface-Grafting PEG Onto Segmented Poly(Ester-Urethane) Film with High Grafting Density for Biomedical Purpose. *Polymers* **2018**, *10*, 1125. [CrossRef]
8. Zhang, J.; Yang, B.; Jia, Q.; Xiao, M.; Hou, Z. Preparation, Physicochemical Properties, and Hemocompatibility of the Composites Based on Biodegradable Poly(Ether-Ester-Urethane) and Phosphorylcholine-Containing Copolymer. *Polymers* **2019**, *11*, 860. [CrossRef]
9. Xiao, M.; Zhang, N.; Zhuang, J.; Sun, Y.; Ren, F.; Zhang, W.; Hou, Z. Degradable Poly(ether-ester-urethane)s Based on Well-Defined Aliphatic Diurethane Diisocyanate with Excellent Shape Recovery Properties at Body Temperature for Biomedical Application. *Polymers* **2019**, *11*, 1002. [CrossRef]
10. Nagy, L.; Nagy, M.; Vadkerti, B.; Daróczi, L.; Deák, G.; Zsuga, M.; Kéki, S. Designed Polyurethanes for Potential Biomedical and Pharmaceutical Applications: Novel Synthetic Strategy for Preparing Sucrose Containing Biocompatible and Biodegradable Polyurethane Networks. *Polymers* **2019**, *11*, 825. [CrossRef]
11. Worsley, A.; Vassileva, K.; Tsui, J.; Song, W.; Good, L. Polyhexamethylene Biguanide:Polyurethane Blend Nanofibrous Membranes for Wound Infection Control. *Polymers* **2019**, *11*, 915. [CrossRef] [PubMed]
12. Comuzzi, L.; Iezzi, G.; Piattelli, A.; Tumedei, M. An In Vitro Evaluation, on Polyurethane Foam Sheets, of the Insertion Torque (IT) Values, Pull-Out Torque Values, and Resonance Frequency Analysis (RFA) of NanoShort Dental Implants. *Polymers* **2019**, *11*, 1020. [CrossRef] [PubMed]
13. Gregorí Valdés, B.S.; Gomes, C.S.B.; Gomes, P.T.; Ascenso, J.R.; Diogo, H.P.; Gonçalves, L.M.; Galhano dos Santos, R.; Ribeiro, H.M.; Bordado, J.C. Synthesis and Characterization of Isosorbide-Based Polyurethanes Exhibiting Low Cytotoxicity Towards HaCaT Human Skin Cells. *Polymers* **2018**, *10*, 1170. [CrossRef] [PubMed]
14. Bigall, N.C.; Parak, W.J.; Dorfs, D. Fluorescent, magnetic and plasmonic-Hybrid multifunctional colloidal nano objects. *Nano Today* **2012**, *7*, 282–296. [CrossRef]
15. Li, B.; Shao, W.; Wang, Y.; Xiao, D.; Xiong, Y.; Ye, H.; Zhou, Q.; Jin, Q. Synthesis and Morphological Control of Biocompatible Fluorescent/Magnetic Janus Nanoparticles Based on the Self-Assembly of Fluorescent Polyurethane and Fe_3O_4 Nanoparticles. *Polymers* **2019**, *11*, 272. [CrossRef]
16. Le, T.C.; Zhai, J.L.; Chiu, W.H.; Tran, P.A.; Tran, N. Janus particles: Recent advances in the biomedical applications. *Int. J. Nanomed.* **2019**, *14*, 6749–6777. [CrossRef]
17. Brzeska, J.; Elert, A.M.; Morawska, M.; Sikorska, W.; Kowalczuk, M.; Rutkowska, M. Branched Polyurethanes Based on Synthetic Polyhydroxybutyrate with Tunable Structure and Properties. *Polymers* **2018**, *10*, 826. [CrossRef]
18. Li, J.-W.; Lee, H.-T.; Tsai, H.-A.; Suen, M.-C.; Chiu, C.-W. Synthesis and Properties of Novel Polyurethanes Containing Long-Segment Fluorinated Chain Extenders. *Polymers* **2018**, *10*, 1292. [CrossRef]
19. Kohári, A.; Halász, I.Z.; Bárány, T. Thermoplastic Dynamic Vulcanizates with In Situ Synthesized Segmented Polyurethane Matrix. *Polymers* **2019**, *11*, 1663. [CrossRef]
20. Strąkowska, A.; Członka, S.; Strzelec, K. POSS Compounds as Modifiers for Rigid Polyurethane Foams (Composites). *Polymers* **2019**, *11*, 1092. [CrossRef]

21. Niedziela, D.; Ireka, I.E.; Steiner, K. Computational Analysis of Nonuniform Expansion in Polyurethane Foams. *Polymers* **2019**, *11*, 100. [CrossRef] [PubMed]

22. Cheikh, W.; Rózsa, Z.B.; Camacho López, C.O.; Mizsey, P.; Viskolcz, B.; Szőri, M.; Fejes, Z. Urethane Formation with an Excess of Isocyanate or Alcohol: Experimental and Ab Initio Study. *Polymers* **2019**, *11*, 1543. [CrossRef] [PubMed]

23. Boros, R.Z.; Farkas, L.; Nehéz, K.; Viskolcz, B.; Szőri, M. An Ab Initio Investigation of the 4,4'-Methlylene Diphenyl Diamine (4,4'-MDA) Formation from the Reaction of Aniline with Formaldehyde. *Polymers* **2019**, *11*, 398. [CrossRef] [PubMed]

24. Zheng, J.; Suh, S. Strategies to reduce the global carbon footprint of plastics. *Nat. Clim. Chang.* **2019**, *9*, 374–375. [CrossRef]

25. Li, J.-W.; Tsen, W.-C.; Tsou, C.-H.; Suen, M.-C.; Chiu, C.-W. Synthetic Environmentally Friendly Castor Oil Based-Polyurethane with Carbon Black as a Microphase Separation Promoter. *Polymers* **2019**, *11*, 1333. [CrossRef] [PubMed]

26. Paciorek-Sadowska, J.; Borowicz, M.; Czupryński, B.; Isbrandt, M. Effect of Evening Primrose Oil-Based Polyol on the Properties of Rigid Polyurethane–Polyisocyanurate Foams for Thermal Insulation. *Polymers* **2018**, *10*, 1334. [CrossRef]

27. Zhou, W.; Bo, C.; Jia, P.; Zhou, Y.; Zhang, M. Effects of Tung Oil-Based Polyols on the Thermal Stability, Flame Retardancy, and Mechanical Properties of Rigid Polyurethane Foam. *Polymers* **2019**, *11*, 45. [CrossRef]

28. Choi, S.M.; Lee, M.W.; Shin, E.J. One-Pot Processing of Regenerated Cellulose Nanoparticles/Waterborne Polyurethane Nanocomposite for Eco-friendly Polyurethane Matrix. *Polymers* **2019**, *11*, 356. [CrossRef]

29. Nečasová, B.; Liška, P.; Kelar, J.; Šlanhof, J. Comparison of Adhesive Properties of Polyurethane Adhesive System and Wood-plastic Composites with Different Polymers after Mechanical, Chemical and Physical Surface Treatment. *Polymers* **2019**, *11*, 397. [CrossRef]

30. Liu, X.; Song, Q.; Tang, Y.; Li, W.; Xu, J.; Wu, J.; Wang, F.; Brookes, P.C. Human health risk assessment of heavy metals in soil–vegetable system: A multi-medium analysis. *Sci. Total Environ.* **2013**, *463*, 530–540. [CrossRef]

31. Xue, D.; Li, T.; Chen, G.; Liu, Y.; Zhang, D.; Guo, Q.; Guo, J.; Yang, Y.; Sun, J.; Su, B.; et al. Sequential Recovery of Heavy and Noble Metals by Mussel-Inspired Polydopamine-Polyethyleneimine Conjugated Polyurethane Composite Bearing Dithiocarbamate Moieties. *Polymers* **2019**, *11*, 1125. [CrossRef] [PubMed]

32. Yu, R.; Zhu, X.; Zhang, M.; Fang, C. Investigation on the Short-Term Aging-Resistance of Thermoplastic Polyurethane-Modified Asphalt Binders. *Polymers* **2018**, *10*, 1189. [CrossRef] [PubMed]

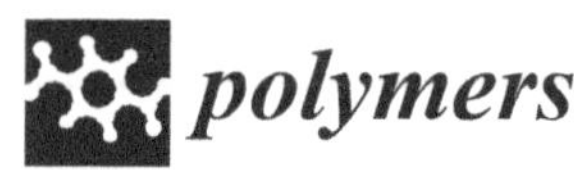 *polymers*

Article

Thermoplastic Dynamic Vulcanizates with In Situ Synthesized Segmented Polyurethane Matrix

Andrea Kohári, István Zoltán Halász and Tamás Bárány *

Department of Polymer Engineering, Faculty of Mechanical Engineering, Budapest University of Technology and Economics, Műegyetem rkp. 3., H-1111 Budapest, Hungary; koharia@pt.bme.hu (A.K.); halaszi@pt.bme.hu (I.Z.H.)
* Correspondence: barany@pt.bme.hu; Tel.: +36-1-463-3740

Received: 24 September 2019; Accepted: 9 October 2019; Published: 12 October 2019

Abstract: The aim of this paper was the detailed investigation of the properties of one-shot bulk polymerized thermoplastic polyurethanes (TPUs) produced with different processing temperatures and the properties of thermoplastic dynamic vulcanizates (TDVs) made by utilizing such in situ synthetized TPUs as their matrix polymer. We combined TPUs and conventional crosslinked rubbers in order to create TDVs by dynamic vulcanization in an internal mixer. The rubber phase was based on three different rubber types: acrylonitrile butadiene rubber (NBR), carboxylated acrylonitrile butadiene rubber (XNBR), and epoxidized natural rubber (ENR). Our goal was to investigate the effect of different processing conditions and material combinations on the properties of the resulting TDVs with the opportunity of improving the interfacial connection between the two phases by chemically bonding the crosslinked rubber phase to the TPU matrix. Therefore, the matrix TPU was synthesized in situ during compounding from diisocyanate, diol, and polyol in parallel with the dynamic vulcanization of the rubber mixture. The mechanical properties were examined by tensile and dynamical mechanical analysis (DMTA) tests. The morphology of the resulting TDVs was studied by atomic force microscopy (AFM) and scanning electron microscopy (SEM) and the thermal properties by differential scanning calorimetry (DSC). Based on these results, the initial temperature of 125 °C is the most suitable for the production of TDVs. Based on the atomic force micrographs, it can be assumed that phase separation occurred in the TPU matrix and we managed to evenly distribute the rubber phase in the TDVs. However, based on the SEM images, these dispersed rubber particles tended to agglomerate and form a quasi-continuous secondary phase where rubber particles were held together by secondary forces (dipole–dipole and hydrogen bonding) and can be broken up reversibly by heat and/or shear. In terms of mechanical properties, the TDVs we produced are on a par with commercially available TDVs with similar hardness.

Keywords: thermoplastic dynamic vulcanizates; TDV; thermoplastic polyurethane; TPU; in situ produced matrix

---

1. Introduction

One of the most important reasons for the successful and continuously growing market penetration of thermoplastics is the diverse characteristics of their blends. By melt blending two (or more) different polymers, a material with new or improved properties can be produced. Blending can enhance mechanical performance (especially toughness), resistance to thermal degradation, improve processability, support cost efficiency, etc., or even result in novel properties. One emerging family of polymer blends is the group of thermoplastic dynamic vulcanizates (TDVs). These TDVs, composed of a continuous thermoplastic phase in which a dynamically cured rubber phase is dispersed, combine the elasticity of rubbers with the easy processing and recyclability of thermoplastics. The term "dynamically" refers to the fact that the rubber component is cured simultaneously with its dispersion in the molten thermoplastic resin by intensive mixing/kneading [1–3]. With this technique, a fine dispersion of the rubber phase in the thermoplastic matrix can be achieved, which is often referred to as a "sea-island" structure with submicron–micron-sized rubber "islands" in the thermoplastic matrix "sea". This structure combines the beneficial properties of the components, namely, the elastic behavior of rubbers and the simple processing (and possible reprocessing as well) of thermoplastic polymers [4].

The development of TDVs started with research on the impact modification possibilities of isotactic polypropylenes (iPP). The results of these studies clarified that the impact resistance of iPP homopolymers could be highly improved through the incorporation of uncured ethylene-propylene-based rubbers. This toughening was even more pronounced below the glass transition temperature range of iPP and was strongly dependent on the dispersion state of the rubber phase. However, the dispersibility of the rubber phase was limited by the recurring agglomeration and coalescence of the rubber particles, which happened during compounding. As a solution for this issue, partial curing of the rubber phase was recommended and attempted [5–7]. The breakthrough was achieved in the late 70s and early 80s when several patents were filed and the first commercialized TDV appeared under the trademark of Santoprene® by Monsanto. It was composed of iPP and ethylene-propylene-diene terpolymer rubber (EPDM). Based on intensive research, the requirements for the optimal mechanical performance of TDVs were deduced, which can be summarized as follows [8–10]:

- the interfacial tension of the constituents is small,
- the crosslink density of the rubber phase is relatively high,
- the matrix polymer is semi-crystalline [11,12].

Similar to blends of various polymers, polyurethanes (PUs) can satisfy an enormously wide range of application requirements due to the great structural diversity of its constituents: isocyanates, polyols, and chain extenders. These constituents may be bifunctional or polyfunctional, allowing the production of both linear and crosslinked molecular structures. Through the appropriate selection of the properties of the precursors such as the polarity of the chains, the mechanical, thermal, and optical behavior of the final PU can be tailored according to the requirements, and through the modification of the constituents, reactive moieties can be introduced into the final chain [13–16]. PUs can be synthetized in one shot or via the pre-polymer route. The former means the simultaneous mixing of all three main PU components, while the latter is composed of two separate steps. In the first step, the polyol reacts with an excess of isocyanate to form an isocyanate-terminated urethane prepolymer, then this prepolymer reacts with the chain extender and the PU is obtained [17]. The one-shot method leads to a relatively more random polymer chain segment structure, whereas the prepolymer route delivers a more regular sequence of the chain segments [18].

Several studies can be found on compounding PUs with rubbers (particularly with NBR rubbers, due to their similar polar character) with the aim of producing polymer blends with novel performance characteristics, although these usually focus on solution and high-temperature melt blending techniques, which are less economical as they need more time and energy. For this reason, dynamic curing is seldom, if ever, used. It was reported that the quasi-static and dynamic mechanical properties were improved (sometimes even outperforming the neat PU), although this reinforcing effect was limited by the melting of the PU phase at higher temperatures as well as by the breakdown of secondary structures between the PU and the NBR at higher strain amplitudes [19]. The same reinforcing effect was observed in blends produced with the solution method [20]. It was reported that if the acrylonitrile content of the NBR was increased, the properties of the blends improved due to the better compatibility between the PU and NBR phases [21]. The incorporation of PU in NBR and hydrogenated NBR (HNBR) rubbers were also researched with the aim of enhancing the wear resistance of the rubber. This technique may have application potential in areas where elastic behavior and good wear resistance are simultaneously required [11,12].

As introduced above, materials with promising novel performance characteristics can be created by blending thermoplastic polymers with rubbers. This potential can be augmented by the utilization of thermoplastic polyurethanes as matrix materials due to their relatively simple in situ synthesis.

Note that a wide range of properties can be obtained within the thermoplastic PU family due to the diversity of the available precursors. Additional benefits can be expected from the dynamic curing introduced above. Our present paper represents the first steps of our research in this field and is aimed at investigating the properties of one-shot bulk polymerized TPUs produced with different processing temperatures as well as the properties of TDVs produced by the dynamic vulcanization of various rubber compounds accompanied by the in situ polymerization of the TPU matrix. The target of this research work was the exploration of the effect of different processing conditions and the material combination on the properties of the resulting TDV, keeping the mechanical performance in focus.

2. Materials and Methods

2.1. Materials and Processing

The tested polyurethanes were prepared from 4,4′-methylenebis(phenyl isocyanate) (MDI), 1,4-butanediol chain extender (BD), and polyether polyol poly(tetrahydrofuran) (PTHF) with a molecular mass of 1000 g/mol and a functionality of 2.0. The purity of all components was over 99%; the components were supplied by Sigma-AldrichDarmstadt, Germany. MDI was used as received, and butanediol and PTHF were dried at 90 °C for 4 h in vacuum. The composition of the materials was described by the ratio of the –NCO (isocyanate) and –OH (hydroxyl) functional groups at the start of the reaction (NCO/OH ratio) and the ratio of the –OH functional groups of the polyol to the total diol (pOH/OH ratio). The NCO/OH ratio was kept constant at 1.05, while the pOH/OH ratio of the TPUs was set to 0.5.

As a base rubber for the rubber phase of the TDVs acrylonitrile-butadiene (NBR), carboxylated acrylonitrile-butadiene (XNBR), and epoxidized natural rubber (ENR) were used. The NBR rubber was produced by Lanxess under the name of Perbunan 3445F. Its Mooney viscosity (ML, 1 + 4, 100 °C) was 45 ± 5, and its bound acrylonitrile content was 34 ± 1 wt%. The XNBR rubber was produced by Lanxess under the name of Krynac X146 with a Mooney viscosity (ML, 1 + 4, 100 °C) of 45 ± 5 and a bound acrylonitrile content of 32.5 ± 1.5 wt%. The ENR rubber was produced by Muang Mai Guthrie Company Limited under the name of Dynathai Epoxyprene 50. Its Mooney viscosity (ML, 1 + 4, 100 °C) is 70–90 and its level of epoxidation is 50 ± 2%. As a curative, dicumyl peroxide with 40 wt% active peroxide content (DCP40) was used (Norac, Norox DCP-40BK). The formulation of the rubber mixtures was as follows: base rubber 100 phr, DCP40 3.75 phr (1.5 phr active peroxide). The rubber mixtures were prepared on a laboratory two-roll mill (Labtech LRM-SC-110, Labtech Engineering Co. Ltd., Samutprakarn, Thailand) at roll temperatures of 70 and 50 °C (front and rear, respectively), and friction of 1.3.

One-step bulk polymerization of the TPUs and dynamic vulcanization of the rubber phase was carried out in a Brabender internal mixer (Brabender Plasti-Corder equipped with a W 50 EHT chamber, Brabender GmbH., Duisburg, Germany) at initial temperatures of 100, 110, 125, and 150 °C for the TPU production and 125 °C for the TDV production with a rotor speed of 50 rpm and a mixing time of 30 min. The reaction scheme is shown in Figure 1. Torque and melt temperature were recorded during the process. TDVs contained 50 wt% TPU and 50 wt% rubber mixture. The initial temperature was 125 °C. First, the components of the TPU were added, then the rubber blend after 13 min.

The 0.5 mm thick TPU and TDV sheets were compression molded at 190 °C under a pressure of 2 MPa for 3 min in a Collin Teach-Line Platen Press 200E laboratory press (Dr. Collin GmbH, Ebersberg, Germany). Specimens for further tests were cut from the sheets produced.

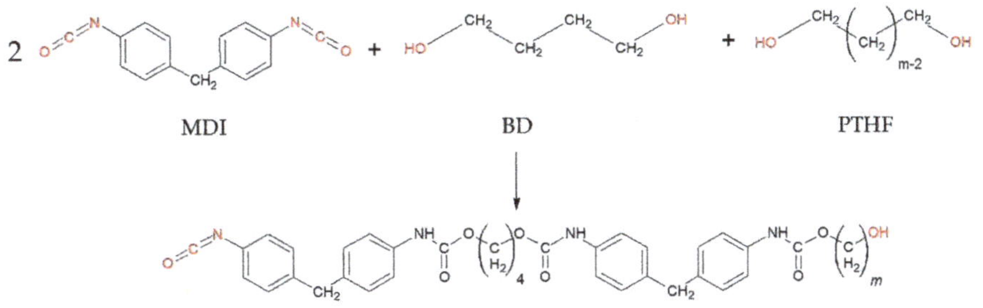

Figure 1. The reaction scheme of the synthesis of thermoplastic polyurethane.

2.2. Testing Methods

Hardness was determined according to the ISO 7619-1 Shore A method with a Zwick H04.3150.000 hardness tester (Zwick GmbH, Ulm, Germany) on the compression molded sheets. Ten tests were performed on each compound.

The tensile mechanical properties of the compounds were investigated according to the ISO 37 standard on a Zwick Z005 universal testing machine with a 5 kN load cell (Zwick GmbH, Ulm, Germany) at room temperature at a crosshead speed of 500 mm/min. Five tests were performed on each compound.

Curing properties of the rubber compounds were tested on a MonTech D-RPA 3000 Dynamic Rubber Process Analyzer, (MonTech GmbH, Buchen, Germany). The curing curves were recorded both in isothermal conditions at 170 °C (1.67 Hz and 1° amplitude) and non-isothermal conditions. The purpose of the latter was to verify the proper vulcanization of the rubber phase during dynamic vulcanization, therefore the temperature profile of the vulcanization test was set according to the temperature curves recorded during the production of the corresponding TDV.

Polymerization of the TPU phase was investigated with a TA AR 2000 parallel plate rheometer (TA Instruments Ltd., New Castle, DE, USA) with a sinusoidal oscillation. We set the temperature profiles of each test according to the temperature curves recorded during the production of the TDVs in order to confirm the proper polymerization of the TPU. The strain amplitude and oscillation frequency were 1% and 10 rad/s. Curves were recorded on a time scale of 30 min and the gap was set to 0.5 mm between the parallel plates, which had a diameter of 25 mm.

Thermal analysis of the samples was carried out on a TA Q2000 DSC machine (TA Instruments Ltd., New Castle, DE, USA) in a N_2 atmosphere with a heat–cool–heat cycle. The heating rate was 10 °C/min in the −90 … .250 °C range.

Dynamic mechanical tests of the compounds were performed with a TA Q800 DMTA machine (TA Instruments Ltd., New Castle, DE, USA) in tensile mode on rectangular specimens with dimensions of 0.5 mm × 2.5 mm × 10 mm (thickness × width × clamped length). Tests were run in the range of −100 … 200 °C with a 3 °C/min heating rate at a frequency of 10 Hz with a preload of 0.01 N and superimposed 0.1% sinusoidal strain.

The structure of the TPUs and TDVs were analyzed by attenuated total reflectance Fourier transform infrared spectroscopy (ATR-FTIR). A Bruker Tensor type FTIR machine (Bruker Optics Inc., Billerica, MA, USA) was used equipped with a Specac Golden Gate ATR unit (Specac Ltd., Orpington, UK) in the wavelength range of 4000 to 600 cm^{-1} with a resolution of 4 cm^{-1}, accumulating 16 scans.

The morphology of the TDVs was characterized with the use of AFM micrographs of cryo-microtomed surfaces. Samples were cooled down to −80 °C and cut with a glass knife with the use of a Leica EM UC6/FC7 ultramicrotome (Leica Microsystems GmbH, Wien, Austria). AFM images were taken with a Nanosurf FlexAFM 5 (Nanosurf AG, Liestal, Switzerland) type AFM in tapping mode in air at room temperature. A single-beam silicon cantilever was used with a nominal force constant of 48 N/m and a resonance frequency of 190 kHz. It was a TAP 190Al-G cantilever (Budget

Sensors, Innovative Solutions Bulgaria Ltd., Sofia, Bulgaria). The scan rate was between 0.5 and 1.0 Hz. The morphology of the TDVs was also investigated with a JEOL JSM 6380LA (Jeol Ltd., Tokyo, Japan) type scanning electron microscope (SEM). The samples were immersed in liquid nitrogen for 3 min and then fractured. The fractured surfaces were etched with boiling tetrahydrofuran (good solvent for TPU) for 1 h. After that, the samples were dried under vacuum for a day [22]. Before the test, the etched surfaces were coated with a thin layer of gold.

3. Results

3.1. Synthesis of Thermoplastic Polyurethanes (TPUs)

First, we evaluated the torque and temperature curves that were recorded during the production of the specimens. The temperature profiles are very important because we needed to select an initial temperature in which the TPU could be polymerized and the rubber phase could be vulcanized at a sufficient rate, but without the degradation of the materials. The torque curves measured during the reactions also carry important information. Torque is proportional to melt viscosity, which depends on molecular weight. Thus, the torque change in the internal mixer is a good indicator of the change in molecular weight [23]. Figure 2 shows that in the case of the samples with initial temperatures of 100 and 110 °C, the torque increased until the end of the mixing time and did not reach an equilibrium. From this, we can conclude that polymerization was too slow at these initial temperatures, and did not finish within the time span of compounding. In contrast, at the initial temperatures of 125 and 150 °C, the torque reached a plateau. Due to the nature of the step polymerization in the first two cases (100 and 110 °C), oligomers with various molecular weights or perhaps short polymers chains were produced, while in the second two cases (125 and 150 °C), high molecular weight polymer chains were generated at the end of the mixing time. Figure 2 also shows the recorded temperature curves. In the case of the TPU100 and TPU110 samples, the temperature curves also did not reach a steady state. The maximum temperatures (114 and 137 °C) were low for the vulcanization of the rubber phase, and polymerization was also slow at these temperatures. In contrast, the 186 °C of the TPU150 sample could lead to the degradation of the rubber phase. The final temperature of the TPU125 sample was 175 °C, which is suitable for both polymerization and vulcanization, so we chose this initial temperature (125 °C) for preparing the TDVs.

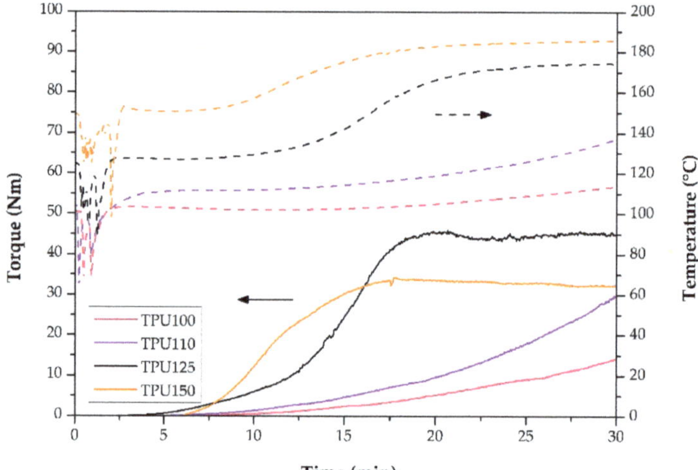

Figure 2. The parameters recorded during the synthesis of the thermoplastic polyurethanes (TPUs): torque curves (——) and temperature curves (– – –).

3.2. Thermoplastic Dynamic Vulcanizates (TDVs)

3.2.1. Production of TDVs

Isothermal Curing Properties

Before preparing the TDVs, we determined the isothermal curing properties of the rubber mixtures (continuous lines in Figure 3). It can be observed that in the case of NBR and XNBR, the curves almost ran together, but in the case of ENR, the torque was far lower during the tests. We determined the vulcanization times (t_{90}) of the mixtures, which was 7 min for NBR and XNBR, and 8 min for ENR.

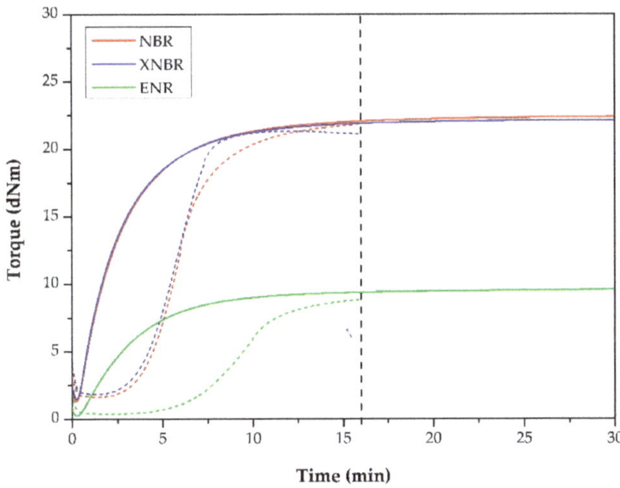

Figure 3. The curing curves of the rubber mixtures in isothermal (——) and in non-isothermal conditions (– – –).

On the basis of the torque curves of the TPUs (Figure 2) and the vulcanization times, we determined that we had to add the rubber mixtures in the thirteenth minute of mixing to have in situ polymerization and vulcanization occur simultaneously.

Torque and Temperature Curves

The best way to keep track of the processes in the internal mixer is to analyze the torque and the temperature curves. In the torque curves recorded during the production of TDVs (Figure 4), there was a sharp and steep increase when the rubber mixture was added, followed by a sudden decrease. This decrease was due to the mastication of the rubber. Thereafter, the curves showed a pronounced increase again, which was due to the polymerization of the TPU and the vulcanization of the rubber phase. Finally, in the last stage, there was another decrease due to the fragmentation and dispersion of the vulcanized rubber islands.

The temperature curves (Figure 4) decreased at two points where we added the components. Figure 4 clearly shows that the torque and temperature values of the TDV sample containing ENR were well below that of the other two samples as well as the matrix TPU. The ENR rubber mixture was much softer than the others, consequently, even after the addition of ENR, the torque did not increase as intensely as in the internal mixer. At the same time, the temperature increase due to friction was also reduced, therefore we did not achieve the optimal conditions for the vulcanization of the rubber phase.

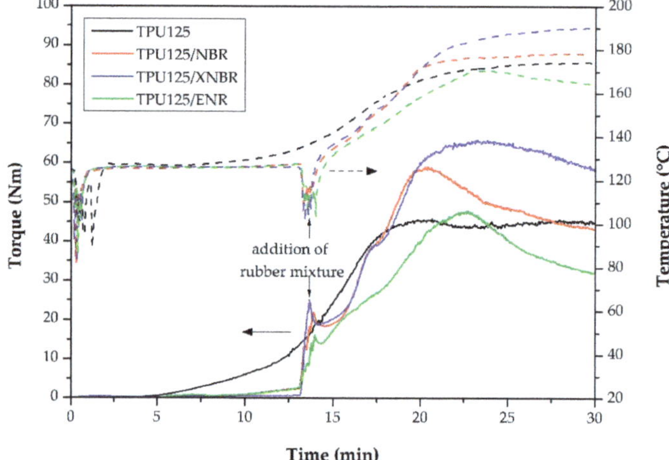

Figure 4. The parameters recorded during the synthesis of the thermoplastic dynamic vulcanizates (TDVs): torque curves (——) and temperature curves (– – –).

Non-Isothermal Curing Properties

Later, we also recorded the non-isothermal curing curves of the mixtures (Figure 3, dashed lines). The purpose of the test was to verify whether the rubber phase was properly vulcanized, therefore the temperature profile of the test was set according to the temperature curves recorded during the production of the corresponding TDV. We approximated these temperature curves with linear sections as shown in Figure 5.

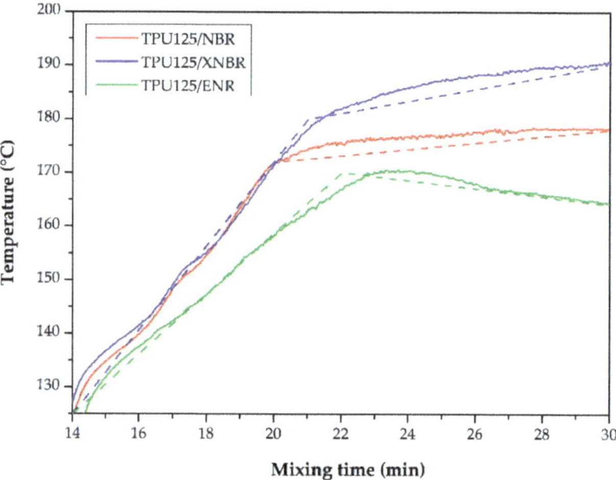

Figure 5. The real temperature profiles recorded during the production of TDVs (——) and their approximate curves (– – –) that were used for the non-isothermal curing test for the rubber compounds.

About 16 min were available for the vulcanization of the rubber mixtures during the production of TDVs, which we indicated with a vertical line in Figure 3. The non-isothermal curing test indicated that we achieved quite a good degree of vulcanization in TDVs containing NBR and XNBR. As shown in Figure 3 (dashed line), the torque curve of the ENR sample did not show saturation. This is because the maximum temperature that the TDV containing ENR reached was not sufficient to completely vulcanize the rubber phase during compounding. This is also supported by the fact that the material removed from the internal mixer was in this case slightly tacky.

The Investigation of Matrix Polymerization

In a rheometer, we modeled the processes in the internal mixer in order to investigate the polymerization of the matrix. Our aim was verify that the polymerization of TPU is possible with the temperature profiles of the production of TDVs. We approached these temperature curves with straight sections as shown in Figure 6 (dashed lines).

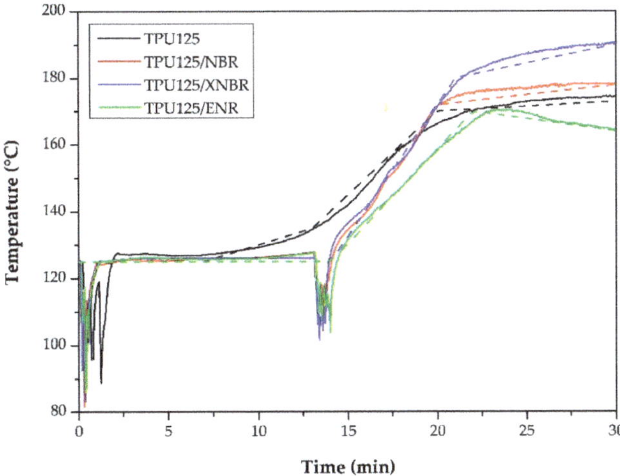

Figure 6. The real temperature profiles recorded during the production of TDVs (——) and the approximate curves (– – –) that were used for the investigation of the polymerization tests of the matrix TPU.

During the evaluation of the curves, we examined the complex viscosity (η^*) of the TPU. Based on this, we compared the processes during the synthesis of pure TPU and those during the production of TDVs. The curves (Figure 7) show that the thermoplastic polyurethane polymerized in each TDV because the final complex viscosity values are similar or greater than those of the TPU.

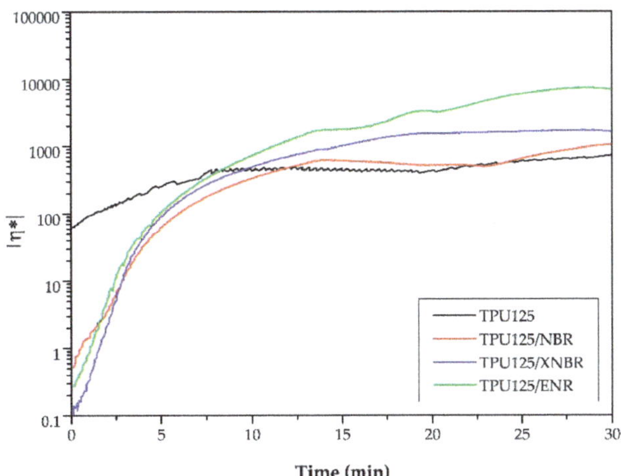

Figure 7. The simulation of the complex viscosity of the matrices in a rheometer during the production of pure TPU and TDVs.

3.2.2. Mechanical Properties

The tensile test results show that the modulus (Figure 8), the tensile strength, and the elongation at break of the TDVs are lower than those of the pure TPU (Table 1). The hardness of the materials was reduced by the rubber phase, which is consistent with the decrease in modulus. TDVs containing NBR and XNBR have similar strength properties and Shore A hardness. The tensile strength and hardness of the TPU125/ENR sample are lower, while the elongation at break is higher than for other TDVs.

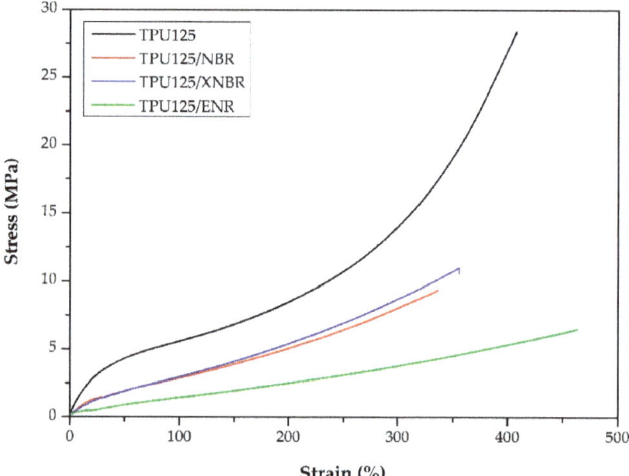

Figure 8. Typical tensile curves of the TPU matrix and the TDVs.

Table 1. Mechanical properties of the TPU matrix and the TDVs.

	TPU125	TPU125/NBR	TPU125/XNBR	TPU125/ENR
Elongation at break (%)	377 ± 52	316 ± 35	328 ± 38	447 ± 21
Modulus at 10% (MPa)	1.73 ± 0.05	0.84 ± 0.05	0.77 ± 0.08	0.45 ± 0.08
50% (MPa)	4.35 ± 0.06	1.98 ± 0.06	2.23 ± 0.22	1.04 ± 0.10
100% (MPa)	5.64 ± 0.05	2.92 ± 0.14	3.18 ± 0.23	1.60 ± 0.15
200% (MPa)	8.63 ± 0.09	5.04 ± 0.17	5.79 ± 0.38	2.74 ± 0.26
300% (MPa)	14.24 ± 0.17	7.92 ± 0.16	9.06 ± 0.48	4.09 ± 0.39
Tensile strength (MPa)	24.4 ± 7.8	8.9 ± 0.9	10.8 ± 1.7	6.6 ± 0.9
Hardness (Shore A°)	81.3 ± 0.9	66.4 ± 0.5	64.5 ± 0.7	55.9 ± 0.9

These results are due to the poor properties of the cured rubbers. Our rubber blends were only model materials that did not contain any filler (e.g., carbon black, silica) and thus had poor mechanical properties. We used such model mixtures because at this stage of our research, we sought simplicity to minimize the effects of possible variables and influencing parameters.

3.2.3. Dynamical Mechanical Analysis

Figure 9 and Table 2 show that the T_g values of the TDVs are higher than those of the matrix. On the tan delta curves of the TDVs, there is a shoulder on the damping peak, which belongs to the glass transition temperature of the soft segments of the TPU phase, while the well-developed peak represents the glass transition of the rubber phase.

In accordance with the results of the tensile test, the storage moduli of the TDVs are smaller than that of the TPU matrix, which is caused by the inherently lower storage modulus of the rubbers when compared to the TPU matrix.

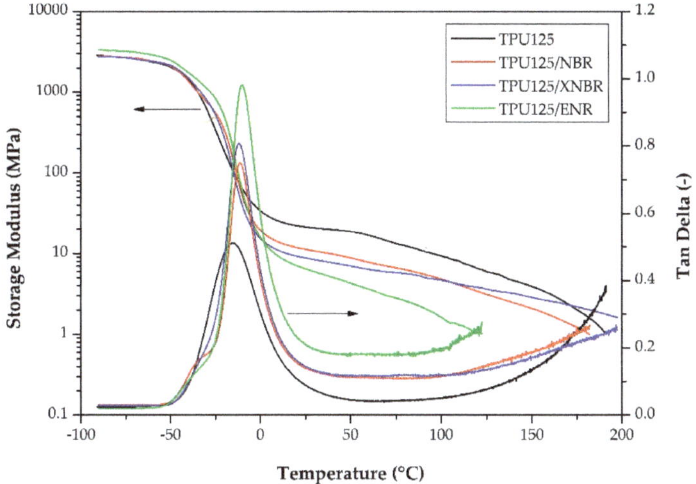

Figure 9. Thermomechanical curves of the TPU matrix and the TDVs.

Table 2. The glass transition temperatures (T_g) and room temperature storage modulus ($E'_{23\,°C}$) of the TDVs.

Sample	T_g (°C)	$E'_{23°C}$ (MPa)
TPU125	−15.82	21.82
TPU125/NBR	−11.36	11.51
TPU125/XNBR	−12.05	8.98
TPU125/ENR	−10.23	6.70

3.2.4. Differential Scanning Calorimetry

On the thermograms of the TDVs, there are three different glass transitions. The first one belongs to the soft segments of the TPU, the second to the rubber phase, and the third to the TPU's hard segment (Figure 10 and Table 3). On the curve of TPU125 and TPU125/ENR, there is also a melting peak, which belongs to the crystalline part of the hard phase. The melting temperature and the peak intensity of the TPU125/ENR sample are smaller than those of the TPU125. For the other two samples, the melting peak is not visible. This is due to the rubber phase, which did not allow the TPU chains to organize and crystallize because the hard segments can form strong hydrogen bonds with the –C≡N and >C=O groups of the rubber [24].

Table 3. The glass transition (T_g) and the melting (T_m) temperature of the different phases in the TDVs.

Sample	$T_{g,\,soft}$ (°C)	$T_{g,\,rubber}$ (°C)	$T_{g,\,hard}$ (°C)	T_m (°C)
TPU125	−42.1	—	56.5	211.3
NBR	—	−23.0	—	—
XNBR	—	−22.7	—	—
ENR	—	−24.6	—	—
TPU125/NBR	−43.1	−21.5	53.7	not visible
TPU125/XNBR	−43.0	−20.2	57.6	not visible
TPU125/ENR	−45.5	−22.3	49.9	207.0

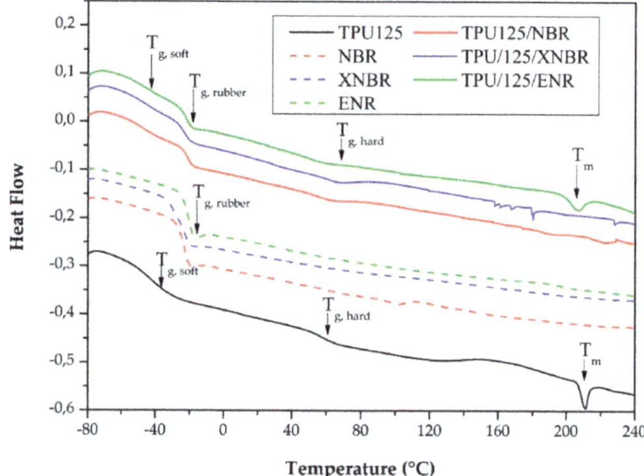

Figure 10. DSC thermograms of the TPU matrix and the TDVs.

3.2.5. Fourier-Transform Infrared Spectroscopy

We used ATR-FTIR spectroscopy to study the chemical structure of the polymers produced. The recorded spectrum of pure TPU125 and the TDVs containing different rubbers are shown in Figure 11. The spectra contain all of the characteristic peaks (i.e., the >N–H peak at 3318 and 1530 cm^{-1} and the >C=O peaks at 1730 and 1702 cm^{-1}) from the urethane linkage. The peak at 1730 cm^{-1} represents the stretching vibration of free >C=O groups, and the peak at 1702 cm^{-1} shows the hydrogen-bonded version. The peak of the isocyanate groups (2275–2250 cm^{-1}) was not visible in the spectrum, which indicates that the diisocyanate reacted completely. The peaks at 2938 and 2854 cm^{-1} belong to the >CH$_2$ groups of the polymer chains. The aromatic ring in the diisocyanate has an absorption peak at 1597 cm^{-1}. On the spectrum of the TPU125/NBR and the TPU125/XNBR, three new peaks appeared, which belong to the rubber phase. The –C≡N peak appeared at 2238 cm^{-1}, the CH peak at 1447 cm^{-1}, and the =CH$_2$ peak at 968 cm^{-1}. In the case of TPU125/ENR, one new peak appeared at 869 cm^{-1}, which belongs to the epoxy group [24–26].

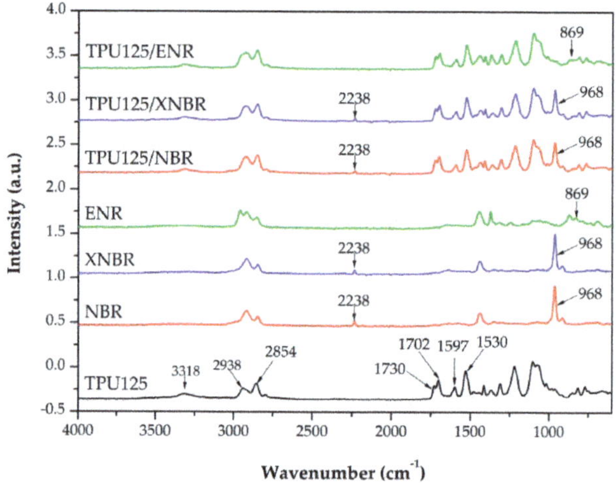

Figure 11. Fourier-transform infrared spectra of the TDVs and TPU125.

3.2.6. Morphology

Atomic Force Microscopy (AFM)

Figure 12 shows the atomic force microscopy phase images of the TDVs and the neat TPU. The bright parts indicate the hard segments of the TPU due to its high surface energy and modulus, and the dark parts are the soft segment of the matrix [27,28]. Figure 12d shows that phase separation took place in the material. In Figure 12a–c (marked by black arrows), there are 1 to 2 μm morphological units, which are probably the rubber islands distributed in the matrix.

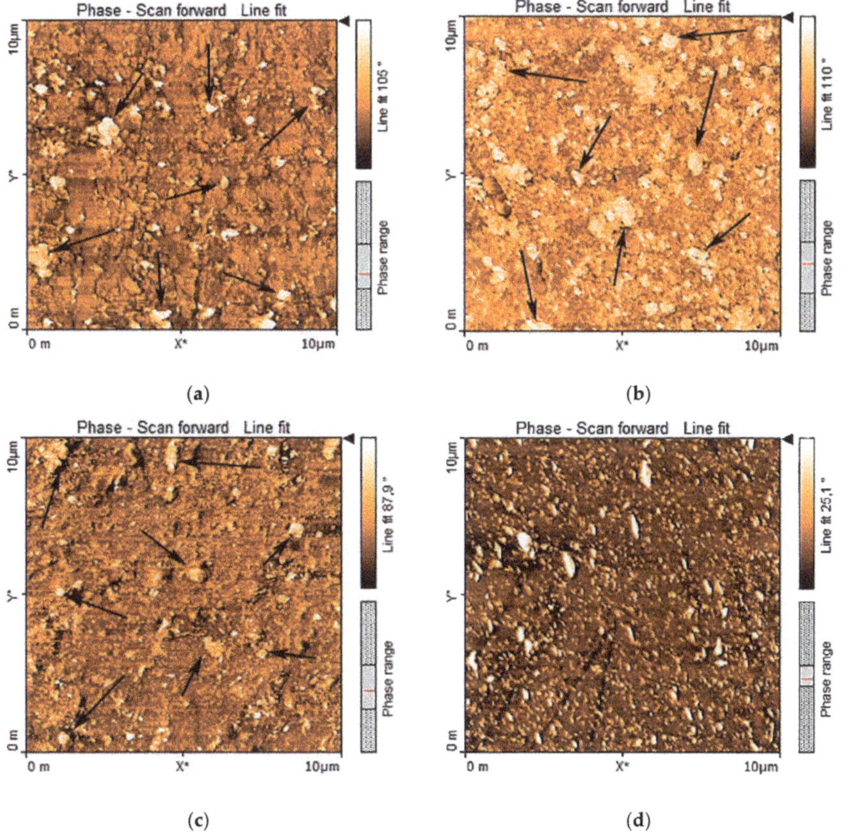

Figure 12. Atomic force microscopy (AFM) phase images (10 μm × 10 μm) of (**a**) TPU125/NBR, (**b**) TPU125/XNBR, (**c**) TPU125/ENR, and (**d**) TPU125.

The pictures show that the phase separation occurred in the thermoplastic polyurethane we produced and that we managed to evenly distribute the rubber phase in the matrix.

Scanning Electron Microscopy (SEM)

Figure 13 shows the SEM images of the cryo-fractured (a, c, e, g) and the etched, cryo-fractured (b, d, f, h) surfaces of the samples. Etching with tetrahydrofuran (THF) was successful since there were significant differences between the pairs of pictures. In the case of the neat TPU (Figure 13h), holes with an elongated shape appeared on the surface as a result of etching. This may suggest that the polymerization of the TPU did not finish completely, therefore small regions with lower molecular weights were present in the material. THF dissolved these regions faster than the rest of the TPU, which explains the presence of the holes. Figure 13b shows the TDV containing NBR. The agglomerated particles of the NBR phase can be observed on the surface, and the holes indicate the dissolved TPU matrix. In the case of the TDV containing XNBR (Figure 13d), a stronger interaction may have developed between the XNBR and the TPU phase as a result of a reaction between the carboxyl groups of the XNBR and the isocyanate groups of the MDI component of the TPU, resulting in a chemically bonded thin TPU layer on the surface of the XNBR phase. This assumption is supported by Figure 13d, which shows that a thin layer of TPU remained on the etched surface due to the above-mentioned strong interaction between the XNBR phase and the TPU. Figure 13b,d suggest that the dispersed vulcanized rubber particles are agglomerated, creating a secondary structure. There are weak secondary bonds between these particles, which can reversibly dissociate at elevated temperatures and/or due to high shear. This phenomenon has already been demonstrated for other material pairs (PP/EPDM) [29]. In the case of the TDV containing ENR (Figure 13f), the rubber phase was not visible, which suggests that it was also dissolved. This is most probably because its degree of vulcanization was insufficient to make it insoluble in THF.

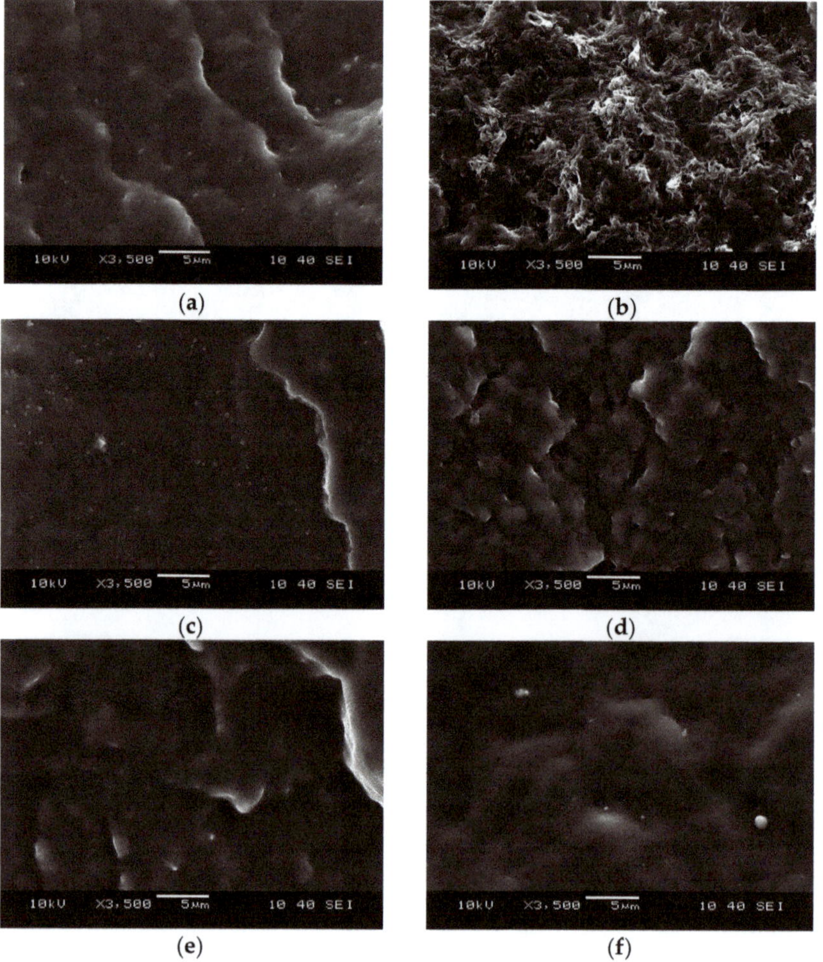

Figure 13. *Cont.*

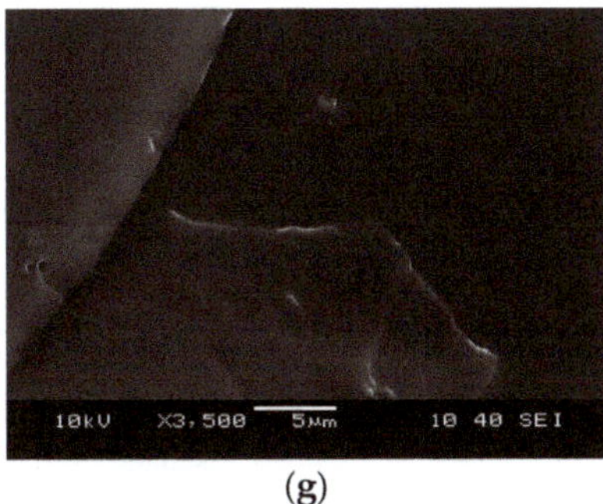 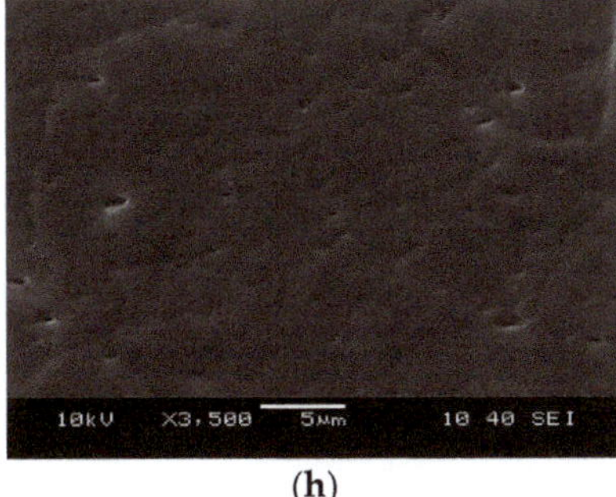

(g)   (h)

Figure 13. SEM images of the cryo-fractured, and cryo-fractured and etched surfaces of (**a**,**b**) TPU125/NBR, (**c**,**d**) TPU125/XNBR, (**e**,**f**) TPU125/ENR, and (**g**,**h**) neat TPU125.

4. Conclusions

We prepared thermoplastic dynamic vulcanizates with an in situ synthesized TPU matrix and with polar rubbers NBR, XNBR, and ENR. The optimum initial temperature for the internal mixer at which the TPU polymerized and the rubber phase was vulcanized at a sufficient rate without degradation of the materials was 125 °C. The non-isothermal vulcanization and rheological tests confirmed the successful in situ polymerization of the matrix and the dynamic vulcanization of the rubber phase. Based on the AFM phase images, it is likely that the rubber segments were successfully dispersed in the TPU matrix. However, based on the SEM images, these dispersed rubber particles tend to agglomerate and form a secondary quasi-continuous phase structure formed by rubber particles held together only by secondary forces. TDVs have lower hardness and weaker strength characteristics than pure TPU. These TDVs have a higher tensile strength, while their elongation at break is only slightly below that of commercial thermoplastic vulcanizates (TPVs) with similar hardness.

Author Contributions: Conceptualization, A.K., I.Z.H., and T.B.; Methodology, A.K. and I.Z.H.; Validation, A.K., I.Z.H., and T.B.; Investigation, A.K. and I.Z.H.; Resources, I.Z.H. and T.B.; Data Curation, A.K. and I.Z.H.; Writing—Original Draft Preparation, A.K.; Writing, Review & Editing, A.K.; I.Z.H., and T.B.; Visualization, A.K.; Supervision, T.B. and I.Z.H.; Project Administration, T.B.; Funding Acquisition, T.B. and I.Z.H.

Funding: This work was supported by the National Research, Development, and Innovation Office, Hungary (K_18128268) and by the Higher Education Excellence Program of the Ministry of Human Capacities in the Framework of the Nanotechnology research area of the Budapest University of Technology and Economics (BME FIKP-NANO).

Acknowledgments: We would like to sincerely thank the late h.c. mult. József Karger-Kocsis for his support and valuable comments which serve as the solid foundation of our research.

Conflicts of Interest: The authors declare no conflicts of interest.

References

1. Karger-Kocsis, J. Thermoplastic dynamic vulcanizates. In *Polypropylene: An A-Z Reference*; Kluwer Publisher: Dordrecht, The Netherlands, 1999; Volume 2, pp. 853–858.
2. Legge, N.R.; Schroder, E.H. *Thermoplastic Elastomers*; Hanser University Press: New York, NY, USA, 1987.
3. Mülhaupt, R. Metallocene catalyst and tailor-made polyolefins. In *Polypropylene: An A–Z Reference*; Karger-Kocsis, J., Ed.; Kluwer Publisher: Dodrecht, The Netherlands, 1999; pp. 454–475.
4. Karger-Kocsis, J. Thermoplastic rubbers via dynamic vulcanization. In *Polymer Blends and Alloys*; Shonaike, G.O., Simon, G.P., Eds.; Marcel Dekker: New York, NY, USA, 1999; pp. 125–153.
5. Karger-Kocsis, J.; Kalló, A.; Kuleznev, V.N. Phase structure of impact-modified polypropylene blends. *Polymer* **1984**, *25*, 279–286. [CrossRef]
6. Wu, S. A generalized criterion for rubber toughening: The critical matrix ligament thickness. *J. Appl. Polym. Sci.* **1988**, *35*, 549–561. [CrossRef]
7. Dao, K.C. Mechanical properties of polypropylene/crosslinked rubber blends. *J. Appl. Polym. Sci.* **1982**, *27*, 4799–4806. [CrossRef]

8. Coran, A.Y.; Patel, R. Rubber-thermoplastic compositions. Part IV. Thermoplastic vulcanizates from various rubber-plastic combinations. *Rubber. Chem. Technol.* **1981**, *54*, 892–903. [CrossRef]
9. Coran, A.Y.; Patel, R.; Williams, D. Rubber-Thermoplastic Compositions. Part V. Selecting Polymers for Thermoplastic Vulcanizates. *Rubber. Chem. Technol.* **1982**, *55*, 116–136. [CrossRef]
10. Akiba, M.; Hashim, A.S. Vulcanization and crosslinking in elastomers. *Prog. Polym. Sci.* **1997**, *22*, 475–521. [CrossRef]
11. Tahir, M.; Stöckelhuber, K.W.; Mahmood, N.; Komber, H.; Heinrich, G. Reactive blending of nitrile butadiene rubber and in situ synthesized thermoplastic polyurethane-urea: Novel preparation method and characterization. *Macromol. Mater. Eng.* **2015**, *300*, 242–250. [CrossRef]
12. Tahir, M.; Stöckelhuber, K.W.; Mahmood, N.; Komber, H.; Formanek, P.; Wiessner, S.; Heinrich, G. Highly reinforced blends of nitrile butadiene rubber and in-situ synthesized polyurethane–urea. *Europ. Polym. J.* **2015**, *73*, 75–87. [CrossRef]
13. Berezkin, Y.; Urick, M. Modern Polyurethanes: Overview of Structure Property Relationship. In *Polymers for Personal Care and Cosmetics*; American Chemical Society: Washington, WA, USA, 2013; Volume 1148, pp. 65–81.
14. Oertal, G. *Polyurethane Handbook*; Hanser University Press: New York, NY, USA, 1994.
15. Petrovic, Z.S. Polyurethanes. In *Handbook of Polymer Synthesis*, 2nd ed.; Kircheldorf, H.R.N.O., Swift, G., Eds.; Marcel Dekker: New York, NY, USA, 1994; pp. 503–539.
16. Woods, G.; Pattison, J.B. *The ICI Polyurethanes Book*, 2nd ed.; John Wiley & Sons: Hoboken, NJ, USA, 1990.
17. Szycher, M. *Handbook of Polyurethanes*; CRC Press: New York, NY, USA, 2013.
18. Prisacariu, C. *Polyurethane Elastomers: From Morphology to Mechanical Aspects*; Springer: New York, NY, USA, 2011.
19. Dimitrievski, I.S.; Marinovic, T. Effect of PU-NBR interactions on blends' dynamic properties. In Proceedings of the 5th European Rheology Conference, Portoroz, Slovenia, 8–11 April 2019; pp. 73–74.
20. Desai, S.; Thakore, I.M.; Brennan, A.; Devi, S. Polyurethane-Nitrile Rubber Blends. *J. Macromol. Sci. A* **2001**, *38*, 711–729. [CrossRef]
21. Im, H.G.; Ka, K.R.; Kim, C.K. Characteristics of polyurethane elastomer blends with poly(acrylonitrile-*co*-butadiene) rubber as an encapsulant for underwater sonar devices. *Ind. Eng. Chem. Res.* **2010**, *49*, 7336–7342. [CrossRef]
22. Demma, G.; Martuscelli, E.; Zanetti, A.; Zorzetto, M. Morphology and properties of polyurethane-based blends. *J. Mater. Sci.* **1983**, *18*, 89–102. [CrossRef]
23. Pukánszky, B.; Bagdi, K.; Tóvölgyi, Z.; Varga, J.; Botz, L.; Hudak, S.; Dóczi, T.; Pukánszky, B. Nanophase separation in segmented polyurethane elastomers: Effect of specific interactions on structure and properties. *Europ. Polym. J.* **2008**, *44*, 2431–2438. [CrossRef]
24. Ning, N.; Qin, H.; Wang, M.; Sun, H.; Tian, M.; Zhang, L. Improved dielectric and actuated performance of thermoplastic polyurethane by blending with XNBR as macromolecular dielectrics. *Polymer* **2019**, *179*, 121646. [CrossRef]
25. Guénaëlle, A.L.V.; Noël, P.G.R.; Marc, O.D.M. *Easy Identification of Plastics and Rubbers*; Rapra Technology Limited: Shawbury, UK, 2001.
26. Kalkornsurapranee, E.; Vennemann, N.; Kummerlöwe, C.; Nakason, C. Novel thermoplastic natural rubber based on thermoplastic polyurethane blends: Influence of modified natural rubbers on properties of the blends. *Iran. Polym. J.* **2012**, *21*, 689–700. [CrossRef]
27. Hossieny, N.; Shaayegan, V.; Ameli, A.; Saniei, M.; Park, C.B. Characterization of hard-segment crystalline phase of thermoplastic polyurethane in the presence of butane and glycerol monosterate and its impact on mechanical property and microcellular morphology. *Polymer* **2017**, *112*, 208–218. [CrossRef]
28. Eceiza, A.; Larrañaga, M.; de la Caba, K.; Kortaberria, G.; Marieta, C.; Corcuera, M.A.; Mondragon, I. Structure–property relationships of thermoplastic polyurethane elastomers based on polycarbonate diols. *J. Appl. Polym. Sci.* **2008**, *108*, 3092–3103. [CrossRef]
29. Wu, H.; Tian, M.; Zhang, L.; Tian, H.; Wu, Y.; Ning, N.; Chan, T.W. New understanding of morphology evolution of thermoplastic vulcanizate (TPV) during dynamic vulcanization. *ACS Sustainable Chem. Eng.* **2014**, *3*, 26–32. [CrossRef]

Article

Urethane Formation with an Excess of Isocyanate or Alcohol: Experimental and Ab Initio Study

Wafaa Cheikh, Zsófia Borbála Rózsa, Christian Orlando Camacho López, Péter Mizsey, Béla Viskolcz, Milán Szőri * and Zsolt Fejes *

Institute of Chemistry, University of Miskolc, Miskolc-Egyetemváros A/2, H-3515 Miskolc, Hungary; cheikhwafaa.92@gmail.com (W.C.); kemzsofi@uni-miskolc.hu (Z.B.R.); chrisscamacho@gmail.com (C.O.C.L.); kemizsey@uni-miskolc.hu (P.M.); bela.viskolcz@uni-miskolc.hu (B.V.)
* Correspondence: milan.szori@uni-miskolc.hu (M.S.); kemfejes@uni-miskolc.hu (Z.F.); Tel.: +36-46-565-111/1337 (M.S.); +36-46-565-111/1911 (Z.F.)

Received: 29 August 2019; Accepted: 19 September 2019; Published: 22 September 2019

Abstract: A kinetic and mechanistic investigation of the alcoholysis of phenyl isocyanate using 1-propanol as the alcohol was undertaken. A molecular mechanism of urethane formation in both alcohol and isocyanate excess is explored using a combination of an accurate fourth generation Gaussian thermochemistry (G4MP2) with the Solvent Model Density (SMD) implicit solvent model. These mechanisms were analyzed from an energetic point of view. According to the newly proposed two-step mechanism for isocyanate excess, allophanate is an intermediate towards urethane formation via six-centered transition state (TS) with a reaction barrier of 62.6 kJ/mol in the THF model. In the next step, synchronous 1,3-H shift between the nitrogens of allophanate and the cleavage of the C–N bond resulted in the release of the isocyanate and the formation of a urethane bond via a low-lying TS with 49.0 kJ/mol energy relative to the reactants. Arrhenius activation energies of the stoichiometric, alcohol excess and the isocyanate excess reactions were experimentally determined by means of HPLC technique. The activation energies for both the alcohol (measured in our recent work) and the isocyanate excess reactions were lower compared to that of the stoichiometric ratio, in agreement with the theoretical calculations.

Keywords: urethane formation; isocyanate excess; mechanism; ab initio; allophanate; kinetics

---

1. Introduction

Isocyanates are among the most valued synthetic intermediates [1]. Their reactions with various nucleophiles give rise to important classes of compounds, such as urethanes, thiouretanes and ureas. These reactions are of industrial importance because they provide the basis of the very versatile class of polymers, polyurethanes, where the main process is the reaction of di-isocyanates with polyols.

From a kinetic and mechanistic point of view, the addition reaction between the isocyanato and the hydroxyl group has been of interest since the 1930 [2]. The first detailed kinetic investigations of uncatalyzed and catalyzed reactions was made by Baker and co-workers in the 1940s [3–5]. Their studies concluded that the apparently bimolecular addition is catalyzed by both the alcohol reactant and the urethane product [4]. Later, the possibility of having alcohol associates as the active reacting partner was pointed out [6–8]. The rate constant of the reaction strongly depends on the solvent [9].

The experimental activation energies for the reactions of aryl isocyanates with alcohols are generally in the range of 17–54 kJ/mol ([10,11] and references cited in each). For a given reaction the activation energy depends on the solvent and the ratio of the reactants. Theoretical calculations showed that the rather high energy barrier (>100 kJ/mol) [8,12,13] needed for reaching the bimolecular transition state (direct addition) becomes substantially lower if one or two additional alcohol molecules (alcohol catalysis), or a urethane molecule (autocatalysis) are also incorporated into the transition state [8,14,15]. A schematic mechanism of such alcohol catalysis is presented in the upper part of Figure 1.

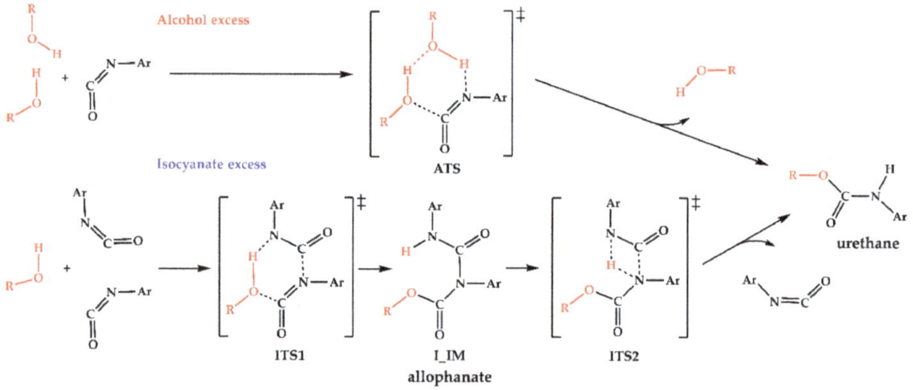

Figure 1. Elementary reaction mechanism for urethane bond formation. The alcohol excess mechanism (top) involves a hydrogen-bonded alcohol associate as the reactant, while the isocyanate excess mechanism (bottom) starts with dipole-dipole stabilized intermolecular isocyanate dimer. In the present study R = Pr and Ar = Ph.

Strong intermolecular hydrogen bonds can stabilize these alcohol associates which is also confirmed by consistent molecular dynamic simulation and X-ray experiment study on liquid 1-propanol by Akiyama and co-workers [16]. This fact makes the above-mentioned mechanism plausible in the condition of excess alcohol. On the other hand, isocyanates also have potential to form associates due to its large permanent electric dipole moment ($|\mu_{tot,MP2/aug-cc-pVTZ}|$ = 2.78 D [17], $|\mu_{tot,MW}|$ = 2.81 D [18]). Indeed, Lenzi et al. reported interaction energy of 24.3–32.8 kJ/mol for alkyl-isocyanates dimers using density functional theory (DFT) calculation [19]; therefore, these isocyanate associates can also be formed in isocyanate excess and can provide a starting point for urethane formation. The proposed reaction mechanism can be seen in the bottom of Figure 1. In this paper, we present this new possible reaction mechanism, supported by both theoretical and experimental findings, in which two isocyanate molecules facilitate the urethane formation process. The theoretical investigation of the reaction mechanism requires an adequate and robust quantum chemical protocol. The fourth generation G4MP2 quantum chemical protocol had been demonstrated several times [20–22] to provide overall thermodynamic results with chemical accuracy [23].

2. Materials and Methods

2.1. Materials

The reaction of phenyl isocyanate (PhNCO) and 1-propanol (PrOH) was conducted at a stoichiometric ratio and at 20-fold isocyanate molar excess. PhNCO (≥99%, Acros Organics BVBA, Geel, Belgium) was used as received. Acetonitrile (ACN) was HPLC grade (VWR International LLC, Debrecen, Hungary). To achieve low water content, PrOH (≥99%, VWR International LLC, Debrecen, Hungary) and tetrahydrofuran (THF) (≥99%, VWR International LLC, Debrecen, Hungary) were stored over 20%(m/V) activated molecular sieves (3Å, beads, VWR International LLC, Debrecen, Hungary) for at least two days [24]. *n*-Butylamine (≥99%) was purchased from Merck Kft. (Budapest, Hungary). *N*,*N*′-diphenylurea (≥98%) was purchased from Alfa-Aesar (Ward Hill, MA, USA).

2.2. Kinetic Experiments

Stock solutions of 2.0 M PhNCO and 2.0 M PrOH in THF (for the stoichiometric runs), and 4.0 M PhNCO and 0.2 M PrOH in THF (for the NCO excess runs) were prepared in volumetric flasks. From the prethermostated ($\pm$0.1 °C) stock solutions, 5.0 mL of PhNCO and 5.0 mL of PrOH solutions were pipetted into a prethermostated glass vial, which was then capped. The experiments were conducted at 303,313 and 323 K. At different time intervals a sample of 10 μL was withdrawn from the reaction mixture and mixed into 990 μL ACN containing 30 μL of *n*-butylamine in order to quench the reaction. The amine reacted spontaneously with the isocyanate to form the adduct *N*-butylphenylurea. The quenched samples were further diluted by a factor of 50 (for the PhNCO excess runs) or 5 (for the stoichiometric runs) with an ACN:H_2O = 1:1 mixture and were subjected to HPLC analysis. The concentration of the *N,N'*-diphenylurea side-product (originating from the hydrolysis of PhNCO) was also determined and was found to be a maximum of 5.6% of the starting PhNCO concentration.

2.3. Analysis Method

Analysis of the quenched and diluted samples was done using a Shimadzu HPLC (Shimadzu Corporation, Kyoto, Japan) equipped with LC-20AD pumps, SIL-20AC autosampler, DGU-20A3R degassing unit, CTO-20A column oven and a SPD-M20A photodiode array detector. A SunShell C8 column (2.6 μm, 150 $\times$ 3.0 mm; ChromaNik Technologies Inc., Osaka, Japan) thermostated at 40 °C was used for the separation. The injection volume was 25 μL. The eluent was ACN:H_2O with a gradient as follows: 0–3.50 min, 42% ACN; 3.51–4.50 min, 82% ACN; 4.51–9.00 min and 42% ACN, at a flow rate of 0.6 mL/min. The product *n*-propyl phenylcarbamate was quantified at 239 nm. For calibration, the reference compound was synthesized from PhNCO in PrOH and purified by flash chromatography.

2.4. Theoretical Method

G4MP2 composite method [23] was applied for obtaining accurate thermodynamic properties, such as zero-point corrected relative energy (ΔE_0), relative enthalpy ($\Delta H(T)$) and relative molar Gibbs free energy ($\Delta G(T,P)$) for the species involved in the studied reaction mechanisms. As part of G4MP2 protocol B3LYP [25], functional was applied in combination with the 6–31G(2df,p) (this basis set is noted as GTBas3 in Gaussian09 [26]) basis set for Berny algorithm driven geometry optimizations (using "tight" convergence criteria with the following thresholds: maximum force = 0.000015, RMS force = 0.000010, maximum displacement = 0.000060 and RMS displacement = 0.000040) and frequency calculations. Normal mode analysis was performed on the optimized structures at the same level of theory to characterize their identities on the potential energy surface (PES). TS structures were also checked by visual inspection of the intramolecular motions corresponding to the imaginary wavenumber using GaussView05 [27] and were confirmed by intrinsic reaction coordinate (IRC) calculations [28] for mapping out the minimal energy pathways (MEP).

For each step of the G4MP2 protocol, including geometry optimization and single point calculations, the SMD polarizable continuum model [29] was used to mimic the effect of the surrounding solvent of 1-propanol (PrOH, ε_r = 20.524) as well as that of tetrahydrofuran (THF, ε_r = 7.4257). It is worthy to note that the static relative permittivity for phenyl isocyanate (PhNCO, ε_r = 8.940 [30]) is close to that of THF; therefore, the potential energy surface (PES) obtained in PhNCO and in THF can be expected to be similar. The SMD model is considered highly accurate, since it achieves mean unsigned errors of 2.5–4.1 kJ/mol in the solvation free energies of neutral species [29] for the reported test set. All quantum chemical calculations were performed by the Gaussian09 [26] software package. The optimized structures and calculated G4MP2 thermochemical properties (E_0, H(298.15 K) and G(298.15 K, 1 atm)) are collected in the Tables S1 and S2 of the Supporting Information.

3. Results and Discussion

3.1. Results of the Kinetic Experiments

The rate constants (k_S for the stoichiometric reaction, $k_{I,obs}$ for the reaction running at 20-fold isocyanate excess) at different temperatures were determined by plotting the urethane concentration against time (Figure 1) and applying a non-linear regression using the kinetic Equation (1) for second order and Equation (2) for pseudo first-order reactions. For the latter, because of the 20-fold isocyanate excess, the isocyanate concentration during the reaction was regarded to be constant ($[PhNCO]_0$). In like manner, the rate constant k_I can be calculated from the observed rate constant k_{obs} (Equation (3)).

$$[urethane] = [PrOH]_0 \times \left(1 - \frac{1}{1 + [PrOH]_0 \times k_S \times t}\right) \tag{1}$$

$$[urethane] = [PrOH]_0 \times \left(1 - e^{-k_{I,obs} \times t}\right) \tag{2}$$

$$k_{I,obs} = k_I \times [PhNCO]_0 \tag{3}$$

It is apparent from Figure 2 that the first few data points fit well for the appropriate equations, namely, the second order one (Equation (1)) for Figure 2a and the pseudo first-order one (Equation (2)) for Figure 2b, but at later reaction stages a positive deviation occurs which possibly accounts from urethane autocatalysis. In case of a stoichiometric NCO/OH ratio, the addition can be described with second-order kinetics up to 50–60% conversion. When the isocyanate is in 20-fold excess, the reaction follows pseudo first-order kinetics only up to a conversion of 25–30%. Therefore, only the initial domain of the data (see Figure 2) were used for non-linear regressions and reaction rate constant calculations.

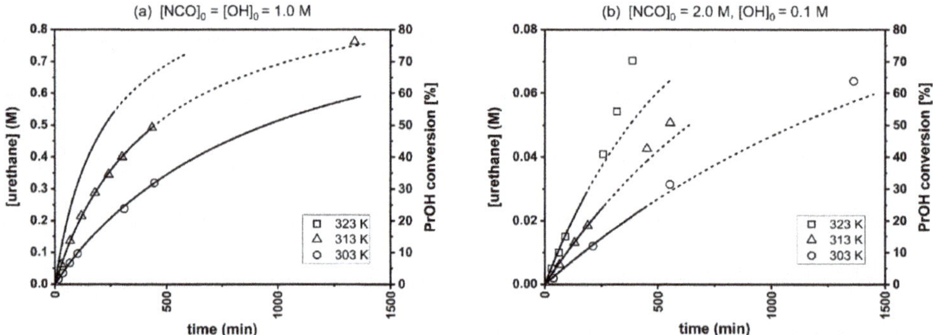

Figure 2. Experimental kinetic curves. (**a**) Second-order kinetics for the stoichiometric ratio. (**b**) pseudo first-order kinetics for the 20-fold PhNCO excess. Data points used for fitting and reaction rate constants' determinations are indicated by solid curve segments.

Table 1 summarizes the kinetic parameters of the reactions. For the alcohol excess reaction, the rate constants (k_A) and the activation energies were measured in our previous work [11]. Both at alcohol excess and at isocyanate excess the Arrhenius activation energies are lower than that of the stoichiometric reaction. (For the Arrhenius plots see Figure S1 in Supporting Information.) From this it is assumed that not only alcohol, but isocyanate molecules can also exert a catalytic effect and facilitate the reaction. At or near stoichiometric ratios, both self-catalytic pathways can occur.

Table 1. Experimental reaction rate constants (k_A, k_S and k_I) at different temperatures, Arrhenius activation energies (E_a) and pre-exponential factors (A). E_a and A values were obtained by the method of least squares. For $[NCO]_0/[OH]_0 = 0.005$, data are taken from [11]. (n.m. = not measured).

Temperature, K	Alcohol Excess $[NCO]_0/[OH]_0 = 0.005$	Stoichiometric Ratio $[NCO]_0/[OH]_0 = 1$	Isocyanate Excess $[NCO]_0/[OH]_0 = 20$
	$k_A \times 10^5$, $M^{-1}\,s^{-1}$	$k_S \times 10^5$, $M^{-1}\,s^{-1}$	$k_I \times 10^5$, $M^{-1}\,s^{-1}$
303	*n.m.*	1.76 ± 0.18	0.52 ± 0.04
313	0.16 ± 0.01	3.72 ± 0.32	0.91 ± 0.07
323	0.23 ± 0.01	7.41 ± 0.60	1.55 ± 0.11
333	0.33 ± 0.02	*n.m.*	*n.m.*
E_a, kJ mol^{-1}	30.4 ± 1.6	58.6 ± 6.0	44.2 ± 4.5
A, M^{-1} s^{-1}	18.8 ± 1.0	234113 ± 23971	214.9 ± 21.9

Rate constants in Table 1 are apparent rate constants, as the values depends on reaction conditions, such as the applied solvent and the concentrations of the reactants.

3.2. Results of the Theoretical Calculations

Hydrogen bond stabilized alcohol associates have been confirmed [16] and their role of reduction of the activation barrier in the urethane formation is already accepted [8]. Therefore, the energies of the PrOH dimer and PhNCO were used as the references in this G4MP2 model calculation. Thermodynamic values for the stationary points of the reactive potential energy surface are summarized in Table 2 and relative zero-point corrected energies in PrOH and THF are also displayed in Figure 3.

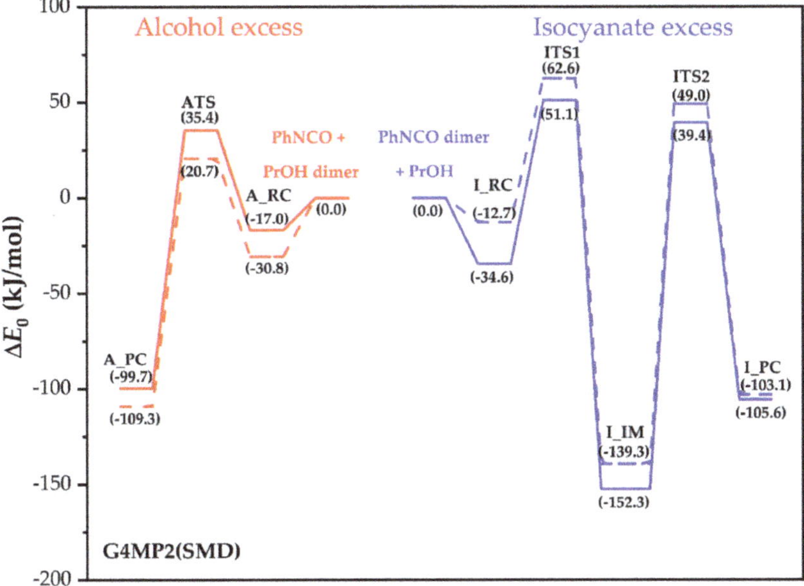

Figure 3. G4MP2 energy profiles (zero-point corrected) for the alcoholic route in solvents 1-PrOH (red solid line) and THF (red dashed line), and for the isocyanate route with 1-PrOH (blue solid line) and THF (blue dashed line).

Table 2. G4MP2 thermochemical properties calculated in 1-propanol (PrOH) and in tetrahydrofuran (THF), including zero-point corrected relative energies (ΔE_0), relative enthalpies ($\Delta H(T)$) and relative Gibbs free energies ($\Delta G(T,P)$) at T = 298.15 K, and P = 1 atm. (A) according to alcohol excess, and according to isocyanate excess (I). All values are in kJ/mol.

Pathway	Species	ΔE_0		$\Delta H(T)$		$\Delta G(T,P)$	
		PrOH	**THF**	**PrOH**	**THF**	**PrOH**	**THF**
	PhNCO + 2 PrOH	0	0	0	0	0	0
Alcohol	A_RC	−17.0	−30.8	−14.0	−27.2	25.1	50.5
Excess (A)	ATS	35.4	20.7	32.7	17.0	91.4	119.2
	A_PC	−99.7	−109.3	−100.9	−109.8	−47.4	−17.2
	2 PhNCO + PrOH	0	0	0	0	0	0
	I_RC	−34.6	−12.7	−33.0	−11.1	−12.7	36.9
Isocyanate	ITS1	51.1	62.6	44.0	55.7	62.6	141.3
Excess (I)	I_IM	−152.3	−139.3	−160.1	−147.2	−139.3	−59.0
	ITS2	39.4	49.0	31.5	41.2	49.0	129.4
	I_PC	−105.6	−103.1	−109.2	−106.4	−103.1	−38.4

In line with the theoretical and experimental work of Raspoet et al. [8], a reactive complex of the alcohol excess reaction (A_RC) had been characterized and its structure is shown in Figure 4. This structure is stabilized by three strong hydrogen bonds between the molecular moieties, and the energy gain of the complex formation is 16.9 kJ/mol in PrOH medium (values obtained in propanol solvent will be discussed further). In this concerted mechanism, the transition state structure (ATS in Figure 4) is a six centered structure. In ATS, the positively charged hydrogen of PrOH shifts to the electron rich nitrogen of the PhNCO, while the NCO group is being bent, activating the carbon for the formation of a new C–O bond, while the other PrOH and the hydrogen of this alcohol's oxygen is transferred to the other alcohol in the same time. Due to the complex interaction network, the transition state energy is only 35.4 kJ/mol above the reactant level, which is consistent with the theoretical value of 27.0 kJ/mol (obtained at the MP2/6-311++G(d,p) or MP2/6-31G(d,p) level of theory) reported by Raspoet et al. [8] for methanol and hydrogen isocyanate. As a result of the IRC calculation, the product complex (A_PC) was also localized and the relevant structural parameters are displayed in Figure 4. As is seen, the urethane bond formed is strongly hydrogen bonded to the oxygen of the remaining PrOH. This exothermic reaction releases 99.7 kJ/mol energy to form A_PC. Interestingly, the relative energies of these stationary points become significantly lower by the replacement of the solvent of PrOH to THF. Obviously, the catalytic effect of the second alcohol can only be manifested when enough PrOH dimer is accessible for the urethane formation reaction.

Despite of intensive use of PhNCO as a proxy in the mechanistic studies for the urethane formation, the physicochemical properties of liquid PhNCO are scarcely mentioned in the literature. For example, only a schematic representation of the intermolecular interactions between PhNCO molecules can be found in the work of Baev [31], with an enthalpy of vaporization value ($\Delta_{vap}H° = 46.5 \pm 0.3$ kJ/mol), while neither the viscosity or liquid structure of PhNCO were never reported to the best of our knowledge. This $\Delta_{vap}H$ value is similar to that of 1-propanol ($\Delta_{vap}H° = 47.5$ kJ/mol) [32]. On the other hand, the kinematic viscosity of PhNCO is 0.96 mm^2/s (298 K) according to our measurement, which is about 2.76 times smaller than that of 1-PrOH (2.65 mm^2/s at 298 K). Due to the recent development of an accurate GAFF-based force field [33] for isocyanate compounds, the structural elucidation of PhNCO liquid is expected to come. Until then, as supported by the above-mentioned $\Delta_{vap}H°$ [31] and interaction energy [19] of PhNCO being similar to those of propanol, one might hypothesize that the PhNCO dimers are stable enough to act as a reactant for the urethane formation under isocyanate excess.

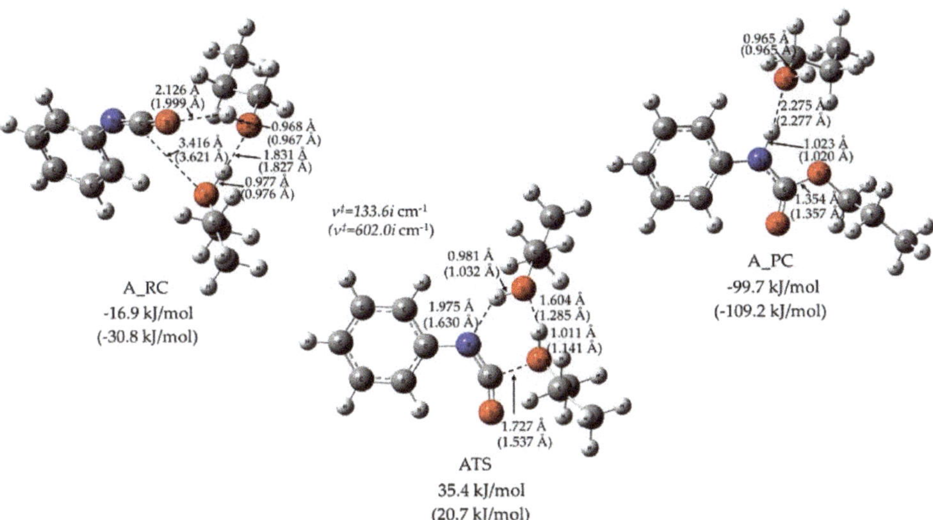

Figure 4. Reactive complex (RC), transition state structure (TS) and product complex (PC) structures (obtained at a B3LYP/6-31G(2df,p) level of theory from the G4MP2 calculation) for the alcohol excess reaction mechanism of urethane bond formation in solvent 1-PrOH or THF (in parenthesis). The relative zero-point corrected energies are also presented in kJ/mol.

The reactive potential energy profile of two phenyl isocyanate molecules with PrOH is shown in Figure 3. The reactive complex (I_RC) is stabilized with a hydrogen bond between the nitrogen of one of the PhNCOs and the hydroxyl of the PrOH molecule, as shown in Figure 5. In addition, the lone pairs of the hydroxyl point towards the positively charged carbon atom of the NCO group in the second PhNCO with a distance of 2.992 Å. These interactions can significantly reduce the relative energy of the reactive complex (−34.6 kJ/mol) compared to that of the reactants. The six-centered transition state structure (ITS1) resulted in the formation of allophanate (I_IM), which has two synchronized bond forming components that are combined with hydrogen-abstraction motion, as shown in Figure 5. In that case, both isocyanate groups are bent, and a long, new C–N bond is being formed between the isocyanate groups (2.320 Å), while the critical distance between the alcohol's oxygen and the isocyanato carbon is extremely small (1.519 Å). In the hydrogen abstraction component of the motion along the reaction coordinate, the moving hydrogen is attacked by the nitrogen of the isocyanato group from relatively large distance ($r_{H\text{-}N}$ = 1.474 Å) and the O–H length is slightly elongated ($r_{O\text{-}H}$ = 1.067 Å). This motion also leads to the formation of a new C=O bond with a distance of 1.337 Å. ITS1 is 51.1 kJ/mol higher in energy compared to the energy level of the reactants (PhNCO dimer and PrOH) and it is 15.7 kJ/mol higher in relative energy than ATS in the case of the alcohol excess mechanism.

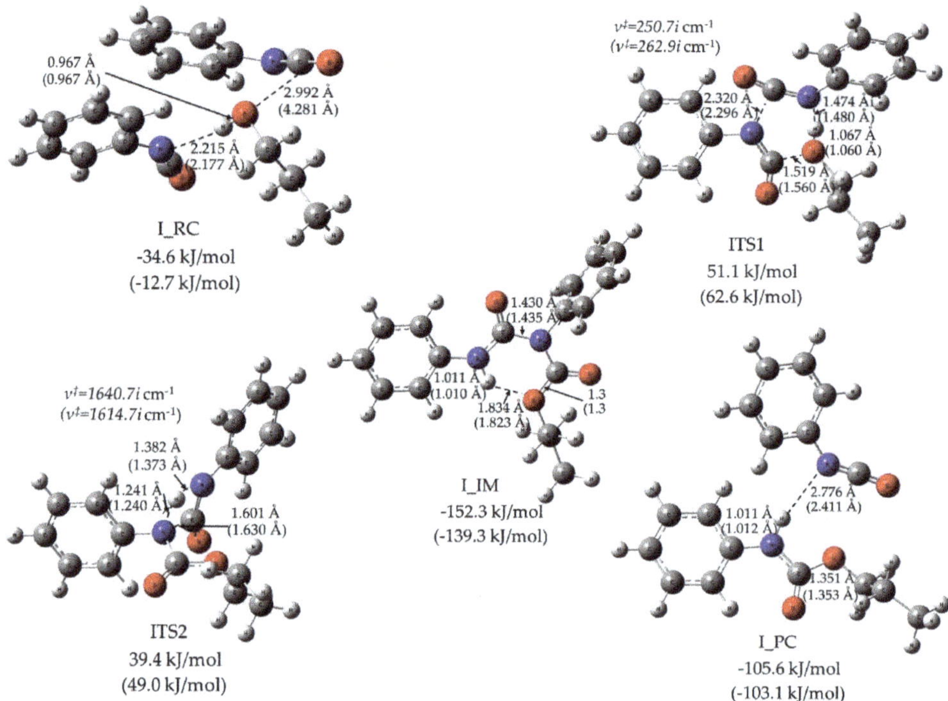

Figure 5. Reactive complex (RC), transition state structure (TS), intermediate (IM) and product complex (PC) structures (obtained at the B3LYP/6 31G(2df,p) level of theory from the G4MP2 calculation) for the isocyanate excess reaction mechanism of urethane bond formation in solvent 1-PrOH or THF (in parenthesis). The relative zero-point corrected energies are also presented in kJ/mol.

As IRC calculation started from ITS1 confirmed, I_RC and I_IM are connected through ITS1. The allophanate formed (I_IM, propyl *N,N'*-diphenylallophanate) is a thermodynamically stable intermediate of this potential energy surface with the corresponding relative zero-point energy of—152.3 kJ/mol. In its planar central structure, a strong intramolecular hydrogen bond can be found with a short H–O distance (r_{NH-O} = 1.834 Å). According to our B3LYP/6-31G(2df,p) calculation, N–H bond stretching mode and its rocking mode can be seen as intensive IR peaks at 3547.7 cm^{-1} and 1574.2 cm^{-1}, respectively. Furthermore, an additional four IR wavenumbers with high intensities can be assigned to the allophanate functional group. Symmetric and asymmetric C=O stretch modes are at 1761.2 cm^{-1} and 1713.7 cm^{-1}, respectively. The remaining two complex vibrational motions of the allophanate ire at 1367.1 cm^{-1} and 1213.1 cm^{-1}. These IR spectral data may be used to monitor the components that take part in the reaction [34], although assignment of these peaks can be difficult due to the multicomponent reaction mixtures, as well as the overlap amongst the IR peaks corresponding to similar functional groups (e.g., allophanate, biuret and urethane). Proper peak assignment for allophanates is still under debate [35].

Nevertheless, the allophanate intermediate can further react through transition state ITS2, leading to the urethane–phenyl isocyanate complex I_PC. As can be seen from Figure 5, ITS2 is a tight, four-centered transition state corresponding to a hydrogen shift from one of the allophanate nitrogens to the other. Comparing the relative energy of ITS1 and ITS2, ITS1 is found to be an energetic bottleneck of this reaction's channel, since all thermodynamic parameters are higher for ITS1 than for ITS2 by at least 11.8 kJ/mol, as shown in Table 2. In contrast to the propanol excess mechanism, solvent change (from PrOH to THF) increased the relative energy, enthalpy and Gibbs free energy values in the isocyanate excess mechanism, as also seen from Table 2.

Allophanate formation in isocyanate excess has been already reported [36], although the previously proposed reaction mechanism starts from a covalently bonded, cyclic isocyanate dimer (uretdione) which then reacts with alcohol to give allophanate. Allophanate can then decompose to urethane and isocyanate. In contrast to that, our proposed mechanism only assumes the formation of the non-covalent dimer, which can react with alcohol through a low-lying, six-centered transition state to form an allophanate intermediate. This transition state is structurally similar to the proposed one at alcohol excess.

4. Conclusions

We conclude, that based on theoretical and experimental results, urethane formation can occur with the active participation of three molecules. One of these molecules, originating from either the excess alcohol or isocyanate, corresponds to self-catalysis. Our new findings indicate that, besides the alcohol-catalyzed route which had already been discussed in the literature and verified by this study, an isocyanate-catalyzed mechanism can also exist. This route, in contrast to the one-step alcohol-catalyzed mechanism, includes two consecutive reactions and the formation of an allophanate intermediate. The key step of the new mechanism is the 1,3-H shift between the nitrogens of the allophanate. The potential energy surface (PES) highly depends on the applied solvent. This agrees with the well-known solvent dependence of the reaction kinetics of urethane formation. The experimental findings, i.e., lower activation energies for either the alcohol or the isocyanate-excess reactions compared to the stoichiometric reaction, also suggest that both self-catalytical pathways could be feasible. Considering the importance of catalysis in polyurethane synthesis, molecular understanding the role of the third molecule in the reaction mechanism of urethane formation gives a new direction to the design of a better catalyst.

Supplementary Materials: The following are available online at http://www.mdpi.com/2073-4360/11/10/1543/s1.

Author Contributions: Conceptualization, M.S., B.V. and P.M.; methodology, Z.F. and M.S.; software, M.S.; validation, M.S. and Z.F.; formal analysis, W.C., Z.B.R. and C.O.C.L.; investigation, W.C., Z.B.R. and C.O.C.L.; resources, B.V. and P.M.; data curation, Z.F. and M.S.; writing—original draft preparation, W.C., C.O.C.L., Z.F. and M.S.; writing—review and editing, Z.B.R.; visualization, W.C. and C.O.C.L.; supervision, Z.F. and M.S.; project administration, B.V.; funding acquisition, B.V. and P.M.

Funding: This research was supported by OTKA 128543, the European Union and the Hungarian State, co-financed by the European Regional Development Fund in the framework of the GINOP-2.3.4-15-2016-00004 project, aimed to promote the cooperation between higher education and industry. Milán Szőri gratefully acknowledges the financial support of the János Bolyai Research Scholarship of the Hungarian Academy of Sciences (BO/00113/15/7) and the additional financial support (Bolyai+) from the New National Excellence Program of the Ministry of Human Capacities (ÚNKP-18-4-ME/4).

Acknowledgments: The GITDA (Governmental Information-Technology Development Agency, Hungary) is also gratefully acknowledged for allocating computing resources used in this work.

Conflicts of Interest: The authors declare no conflict of interest. The funders had no role in the design of the study; in the collection, analyses, or interpretation of data; in the writing of the manuscript, or in the decision to publish the results.

References

1. Boros, R.Z.; Farkas, L.; Nehéz, K.; Viskolcz, B.; Szőri, M. An Ab Initio Investigation of the 4,4'-Methylene Diphenyl Diamine (4,4'-MDA) Formation from the Reaction of Aniline with Formaldehyde. *Polymers* **2019**, *11*, 398. [CrossRef] [PubMed]
2. Davis, T.L.; Farnum, J.M. Relative Velocities of Reaction of Alcohols with Phenyl Isocyanate. *J. Am. Chem. Soc.* **1934**, *56*, 883–885. [CrossRef]

3. Baker, J.W.; Holdsworth, J.B. The Mechanism of Aromatic Side-chain Reactions with Special Reference to the Polar Effects of Substituents. Part XIII. Kinetic Examination of the Reaction of Aryl Isocyanates with Methyl Alcohol. *J. Chem. Soc.* **1947**, 713–726. [CrossRef]

4. Baker, J.W.; Gaunt, J. The Mechanism of the Reaction of Aryl Isocyanates with Alcohols and Amines. Part III. The "Spontaneous" Reaction of Phenyl Isocyanate with Various Alcohols. Further Evidence Relating to the Anomalous Effect of Dialkylanilines in the Base-catalysed Reaction. *J. Chem. Soc.* **1949**, 19–24. [CrossRef]

5. Baker, J.W.; Gaunt, J. The Mechanism of the Reaction of Aryl Isocyanates with Alcohols and Amines. Part, V. Kinetic Investigations of the Reaction between Phenyl Isocyanate and Methyl and Ethyl Alcohols in Benzene Solution. *J. Chem. Soc.* **1949**, 27–31. [CrossRef]

6. Ephraim, S.; Woodward, A.E.; Mesrobian, R.B. Kinetic Studies of the Reaction of Phenyl Isocyanate with Alcohols in Various Solvents. *J. Am. Chem. Soc.* **1958**, *80*, 1326–1328. [CrossRef]

7. Lammiman, S.A.; Satchell, R.S. The Kinetics and Mechanism of the Spontaneous Alcoholysis of *p*-Chlorophenyl Isocyanate in Diethyl Ether. The Association of Alcohols in Diethyl Ether. *J. Chem. Soc. Perkin Trans.* **1972**, *2*, 2300–2305. [CrossRef]

8. Raspoet, G.; Nguyen, M.T. The Alcoholysis Reaction of Isocyanates Giving Urethanes: Evidence for a Multimolecular Mechanism. *J. Org. Chem.* **1998**, *63*, 6878–6885. [CrossRef]

9. Makitra, R.G.; Midyana, G.G.; Pal'chikova, E.Ya.; Romanyuk, A.V. Solvent Effect on the Kinetics of Carbamoylation of Alcohols. *Russ. J. Org. Chem.* **2012**, *48*, 25–31. [CrossRef]

10. Król, P.; Wojturska, J. Kinetic Study on the Reaction of 2,4-and 2,6-Tolylene Diisocyanate with 1-Butanol in the Presence of Styrene, as a Model Reaction for the Process that Yields Interpenetrating Polyurethane–Polyester Networks. *J. Appl. Polym. Sci.* **2003**, *88*, 327–336. [CrossRef]

11. López, C.O.C.; Fejes, Z.; Viskolcz, B. Microreactor Assisted Method for Studying Isocyanate–Alcohol Reaction Kinetics. *J. Flow Chem.* **2019**, *9*, 199–204. [CrossRef]

12. Çoban, M.; Aylin, F.; Konuklar, S. A Computational Study on the Mechanism and the Kinetics of Urethane Formation. *Comput. Theor. Chem.* **2011**, *963*, 168–175. [CrossRef]

13. Kössl, F.; Lisaj, M.; Kozich, V.; Heyne, K.; Kühn, O. Monitoring the Alcoholysis of Isocyanates with Infrared Spectroscopy. *Chem. Phys. Lett.* **2015**, *621*, 41–45. [CrossRef]

14. Wang, X.; Hu, W.; Gui, D.; Chi, X.; Wang, M.; Tian, D.; Liu, J.; Ma, X.; Pang, A. DFT Study of the Proton Transfer in the Urethane Formation between 2,4-Diisocyanatotoluene and Methanol. *Bull. Chem. Soc. Jpn.* **2013**, *86*, 255–265. [CrossRef]

15. Somekawa, K.; Mitsushio, M.; Ueda, T. Molecular Simulation of Potential Energies, Steric Changes and Substituent Effects in Urethane Formation Reactions from Isocyanates. *J. Comput. Chem. Jpn.* **2016**, *15*, 32–40. [CrossRef]

16. Akiyama, I.; Ogawa, M.; Takase, K.; Takamuku, T.; Yamaguchi, T.; Ohtori, N. Liquid Structure of 1-Propanol by Molecular Dynamics Simulations and X-Ray Scattering. *J. Solution Chem.* **2004**, *33*, 797–809. [CrossRef]

17. Sun, W.; Silva, W.G.D.P.; van Wijngaarden, J. Rotational Spectra and Structures of Phenyl Isocyanate and Phenyl Isothiocyanate. *J. Phys. Chem. A* **2019**, *123*, 2351–2360. [CrossRef]

18. Partington, J.R.; Cowley, E.G. Dipole Moments of Ethyl and Phenyl Isocyanates. *Nature* **1935**, *135*, 1038. [CrossRef]

19. Lenzi, V.; Driest, P.J.; Dijkstra, D.J.; Ramos, M.M.D.; Marques, L.S.A. Investigation on the Intermolecular Interactions in Aliphatic Isocyanurate Liquids: Revealing the Importance of Dispersion. *J. Mol. Liquids* **2019**, *280*, 25–33. [CrossRef]

20. Ramakrishnan, R.; Dral, P.O.; Rupp, M.; von Lilienfeld, O.A. Quantum Chemistry Structures and Properties of 134 Kilo Molecules. *Sci. Data* **2015**, *1*, 140022. [CrossRef]

21. Boros, R.Zs.; Koós, T.; Cheikh, W.; Nehéz, K.; Farkas, L.; Viskolcz, B.; Szőri, M. A Theoretical Study on the Phosgenation of Methylene Diphenyl Diamine (MDA). *Chem. Phys. Lett.* **2018**, *706*, 568–576. [CrossRef]

22. Schalk, O.; Townsend, D.; Wolf, T.J.A.; Holland, D.M.P.; Boguslavskiy, A.E.; Szőri, M. Albert Stolow Time-Resolved Photoelectron Spectroscopy of Nitrobenzene and its Aldehydes. *Chem. Phys. Lett.* **2018**, *691*, 379–387. [CrossRef]

23. Curtiss, L.A.; Redfern, P.C.; Raghavachari, K. Gaussian-4 Theory Using Reduced Order Perturbation Theory. *J. Chem. Phys.* **2007**, *127*, 124105. [CrossRef] [PubMed]

24. Williams, D.B.G.; Lawton, M. Drying of Organic Solvents: Quantitative Evaluation of the Efficiency of Several Desiccants. *J. Org. Chem.* **2010**, *75*, 8351–8354. [CrossRef] [PubMed]

25. Becke, A.D. Density-Functional Thermochemistry. III. The Role of Exact Exchange. *J. Chem. Phys.* **1993**, *98*, 5648–5652. [CrossRef]
26. Frisch, M.J.; Trucks, G.W.; Schlegel, H.B.; Scuseria, G.E.; Robb, M.A.; Cheeseman, J.R.; Scalmani, G.; Barone, V.; Mennucci, B.; Petersson, G.A.; et al. *Gaussian 09, Revision, E.01*; Gaussian, Inc.: Wallingford, CT, USA, 2009.
27. Dennington, R.D.; Keith, T.A.; Millam, J.M. *GaussView05*; Semichem Inc.: Shawnee Mission, KS, USA, 2009.
28. Gonzalez, C.; Schlegel, H.B. An Improved Algorithm for Reaction Path Following. *J. Chem. Phys.* **1989**, *90*, 2154–2161. [CrossRef]
29. Marenich, A.V.; Cramer, C.J.; Truhlar, D.G. Universal Solvation Model Based on the Generalized Born Approximation with Asymmetric Descreening. *J. Chem. Theory Comput.* **2009**, *5*, 2447–2464. [CrossRef]
30. Lide, D.R. (Ed.) *CRC Handbook of Chemistry and Physics*, 90th ed.; CRC Press/Taylor and Francis: Boca Raton, FL, USA, 2010.
31. Baev, A.K. *Specific Intermolecular Interactions of Nitrogenated and Bioorganic Compounds*; Springer: Heidelberg, Germany, 2013; p. 318.
32. Majer, V.; Svoboda, V. *Enthalpies of Vaporization of Organic Compounds: A Critical Review and Data Compilation*; Blackwell Scientific Publications: Oxford, UK, 1985; p. 300.
33. Lenzi, V.; Driest, P.J.; Dijkstra, D.J.; Ramos, M.M.; Marques, L.S. GAFF-IC: Realistic Viscosities for Isocyanate Molecules with a GAFF-based Force Field. *Mol. Simulat.* **2019**, *45*, 207–214. [CrossRef]
34. Al Nabulsi, A.; Cozzula, D.; Hagen, T.; Leitner, W.; Müller, T.E. Isocyanurate Formation during Rigid Polyurethane Foam Assembly: A Mechanistic Study Based on in situ IR and NMR Spectroscopy. *Polym. Chem.* **2018**, *9*, 4891–4899. [CrossRef]
35. Stern, T. Hierarchical fractal-structured allophanate-derived network formation in bulk polyurethane synthesis. *Polym. Adv. Technol.* **2017**, *29*, 1–12. [CrossRef]
36. Delebecq, E.; Pascault, J.-P.; Boutevin, B.; Ganachaud, F. On the Versatility of Urethane/Urea Bonds: Reversibility, Blocked Isocyanate, and Non-isocyanate Polyurethane. *Chem. Rev.* **2013**, *113*, 80–118. [CrossRef] [PubMed]

 *polymers*

Article

Synthetic Environmentally Friendly Castor Oil Based-Polyurethane with Carbon Black as a Microphase Separation Promoter

Jia-Wun Li [1], Wen-Chin Tsen [2], Chi-Hui Tsou [3], Maw-Cherng Suen [4,*] and Chih-Wei Chiu [1,*]

[1] Department of Materials Science and Engineering, National Taiwan University of Science and Technology, Taipei 10607, Taiwan

[2] Department of Fashion and Design, LEE-MING Institute of Technology, No. 22, Sec. 3, Tailin. Rd., New Taipei City 24305, Taiwan

[3] Sichuan Provincial Key Lab of Process Equipment and Control, Material Corrosion and Protection Key Laboratory of Sichuan Province, College of Materials Science and Engineering, Sichuan University of Science and Engineering, Zigong 643000, China

[4] Department of Fashion Business Administration, LEE-MING Institute of Technology, New Taipei City 24305, Taiwan

* Correspondence: sunmc0414@gmail.com (M.-C.S.); cwchiu@mail.ntust.edu.tw (C.-W.C.)

Received: 9 July 2019; Accepted: 8 August 2019; Published: 12 August 2019

Abstract: This study created water polyurethane (WPU) prepolymer by using isophorone diisocyanate, castor oil, dimethylolpropionic acid, and triethanolamine (TEA) as the hard segment, soft segment, hydrophilic group, and neutralizer, respectively. TEA, deionized water, and carbon black (CB) were added to the prepolymer under high-speed rotation to create an environmentally friendly vegetable-oil-based polyurethane. CB served as the fortifier and promoter of microphase separation. Fourier transform infrared spectroscopy was performed to elucidate the role of H-bond interactions within the CB/WPUs. Additionally, atomic force microscopy was conducted to determine the influence of H-bond interactions on the degree of microphase separation in the WPU. Furthermore, this study used four-point probe observation to discover the materials' conductivity of CB in the WPU. Thermogravimetric analysis and dynamic mechanical analysis were performed to measure the thermal properties of the CB/WPUs. The mechanical properties of CB/WPUs were measured using a tensile testing machine. The CB/WPUs were also soaked in 1 wt.% NaOH solution for different amounts of time to determine the degradation properties of the CB/WPUs. Finally, scanning electron microscopy was performed to observe the topography of the CB/WPUs after degradation.

Keywords: carbon black; microphase separation promoter; polyurethane; vegetable oil; environmentally friendly

---

1. Introduction

The promotion of green materials has attracted global attention to the development of environmentally friendly polymers. Numerous studies have investigated vegetable oils [1,2], such as castor oil (CO) [3,4] and soybean oil [5]. CO is extracted from the seeds of caster plants. Because every CO molecule consists of three –OH groups, CO can serve as a cross-link agent [6]. Scholars have imported CO into polyurethane (PU) structures [7,8], and even implemented the postmodified ester [9,10] and alkenyl [11,12] groups of CO in PU. PU has outstanding mechanical properties, abrasion resistance, and stiffness [13,14]. Therefore, this study synthesized green PU by using natural materials and isocyanate [15]. Water polyurethane (WPU) has inferior mechanical properties 5 to oil PU [16]. The mechanical properties of PU are mainly determined by the degree of microphase separation during PU formation. Microphase separation is generated by the polarity between hard and soft segments, which also causes the clustering of hard segments [17,18]. The formation of microphase separation mainly arises from the generation of hydrogen bond between the -NH group on urethane group (–COONH–) hard segment and C=O group, which promotes the aggregation of hard segment.

Scholars have employed 1-octadecanol as a promoter to improve the degree of microphase separation and mechanical properties in PU [19]. In other studies, the H-bond interaction between inorganic substances and PU has been used to increase the degree of microphase separation. Xia et al. discovered that adding a suitable amount of silicate clay increased the degree of completion of microphase separation during the PU production process. However, the addition of excessive amounts of clay lowered the degree of microphase separation [20]. Studies have also reported that the addition of multiwalled nanotubes accelerated the microphase separation of structures within the thermoplastic PU matrix [21,22]. This type of microphase separation is mainly caused by the H-bond interaction formed by the –COOH or –OH group of the inorganic substance and the C=O group in the thermoplastic PU. The increase in the degree of microphase separation also improved the mechanical properties of PU. Although WPU has poorer mechanical properties than PU, the WPU synthesis process does not require the use of solvents. Thus, the production of WPU creates fewer volatile organic compounds and is thus more environmentally friendly [23]. In one study, scholars used isophorone diisocyanate (IPDI) and CO to successfully synthesize WPU with outstanding mechanical properties [24]. The present study aimed to synthesize the aforementioned type of green PU and employ inorganic substances as the microphase separation accelerator for producing environmentally friendly PU nanocomposites. The inclusion of inorganic substances in PU can endow PU with various functions [25,26]. For example, the blending of PU and Ag produced an antibacterial material [27]; the use of oyster shell powder as a nucleating agent in PU imbued PU with the antibacterial properties of the oyster shell powder, producing a biomedical material [28]; and the addition of carbon material to PU produced a conductive material [29]. Carbon black (CB) is a natural product produced when hydrocarbon substances are burned in insufficient air. The surface layer of CB comprises –OH groups, which can form H-bond interactions with C=O groups in PU. These groups not only can attract the hard segments to around CB to increase the microphase separation, but also can promote the spread of CB within PU, creating PU with conductivity and adding to the functionality of PU to enable production of static conductive materials. This study employed the –OH group in CB to promote microphase separation in CO-based PU and endow PU with static conductive functionality.

This study employed IPDI, CO, dimethylolpropionic acid (DMPA), and triethanolamine (TEA) as the hard segment, soft segment, hydrophilic group, and neutralizer, respectively, to create a WPU prepolymer. Scheme 1 displays the process of adding TEA, deionized water, and CB to the PU under high-speed rotation to produce environmentally friendly CO-based WPU. Figure 1 shows the cycle of environmental CB/WPUs. Fourier transform infrared spectroscopy (FTIR) was used to determine the functional group changes. Thermogravimetric analysis (TGA) and dynamic mechanical analysis (DMA) were used to investigate the thermal properties of the CB/WPUs. Additionally, the mechanical properties and surface topography of the CB/WPUs were obtained using a tensile testing machine and atomic force microscopy (AFM). A four-point probe was employed to measure the conductivity of the CB/WPUs. Subsequently, this study analyzed the degradation properties of the CB/WPUs after they were soaked in NaOH for different periods of time and employed scanning electron microscopy (SEM) to observe the topography of the hydrolyzed CB/WPUs.

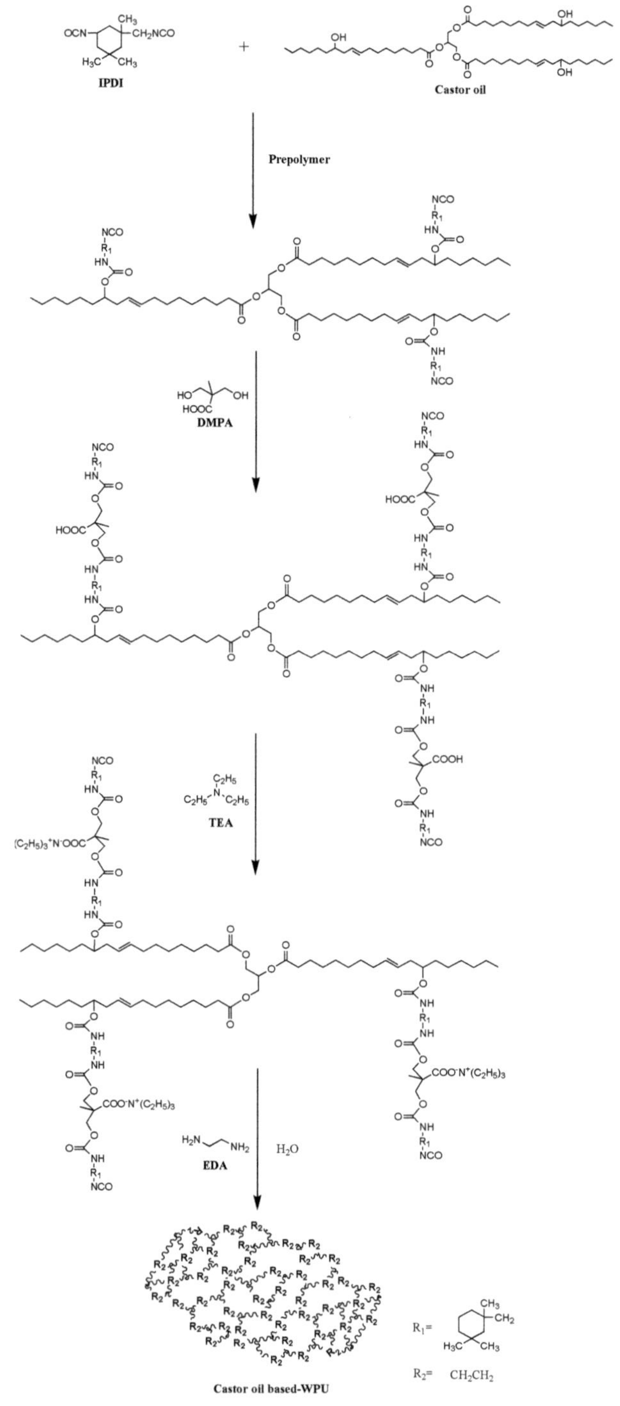

Scheme 1. Formula for the castor oil based-WPU.

due to the H-bond interaction produced by the –OH group of CB and –C=O group of PU. The FTIR results revealed that CB/WPU-02 had the most H-bonds, causing an unstable film-forming process and increasing the average surface roughness. The phase figure indicates that WPU had similar microphase separation conditions as PU. When 1 wt.% of CB was added, the number of surface hard segments was increased. The hard segments (i.e., the white dots in Figure 3) showed the presence of CB. Because in the phase diagram of AFM, it is able to identify the soft or hard degree of the film, and when the values are higher (white dots), it represents the film is harder. The microphase separation in PU can promote the separation of soft segment and hard segment. The hard segment belongs to the harder area in the film. So the white dots represent the hard segment (IPDI), but they can also represent CB (because CB is relatively harder compared to the film). This indicates that CB promotes microphase separation. When 2 wt.% of CB was added, the hard segments began to cluster and the soft segments (i.e., darker area) were evenly distributed. This result indicated that the degree of microphase separation was greater in CB/WPU-02 than in CB/WPU-01. The phase figure of CB/WPU-03 indicates that when the amount of CB added was 3 wt.%, the white dots were clustered. Microphase separation in PU is caused by the different polarity of hard and soft segments. Additionally, hard segment clustering is caused by H-bond interactions between hard segments. Because the CB/WPU-02 film exhibited the most H-bond interactions, it had the greatest degree of microphase separation.

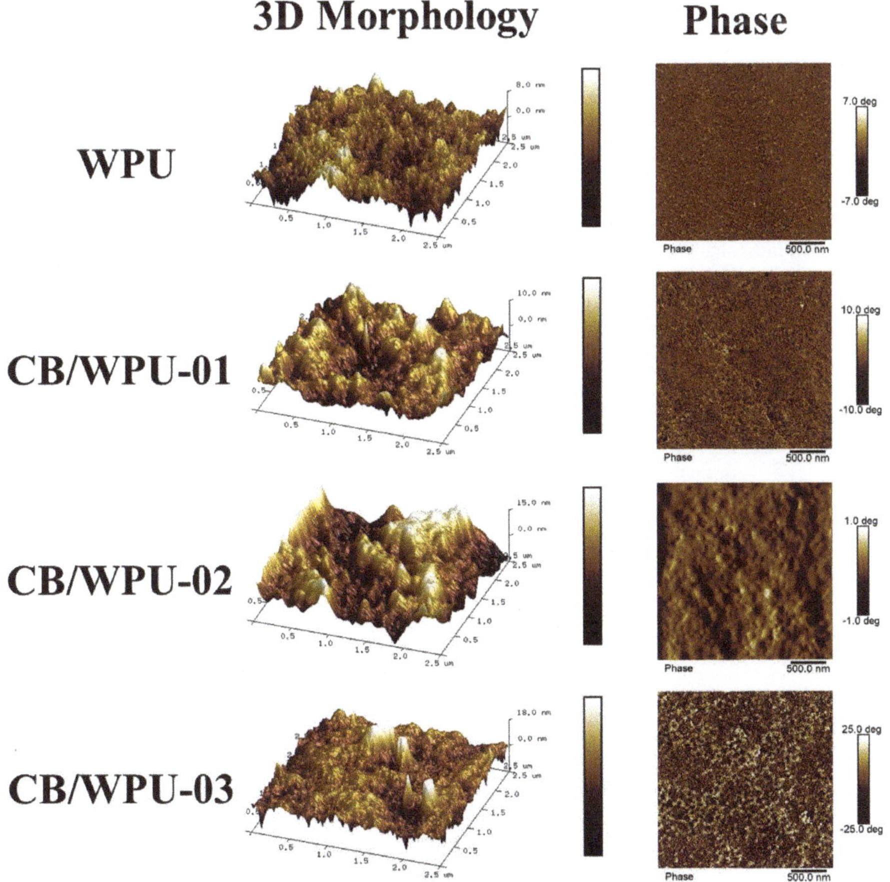

Figure 3. AFM 3D Morphology and phase images of the CB/WPUs.

The amount of CB substantially influences the conductivity of CB/WPUs. The resistive properties of the CB/WPUs are displayed in Table 3. The standard WPU did not exhibit conductivity. When 1 wt.% CB was added, the resistive value was 2.01×10^6, indicating that the addition of 1 wt.% CB gave the WPU static conductivity. CB/WPU-02 and CB/WPU-03 had resistive values of 8.41×10^5 and 3.99×10^4, respectively. This was because more CB was present in the WPU and indicates that all samples had static conductivity.

Table 3. DMA and electrical resistance results of the CB/WPUs.

Sample	T_{gd} from Tanδ (°C)	Tan δ_{max}	Electrical Resistance (ohm)
WPU	52.66	0.403	– [a]
CB/WPU-01	59.36	0.400	2.01×10^6
CB/WPU-02	59.41	0.398	8.41×10^5
CB/WPU-03	55.82	0.384	3.99×10^4

[a] Indicates that the resistance value is not detected.

3.3. Thermal Properties

Figure 4 displays the TGA curves of WPUs containing different amounts of CB. Table 2 records the following TGA values of the samples: T_{onset}, T_{50}, and 700 °C residue. The results indicated that the T_{onset} of WPU, CB/WPU-01, CB/WPU-02, and CB/WPU-03 were 290.49 °C, 294.89 °C, 297.13 °C, and 299.58 °C, respectively. Additionally, the T_{50} of WPU, CB/WPU-01, CB/WPU-02, and CB/WPU-03 were 337.05 °C, 344.55 °C, 347.21 °C, and 348.43 °C. Thus, when 3 wt.% CB was added to the WPU, the temperature required to reach T_{onset} or T_{50} was increased by 10 °C. This was because CB has greater thermostability than WPU. Therefore, the thermostability of WPU was increased by adding more CB. Furthermore, the 700 °C residue masses of WPU, CB/WPU-01, CB/WPU-02, and CB/WPU-03 were 1.81%, 2.24%, 2.30%, and 2.45%, respectively. The increase of residue mass in CB/WPUs with CB content again indicates that more CB has remained in WPU.

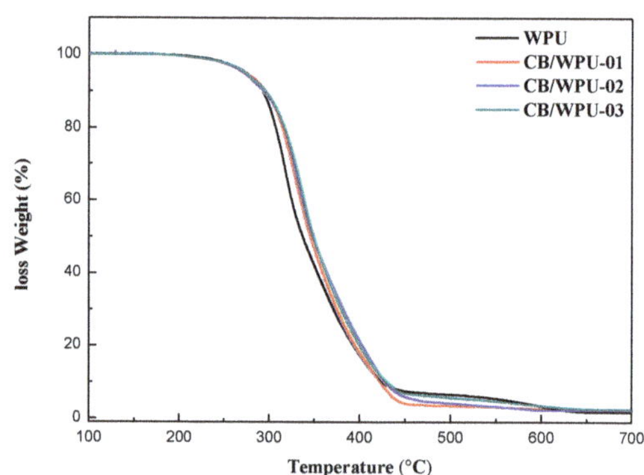

Figure 4. TGA curves of the CB/WPUs.

3.4. Dynamic Mechanical Analysis (DMA)

Figure 5 displays the tanδ of the WPUs obtained when different amounts of CB were added. The dynamic glass transition temperature is denoted T_{gd}. Table 3 lists the maximum T_{gd} and tanδ for the different CB/WPUs. The T_{gd} of the standard WPU was 52.66 °C. Adding 2 wt.% CB to WPU resulted in a higher T_{gd} of 59.41 °C. This was possibly due to H-bond interactions between the –OH groups in CB and C=O groups in WPU. This H-bond interaction prevents WPU molecular chains from moving, causing a higher T_{gd}. However, when 3 wt.% CB was added, the T_{gd} was lower. This may have been caused by the extensive clustering of CB, which restrained the –OH groups of CB and prevented them from forming H-bonds with the C=O groups, as was demonstrated in the FTIR analysis. Additionally, the maximum tanδ of WPU, CB/WPU-01, CB/WPU-02, and CB/WPU-03 was 0.403, 0.400, 0.398, and 0.384, respectively. The tanδ is obtained from the loss modulus divided by the storage module. So, lower tanδ value means harder membrane. Thus, the maximum tanδ decreased with an increase of CB. This was because the addition of CB increased the hardness of the CB/WPUs, reducing the flexibility and increasing the stickiness of the WPUs.

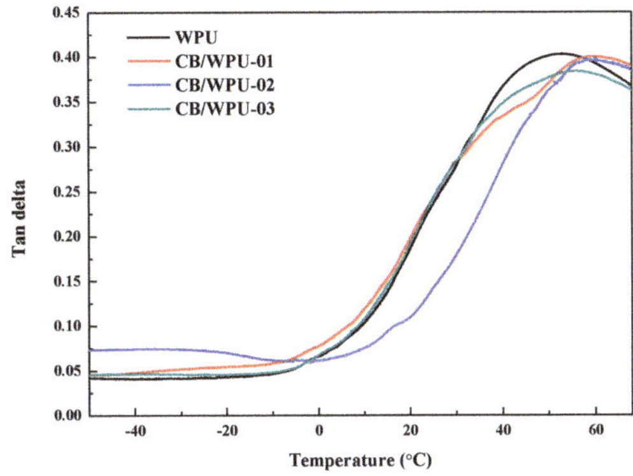

Figure 5. DMA Tan δ curves of the CB/WPUs.

3.5. Tensile Properties

Figure 6 displays the stress–strain curve of the various CB/WPUs. Table 4 records the maximum tensile strength, breaking strain, and Young's modulus of each CB/WPU. The maximum tensile strength of WPU, CB/WPU-01, CB/WPU-02, and CB/WPU-03 was 9.7, 12.9, 13.8, and 10.4 MPa, respectively, whereas their breaking strains were 123%, 108%, 69%, and 60%. CB/WPU-02 thus had the greatest tensile strength. Possible reasons for this include the dispersibility of CB when 2 wt.% CB was added to WPU and the H-bond interactions between CB and WPU. In these two situations, the degree of microphase separation was increased. This microphase separation result was also shown in the AFM results. When an excessive amount of CB (i.e., 3 wt.%) was added, the CB in the WPU was severely clustered. This clustering increased the stress concentration point of the WPU and caused more defects in the material, reducing the maximum tensile strength and breaking strain. The Young's modulus of WPU, CB/WPU-01, CB/WPU-02, and CB/WPU-03 was 0.85, 1.39, 3.64, and 1.77 MPa, respectively. CB/WPU-02 thus exhibited the greatest Young's modulus. This result is consistent with the aforementioned description.

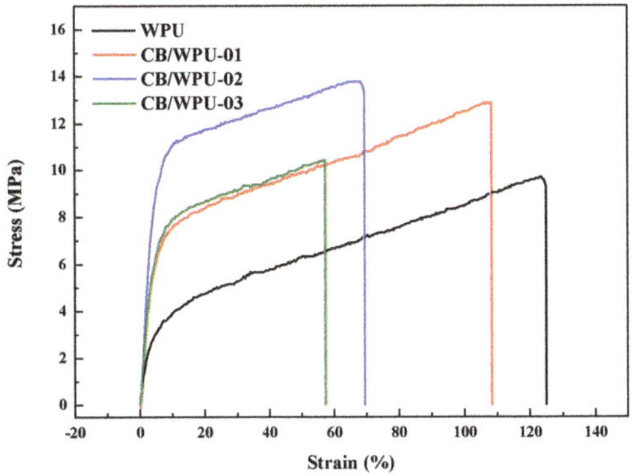

Figure 6. Tensile properties of the CB/WPUs.

Table 4. Tensile properties of the CB/WPUs.

Sample	Tensile Strength (MPa)	Young's Modulus (MPa)	Elongation at Break (%)
WPU	9.7	0.85	123
CB/WPU-01	12.9	1.39	108
CB/WPU-02	13.8	3.64	69
CB/WPU-03	10.4	1.77	60

3.6. Hydrolytic Degradation

This study sought to develop an environmentally friendly and degradable material. Figure 7 displays the mass loss of the CB/WPUs after degradation in 1 wt.% NaOH. The mass loss of all samples exceeded 15% after 12 h of degradation. After 16 h of degradation, most of the polymer structure was fractured, causing difficulties in sampling. CB/WPU-02 exhibited the lowest degradation speed. This was because the CB/WPU-02 had the most H-bond interactions. When the amount of CB added was 3 wt.%, the CB was severely clustered within the WPU, increasing the number of defects in the material. This enabled 1 wt.% NaOH to enter the CB/WPU-03 film more easily and caused the CB/WPU-03 to lose more mass.

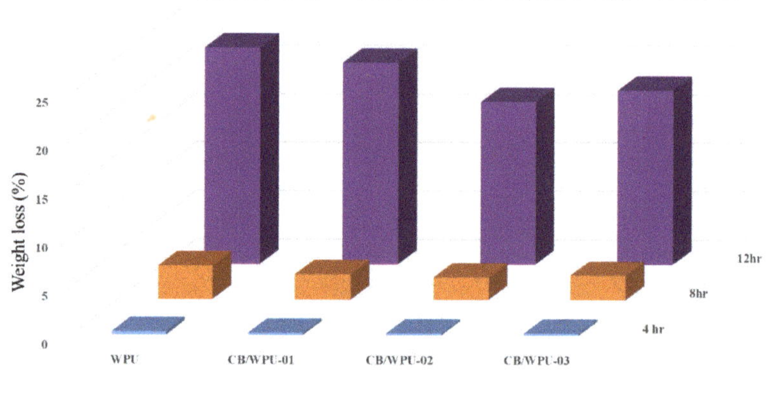

Figure 7. Weight loss of hydrolytic degradation results of the CB/WPUs.

3.7. SEM Morphology Analysis

Figure 8 displays the SEM surface topology of CB/WPUs after degradation in 1 wt.% NaOH for different amounts of time. After 12 h, numerous holes had developed on the WPU sample containing no CB, whereas CB/WPU-01 had considerably fewer holes. This was because the H-bond interactions between CB and WPU prevented the 1 wt.% NaOH solution from infiltrating the WPU film. The surface topography of CB/WPU-02 is rougher than the other samples. This was possibly because more H-bond interactions were causing the greatest degree of microphase separation within the CB/WPU-02 (such as AFM). The high degree of microphase separation promoted the clustering of degradable casotor oil segments, so after CB/WPU-02 was immersed in 1 wt.% NaOH solution, the clustered soft segments on CB/WPU-02 surface were degraded and the shape was changed. After being degraded for 12 h, large cavities formed on CB/WPU-03. This was because of severe CB clustering in the WPU, which caused numerous defects in the CB/WPU-03 film. In summary, from the degradation loss rate and SEM, it was known that CB/WPU-02 had less degradation loss rate and no formation of cavities, and this was possibly because the H-bond interactions inhibited 1 wt.% NaOH, permeating into CB/WPUs film.

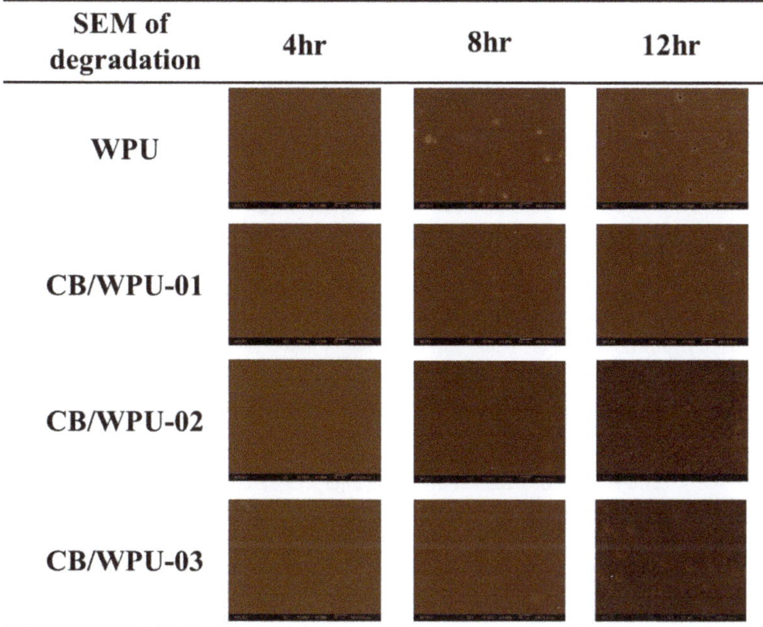

Figure 8. Scanning electron microscope micrographs of the CB/WPUs with hydrolytic degradation for 4, 8, and 12 h at 45 °C.

4. Conclusions

This study successfully synthesized an environmentally friendly vegetable-oil-based PU. FTIR analyses revealed that CB/WPU-02 had the most H-bond interactions. The AFM and conductivity test results revealed that 2 wt.% CB was the optimal amount of CB to add to promote microphase separation. This was due to CB microclustering when 2 wt.% CB was employed. The CB microclustering and H-bond interactions caused CB/WPU-02 to exhibit the highest degree of microphase separation. Additionally, TGA revealed that greater amounts of CB in the WPU resulted in higher initial degradation temperatures. The DMA and tensile test revealed that the CB/WPU-02 film had the highest T_{gd} and most favorable mechanical properties. When the amount of CB added was 3 wt.%, the CB clustered severely and numerous defects were created. The degradation experiment and SEM results revealed that CB/WPU-02 exhibited the greatest degradation stability of all samples. NaOH was found to not easily infiltrate the CB/WPU-02 film, causing the degradation process to commence from the film surface. In summary, when 1 wt.% of CB is added (CB/WPU-01), it has already become the static conductive material (such as Table 3). So the cost has already been reduced, it is much more suitable to be used in the application of operation room and weapon storehouse. As for the raise of benefit, 2 wt.% of CB addition (CB/WPU-02) will be the most appropriate addition level.

Author Contributions: Formal analysis, J.-W.L.; Project administration, M.-C.S. and C.-W.C.; Resources, C.-H.T.; Writing—review and editing, W.-C.T.

Funding: This research received no external funding.

Conflicts of Interest: The authors declare no conflict of interest.

References

1. Zhu, Y.; Romain, C.; Williams, C.K. Sustainable polymers from renewable resources. *Nature* **2016**, *540*, 354–362. [CrossRef] [PubMed]

2.   Samarth, N.B.; Mahanwar, P.A. Modified vegetable oil based additives as a future polymeric material. *Open. J. Org. Polym. Mater.* **2015**, *5*, 1–22. [CrossRef]

3.   Díez-Pascual, A.M.; Díez-Vicente, A.L. Wound healing bionanocomposites based on castor oil polymeric films reinforced with chitosan-modified ZnO nanoparticles. *Biomacromolecules* **2015**, *16*, 2631–2644. [CrossRef] [PubMed]

4.   Ibrahim, S.; Ahmad, A.; Mohamed, N. Characterization of novel castor oil-based polyurethane polymer electrolytes. *Polymers* **2015**, *7*, 747–759. [CrossRef]

5.   Xu, Y.; Yuan, L.; Wang, Z.; Wilbon, P.A.; Wang, C.; Chu, F.; Tang, C. Lignin and soy oil-derived polymeric biocomposites by "grafting from" RAFT polymerization. *Green Chem.* **2016**, *18*, 4974–4981. [CrossRef]

6.   Gurunathan, T.; Mohanty, S.; Nayak, S.K. Isocyanate terminated castor oil-based polyurethane prepolymer: Synthesis and characterization. *Prog. Org. Coat.* **2015**, *80*, 39–48. [CrossRef]

7.   Rojas, M.F.; Miranda, L.P.; Ramirez, A.M.; Quintero, K.P.; Bernard, F.; Einloft, S.; Carreño Díaz, L.A. New biocomposites based on castor oil polyurethane foams and ionic liquids for CO_2 capture. *Fluid Phase Equilib.* **2017**, *452*, 103–112. [CrossRef]

8.   Nguyen Dang, L.; Le Hoang, S.; Malin, M.; Weisser, J.; Walter, T.; Schnabelrauch, M.; Seppälä, J. Synthesis and characterization of castor oil-segmented thermoplastic polyurethane with controlled mechanical properties. *Eur. Polym. J.* **2016**, *81*, 129–137. [CrossRef]

9.   Dave, V.J.; Patel, H.S. Synthesis and characterization of interpenetrating polymer networks from transesterified castor oil based polyurethane and polystyrene. *J. Saudi Chem. Soc.* **2017**, *21*, 18–24. [CrossRef]

10.   Li, S.; Xu, C.; Yang, W.; Tang, Q. Thermoplastic Polyurethanes Stemming from Castor Oil: Green Synthesis and Their Application in Wood Bonding. *Coatings* **2017**, *7*, 159.

11.   Ionescu, M.; Radojčić, D.; Wan, X.; Shrestha, M.L.; Petrović, Z.S.; Upshaw, T.A. Highly functional polyols from castor oil for rigid polyurethanes. *Eur. Polym. J.* **2016**, *84*, 736–749. [CrossRef]

12.   Chen, G.; Guan, X.; Xu, R.; Tian, J.; He, M.; Shen, W.; Yang, J. Synthesis and characterization of UV-curable castor oil-based polyfunctional polyurethane acrylate via photo-click chemistry and isocyanate polyurethane reaction. *Prog. Org. Coat.* **2016**, *93*, 11–16. [CrossRef]

13.   Otto, G.P.; Moisés, M.P.; Carvalho, G.; Rinaldi, A.W.; Garcia, J.C.; Radovanovic, E.; Fávaro, S.L. Mechanical properties of a polyurethane hybrid composite with natural lignocellulosic fibers. *Compos. Part B* **2017**, *110*, 459–465. [CrossRef]

14.   Ashrafizadeh, H.; Mertiny, P.; McDonald, A. Evaluation of the effect of temperature on mechanical properties and wear resistance of polyurethane elastomers. *Wear* **2016**, *368*, 26–38. [CrossRef]

15.   Zia, F.; Zia, K.M.; Zuber, M.; Kamal, S.; Aslam, N. Starch based polyurethanes: A critical review updating recent literature. *Carbohydr. Polym.* **2015**, *134*, 784–798. [CrossRef] [PubMed]

16.   Wu, Y.; Du, Z.; Wang, H.; Cheng, X. Preparation of waterborne polyurethane nanocomposite reinforced with halloysite nanotubes for coating applications. *J. Appl. Polym. Sci.* **2016**, *133*, 43949. [CrossRef]

17.   Liu, H.; Huang, W.; Yang, X.; Dai, K.; Zheng, G.; Liu, C.; Shen, C.; Yan, X.; Guo, J.; Guo, Z. Organic vapor sensing behaviors of conductive thermoplastic polyurethane–graphene nanocomposites. *J. Mater. Chem. C* **2016**, *4*, 4459–4469. [CrossRef]

18.   Li, J.W.; Lee, H.T.; Tsai, H.A.; Suen, M.C.; Chiu, C.W. Synthesis and Properties of Novel Polyurethanes Containing Long-Segment Fluorinated Chain Extenders. *Polymers* **2018**, *10*, 1292. [CrossRef]

19.   Chen, S.; Cao, Q.; Jing, B.; Cai, Y.; Liu, P.; Hu, J. Effect of microphase-separation promoters on the shape-memory behavior of polyurethane. *J. Appl. Polym. Sci.* **2006**, *102*, 5224–5231. [CrossRef]

20.   Xia, H.; Song, M.; Zhang, Z.; Richardson, M. Microphase separation, stress relaxation, and creep behavior of polyurethane nanocomposites. *J. Appl. Polym. Sci.* **2007**, *103*, 2992–3002. [CrossRef]

21.   Hosseini-Sianaki, T.; Nazockdast, H.; Salehnia, B.; Nazockdast, E. Microphase separation and hard domain assembly in thermoplastic polyurethane/multiwalled carbon nanotube nanocomposites. *Polym. Eng. Sci.* **2015**, *55*, 2163–2173. [CrossRef]

22.   Landa, M.; Canales, J.; Fernández, M.; Muñoz, M.E.; Santamaría, A. Effect of MWCNTs and graphene on the crystallization of polyurethane based nanocomposites, analyzed via calorimetry, rheology and AFM microscopy. *Polym. Test.* **2014**, *35*, 101–108. [CrossRef]

23.   Ghosh, B.; Gogoi, S.; Thakur, S.; Karak, N. Bio-based waterborne polyurethane/carbon dot nanocomposite as a surface coating material. *Prog. Org. Coat.* **2016**, *90*, 324–330. [CrossRef]

24. Liu, K.; Miao, S.; Su, Z.; Sun, L.; Ma, G.; Zhang, S. Castor oil-based waterborne polyurethanes with tunable properties and excellent biocompatibility. *Eur. J. Lipid Sci. Technol.* **2016**, *118*, 1512–1520. [CrossRef]
25. Wang, G.; Fu, Y.; Guo, A.; Mei, T.; Wang, J.; Li, J.; Wang, X. Reduced graphene oxide–polyurethane nanocomposite foam as a reusable photoreceiver for efficient solar steam generation. *Chem. Mater.* **2017**, *29*, 5629–5635. [CrossRef]
26. Dong, M.; Li, Q.; Liu, H.; Liu, C.; Wujcik, E.K.; Shao, Q.; Ding, T.; Mai, X.; Shen, C.; Guo, Z. Thermoplastic polyurethane-carbon black nanocomposite coating: Fabrication and solid particle erosion resistance. *Polymer* **2018**, *158*, 381–390. [CrossRef]
27. Fu, H.; Wang, Y.; Li, X.; Chen, W. Synthesis of vegetable oil-based waterborne polyurethane/silver-halloysite antibacterial nanocomposites. *Compos. Sci. Technol.* **2016**, *126*, 86–93. [CrossRef]
28. Tsou, C.H.; Wu, C.S.; Hung, W.S.; De Guzman, M.R.; Gao, C.; Wang, R.Y.; Chen, J.; Wan, N.; Peng, Y.J.; Suen, M.C. Rendering polypropylene biocomposites antibacterial through modification with oyster shell powder. *Polymer* **2019**, *160*, 265–271. [CrossRef]
29. Liu, H.; Huang, W.; Gao, J.; Dai, K.; Zheng, G.; Liu, C.; Shen, C.; Yan, X.; Guo, J.; Guo, Z. Piezoresistive behavior of porous carbon nanotube-thermoplastic polyurethane conductive nanocomposites with ultrahigh compressibility. *Appl. Phys. Lett.* **2016**, *108*, 011904.
30. Shearouse, W.C.; Lillie, L.M.; Reineke, T.M.; Tolman, W.B. Sustainable polyesters derived from glucose and castor oil: Building block structure impacts properties. *ACS Macro Lett.* **2015**, *4*, 284–288. [CrossRef]
31. Lillie, L.M.; Tolman, W.B.; Reineke, T.M. Degradable and renewably-sourced poly(ester-thioethers) by photo-initiated thiol–ene polymerization. *Polym. Chem.* **2018**, *9*, 3272–3278. [CrossRef]
32. Yang, W.; Cheng, X.; Wang, H.; Liu, Y.; Du, Z. Surface and mechanical properties of waterborne polyurethane films reinforced by hydroxyl-terminated poly(fluoroalkyl methacrylates). *Polymer* **2017**, *133*, 68–77. [CrossRef]

Communication

Sequential Recovery of Heavy and Noble Metals by Mussel-Inspired Polydopamine-Polyethyleneimine Conjugated Polyurethane Composite Bearing Dithiocarbamate Moieties

Dingshuai Xue [1],*, **Ting Li [2]**, **Guoju Chen [3]**, **Yanhong Liu [1]**, **Danping Zhang [1]**, **Qian Guo [1]**, **Jujie Guo [1]**, **Yueheng Yang [1]**, **Jiefang Sun [4]**, **Benxun Su [1]**, **Lei Sun [5]** and **Bing Shao [2]**

[1]   State Key Laboratory of Lithospheric Evolution, Institute of Geology and Geophysics, Chinese Academy of Sciences, Beijing 100029, China

[2]   School of Public Health, Capital Medical University, Beijing 100069, China

[3]   State Key Laboratory for Comprehensive Utilization of Nickel and Cobalt Resources, Jinchang 737100, China

[4]   Beijing Key Laboratory of Diagnostic and Traceability Technologies for Food poisoning, Beijing Center for Disease Prevention and Control, Beijing 100013, China

[5]   Center for Biological Imaging, Institute of Biophysics, Chinese Academy of Sciences, Beijing 100101, China

*   Correspondence: xuedingshuai@mail.iggcas.ac.cn; Tel.: +86-10-8299-8487

Received: 7 February 2019; Accepted: 19 June 2019; Published: 2 July 2019

Abstract: Dithiocarbamate-grafted polyurethane (PU) composites were synthesized by anchoring dithiocarbamate (DTC) as a chelating agent to the polyethyleneimine-polydopamine (PE-DA)-functionalized graphene-based PU matrix (PE-DA@GB@PU), as a new adsorbent material for the recovery of Cu^{2+}, Pb^{2+}, and Cd^{2+} from industrial effluents. After leaching with acidic media to recover Cu^{2+}, Pb^{2+}, and Cd^{2+}, dithiocarbamate-grafted PE-DA@GB@PU (DTC-g-PE-DA@GB@PU) was decomposed and PE-DA@GP was regenerated. The latter was used to recover Pd^{2+}, Pt^{4+}, and Au^{3+} from the copper leaching residue and anode slime. The present DTC-g-PE-DA@GB@PU and PE-DA@GB@PU composites show high adsorption performance, effective separation, and quick adsorption of the target ions. The morphologies of the composites were studied by scanning electron microscopy and their structures were investigated by Fourier transform infrared (FT-IR) spectroscopy and Raman spectroscopy. The effects of pH values, contact time, and initial metal ion concentration conditions were also studied. An adsorption mechanism was proposed and discussed in terms of the FT-IR results.

Keywords: polyurethane; graphene; metallurgy; polymeric composites; thiocarbamate

---

1. Introduction

Heavy metal ions (Cu^{2+}, Pb^{2+}, and Cd^{2+}) are non-biodegradable, accumulative, and persistent environmental contaminants. In recent years, the surge of industrial effluents and mine wastewater has intensified environmental problems, posing a severe harm to ecological and global public health because of the carcinogenicity, high toxicity, and biological accumulation of such ions [1]. Pd^{2+}, Pt^{4+}, and Au^{3+} are well-known noble metals (NMs) used for coins, jewelry, catalysts, and electric and corrosion-resistant materials. As NM resources are limited, their demand in the jewelry market and for medical applications is generally quite high [2]. Consequently, considerable research efforts have been made toward the removal of heavy metal ions from industrial wastewater and recovery of NMs from industrial waste, which usually involve using adsorbents derived from polyurethane (PU) by virtue of its extremely low cost, 3D porous structure, easy handling and storage [3]. Jinchuan Group Ltd. is China's biggest platinum group metals producer. During the electrorefining process, some heavy metals (HMs) in industrial effluents are discharged to the tailings dam and some NMs accumulate in the byproduct copper leaching residue and anode slime. The latter is an important source for recycling and recovery of Au, Pd, and Pt [4]. These sludges are categorized as hazardous wastes, and their storage and disposal are very costly. However, as these wastes contain valuable metals in abundance, it is extremely important to maximize the utilization and recycling of HMs and NMs economically to achieve both environmental protection and sustainable development.

For this purpose, many kind of adsorbents have been synthesized and various techniques such as nanoparticles [5], nanofibers [6,7], ion exchange [8,9], membrane separation [10,11], and solvent extraction [12,13] have been investigated over the past few years. However, the extraction of HM and NM ions from wastes is still challenging. In this respect, the critical issues are low selectivity and insufficient adsorption capacity of the adsorbents. PU is the most practical adsorbent material for NMs. Graphene-based composite materials have attracted considerable attention because of the large specific surface area of graphene [14]. It has been shown that the fusion of graphene oxide (GO) with PU polymers can enhance the mechanical properties and surface area of graphene-based PU (GB@PU) [15].

In the present work, a polyethyleneimine-polydopamine (PE-PDA) hybrid coating was spontaneously co-deposited onto the surface of GB@PU [16]. By taking advantage of the self-polymerization of catechol-inspired natural PDA, PE-DA coated GB@PU (PE-DA@GB@PU) hybrid was prepared via Michael addition or Schiff base reactions [17]. The homogeneous and well-controlled coating layer of PDA on GB@PU improved the hydrophilicity of the GO sheets. Simultaneously, the PDA layer offered abundant active sites to graft the high-density PE molecule. The prepared PE-DA@GB@PU composite exhibited excellent selectivity and high adsorption capacity for NMs. In order to further improve the adsorbent performance, dithiocarbamate (DTC) was anchored on the surface of PE-DA@GB@PU. DTC could strongly chelate with various metal ions owing to its sulfur and nitrogen groups, which provided lone pair electrons form chemical bonding [18,19]. First, DTC-g-PE-DA@GB@PU was used to recover HMs from industrial effluents. When the adsorption efficiency of DTC-g-PE-DA@GB@PU decreased, 0.1 M HCl was used to leach the HMs and regenerate PE-DA@GB@PU, which was used to recover NMs from the copper leaching residue and anode slime of Jinchuan Group Ltd. This new design strategy greatly improves sorbent utilization efficiency and saves the raw material. Continuous recycling of two different types of metal ions is more environmentally-friendly than other sorbents. In addition, the adsorbents are readily regenerated and exhibit excellent stability and recyclability.

2. Materials and Methods

2.1. Materials

Open-cell type polyether PU was purchased from Taobao (China). Graphene oxide solution was acquired from XFNANO Materials Tech Co., Ltd. (China). Dopamine hydrochloride (DA) and PE (M_w = 600 Da) were obtained from Sigma-Aldrich (USA) and Aladdin (China), respectively. Other chemicals, including dopamine hydrochloride (DA), PE (M_w = 600 Da), ethanol (EtOH), sodium hydroxide (NaOH), ammonium hydroxide, and carbon disulfide (CS_2), were obtained from Sinopharm Chemical Reagent Co., Ltd. The stock standard solutions for Cu^{2+}, Pb^{2+}, Cd^{2+}, Pd^{2+}, Pt^{4+}, and Au^{3+} were procured from Spex CertiPrep Inc. (Metuchen, NJ, USA).

2.2. Instrumentation

An inductively coupled plasma optical emission spectrometer (ICP-OES) was used to determine the concentration of metal ions (Table S1, Supplementary Materials). A 540 Zeiss scanning electron microscope (SEM, Merlin, Oberkochen, Germany) with energy dispersive X-ray spectroscopy (EDS, Oxford Instruments, Oxford, the UK), Fourier-transform infrared spectrometer (FTIR, Bruker VERTEX 70v, Karlsruhe, Germany), Raman spectrometer (HR800, HORIBA Jobin Yvon, Paris, France) and wide-angle X-ray diffraction (WAXD, Panalytical X'Pert PRO, Almelo, The Netherlands) were used for the characterization of the new materials.

2.3. Synthesis of PE-DA@GB@PU and DTC-g-PE-DA@GB@PU

The PE-DA@GB@PU and DTC-g-PE-DA@GB@PU materials were prepared from PU, as illustrated in Scheme 1. Raw PU (3.0 g) was boiled in 2 M HCl solution for 2 h for the amination of PU (APU). GB@PU was prepared by a dip coating and solvothermal synthesis method [20,21]. First, AP was immersed in GO (2.0 mg mL^{-1}) solution, then transferred into a Tetrafluoro autoclave. After ultrasonication for 1 h, the Tetrafluoro autoclave was heated to 90 °C and maintained at this temperature for 12 h. Next, GB@PU was dried in a vacuum oven at 50 °C. Finally, GB@PU was immersed in Tris buffer solution (pH = 8.5) of DA (2 mg/mL) and PE (4 mg/mL) in a mass ratio of 1:1 (25 °C, 24 h). After modification, the samples were withdrawn and washed several times using deionized water, then dried in a vacuum oven at 50 °C overnight. Subsequently, PE-DA@GB@PU was immersed in 20 mL of 1.0 M NaOH solution, and 10 mL of 0.1 M CS$_2$ solution in ethanol was added dropwise to the reaction system. The reaction mixture was then stirred at room temperature for 24 h. Finally, DTC-g-PE-DA@GB@PU was filtered and rinsed four times with water [22].

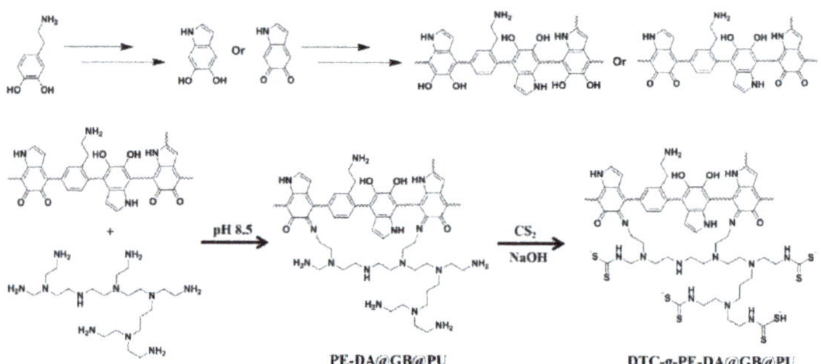

Scheme 1. Synthetic routes of PE-DA@GB@PU and DTC-g-PE-DA@GB@PU.

2.4. Static Adsorption Experiments

Adsorption experiments were carried out by batch method. A series of standards or sample solutions containing Cu^{2+}, Pb^{2+}, Cd^{2+}, Pd^{2+}, Pt^{4+}, and Au^{3+} were transferred into a 50-mL triangular bottle with a grinding plug. The pH of each solution was adjusted to the desired value and the volume was adjusted to 20 mL. Either PE-DA@GB@PU or DTC-g-PE-DA@GB@PU (0.02 g) was added, and the mixture was shaken vigorously to attain equilibrium by a vibrating machine. The remaining metal ions in the solution were determined by ICP-OES. The amounts of HM and NM ions adsorbed by the adsorbent were calculated by the difference.

The percentage of each metal ion adsorbed on the adsorbent (A) was calculated according to Equation (1):

$$A = \frac{(D_i - D_e) * 100\%}{D_i} \tag{1}$$

After equilibration, the extent of adsorption (F_e (mg g^{-1})) was calculated according to Equation (2):

$$F_e = \frac{(D_i - D_e)B}{w} \tag{2}$$

where D_i and D_e are the initial and equilibrium concentrations of the metal ions in the solution, respectively; and B and w are the volume (mL) and mass (g) of the solution and adsorbent, respectively.

3. Results

Dopamine can easily form insoluble PDA in alkaline solutions via versatile reactions with molecules containing thiol or amine groups by Michael addition and/or Schiff base reactions. Herein, PE was introduced on the GB@PU surface with PE-DA hybrid coating because it had a larger amine density, which proved advantageous for interaction with CS_2 to form dithiocarbamate composites. The reaction scheme is represented in Scheme 1 [23,24].

3.1. Characterization of PE-DA@GB@PU and DTC-g-PE-DA@GB@PU

Raw PU (Figure 1a) at 70× magnification exhibited a smooth reticular skeleton, whereas the surface of the APU was denuded (Figure 1b). As shown in Figure 1c, the appearance of GB@PU was markedly different from that of raw PU and APU. Raman scattering revealed that the coating observed on the skeleton of GB@PU was composed of graphene sheets (Figure S1, Supplementary Materials). The skeleton of PE-DA@GB@PU was covered with a large number of microscale protrusions (Figure 1d). As shown in Figure 1e,f the skeleton of DTC-g-PE-DA@GB@PU was more smooth without protrusions as compared with PE-DA@GB@PU.

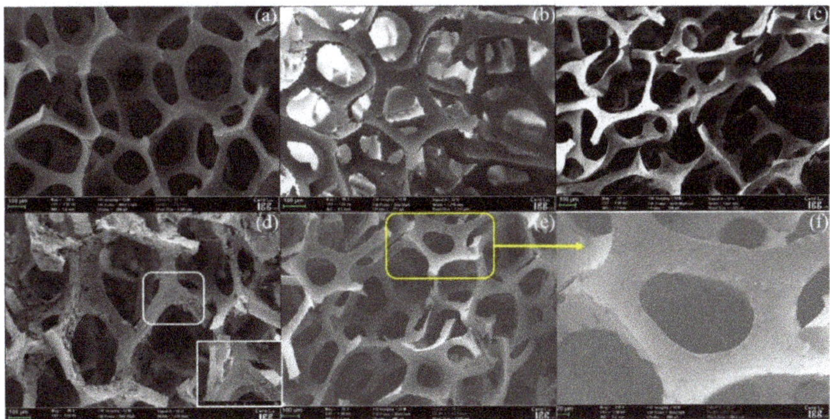

Figure 1. SEM image of (**a**) raw PU, (**b**) APU, (**c**) GB@PU, (**d**) PE-DA@GB@PU, and (**e**) DTC-g-PE-DA@GB@PU, (**f**) magnification SEM image of (**e**).

The chemical structures of PU, APU, GB@PU, PE-DA@GB@PU and DTC-g-PE-DA@GB@PU were analyzed by FTIR spectroscopy, and their FTIR spectra are shown in Figure 2. The broadening of the band at 3308 cm^{-1} in the spectrum of APU was ascribed to stretching vibration of N–H for PU shifted to 3293 cm^{-1} for APU [25]. After graphene coated on the APU skeleton, the peaks of GB@PU between 2700 and 3500 cm^{-1} became weaker, because the graphene coating inhibited spectral absorption. Compared to APU and GB@PU, PE-DA@GB@PU showed a new absorption peak at ~1650 cm^{-1} and a broad band from 3200 to 3600 cm^{-1}. The former could be assigned to the carbonyl groups derived from dopamine oxides, whereas the latter likely corresponded to the stretching vibration of hydroxy or amino groups of dopamine and PE [17]. For DTC-g-PE-DA@GB@PU, the characteristic peak of NH_2 group at 3216 cm^{-1} disappeared and a new adsorption band could be observed at 958 cm^{-1}, which indicated that the amino groups had reacted with CS_2. However, the characteristic peaks of the DTC group (N–CS_2 vibration and C=S vibration) were obscured by the strong absorption peaks of the PU matrix at ~1490 and 1100 cm^{-1}.

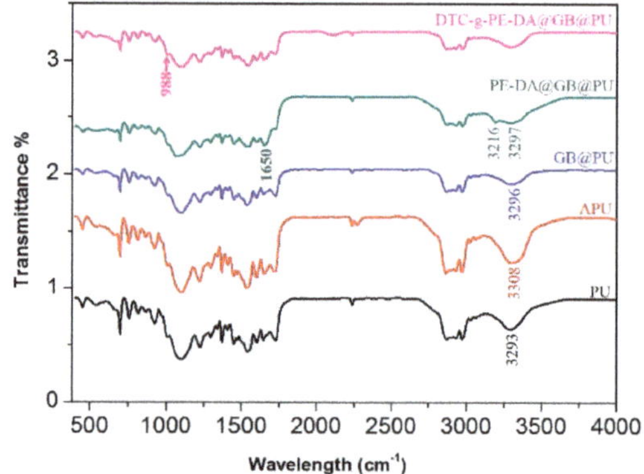

Figure 2. FTIR spectra of (**a**) raw PU, (**b**) APU, (**c**) GB@PU, (**d**) PE-DA@GB@PU, and (**e**) DTC-g-PE-DA@GB@PU.

3.2. pH Effect

Solution pH is one of the most important factors for the metal ions adsorption [26]. The influence of pH on the sorption of HM and NM ions on DTC-g-PE-DA@GB@PU and PE-DA@GB@PU, respectively, was studied over a pH range (Figure 3a,b). As shown in Figure 3a, there was a general increase in the recoveries of these ions with increasing pH of the solution. At low pH, the H^+ ions competed with HM ions for the ion-exchange reaction at the adsorption sites, and accordingly, low adsorption recoveries were observed. With increasing pH, the concentration of H^+ ions decreased and the active adsorption sites mainly turned into dissociated states, which were sufficient for catching the HMs by coordination interaction between dithiocarbamate groups and HM ions. According to the literature [27], DTC compounds are generally decomposed into the constituent amine and carbon disulfide in acidic media (presented in Figure S2, Supplementary Materials). After 0.1 M HCl was used to elute the HM ions from DTC-g-PE-DA@GB@PU, PE-DA@GB@PU is regenerated. PE-DA@GB@PU exhibited excellent extraction yields of Pd^{2+}, Pt^{4+}, and Au^{3+} over a wide pH range because of the rich amino groups and flexible chains of PE [17] (Figure 3b).

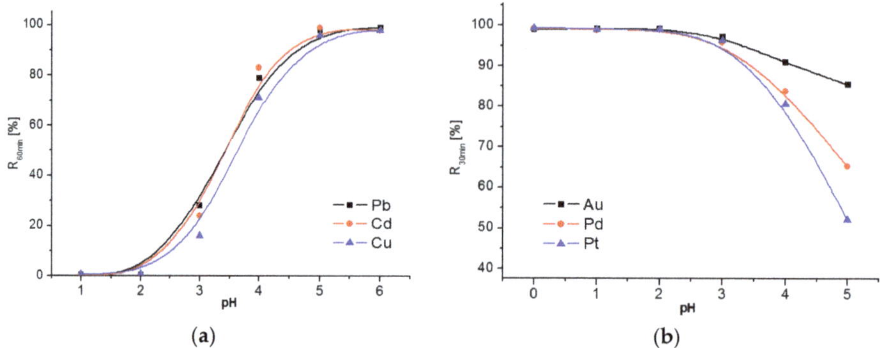

Figure 3. Effect of pH on the adsorption of Cu^{2+}, Pb^{2+}, and Cd^{2+} monometallic solutions (10.0 µg mL^{-1}) by DTC-g-PE-DA@GB@PU (**a**) and the adsorption of Pd^{2+}, Pt^{4+}, and Au^{3+} monometallic solutions (1.0 µg mL^{-1}) by PE-DA@GB@PU (**b**). Other conditions: adsorbent, 20 mg; shaking time: heavy metal ions 60 min, noble metal ions 30 min; and temperature, 25 °C.

3.3. Adsorption Isotherms

The adsorption capacities of DTC-g-PE-DA@GB@PU for HM ions and PE-DA@GB@PU for NM ions were assessed using the equilibrium adsorption isotherm by varying the initial concentrations (The concentrations of Cu, Pb, Cd, Pd, Pt, and Au were in the ranges 20–100, 20–180, 20–120, 20–800, 20–600, and 20–800 µg mL^{-1}, respectively), as shown in Figure 4.

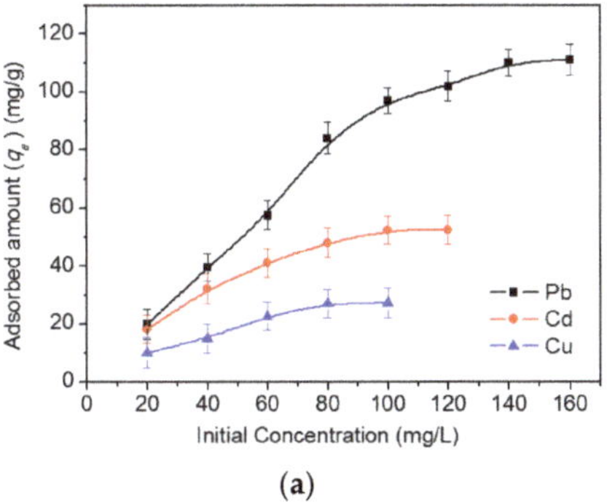
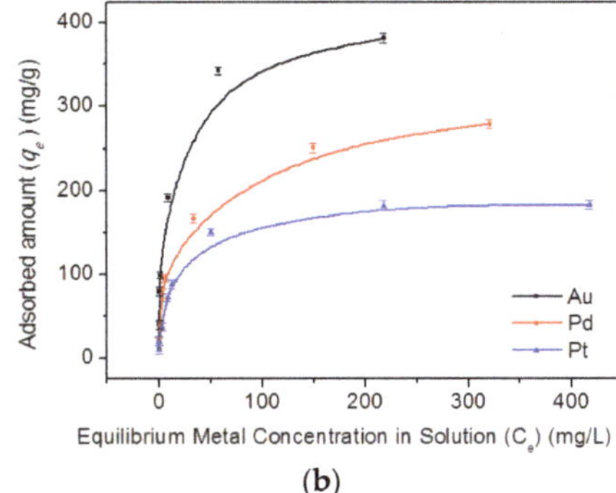

(a)   (b)

Figure 4. Adsorption isotherms of Cu^{2+}, Pb^{2+}, and Cd^{2+} onto DTC-g-PE-DA@GB@PU (a) and Pd^{2+}, Pt^{4+}, and Au^{3+} onto PE-DA@GB@PU (b). Other conditions: solution pH = 6 for DTC-g-PE-DA@GB@PU, solution pH = 0 for PE-DA@GB@PU; volume, 20 mL; adsorbent dose, 20 mg; contact time: 30 min for PE-DA@GB@PU; 60 min for DTC-g-PE-DA@GB@PU, temperature, 25 °C; n = 3.

The equations of Langmuir, Freundlich, and Dubinin–Radushkevich isotherm models were used to analyze the experimental data. The experimental parameters are presented in Table 1.

Table 1. Parameters of the Langmuir, Freundlich, and Dubinin–Radushkevich isotherms for the adsorption of HM and NM ions onto DTC-g-PE-DA@GB@PU and PE-DA@GB@PU, respectively, at 25 °C.

Isotherms	Isotherm Constants	Heavy Metal Ions			Noble Metal Ions		
		Cu^{2+}	Pb^{2+}	Cd^{2+}	Pd^{2+}	Pt^{4+}	Au^{3+}
Langmuir	F_{max} (mg g^{-1})	41.2	113.9	57.1	285.7	185.2	384.6
	K (L mg^{-1})	0.02915	0.3776	0.1726	0.0986	0.1192	0.2524
	R^2	0.9164	0.9935	0.9971	0.9985	0.9988	0.9997
	R_L	0.2554–0.6317	0.0145–0.1169	0.0461–0.2246	0.0125–0.9103	0.0138–0.8935	0.0049–0.7985
Freundlich	K_f (mg g^{-1}) (L mg^{-1})$^{1/n}$	2.816	44.17	15.57	37.025	32.946	58.844
	n	1.823	4.200	3.207	2.682	3.095	2.597
	R^2	0.9431	0.9891	0.9635	0.9653	0.9614	0.9102
Dubinin–Radushkevich	F_{D-R} (mg g^{-1})	30.6	98.2	45.6	128.14	101.64	195.43
	E	0.082	0.780	0.728	2.946	3.402	2.884
	R^2	0.9832	0.7024	0.8152	0.6336	0.7531	0.7947

The adsorption isotherms of Pb^{2+}, Cd^{2+}, Pd^{2+}, Pt^{4+}, and Au^{3+} fitted well with the Langmuir model, which implied monolayer adsorption (linear correlation coefficients values (R^2) for this model were greater than 0.9935). The theoretical F_{max} values of Pb^{2+}, Cd^{2+}, Pd^{2+}, Pt^{4+}, and Au^{3+} (113.9, 57.1, 285.7, 185.2, and 384.6 mg g^{-1}, respectively) acquired from the Langmuir model were close to the experimental values (111.9, 53.8, 279.6, 175.3, and 380.5, mg g^{-1}, respectively) at 25 °C. As for Cu(II), the D–R isotherm model fitted best to the experimental data, as concluded when its R^2 value was compared with the R^2 values of the Langmuir and Freundlich models. In addition, the maximum adsorption Q_{D-R} value calculated from the experimental data (30.6 mg g^{-1}) was in good accordance with the experimental Q value (28.7 mg g^{-1}) (Figures S3–S5, Supplementary Materials). Comparison of the maximum capacities of DTC-g-PE-DA@GB@PU and PE-DA@GB@PU with those of other adsorbents is shown in Table S2 (Supplementary Materials). Shannon reported [28] that the ionic radii of the heavy metal ions are in the following order: $Pb^{2+} > Cd^{2+} > Cu^{2+}$. As is all known, dithiocarbamate group can bond to metal ions through its sulfur moiety. Based on the hard and soft acids and bases (HSAB) principle, sulfur is a soft base, which easily forms bonds with large, highly polarizable metal ions (soft-soft interaction). For the three HM ions, Pb^{2+} is much larger and much softer (more polarizable) than the others and interacts more strongly with sulfur. As metal ions become larger and more polarizable ($Cu^{2+} < Cd^{2+} < Pb^{2+}$), the tendency for attachment of the increasingly "soft" metal ions to sulfur increases. The resultant $[Pb(DTC)_4]^{2-}$ ions is a soft–soft combination, while $[Cu(DTC)_4]^{2-}$ is a relatively hard–soft combination, consistent with the HSAB prediction. Therefore, the maximum adsorption capacity order is $Pb^{2+} > Cd^{2+} > Cu^{2+}$. Meanwhile, the adsorption rate has a similar trend. In fact, other researchers have reported the same trend as that obtained in our study [27,29,30].

3.4. Adsorption Kinetics

Figure 5 illustrates the adsorption of HM (Figure 5a) and NM (Figure 5b) ions on the adsorbent as a function of time. The adsorption plots showed that the initial adsorption rate of HM and NM ions was high. With the passage of time, the rate of percent adsorption decreased and then reached equilibrium, which was due to the gradual occupation of the available adsorption sites on DTC-g-PE-DA@GB@PU and PE-DA@GB@PU.

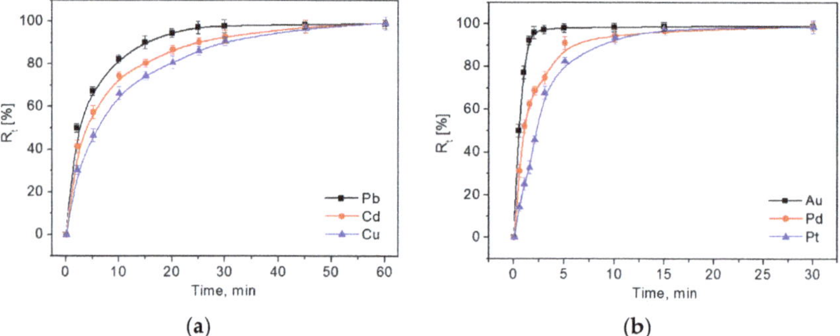

Figure 5. Effect of contact time on the adsorption of 10.0 µg mL^{-1} Cu^{2+}, Pb^{2+}, and Cd^{2+} onto DTC-g-PE-DA@GB@PU (**a**) and 1.0 µg mL^{-1} Pd^{2+}, Pt^{4+}, and Au^{3+} onto PE-DA@GB@PU (**b**). Other conditions: solution pH = 6 for DTC-g-PE-DA@GB@PU, solution pH = 0 for PE-DA@GB@PU; volume, 20 mL; adsorbent dose, 20 mg; temperature, 25 °C; ($n = 3$).

The results of various adsorption parameters obtained from the kinetic models are presented in Table 2. The pseudo-second-order kinetic model fitted well as compared to the pseudo-first-order model. The plots were not linear over the time range, implying that more than one process affected the adsorption process, thereby confirming that adsorption involved multiple stages (Figures S6–S8, Supplementary Materials).

Table 2. Kinetic parameters for the adsorption of HM and NM ions onto DTC-g-PE-DA@GB@PU and PE-DA@GB@PU, respectively (25 °C).

Model	Parameters	Heavy Metal Ions			Noble Metal Ions		
		Cu^{2+}	Pb^{2+}	Cd^{2+}	Pd^{2+}	Pt^{4+}	Au^{3+}
Pseudo-first-order	k_1 (min^{-1})	0.1020	0.1105	0.1021	0.1951	0.2077	0.1967
	R^2	0.9440	0.9649	0.9657	0.9456	0.9240	0.6513
Pseudo-second-order	k_2 (g mg^{-1}min^{-1})	0.6965	2.2519	1.1523	5.1564	1.1750	26.8477
	R^2	0.9988	0.9995	0.9993	0.9997	0.9838	0.9998
Intraparticle diffusion	K_i (mg g^{-1} min$^{-0.5}$)	0.02166	0.01451	0.01773	0.0247	0.0369	0.0118
	R^2	0.8860	0.7155	0.8487	0.6995	0.7628	0.3221

3.5. Adsorption Mechanism

In order to understand the mechanism of adsorption of Cu^{2+}, Pb^{2+}, and Cd^{2+} onto DTC-g-PE-DA@GB@PU, and Pd^{2+}, Pt^{4+}, and Au^{3+} onto PE-DA@GB@PU, FTIR spectroscopy studies were conducted for DTC-g-PE-DA@GB@PU and HM ions loaded DTC-g-PE-DA@GB@PU (abbreviated as DTC-g-H), and PE-DA@GB@PU and NM ions loaded PE-DA@GB@PU (abbreviated as PE-N).

The FT-IR spectrum for DTC-g-PE-DA@GB@PU (Figure 6) showed peaks at 1112, 958, and 1435 cm^{-1}, which were attributed to C=S vibration, C–S stretching vibration, and N–CS$_2$ vibration, respectively [19]. There was a red-shift and decrease in intensity of the characteristic FTIR DTC-g-H signals (C–S and C=S stretching vibrations) after the adsorption of HM ions. In addition, sharp new peaks were observed at 1375 cm^{-1} in the spectra of DTC-g-Cu, DTC-g-Pb, and DTC-g-Cd. These results indicated that a strong metal-ligand bond had formed between the metal ions and chelating groups in DTC-g-PE-DA@GB@PU.

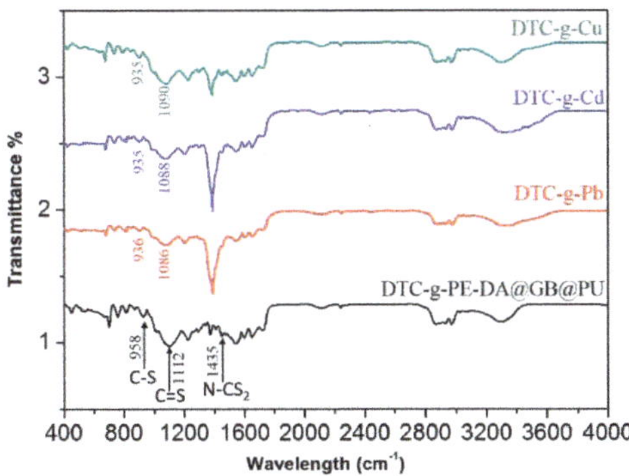

Figure 6. FT-IR spectra of DTC-g-PE-DA@GB@PU adsorbent and the DTC-g-Cu, DTC-g-Pb, and DTC-g-Cd complexes.

As seen in Figure 7, after the adsorption of NM ions, the N–H stretching peak shifted from 3297 to 3319 cm^{-1} and the peak 3216 cm^{-1} disappeared altogether, which was likely attributable to either the electrostatic attraction between the positively charged amine groups in the adsorbent and the negatively charged NM anions [31,32], or because the nitrogen atoms in the amino groups coordinated with the NM ions. Moreover, the interaction between the NM ions and NH group resulted in the broadening of the peaks from 3100 to 3600 cm^{-1}.

The mechanisms of HM and NM ions adsorption by DTC-g-PE-DA@GB@PU and PE-DA@GB@PU were also elucidated by the SEM/EDS measurements before and after loading the HM and NM ions, respectively. DTC-g-PE-DA@GB@PU and PE-DA@GB@PU were shaken in 100-ppm HM ions and 1000-ppm NM ions, respectively, for half an hour. The HM and NM ions were prepared using their nitrates and adjusted to optimal pH. The presence of Cu, Pb, Cd, Pd, Pt, and Au peaks, along with other elemental peaks (where the carbon (C) and oxygen (O) peaks were due to DTC-g-PE-DA@GB@PU and PE-DA@GB@PU) in the EDX analysis spectrum (Figure 8) confirm the adsorption of Cu^{2+}, Pb^{2+}, Cd^{2+} and Pd^{2+}, Pt^{4+}, Au^{3+} onto the surfaces of DTC-g-PE-DA@GB@PU and PE-DA@GB@PU, respectively [33,34].

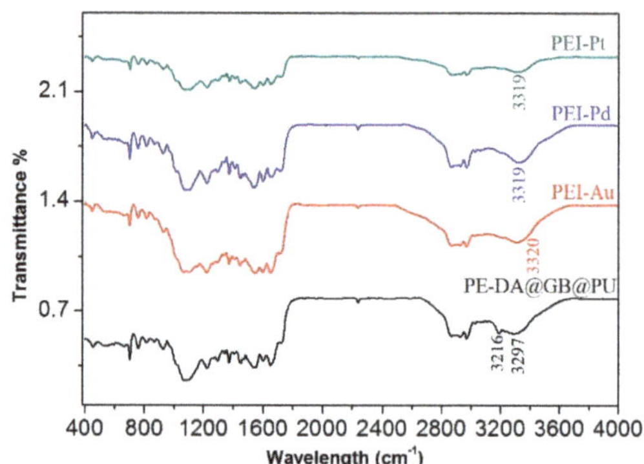

Figure 7. FT-IR spectra of PE-DA@GB@PU adsorbent and the PE-Pd, PE-Pt and PE-Au complexes.

The samples were subjected to WAXD analysis using a PANalytical diffractometer with Cu-Kα radiation (40 kV, 40 mA). The WAXD diffraction patterns of PU, APU, GB@PU, PE-DA@GB@PU, DTC-g-PE-DA@GB@PU, and GO are shown in Figure 9a. The WAXD diffraction pattern of PU exhibited a characteristic broad peak. GO shows a diffraction peak at ~10.8° [20]. After reaction of GO with APU, the diffraction peak of GB@PU was broadened and red shifted to 21.1°, compared with GO, suggesting a reduction in the content of oxygen-containing groups from GO to graphene [35]. For PE-DA@GB@PU and DTC-g-PE-DA@GB@PU, broad and weak peaks were observed in the WAXD patterns, corresponding to the amorphous feature of the crosslinked polymer matrix, implying the presence of intercalated PEI and PDA with GB@PU [36]. To explore the adsorption mechanism, DTC-g-PE-DA@GB@PU was added to copper nitrate, lead nitrate, and cadmium nitrate solution (HM ions ~1000 ppm) separately. Unlike the observations reported by Sharma et al. [33,34], no precipitate or floc appeared. We speculate that nitro-oxidized carboxycellulose nanofibers (NOCNF) constituted a kind of particle material, whereas the DTC-g-PE-DA@GB@PU was bulk material. The surface of the former was negatively charged, which could be induced to aggregate in the presence of cations. However, the form of DTC-g-PE-DA@GB@PU was bulk, and its adsorption characteristics were mainly due to its porous characteristics. In addition, the pH of the HM ion solutions may not be optimum for adsorption because the pH levels of 1000-ppm solutions of Cu^{2+}, Pb^{2+}, and Cd^{2+} were 4.04, 4.05, and 4.87, respectively. However, when some NaOH solution was added to adjust the pH to 6, precipitation appeared (Figure S9, Supplementary Materials). Then we reduced the concentration of the solutions to 100 ppm, and WAXD analysis of DTC-g-PE-DA@GB@PU was performed after adsorption (Figure 9b). Unfortunately, the WAXD patterns show no significant changes. The probable main reason was maybe that the low concentration of HM ion solutions resulted in low adsorption amount and the higher baseline of DTC-g-PE-DA@GB@PU masked the peaks of Cu, Pb, and Cd loaded on the adsorbent. For Pd^{2+}, Pt^{4+}, and Au^{3+}, 1000-ppm solutions were used, because at low pH (pH = 0), no precipitation was observed. As shown in Figure 9b, after the impregnation of PE-DA@GB@PU into gold solution, new peaks appeared at 2θ = 38.4°, 44.5°, 64.8°, and 77.7°, which matched exactly with those of gold, verifying the formation of elemental gold during the adsorption process. Chloroauric acid has high oxidation. Amino groups, which are the major surface constituents of PE-DA@GB@PU, can be easily reduced by $AuCl_4^-$ [37]. For Pd and Pt, there were no obvious changes in the WAXD patterns because the main adsorption mechanism may have been electrostatic attraction. In Figure 8d,e the EDX show the presence of chlorine atom, which could reinforce this assumption.

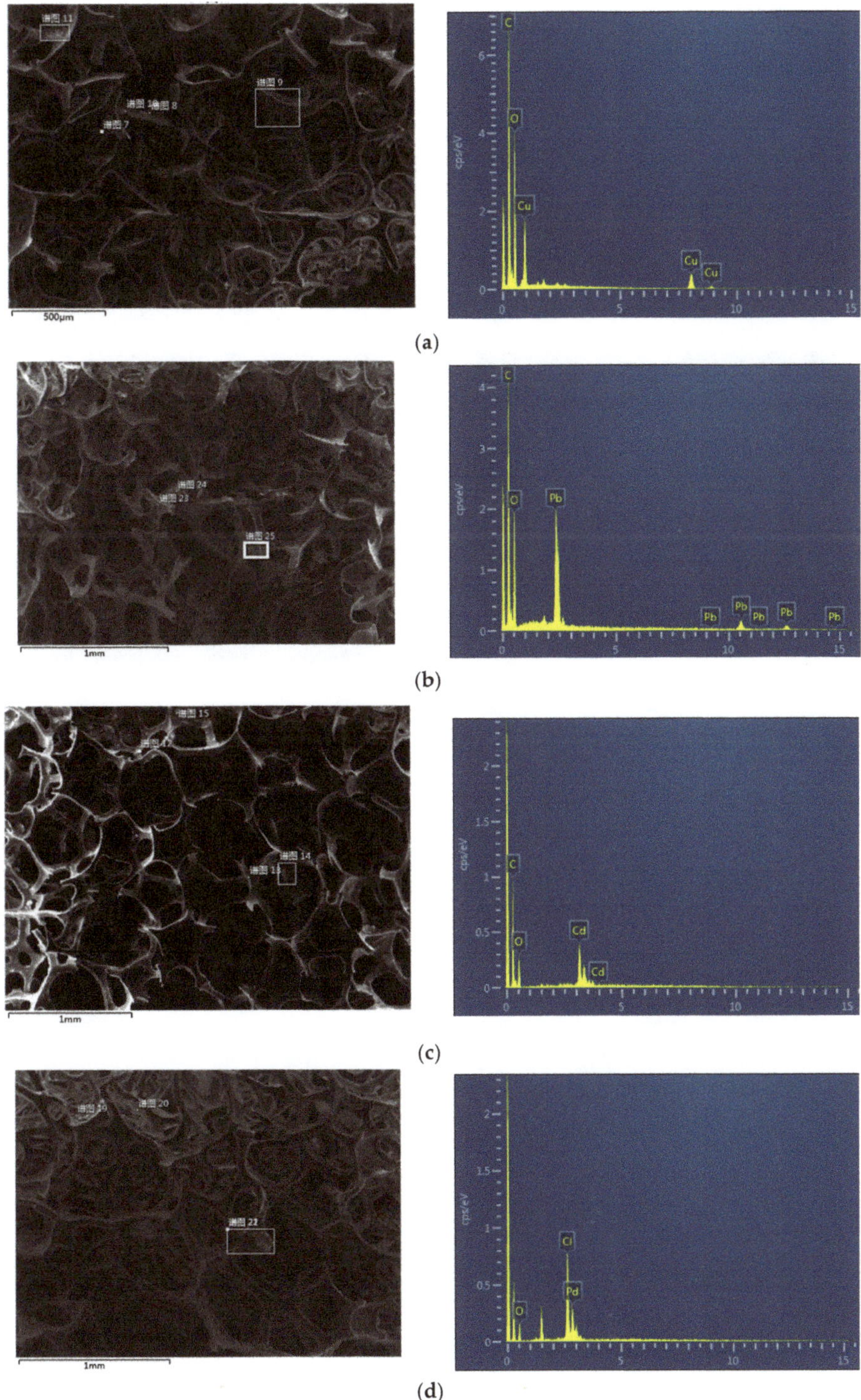

Figure 8. *Cont.*

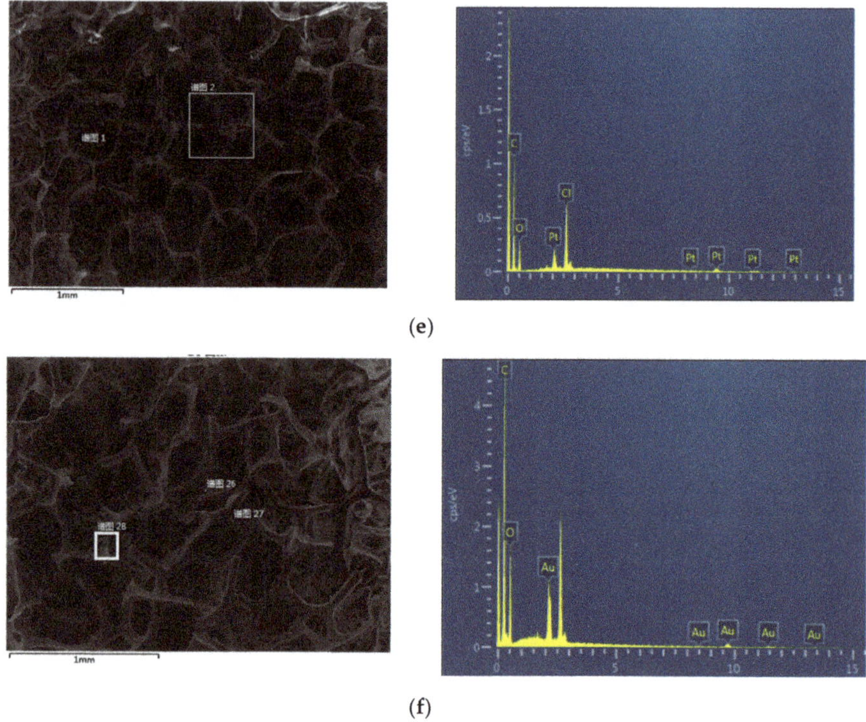

Figure 8. SEM image and energy-dispersive X-ray (EDX) analysis results of (**a**) Cu^{2+}-, (**b**) Pb^{2+}-, (**c**) Cd^{2+}-loaded DTC-g-PE-DA@GB@PU and (**d**) Pd^{2+}-, (**b**) Pt^{4+}-, (**f**) Au^{3+}-loaded PE-DA@GB@PU.

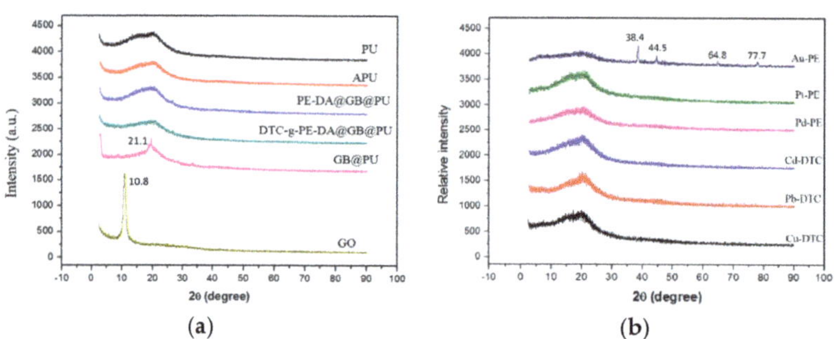

Figure 9. WAXD patterns for (**a**) PU, APU, GB@PU, PE-DA@GB@PU, DTC-g-PE-DA@GB@PU, and GO; (**b**) after Cu^{2+}, Pb^{2+}, and Cd^{2+} adsorption on the DTC-g-PE-DA@GB@PU and Pd^{2+}, Pt^{4+}, and Au^{3+} adsorption on the PE-DA@GB@PU.

3.6. Regeneration of DTC-g-PE-DA@GB@PU and PE-DA@GB@PU

Regeneration of adsorbents is essential in practical applications because of the economic and ecological appeals for sustainability. Thus, the adsorption–desorption ability of DTC-g-PE-DA@GB@PU was examined with of 0.1 M HCl and NaOH solutions. Both of these eluents were effective for leaching the HM ions. However, DTC-g-PE-DA@GB@PU decomposed under strong acid conditions. Therefore, when simple recovery of HM ions is desired, 0.1 M NaOH should be used as the eluent. After the adsorption of the HM ions, DTC-g-PE-DA@GB@PU was soaked in 0.1 M NaOH to remove the adsorbed HM ions. The adsorbent was then dried in air and the adsorption recycling was repeated again. The adsorption–elution processes were repeated for four successive cycles without evident loss of the adsorption capacity (Figure 10a). After four adsorption–desorption processes of using DTC-g-PE-DA@GB@PU to adsorb HM ions, 0.1 M HCl was used in the final desorption step to remove the HM ions. Simultaneously, PE-DA@GB@PU was obtained, which could then be used to directly recover NM ions. After the adsorption of NM ions by PE-DA@GB@PU, the sorbent was soaked in 1.0 M HNO_3 to effectively remove the NM ions. In a similar manner as for DTC-g-PE-DA@GB@PU, the PE-DA@GB@PU sorbent was dried in air, and the adsorption cycle was repeated again. The extraction percentages of NM ions for PE-DA@GB@PU lost nearly 10% of the original efficiency after four cycles (Figure 10b).

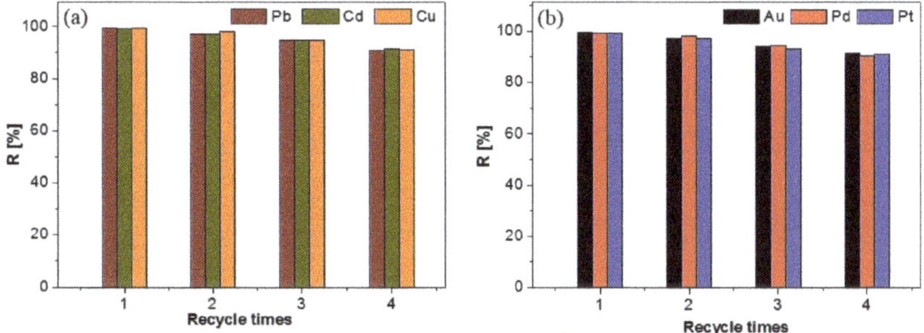

Figure 10. Recyclability of (**a**) DTC-g-PE-DA@GB@PU toward 10.0 µg mL^{-1} Cu^{2+}, Pb^{2+} and Cd^{2+}. Other conditions: solution pH = 6; volume, 20 mL; adsorbent dose, 20 mg; temperature, 25 °C; (**b**) recyclability of PE-DA@GB@PU toward 1.0 µg mL^{-1} Pd^{2+}, Pt^{4+}, and Au^{3+}. Other conditions: solution pH = 0; volume, 20 mL; adsorbent dose, 20 mg; temperature, 25 °C.

The described method was also applied for the removal of Cu^{2+}, Pb^{2+} and Cd^{2+} from industrial effluents and recovery of Pd^{2+}, Pt^{4+}, and Au^{3+} present in copper leaching residue and anode slime, successively. The industrial effluents and the refining wastes (copper leaching residue and anode slime) were obtained from Jinchuan Group International Resources Co. Ltd. The industrial effluents and the refining wastes (copper leaching residue and anode slime) produced in the refining process contained a certain amount of HMs and NMs, respectively. Sorption with these new adsorbents that allowed HM and NM deep concentration and their separation from other base metals proved to be one of the most efficient ways to recover HMs and NMs. Industrial effluents were adjusted to the desired pH values before the adsorption process. After the adsorption of HM ions, 0.1 M HCl was used as an eluent and PE-DA@GB@PU was obtained, which was then used to recover NMs from the copper leaching residue and anode slime. Before the recovery process, copper leaching residue and anode slime were treated with aqua regia by Elci's procedure [38]. The method developed for the determination of HM and NM ions in industrial effluents, copper leaching residue, and anode slime from Jinchuan Group International Resources Co. Ltd. is summarized in Table 3.

Table 3. Standard addition method for HM ions and NM ions determination in industrial effluents and copper leaching residue, anode slime, respectively.

Sample	Analytes	Added (μg g^{-1})	Found [a] (μg g^{-1})	Recovery (%)
Industrial effluents	Cu	—	112.6 ± 5.8	—
		100	209.2 ± 7.6	96.6
	Pb	—	12.7 ± 2.0	—
		10	22.6 ± 2.6	96.0
	Cd	—	5.4 ± 0.4	—
		10	14.6 ± 1.8	92.0
Copper leaching residue	Pd	—	13.2 ± 1.8	—
		20	32.1 ± 3.4	94.5
	Pt	—	20.0 ± 3.7	—
		20	40.7 ± 4.0	104
	Au	—	25.3 ± 2.4	—
		20	46.5 ± 3.0	106
Anode slime	Pd	—	33.4 ± 2.8	—
		30	62.1 ± 2.4	95.7
	Pt	—	8.3 ± 1.6	—
		10	17.7 ± 2.2	94.0
	Au	—	179.8 ± 3.5	—
		100	285.1 ± 2.2	105

[a] Mean value ± standard deviation, $n = 3$.

4. Conclusions

Novel PU composites were synthesized and their adsorptive selectivities for HM and NM ions were investigated. The 'S' donor ligands of DTC-g-PE-DA@GB@PU were responsible for its high selectivity for Cu^{2+}, Pb^{2+} and Cd^{2+}; while PE-DA@GB@PU possessed abundant amine groups that obviously increased its adsorption capacity, and the composite exhibited excellent selectivity properties toward Pd^{2+}, Pt^{4+}, and Au^{3+}. The adsorption processes depended on the pH value, contact time, and initial metal ion concentration. The optimal conditions, adsorption isotherms, and kinetics for recovery were investigated systematically. The proposed method was applied for the removal of HMs from industrial effluents and recovery of NMs from the copper leaching residue and anode slime, with satisfactory results being obtained in all cases.

Supplementary Materials: The following are available online at http://www.mdpi.com/2073-4360/11/7/1125/s1, Figure S1: Raman spectra of PU, APU, GO, GB@PU, PE-DA@GB@PU and DTC-g-PE-DA@GB@PU, Figure S2: The acid decomposition of dithiocarbamate compound, Figure S3: The Langmuir sorption isotherm for batch method of (a) DTC-g-PE-DA@GB@PU for HM ions and (b) PE-DA@GB@PU for NM ions; (25 °C), Figure S4: The Freundlich sorption isotherm for batch method of (a) DTC-g-PE-DA@GB@PU for HM ions and (b) PE-DA@GB@PU for NM ions; (25 °C), Figure S5: The D–R sorption sorption isotherm for batch method of(a) DTC-g-PE-DA@GB@PU for HM ions and (b) PE-DA@GB@PU for NM ions; (25 °C), Figure S6: Pseudo-first-order plots for (a) HM ions on DTC-g- PE-DA@GB@PU and (b) NM ions on PE-DA@GB@PU at 25 °C, Figure S7: Pseudo-second-order plots for (a) HM ions on DTC-g-PE-DA@GB@PU and (b) NM ions on PE-DA@GB@PU at 25 °C, Figure S8: Intraparticle diffusion plots for (a) HM ions on DTC-g-PE-DA@GB@PU and (b) NM ions on PE-DA@GB@PU at 25 °C, Figure S9: The pH levels of 1000-ppm solutions of Cu^{2+}, Pb^{2+}, and Cd^{2+} after some NaOH solution was added to adjust the pH to 6, Table S1: Operation parameters of IRIS Advantage ICP-OES, Table S2: Comparison of the maximum adsorption capacities of DTC-g-PE-DA@GB@PU and PE-DA@GB@PU with other adsorbents.

Author Contributions: Funding acquisition, Y.L. and Y.Y.; Investigation, T.L., D.Z., Q.G. and J.G.; Supervision, J.S. and B.S.; Validation, G.J. and L.S.; Sampling, G.C.; Writing-original draft, D.X.; Writing-review and editing, D.X. and B.S.

Funding: This study was financially supported by the National Natural Science Foundation of China (nos. 41403021, 41703021 and 41525012) and Opening Foundation of the State Key Laboratory of Lithosphere Evolution (SKL-K201604).

Conflicts of Interest: The authors declare no conflict of interest.

References

1. Liu, X.; Song, Q.; Tang, Y.; Li, W.; Xu, J.; Wu, J.; Wang, F.; Brookes, P.C. Human health risk assessment of heavy metals in soil–vegetable system: A multi-medium analysis. *Sci. Total Environ.* **2013**, *463*, 530–540. [CrossRef] [PubMed]
2. Cieszynska, A.; Wieczorek, D. Extraction and separation of palladium(II), platinum(IV), gold(III) and rhodium(III) using piperidine-based extractants. *Hydrometallurgy* **2018**, *175*, 359–366. [CrossRef]
3. Moawed, E.A.; Ishaq, I.; Abdul-Rahman, A.; El-Shahat, M.F. Synthesis, characterization of carbon polyurethane powder and its application for separation and spectrophotometric determination of platinum in pharmaceutical and ore samples. *Talanta* **2014**, *121*, 113–121. [CrossRef] [PubMed]
4. Yang, Z. Key technology research on the efficient exploitation and comprehensive utilization of resources in the deep Jinchuan nickel deposit. *Engineering* **2017**, *3*, 559–566. [CrossRef]
5. Sharma, P.R.; Varma, A.J. Functional nanoparticles obtained from cellulose: Engineering the shape and size of 6-carboxycellulose. *Chem. Commun.* **2013**, *49*, 8818–8820. [CrossRef] [PubMed]
6. Sharma, P.R.; Chattopadhyay, A.; Sharma, S.K.; Hsiao, B.S. Efficient removal of UO_2^{2+} from water using carboxycellulose nanofibers prepared by the nitro-oxidation method. *Ind. Eng. Chem. Res.* **2017**, *56*, 13885–13893. [CrossRef]
7. Sharma, P.R.; Joshi, R.; Sharma, S.K.; Hsiao, B.S. A simple approach to prepare carboxycellulose nanofibers from untreated biomass. *Biomacromolecules* **2017**, *18*, 2333–2342. [CrossRef] [PubMed]
8. Dabrowski, A.; Hubicki, Z.; Podkościelny, P.; Robens, E. Selective removal of the heavy metal ions from waters and industrial wastewaters by ion-exchange method. *Chemosphere* **2004**, *56*, 91–106. [CrossRef]
9. Nikoloski, A.N.; Ang, K.-L.; Li, D. Recovery of platinum, palladium and rhodium from acidic chloride leach solution using ion exchange resins. *Hydrometallurgy* **2015**, *152*, 20–32. [CrossRef]
10. Barakat, M.A. New trends in removing heavy metals from industrial wastewater. *Arabian J. Chem.* **2011**, *4*, 361–377. [CrossRef]
11. Bhandare, A.A.; Argekar, A.P. Separation and recovery of platinum and rhodium by supported liquid membranes using bis (2-ethylhexyl) phosphoric acid (HDEHP) as a mobile carrier. *J. Membr. Sci.* **2002**, *201*, 233–237. [CrossRef]
12. Silva, J.E.D.; Paiva, A.P.; Soares, D.; Labrincha, A.; Castro, F. Solvent extraction applied to the recovery of heavy metals from galvanic sludge. *J. Hazard. Mater.* **2005**, *120*, 113–118. [CrossRef] [PubMed]
13. Pan, L. Solvent extraction and separation of palladium(II) and platinum(IV) from hydrochloric acid medium with dibutyl sulfoxide. *Miner. Eng.* **2009**, *22*, 1271–1276. [CrossRef]
14. Fu, L.; Yan, Z.; Zhao, Q.; Yang, H. Novel 2D nanosheets with potential applications in heavy metal purification: A review. *Adv. Mater. Interfaces* **2018**, *5*, 1801094. [CrossRef]
15. Ibeh, C.C.; Bubacz, M. Current trends in nanocomposite foams. *J. Cell. Plast.* **2008**, *44*, 493–515. [CrossRef]
16. Liu, Y.; Fang, Y.; Liu, X.; Wang, X.; Yang, B. Mussel-inspired modification of carbon fiber via polyethyleneimine/polydopamine co-deposition for the improved interfacial adhesion. *Compos. Sci. Technol.* **2017**, *151*, 164–173. [CrossRef]
17. Liu, Y.; Luo, R.; Shen, F.; Tang, L.; Wang, J.; Huang, N. Construction of mussel-inspired coating via the direct reaction of catechol and polyethyleneimine for efficient heparin immobilization. *Appl. Surf. Sci.* **2015**, *328*, 163–169. [CrossRef]
18. Wang, X.; Jing, S.; Liu, Y.; Qiu, X.; Tan, Y. Preparation of dithiocarbamate polymer brush grafted nanocomposites for rapid and enhanced capture of heavy metal ions. *RSC Adv.* **2017**, *7*, 13112–13122. [CrossRef]
19. Ge, Y.; Xiao, D.; Li, Z.; Cui, X. Dithiocarbamate functionalized lignin for efficient removal of metallic ions and the usage of the metal-loaded bio-sorbents as potential free radical scavengers. *J. Mater. Chem. A* **2014**, *2*, 2136–2145. [CrossRef]
20. Zhou, S.; Hao, G.; Zhou, X.; Jiang, W.; Wang, T.; Zhang, N.; Yu, L. One-pot synthesis of robust superhydrophobic, functionalized graphene/polyurethane sponge for effective continuous oil–water separation. *Chem. Eng. J.* **2016**, *302*, 155–162. [CrossRef]

21. Wu, C.; Huang, X.; Wu, X.; Qian, R.; Jiang, P. Mechanically flexible and multifunctional polymer-based graphene foams for elastic conductors and oil-water separators. *Adv. Mater.* **2013**, *25*, 5658–5662. [CrossRef] [PubMed]

22. McClain, A.; Hsieh, Y.-L. Synthesis of polystyrene-supported dithiocarbamates and their complexation with metal ions. *J. Appl. Polym. Sci.* **2004**, *92*, 218–225. [CrossRef]

23. Lim, M.-Y.; Choi, Y.-S.; Kim, J.; Kim, K.; Shin, H.; Kim, J.-J.; Shin, D.M.; Lee, J.-C. Cross-linked graphene oxide membrane having high ion selectivity and antibacterial activity prepared using tannic acid-functionalized graphene oxide and polyethyleneimine. *J. Membr. Sci.* **2017**, *521*, 1–9. [CrossRef]

24. Zhang, C.; Ma, M.-Q.; Chen, T.-T.; Zhang, H.; Hu, D.-F.; Wu, B.-H.; Ji, J.; Xu, Z.-K. Dopamine-triggered one-step polymerization and codeposition of acrylate monomers for functional coatings. *ACS Appl. Mater. Interfaces* **2017**, *9*, 34356–34366. [CrossRef] [PubMed]

25. Xue, D.; Wang, H.; Liu, Y.; Shen, P.; Sun, J. Cytosine-functionalized polyurethane foam and its use as a sorbent for the determination of gold in geological samples. *Anal. Methods* **2016**, *8*, 29–39. [CrossRef]

26. Fujiwara, K.; Ramesh, A.; Maki, T.; Hasegawa, H.; Ueda, K. Adsorption of platinum (IV), palladium (II) and gold (III) from aqueous solutions onto L-lysine modified crosslinked chitosan resin. *J. Hazard. Mater.* **2007**, *146*, 39–50. [CrossRef] [PubMed]

27. Bai, L.; Hu, H.; Fu, W.; Wan, J.; Cheng, X.; Zhuge, L.; Xiong, L.; Chen, Q. Synthesis of a novel silica-supported dithiocarbamate adsorbent and its properties for the removal of heavy metal ions. *J. Hazard. Mater.* **2011**, *195*, 261–275. [CrossRef]

28. Shannon, R.D. Revised effective ionic radii and systematic studies of interatomic distances in halides and chalcogenides. *Acta Crystallogr. Sect. A* **1976**, *32*, 751–767. [CrossRef]

29. Zhu, Q.; Li, Z. Hydrogel-supported nanosized hydrous manganese dioxide: Synthesis, characterization, and adsorption behavior study for Pb^{2+}, Cu^{2+}, Cd^{2+} and Ni^{2+} removal from water. *Chem. Eng. J.* **2015**, *281*, 69–80. [CrossRef]

30. Xu, H.; Yuan, H.; Yu, J.; Lin, S. Study on the competitive adsorption and correlational mechanism for heavy metal ions using the carboxylated magnetic iron oxide nanoparticles (MNPs-COOH) as efficient adsorbents. *Appl. Surf. Sci.* **2019**, *473*, 960–966. [CrossRef]

31. Choi, H.A.; Park, H.N.; Won, S.W. A reusable adsorbent polyethylenimine/polyvinyl chloride crosslinked fiber for Pd(II) recovery from acidic solutions. *J. Environ. Manag.* **2017**, *204*, 200–206. [CrossRef] [PubMed]

32. Park, H.N.; Choi, H.A.; Won, S.W. Fibrous polyethylenimine/polyvinyl chloride crosslinked adsorbent for the recovery of Pt(IV) from acidic solution: Adsorption, desorption and reuse performances. *J. Clean. Prod.* **2018**, *176*, 360–369. [CrossRef]

33. Sharma, P.R.; Chattopadhyay, A.; Zhan, C.; Sharma, S.K.; Geng, L.; Hsiao, B.S. Lead removal from water using carboxycellulose nanofibers prepared by nitro-oxidation method. *Cellulose* **2018**, *25*, 1961–1973. [CrossRef]

34. Sharma, P.R.; Chattopadhyay, A.; Sharma, S.K.; Geng, L.; Amiralian, N.; Martin, D.; Hsiao, B.S. Nanocellulose from spinifex as an effective adsorbent to remove cadmium(II) from water. *ACS Sustain. Chem. Eng.* **2018**, *6*, 3279–3290. [CrossRef]

35. Zhao, J.; Guo, Q.; Wang, X.; Xie, H.; Chen, Y. Recycle and reusable melamine sponge coated by graphene for highly efficient oil-absorption. *Colloids Surf. A* **2016**, *488*, 93–99. [CrossRef]

36. Liu, Y.; Xu, L.; Liu, J.; Liu, X.; Chen, C.; Li, G.; Meng, Y. Graphene oxides cross-linked with hyperbranched polyethylenimines: Preparation, characterization and their potential as recyclable and highly efficient adsorption materials for lead(II) ions. *Chem. Eng. J.* **2016**, *285*, 698–708. [CrossRef]

37. Parajuli, D.; Kawakita, H.; Inoue, K.; Funaoka, M. Recovery of gold(III), palladium(II), and platinum(IV) by aminated lignin derivatives. *Ind. Eng. Chem. Res.* **2006**, *45*, 6405–6412. [CrossRef]

38. Elci, L.; Soylak, M.; Buyuksekerci, E.B. Separation of gold, palladium and platinum from metallurgical samples using an amberlite XAD-7 resin column prior to their atomic absorption spectrometric determinations. *Anal. Sci.* **2003**, *19*, 1621–1624. [CrossRef]

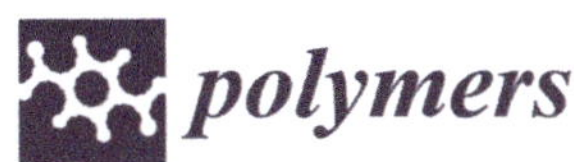 *polymers*

Article

POSS Compounds as Modifiers for Rigid Polyurethane Foams (Composites)

Anna Strąkowska *, Sylwia Członka and Krzysztof Strzelec

Institute of Polymer & Dye Technology, Lodz University of Technology, 90-924 Lodz, Poland
* Correspondence: anna.strakowska@p.lodz.pl

Received: 10 June 2019; Accepted: 25 June 2019; Published: 27 June 2019

Abstract: Three types of polyhedral oligomeric silsesquioxanes (POSSs) with different functional active groups were used to modify rigid polyurethane foams (RPUFs). Aminopropylisobutyl-POSS (AP-POSS), trisilanoisobutyl-POSS (TS-POSS) and octa(3-hydroxy-3-methylbutyldimethylsiloxy-POSS (OH-POSS) were added in an amount of 0.5 wt.% of the polyol weight. The characteristics of fillers including the size of particles, evaluation of the dispersion of particles and their effect on the viscosity of the polyol premixes were performed. Next, the obtained foams were evaluated by their processing parameters, morphology (Scanning Electron Microscopy analysis, SEM), mechanical properties (compressive test, three-point bending test, impact strength), viscoelastic behavior (Dynamic Mechanical Analysis, DMA), thermal properties (Thermogravimetric Analysis, TGA, thermal conductivity) and application properties (contact angle, water absorption). The results showed that the morphology of modified foams is significantly affected by the fillers typology, which resulted in inhomogeneous, irregular, large cell shapes and further affected the physical and mechanical properties of the resulting materials. RPUFs modified with AP-POSS represent better mechanical properties compared to the RPUFs modified with other POSS.

Keywords: RPUF; POSS; reinforcing effect; thermal properties; morphology

---

1. Introduction

Polyurethanes (PUs) are a group of polymers characterized by the most diverse properties, thanks to which they have a very wide range of industrial applications [1–3]. For the synthesis of polyurethanes, isocyanates that were obtained by Wurtz in 1849 are used as the fundamental raw material [4]. Among the synthesized polymeric materials, PUs currently occupy fifth place among the most commonly used plastics in the world and constitute 7.7% of total plastics produced [5,6]. The global PU market is dominated by PU foams, which account for over 65% of global production. Worldwide production of PU foams mainly includes elastic foams, which cover about 37% of world production, making them the most comprehensive group of polyurethane and rigid materials, which immediately after flexible foams, are the second largest group of polyurethane plastics and constitute about 28% of global polyurethane production [7]. Rigid polyurethane foams (RPUFs) are used as high-performance thermal insulation materials in construction, pre-insulated pipelines and in the refrigeration industry due to their properties, such as closed cell structure, low thermal conductivity, and low moisture absorption capacity [8–11]. Open-cell polyurethane foams increase the heat transfer capacity, which means they are characterized by much higher thermal conductivity than RPUFs with a closed-cell structure [12]. The thermal insulation properties of RPUFs also depend on the apparent density of the foam, which has a fundamental influence on the heat conduction coefficient. The values of apparent density for RPUFs with an open cell structure is 10 to 12 kg m^{-3}, while for closed cell foams the value usually lies in the range 25–70 kg m^{-3}. The smallest values of thermal conductivity are observed for RPUFs with closed-cell structure, the apparent density of which is from

25 to 35 kg m^{-3} [13]. A significant impact on the wide range of applications of rigid PUR foams also has mechanical and physical properties. However, PU materials are flammable, giving off toxic gases during combustion that pose a threat to the surrounding environment. Fire resistance should be limited so that it can be successfully used in building applications [14–18].

A review of the related literature indicates that one of the method of increased flammability of PU foams are fire-resistant coatings. For example, a layer-by layer assembly technique of polyethylenimine (PEI), graphene oxide (GO) and synthetic melanin nanoparticles (SMNPs) was constructed on the surface of PU foams [19]. Obtained materials exhibited an excellent radical scavenging capacity and high-temperature stability. Cho et al. employed polydopamine into the flame-retardant surface coating system of the flexible PU foam, realizing a decreased peak heat release rate by 67% [20]. Roberts et al. [21] further analyzed the thermal degradation kinetics of polydopamine-nanocoated PU foam and discovered that polydopamine could consistently decrease the release amount of combustible volatiles [22]. On the other hand, another studies have shown that better improvement of mechanical properties and heat resistance, can be also obtained by use of nanomaterials combined with PU matrix [18,22–28].

It was reported that, in addition to traditional fillers, an interesting group of reactive nanofillers are polyhedral oligomeric silsesquioxanes (POSS), which combine the features of organic and inorganic materials [29–31]. They were discovered at the beginning of the 20th century and are currently in the interest of many research centers due to their unique properties, including well-defined, three-dimensional, chemically and thermally resistant hybrid structure with nanoscopic dimensions [32–35]. Their biggest advantage is the ease of functioning with various organic or inorganic substituents. The most frequently occurring substituents are hydrogen, amine, acrylic and methacrylic, vinyl, allyl, epoxide, hydroxyl or halogen groups [36,37]. Thanks to the hybrid structure and the diversity of function groups, POSS exhibits comprehensive chemical compatibility [38,39]. POSS has a regular polyhedral spatial structure presented as a regular cube. The size of POSS depends on their construction. The size of the octasilsesquioxane core is about 0.5 nm, and the whole particle reaches dimensions of 1 to 3 nm depending on the type of substituents [40]. The oligosilsesquioxanes in the polymer matrix tend to agglomerate so that crystals can form in the polymer matrix [41]. Their dimensions can range from a dozen to several dozen nanometers [42,43]. Due to their properties, POSS compounds are commonly used in multifunctional applications. In contrast to the traditional organic compounds, they do not emit volatile organic substances, thanks to which they are odorless and environmentally friendly [40]. The basic element deciding about the potential application and influencing the final properties of the material are the organofunctional substituents found in the corners of the silicon-oxygen skeleton, and more specifically their number, distribution, and chemical nature. Thanks to the combination of reactive and non-reactive substituents POSS molecules can be easily attached to the polymer by copolymerization, transplantation or simply mixing [44].

Hybrid nature and nanometric size scale make silsesquioxanes extremely attractive materials used as nanofillers [45]. It was shown in previous studies that the introduction of POSS compounds to the polymeric material may increase the temperature of use and decomposition, affect the glass transition temperature, increase the resistance to oxidation of the composite, improve mechanical strength, affect surface hardening, reduce flammability and heat [43,46,47].

The aim of this research is to develop and test PU foams modified with three type of POSS—two of them with closed-cage functionalized by amino- or hydroxyl group (Aminopropylisobutyl-POSS (AP-POSS) and octa(3-hydroxy-3-methylbutyldimethylsiloxy-POSS (OH-POSS) and one with open-cage (trisilanoisobutyl-POSS (TS-POSS)). The influence of the different type of POSSs on thermal properties (Thermogravimetric Analysis, TGA), dynamic mechanical properties (Dynamic Mechanical Analysis, DMA), physico-mechanical properties (compression strength, three-point bending test, impact strength, apparent density, water absorption) and morphology of obtained PU composites was examined in current work. The obtained results, indicate that the addition of POSS influences the morphology of analyzed foams and consequently their further mechanical and thermal properties.

2. Experimental Section

2.1. Materials

The water-blown PUR foams used in this study were obtained from a two-component system supplied by Purinova Sp. z o. o. from Bydgoszcz, Poland, after mixing the polyol (Izopianol 30/10/C) and the diphenylmethane diisocyanate (Purocyn B). The polyol is a mixture of components containing polyester polyol (hydroxyl number ca. 230–250 mgKOH/g, functionality of 2), catalyst (*N,N*-Dimethylcyclohexylamine), flame retardant (Tris(2-chloro-1-methylethyl)phosphate), chain extender (1,2-propanediol) and water as blowing agent [48]. This formulation was selected because of its industrial relevance for the production of PU insulation ensures that the raw materials are available from a range of manufacturers at a competitive cost. AP-POSS, TS-POSS and OH-POSS were provided by Hybrid Plastics Inc. (Hattiesburg, MS, USA). The chemical structures of AP-POSS, TS-POSS and OH-POSS are presented in Figure 1a,b,c respectively.

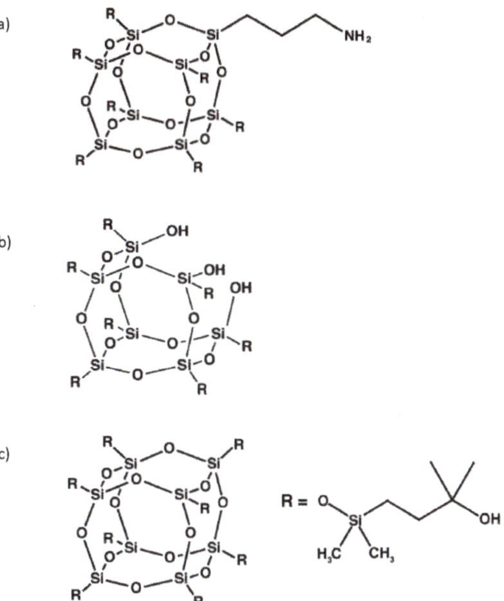

Figure 1. Chemical structure of (**a**) AP-POSS, (**b**) TS-POSS and (**c**) OH-POSS.

2.2. Manufacturing of RPUFs

RPUFs containing 0.5 wt.% of AP-POSS, TS-POSS, and OH-POSS were obtained as follows. The adequate amount of selected POSS was added to the Izopianol 30/10/C and the mixture (component A) was homogenized with an overhead stirrer at 600 RPM under ambient conditions for approximately 60 s. Purocyn B (component B) was added to component A and the mixture was stirred for 10 s at a speed rate of 1800 RPM. Following the information provided by the supplier, the ingredients were mixed in the ratio of 100:160 (ratio of component A to component B). Such prepared system was poured into an open mold and allowed to expand freely in the vertical direction. RPUFs were conditioned at room temperature for 24 h. After this time, samples were cut with a band saw into appropriate shapes (determined by obligatory standards listed below in the *Characterization techniques*) and their physico-mechanical properties were investigated. A schematic figure of the synthesis of RPUFs is presented in Figure 2.

Figure 2. Schematic procedure of synthesis of rigid polyurethane foams (RPUFs).

2.3. Characterization Techniques

The average size of POSS powder particles was measured using a Zetasizer NanoS90 instrument (Malvern Instruments Ltd., UK). The size of particles polyol dispersion (0.04 g/L) was determined with the dynamic light scattering DLS method.

The absolute viscosities of polyol and isocyanate were determined corresponding to ASTM D2930 (equivalent to ISO 2555) using a rotary Viscometer DVII+ (Brookfield, Germany). The torque of samples was measured as a range of shear rate from 0.5 to 100 s^{-1} at room temperature.

The apparent density of foams was determined accordingly to ASTM D1622 (equivalent to ISO 845). The densities of five specimens per sample were measured and averaged.

The morphology and cell size distribution of foams were examined from the cellular structure images of foam which were taken using JEOL JSM-5500 LV scanning electron microscopy (JEOL Ltd., USA). All microscopic observations were made in the high-vacuum mode and at the accelerating voltage of 10 kV. The samples were scanned in the free-rising direction. The average pore diameters, walls thickness, and pore size distribution were calculated using ImageJ software (Media Cybernetics Inc.).

The thermal properties of the synthesized composites were evaluated by TGA measurements performed using a STA 449 F1 Jupiter Analyzer (Netzsch Group, Germany). About 10 mg of the sample was placed in the TG pan and heated in an argon atmosphere at a rate of 10 K min^{-1} up to 600 °C with the sample mass about 10 mg. The decomposition temperatures ($T_{5\%}$, $T_{10\%}$, $T_{50\%}$ and $T_{70\%}$ of mass loss) were determined.

The compressive strength ($\sigma_{10\%}$) of foams was determined accordingly to the ASTM D1621 (equivalent to ISO 844) using Zwick Z100 Testing Machine (Zwick/Roell Group, Germany) with a load cell of 2 kN and the speed of 2 mm min^{-1}. Samples of the specified sizes were cut with a band saw in a direction perpendicular to the foam growth direction. Then, the analyzed sample was placed between two plates and the compression strength was measured as a ratio of the load causing 10% deformation of sample cross-section in the parallel and perpendicular direction to the square surface. The result was averaged of 5 measurements per each sample.

Impact test was carried out in agreement with ASTM D4812 on the pendulum 0.4 kg hammer impact velocity at 2.9 m s^{-1} with the sample dimension of $10 \times 10 \times 100$ mm. All tests were performed at room temperature. At least five samples were prepared for the tests.

Three-point bending test was carried out using Zwick Z100 Testing Machine (Zwick/Roell Group, Germany) at room temperature, according to ASTM D7264 (equivalent to ISO 178). The tested samples were bent with testing speed 2 mm min^{-1}. Obtained flexural stress at break (ε_f) results for each sample were expressed as a mean value. The average of 5 measurements per each type of composition was accepted.

Dynamic mechanical analysis (DMA) was determined using ARES Rheometer (TA Instruments, USA). Torsion geometry was used with samples of thickness 2 mm. Measurements were examined in the temperature range 20–250 °C at a heating rate of 10 °C min^{-1}, using a frequency of 1 Hz and applied deformation at 0.1%.

Surface hydrophobicity was analyzed by contact angle measurements using the sessile-drop method with a manual contact angle goniometer with an optical system OS-45D (Oscar, Taiwan) to capture the profile of a pure liquid on a solid substrate. A water drop of 1 μL was deposited onto the surface using a micrometer syringe fitted with a stainless steel needle. The contact angles reported are the average of at least ten tests on the same sample.

Water absorption of the RPUFs was measured according to ASTM D2842 (equivalent to ISO 2896). Samples were dried for 1 h at 80 °C and then weighed. The samples were immersed in distilled water to a depth of 1 cm for 24 h. Afterward, the samples were removed from the water, held vertically for 10 s, the pendant drop was removed and then blotted between dry filter paper (Fisher Scientific, USA) at 10 s and weighed again. The average of 5 specimens was used.

Changes in the linear dimensions were determined in accordance to the ASTM D2126 (equivalent to ISO 2796). The samples were conditioned at the temperature of 70 °C and −20 °C for 14 days. Change in linear dimensions was calculated in % from Equation (1).

$$\Delta l = ((l - l_o)/l_o) \times 100 \tag{1}$$

where l_o is the length of the sample before thermostating and l is the length of the sample after thermostating. The average of 5 measurements per each type of composition was reported.

3. Results and Discussion

3.1. Average Size of POSS Powder Particles and the Dispersion of POSS-Modified Polyol Premixes

One of the most important parameters determining the behavior of the filler in the polymer matrix is the size of its particles. If the particles are too small, their dispersion may be difficult because they have a greater tendency to aggregate and agglomerate, forming large clusters in the matrix. Too large particles may affect the foaming process and further properties of the obtained materials. The particle size of the POSS powder was measured in a polyol dispersion (0.04 g/L). The results of particle size measurements are given in Figure 3.

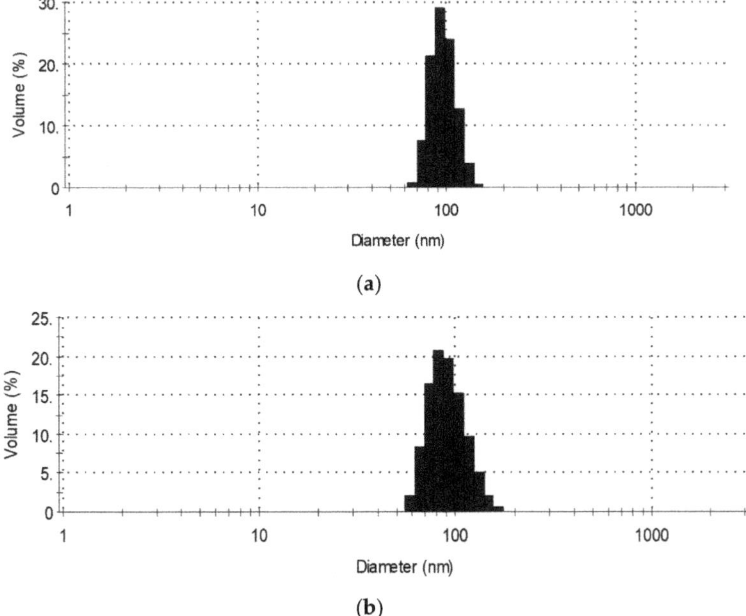

Figure 3. Distribution of particles size of (**a**) AP-POSS, (**b**) TS-POSS.

From the diagram, it follows that the size of AP-POSS particles ranges from 65 to 104 nm, the highest percentage—29% is shown by 82 nm particles. In the case of TS-POSS, the particle size distribution is somewhat larger and ranges from 59 to 108 nm, with the largest 69 nm volume fraction. Such small particle sizes of nanofillers may suggest their tendency to agglomerate in the polyol, which may negatively affect mechanical and functional properties. Figure 4 shows the optical micrographs obtained for the polyol systems with AP-POSS, TS-POSS and additionally OH-POSS. A comparison of the optical images for the sample with AP-POSS (Figure 4a) with that of TS-POSS (Figure 4b) reveals that in both cases the particles are well dispersed in the polyol systems and no aggregates of the POSS's particle are observed. A different trend is observed for the sample with OH-POSS. As presented in Figure 4c, a homogenous dispersion of the polyol system is observed, as a result of the liquid character of the used OH-POSS.

| **(a) AP-POSS** | **(b) TS-POSS** | **(c) OH-POSS** |

Figure 4. Polyol premixes with 0.5 wt.% of (**a**) AP-POSS, (**b**) TS-POSS and (**c**) OH-POSS.

3.2. Impact of POSS on PU Mixture Viscosity

The viscosity of the reactive mixture was measured first since it is a critical parameter affecting the foaming process [49]. Increased viscosity hinders bubble growth, yielding foams with lower cell size. Table 1 presents the results of the change in dynamic viscosity depending on the type of POSS in polyol mixture. The polyol premixes that contained AP-POSS, TS-POSS and OH-POSS are characterized by an increase in their viscosity, as a result of the presence of POSS particles interacting with the polyether polyol through hydrogen bonding and van der Wall's interaction [48]. Compared to control polyol, the greatest dynamic viscosity has AP-POSS modified polyol mixture.

Table 1. Dynamic viscosity and logarithmic plot of the fitting equations for polyol premixes.

Sample Code	Dynamic Viscosity η [mPa·s]			Fitting Equation	Power Law Index (n)	R^2
	0.5 RPM	**5 RPM**	**10 RPM**			
PU-0	628	424	380	y = −0.058 + 0.335	0.335	0.982
PU-AP	2110	1313	1149	y = −0.060 + 0.318	0.298	0.978
PU-TS	1317	1098	957	y = −0.060 + 0.315	0.315	0.978
PU-OH	1149	805	702	y = −0.059 + 0.298	0.318	0.979

The rheological properties of polyol premixes are shown as the viscosity versus shear rate in Figure 5a. In all systems, the viscosity is generally reduced at increased shear rates. Such a phenomenon is typical for non-Newtonian fluids with a pseudoplastic nature and is quite often found in the many previous works [50,51]. To further analyses the data, graph of viscosity versus shear rate is converted to log viscosity versus log shear rate form as shown in Figure 5b. It can be seen that the curvatures of viscosity versus shear rate can be made close to linear using this log-log format with regression of 0.979–0.982. The power law index (*n*) was calculated from the slopes. All results are presented in Table 1. For the system containing AP-POSS, the power law index is lower than that of their TS-POSS and OH-POSS modified system counterparts. It indicates that the effect of the filler on the pseudoplasticity behavior becomes more significant for systems modified with AP-POSS, leading to the highly non-Newtonian behavior.

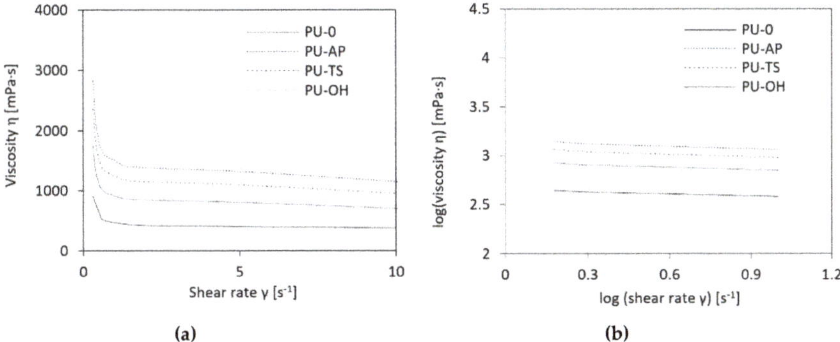

(a) **(b)**

Figure 5. (**a**) Viscosity as a function of shear rate and (**b**) log-log plot of the viscosity vs. the shear rate for polyol premixes.

3.3. The Influence of POSS on the Maximum Temperature (T_{max}) of the Reaction Mixture during the Foaming Process

The reaction of the synthesis of RPUFs is highly exothermic [15,52]. The rate of increase in temperature determines the activity of reaction mixture that is associated with the reactivity of the components of the mixture. As shown in Table 2, the introduction of AP-POSS, TS-POSS, and OH-POSS into the PU system increases the activity of reaction mixture which is confirmed by an increase in the T_{max} during the foaming process in each case. The presence of the additional groups as a result of the incorporation of the filler can lead to the exothermic reaction providing more heat evaporated to the system, and consequently higher temperature of the modified system compared to the PU-0. The T_{max} increases by about 20 °C with the addition of each POSS and appears at longer times compared to the PU-0 (Figure 6). Basically, an analog tendency was observed by other authors in previous works [16,53,54].

Table 2. Selected properties of RPUFs.

Sample Code	Temperature [°C]	Cream Time [s]	Extension Time [s]	Tack-free Time [s]	Cell Size [μm]	Wall Thickness [μm]	Apparent Density [kg m^{-3}]
PU-0	110	43 ± 4	277 ± 10	341 ± 14	472 ± 10	62 ± 4	38
PU-AP	135	49 ± 2	512 ± 11	376 ± 12	390 ± 8	68 ± 2	43
PU-TS	136	47 ± 2	504 ± 8	370 ± 12	402 ± 6	66 ± 3	42
PU-OH	134	46 ± 2	428 ± 9	320 ± 10	410 ± 8	66 ± 2	40

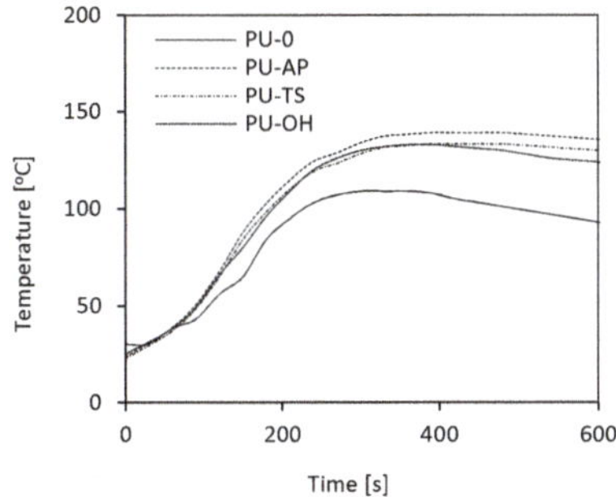

Figure 6. Temperature of reaction mixture in PU formulations.

3.4. Foaming Kinetic of RPUFs

The foaming process was determined by measuring the characteristic processing times like cream, extension and gelation time. The cream time was measured from the start of mixing of components to a visible start of foam growth, extension time elapsing until reaching the highest volume of the foam and gelation time was determined as the time when the foam solidifies completely and the surface is no longer tacky [17]. The results presented in Table 2 indicate a slight increase in cream and extension time for the RPUFs containing AP-POSS, TS-POSS, and OH-POSS. This dependence is mostly related to the fact that well-dispersed filler in the reaction mixture acts as a nucleating agent and higher viscosity of the modified systems is observed. It was reported in a previous work that higher viscosity has a major impact on the growth of RPUFs and causes an increase in reaction time by a few minutes [39]. Also, an increase of filler content affects the kinetics of the reaction and the phase separation. The rate of PU polymerization during foaming and morphology development is slowed down [40]. The addition of the filler into the system decreases the rate of isocyanate conversion during the early stage reaction. Also, due to the presence of the filler, a reduction of the mobility of the molecules takes place [40], leading to prolonged cream and extension time [18,22]. Compared to the PU-0, composites modified with the addition of the fillers are also characterized by a shorter tack-free time, indicating that filler particles act as a curing accelerator. Among studied fillers, the highest values of extension time and tack-free time are determined for PU-AP composites, as a result of higher viscosity, as compared to the PU-TS and PU-OH counterparts.

3.5. Density of RPUFs

Apparent density is one of the most important parameters to control the physical, mechanical and thermal properties of the RPUFs which has influence on their performance and applications. The values of density of prepared foams are presented in Table 2. In general, term, the apparent density tends to increase when the POSS are added. PU-0 is characterized by an apparent density of 38 kg m^{-3}. The apparent density increases to 43, 42 and 40 kg m^{-3} for samples with AP-POSS, TS-POSS, and OH-POSS, respectively. This effect can be explained by an analysis of the role of filler particles on nucleation and cell growth. The POSS particles act as nucleation sites promoting the formation of bubbles, and this is an increasing trend with nanoparticles content, but, at the same time, the growth process of the resulting cells is hindered by an increase of the gelling reaction speed, revealing in bigger viscosity. This results in bubble collapse and higher density foams. Moreover, the reactive groups of POSS particles (such as, hydroxyl and amine groups) would react with isocyanate (–NCO) groups, therefore, the content of isocyanates which reacted with water and produced CO_2 foaming gas would decrease, leading to the decreased foaming ratio and increased density. To sum up, the density of the composite foams increased with the incorporation of POSS filler in the PU system.

3.6. Morphology of RPUFs

The cell morphology is one of the most important factors determining the physico-mechanical properties of RPUFs [23,51]. The foaming process, the formation of cells, and their shape can be explained by a nucleation and growth mechanism [24]. A proper balance of filler concentration, reaction temperature, viscosity and dispersion of the filler in the polymer matrix is the key for optimization of the cellular structure of RPUFs [25]. The cellular structures of RPUFs composites are presented in Figure 7.

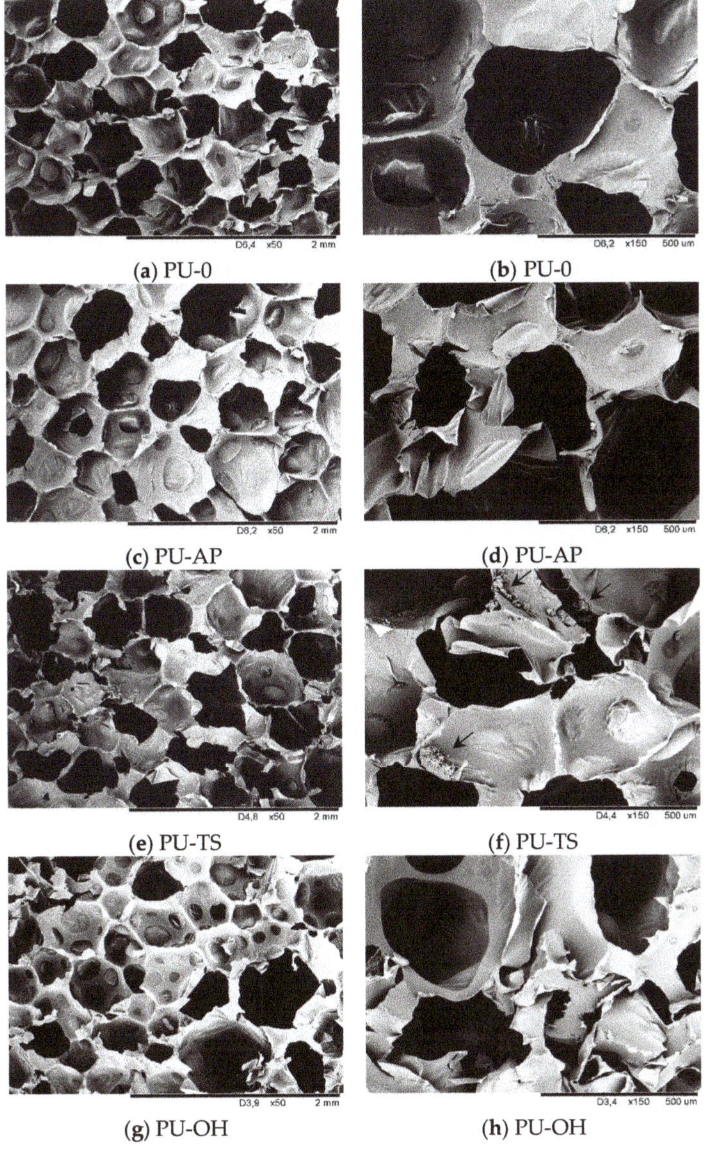

Figure 7. Morphology of (**a,b**) PU-0; (**c,d**) PU-AP, (**e,f**) PU-TS, (**g,h**) PU-OH observed at different magnification.

As observed from the micrograph of the neat PU-0 (Figure 7a,b), the cell size and cell distribution are nearly uniform and the PU-0 consists of closed cells with a negligible amount of cells with broken walls. With the addition of AP-POSS, the overall cell structure becomes less uniform and the number of broken cells increased (Figure 7c,d). A similar trend is observed for sample PU-TS, as shown in Figure 7e,f, although it has a higher content of broken cells compared to PU-AP. The more homogenous structure is observed in Figure 7g,h, which corresponds to the PU-OH. The closed-cell structure is well-preserved, and the number of broken cells is decreased. Higher content of open cells in the case of RPUFs modified with AP-POSS and TS-POSS can be connected with poor interfacial adhesion between the filler surface and the polymer matrix, which promotes earlier cell collapsing phenomena and increases a high possibility of generating open pores [26]. Moreover, the possible interphase interactions between POSS and PU in cell struts disturbed formulation of stable foam structure [27] which results in the coalescence of crowded cells. The alteration of cell morphology as the result of filler incorporation was also observed in previous studies [28,35,55].

The values of the cell size of the RPUFs were statistically analyzed by means of *ImageJ* software from SEM images and the median values are summarized in Table 2. The cell size distribution of the RPUFs is presented in Figure 8. From the table, the PU-0 has fewer cells with a larger cell size than the POSS-modified composites. In general, the PU-0 has an average cell size of 466 μm, and the addition of small amounts of POSS yielded smaller cells. RPUFs with AP-POSS, TS-POSS, and OH-POSS have a cell size of 396 μm, 389 μm and 408 μm, respectively. This means that RPUFs composites containing POSS have a higher cell density and smaller cell size than those of the PU-0. Therefore, it can be concluded that the POSS addition has an effect on reducing the cell size. This may be due to the increased viscosity of the system after POSS addition which restrains the expansion of the cells. Moreover, it has been well established in previous works that filler particles can act as nucleation sites for cell formation and since a higher number of cells starts to nucleate at the same time, thus a higher amount of cells with reduced diameter is present [56–61].

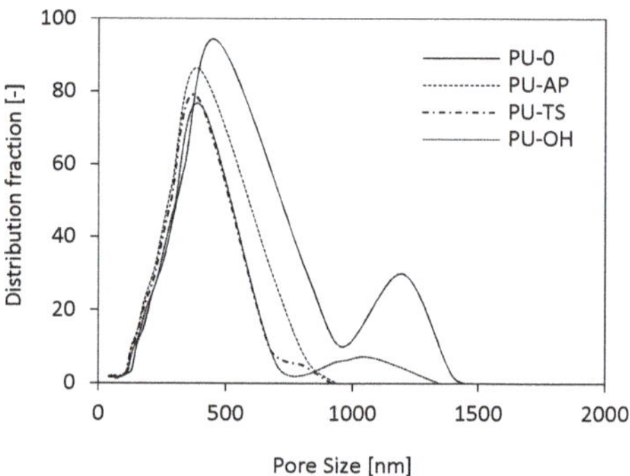

Figure 8. Cell size distributions of RPUFs.

3.7. Compressive Strength of RPUFs

The mechanical properties of RPUFs depend primarily on the cells' morphology with the strength being higher in the direction of foam expansion. In Figure 9 it can be seen that all the compressive stress-strain plots of RPUFs are composed of a first linear region which corresponds to the elastic response of the material and a second region in which the curves present a large plateau due to the plastic deformation and rupture of the cell walls while the stress is constant until the cells are crushed. Nevertheless, some differences can be observed between the samples. The increase in brittleness caused by the reinforcements determines a more abrupt transition from the elastic region to the plateau, in contrast to the smooth transition observed in the case of the PU-0. The elongation at break of the PU composites decreases with POSS incorporation, implying that POSS particles make the PU matrix more rigid. This is a common result in PU composites reinforced by a conventional filler [16,62,63].

The compression modulus and compressive strength of RPUFs are presented in Table 3. The compressive strength of all materials tested in the direction parallel and perpendicular to the direction of foam rise is greater than the strength of the reference foam. The largest increase in compressive strength is observed for the PU-AP and it is about 351 kPa in a parallel direction and 159 kPa in the perpendicular direction. In the foams containing TS-POSS and OH-POSS, there is a slight decrease in compressive strength compared with RPUFs containing AP-POSS; however, it is still larger than for the PU-0. As presented in Figure 10 the mechanical properties are closely related to the apparent density of polymer composites. An increase in density is accompanied by an increase in the mechanical properties of the composites since in compression the stiffness arises from buckling of cell walls. The higher density is related to more compact cellular structures, hence there is more material per unit area and the modulus and strength increase [64]. POSS-modified foams obtained in this study show apparent density values of 40–43 kg m^{-3} and compressive strengths of 309–351 kPa, which are well in the range exhibited by conventional commercial foams that present densities in the 15–130 kg m^{-3} range and compressive strength values in the range 200–220 kPa (for RPUFs at a density of 40 kg m^{-3}) [60,65]. Based on these results, the foams modified with POSS can potentially be used on an industrial scale in the construction and packaging industries.

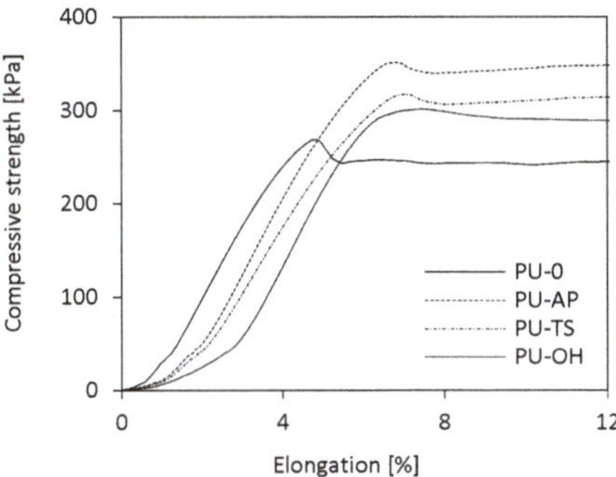

Figure 9. Compression behaviors of RPUFs measured parallel to the foam rise direction.

Table 3. Mechanical properties of RPUFs.

Sample Code	Compressive Strength (Parallel) σ_{10} [kPa]	Compressive Strength (Perpendicular) σ_{10} [kPa]	Young Modulus [MPa]	Flexural Strength σ_f [MPa]	Elongation [%]	Impact Strength [kJ m^{-2}]
PU-0	260	144	5	0.402	11.2	0.35
PU-AP	351	159	6.1	0.469	10.2	0.46
PU-TS	312	144	5.4	0.430	10.4	0.45
PU-OH	309	135	5.2	0.427	10.8	0.42

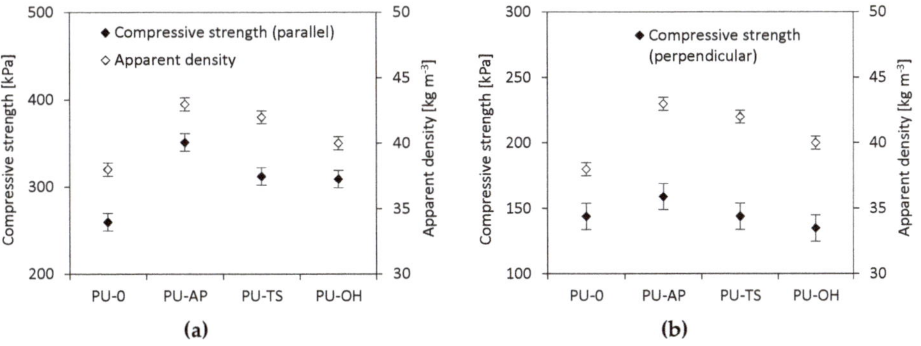

Figure 10. Effect of apparent density on compressive strength of RPUFs of RPUFs measured (**a**) parallel and (**b**) perpendicular to the foam rise direction.

3.8. Flexural Strength of RPUFs

As in the case of compression results presented in Figure 10, the correlation between flexural strength (σ_f) and apparent density is observed as well (Figure 11). It can be also seen that incorporation of POSS filler affects the σ_f of POSS-modified materials. Compared to the PU-0, σ_f is improved by the addition POSS in all cases. The value of tensile strength of PU-AP increases by 38% from 0.402 to 0.469 MPa as compared to the PU-0. Similar trend id observed for RPUFs modified with TS-POSS and OH-POSS. The value of σ_f increases to 0.430 and 0.427 MPa for samples PU-TS and PU-OH. Figure 12 shows the stress-elongation curves for the RPUFs. All samples exhibit a linear elastic behavior in the low-stress region and plastic deformation in the high-stress region, pointing at the comparable mechanical performance of modified foams. The incorporation of POSS reduces the elongation at break (ε_f) of RPUFs in all cases. The reason is due to the presence of POSS aggregates within the PU matrix, which may act as defects during the tensile testing process and decrease ε_f of foam composites.

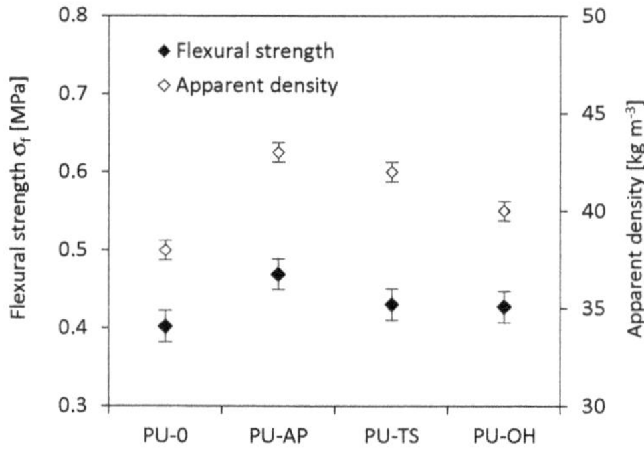

Figure 11. Effect of apparent density on flexural strength (σ_f) of RPUFs.

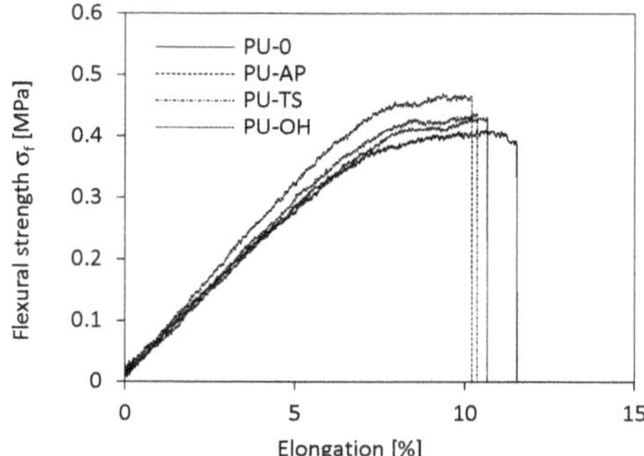

Figure 12. Flexural stress-elongation curves of RPUFs.

3.9. Impact Strength of RPUFs

The correlation between impact strength and apparent density is observed as well (Figure 13). With the incorporation of AP-POSS, TS-POSS, and OH-POSS, the impact strength increases from 0.35 to 0.46, 0.45 and 0.42 kJ m^{-2}, respectively. This behavior is related to the good interface reinforcement matrix and the generation of fracture paths through the POSS-reinforced RPUFs. Thus, the deformability of the RPUFs matrix is reduced, which in turn affects the ductility in the foam surface. With this effect, the foam composite tends to form a more rigid structure and decrease the concentration of POSS, thus reducing the foam's energy absorption, resulting in greater impact strength.

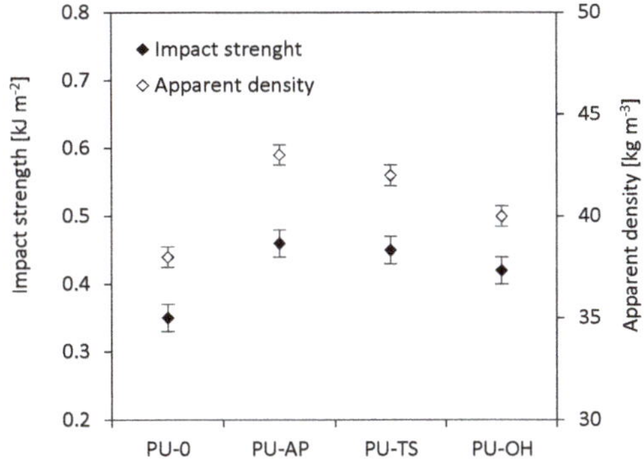

Figure 13. Effect of apparent density on the impact strength of RPUFs.

3.10. Dynamic Mechanical Analysis (DMA) and Thermogravimetric Analysis (TGA)

The dynamic mechanical behavior of RPUFs as a function of the temperature is shown in Figure 14. The results presented in Figure 14a and Table 4, indicate that the incorporation of POSS to the PU matrix affects the value of T_g, which corresponds to the maximum value of the curve loss tangent (*tanδ*) versus temperature. Compared to the RPUFs modified with TS-POSS and OH-POSS, RPUFs containing AP-POSS are characterized by higher T_g. Wu et al. [66] have shown that the T_g of RPUFs reflects the rigidity of the polymer matrix which is a function of the isocyanate index, cross-link density and aromaticity level of the RPUFs. Given that the isocyanate index has been held constant in this study, the increase in the T_g for POSS-modified samples must be a reflection of the increased aromaticity and cross-link density due to the presence of the POSS [67]. Moreover, as shown in Figure 14a, the reference and POSS-modified foams exhibit one wide peak in the range of temperature analyzed. The width of the peak becomes broader with the POSS incorporation due to different relaxation mechanisms appearing in the modified materials as a consequence of the added filler. The broadening of the *tanδ* peak is often assumed to be due to broader distribution in molecular weight between crosslinking points or heterogeneities in the network structures [19].

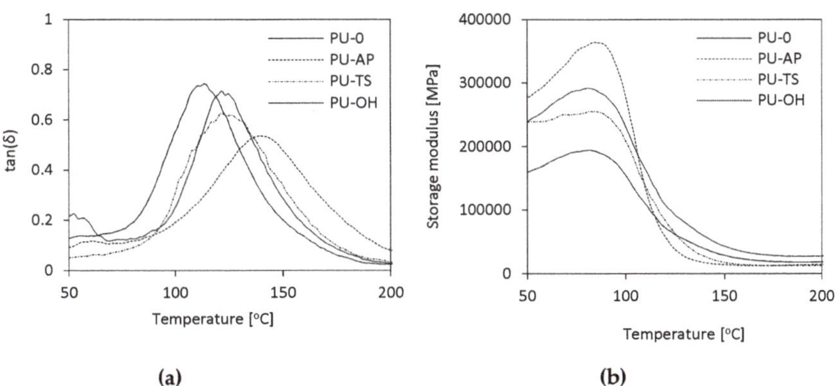

Figure 14. (a) tanδ and (b) storage modulus as a function of temperature plotted for RPUFs.

In Figure 14b, it is also notable that RPUFs modified with POSS are characterized by higher storage modulus (E') as compared to PU-0. It can be concluded that the addition of all POSS has significantly increased the E' of PU and consequently the stiffness of studied composites is also enhanced. This is due to the presence of filler in the PU matrix as well as higher viscosity of the modified systems, which imposes serious limits on the mobility of polymer chains, affecting their higher stiffness. Similar results are reported in the literature [68,69].

Table 4. The results of the thermogravimetric analysis of RPUFs.

Sample Code	T_g [°C]	$T_{5\%}$ [°C]	$T_{10\%}$ [°C]	$T_{50\%}$ [°C]	$T_{70\%}$ [°C]	*Char Residue* [%]
AP-POSS	-	267	280	342	531	21.6
TS-POSS	-	260	287	350	449	18.4
OH-POSS	-	n.d.	n.d.	n.d.	n.d.	n.d.
PU-0	112	220	265	454	591	27.9
PU-AP	137	205	245	418	595	29.0
PU-TS	127	216	261	451	586	28.7
PU-OH	121	210	251	449	585	28.6

The thermal degradation of pure polyurethane foam and hybrid composites was monitored by TGA thermograms as displayed in Figure 15a. The thermo-oxidative decomposition temperatures for 5, 10, 50 and 70% weight loss are evaluated from TGA curves, as listed in Table 4. In the case of PUR foams, thermal degradation occurred in 3 stages. In the first stage of decomposition at about 10% loss of initial mass, dissociation of urethane bonds occurs at a temperature of 150 to 330 °C [70,71]. The second degradation step RPUF corresponding to a weight loss of about 50% occurs at a temperature between 330 and 400 °C and is attributed to the decomposition of the soft polyol segments [70,72]. Then, the third degradation step associated with the degradation of the fragments generated during the second stage occurs at 500 °C, which corresponds to 80% loss of mass [70,72].

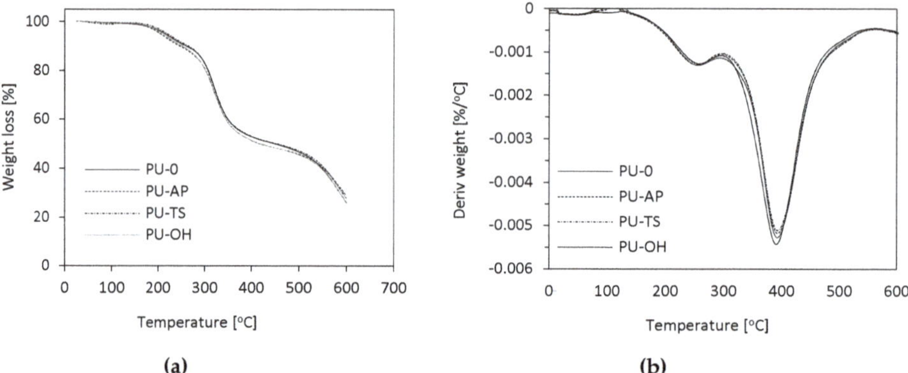

Figure 15. (a) TGA and (b) DTG curves for RPUFs modified with POSS.

It can be observed that the addition of fillers affects the thermal stability of RPUF (Table 3). POSS used as foam modifiers is characterized by higher thermal stability and percentage losses of masses at much higher temperatures than PU-O. However, in the presence of POSS, the acceleration of mass loss at the initial stage of degradation was observed.

The reduction of thermal stability can be attributed to non-homogeneous dispersion of POSS and changes in cross-link density [51]. Confirmation is SEM photos, which clearly show that the presence of POSS increases the heterogeneity of the RPUF morphology. In further stages of degradation, modified foams are slightly more stable and are characterized by weight losses obtained at a similar temperature as pure foam. In further degradation steps, the modified foams are slightly more stable and are characterized by mass losses obtained at a similar temperature as pure foam, with maximum mass losses of approximately 314–322 °C and 551–584 °C, which is related to the reaction of oxygen with hydroperoxides that are themselves unstable and decay, creating more free radicals [73]. In addition, it can be seen that the amount of char residue for POSS filled foams is increased compared to PU-0. This results in more stable char layers that can protect materials from further decomposition and, in turn, increase thermal stability. The change is also visible on the DTG curves, which are the first derivative of TGA that represent the speed of the composite decomposition process during heating. It can be seen in Figure 15b that the degradation rate of POSS modified foams is slightly lower than that of PU-0 foam.

3.11. Dimensional Stability, Contact Angle and Water Absorption

For RPUFs, often used as construction materials, dimensional stability, as well as affinity for water, is a very important parameter. Table 5 and Figure 16 show the dimensional stability of foams at low (−20°C) and high temperature (70 °C), respectively. The variability of dimensions at low temperature was slightly higher than at high temperature for the same foam samples. Furthermore, the % linear changes in length, width, and thickness after exposure indicate that the addition of POSS generally resulted in smaller dimensional changes of the modified foams compared to the reference foam, indicating a stabilizing effect of POSS on the degradation factor. This is particularly evident in the conditions of elevated temperature, where for TS-POSS modified foams, the dimensional stability improved by an average of 20% in comparison with the PU-0. The only exception to this trend is the POSS-OH-modified sample, which shows slightly larger changes in linear dimensions compared to the reference sample, especially at reduced temperatures. However, according to industrial standard, PU panels tested at 70 °C should have less than 3% of linear change. In each case, the dimensional stability of PU foams is thus still considered to be mild and within commercially acceptable limits [74].

Table 5. Dimensional stability and affinity for water of RPUFs.

Sample Code	Dimensional Stability (+70°C) [%]			Dimensional Stability (−20°C) [%]			Water Absorption [%]	Contact Angle [°]
	Width	Length	Thickness	Width	Length	Thickness		
PU-0	1.80	1.6	2.5	1.90	1.8	1.65	12.3	129
PU-AP	1.55	1.5	2.2	1.72	1.74	1.74	11.5	135
PU-TS	1.5	1.4	1.90	1.89	1.75	1.95	11.8	136
PU-OH	1.7	1.9	2.4	2.26	1.82	1.72	11.2	140

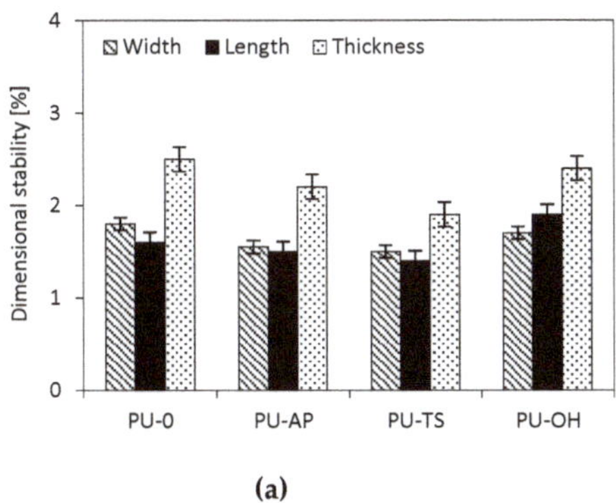

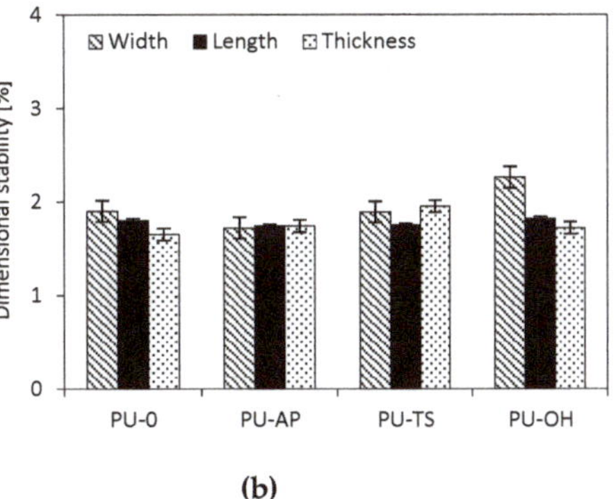

Figure 16. Dimensional stability of RPUFs modified with AP-POSS, TS-POSS and OH-POSS after exposure at 70 °C (**a**) and −20 °C (**b**).

Polyhedral oligomeric silsesquioxanes significantly affected the hydrophobicity of the foams (Figures 17 and 18). Regarding water absorption, it is notable that foams modified by POSS absorb less water than the reference sample. This effect is attributed to the greater surface roughness of foams with smaller pore sizes as well as the lack of large surface pores in which water droplets can be stored. Lower water absorption indicates greater hydrophobicity, which is also well illustrated by the contact angles of foam surfaces with water (Figure 18). The most hydrophobic foam was modified with POSS-OH, which achieved a contact angle of 140° and water absorption at the lowest level (11.2% after 24 h). This is due to the presence of non-polar side chains in the corners of silsesquioxane cages, which reduces the surface energy of the entire system.

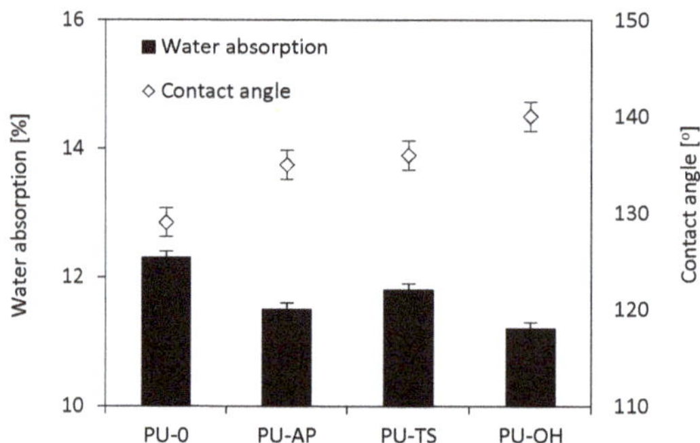

Figure 17. Effect of contact angle on water absorption of RPUFs modified with POSS.

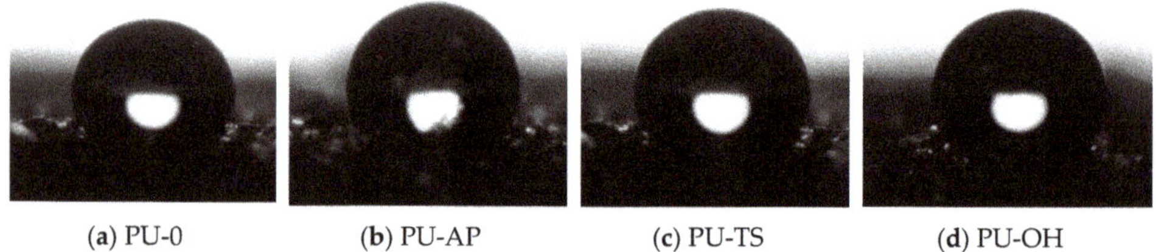

(a) PU-0   (b) PU-AP   (c) PU-TS   (d) PU-OH

Figure 18. The contact angle of the surface of the (a) PU-0, (b) PU-AP, (c) PU-TS, (d) PU-OH.

4. Conclusions

RPUFs were successfully reinforced using POSS with hydroxyl and amino groups. The impact of POSSs on thermal properties, dynamic mechanical properties, physico-mechanical properties (compressive strength, three-point bending test, impact strength apparent density), foaming parameters and morphology of RPUFs was examined. The presented results indicate that the addition of AP-POSS, TS-POSS, and OH-POSS in the range of 0.5 wt.% influences the morphology of analyzed foams and consequently their further mechanical and thermal properties. It was noticed that RPUFs modified with AP-POSS are characterized by smaller and more regular polyurethane cells. This suggests better compatibility between PU foam matrix and AP-POSS compared with other fillers. This results in significant improvement of physico-mechanical properties and thermal stability of composites with AP-POSS. For example, compared to the RPUFs modified with OH-POSS and TS-POSS, composition with 0.5 wt.% of the AP-POSS showed greater compressive strength (351 kPa) and higher flexural strength (0.469 MPa). However, the highest hydrophobicity showed OH-PU foams, which were characterized by the greatest contact angle (140°) and less water uptake (11.2% after 24 h).

Author Contributions: Conceptualization, A.S.; Data curation, S.C.; Formal analysis, A.S. and S.C.; Investigation, A.S. and S.C.; Methodology, A.S. and S.C.; Resources, A.S.; Validation, S.C.; Visualization, S.C.; Writing—original draft, A.S. and S.C.; Writing—review editing, K.S.

Funding: This research received no external funding.

Conflicts of Interest: The authors declare no conflict of interest.

References

1. Nikje, M.M.A.; Noruzian, M.; Moghaddam, S.T. Investigation of Fe3O4/AEAP supermagnetic nanoparticles on the morphological, thermal and magnetite behavior of polyurethane rigid foam nanocomposites. *Polimery* **2015**, *60*, 26–32. [CrossRef]
2. Xie, H.; Yang, W.; Yuen, A.C.Y.; Xie, C.; Xie, J.; Lu, H.; Yeoh, G.H. Study on flame retarded flexible polyurethane foam/alumina aerogel composites with improved fire safety. *Chem. Eng. J.* **2017**, *311*, 310–317. [CrossRef]
3. Yang, C.; Fischer, L.; Maranda, S.; Worlitschek, J. Rigid polyurethane foams incorporated with phase change materials: A state-of-the-art review and future research pathways. *Energy Build.* **2015**, *87*, 25–36. [CrossRef]
4. Tan, S.; Abraham, T.; Ference, D.; Macosko, C.W. Rigid polyurethane foams from a soybean oil-based Polyol. *Polymers* **2011**, *52*, 2840–2846. [CrossRef]
5. Rigid Polyurethane Foam Market - Global Industry Analysis, Size, Share, Growth Trends & Forecasts 2017–2025, PR Newswire, New York. 2017. Available online: https://search.proquest.com/docview/1966284809?rfr_id=info%3Axri%2Fsid%3Aprimo (accessed on 27 May 2018).
6. Global Polyurethane (PU) Foam Market Trends, Analysis & Forecasts 2016–2024, NASDAQ OMX's News Release Distrib. Channel, New York. 2016. Available online: https://search.proquest.com/docview/1793434616?rfr_id=info%3Axri%2Fsid%3Aprimo (accessed on 27 May 2018).
7. Mahajan, N.; Gupta, P. New insights into the microbial degradation of polyurethanes. *RSC Adv.* **2015**, *5*, 41839–41854. [CrossRef]
8. Yang, H.; Wang, X.; Song, L.; Yu, B.; Yuan, Y.; Hu, Y.; Yuen, K.K.R. Aluminum hypophosphite in combination with expandable graphite as a novel flame retardant system for rigid polyurethane foams. *Polym. Adv. Technol.* **2014**, *25*, 1034–1043. [CrossRef]
9. Hu, X.-M.; Wang, D.-M. Enhanced fire behavior of rigid polyurethane foam by intumescent flame retardants. *J. Appl. Polym. Sci.* **2013**, *129*, 238–246. [CrossRef]
10. Shi, Y.; Yuan, Y.; Yang, H.; Yu, B.; Wang, W.; Song, L.; Hu, Y.; Zhang, Y. Phosphorus and Nitrogen-Containing Polyols: Synergistic Effect on the Thermal Property and Flame Retardancy of Rigid Polyurethane Foam Composites. *Ind. Eng. Chem. Res.* **2016**, *55*, 10813–10822.
11. Yang, H.; Liu, H.; Jiang, Y.; Chen, M.; Wan, C. Density Effect on Flame Retardancy, Thermal Degradation, and Combustibility of Rigid Polyurethane Foam Modified by Expandable Graphite or Ammonium Polyphosphate. *Polymers* **2019**, *11*, 668. [CrossRef] [PubMed]
12. Kurańska, M.; Prociak, A. The influence of rapeseed oil-based polyols on the foaming process of rigid polyurethane foams. *Ind. Crop. Prod.* **2016**, *89*, 182–187. [CrossRef]
13. Sobolewski, M.; Błażejczak, A. Izolacyjność cieplna wysokoprężnej pianki poliuretanowej w aerozolu. Cz. 1. Właściwości i zastosowanie pianek poliuretanowych, Izolacje. R. 2014, 19, pp. 11–12. Available online: http://yadda.icm.edu.pl/baztech/element/bwmeta1.element.baztech-5d8e9b3a-a1e9-4bf7-85dd-3adb61863dd1 (accessed on 1 June 2019).
14. Paciorek-Sadowska, J.; Czupryński, B.; Liszkowska, J.; Jaskółowski, W. Nowy poliol boroorganiczny do produkcji sztywnych pianek poliuretanowo-poliizocyjanurowych. Cz. II. Otrzymywanie sztywnych pianek poliuretanowo-poliizocyjanurowych z zastosowaniem nowego poliolu boroorganicznego, Polimery. T. 2010, 55, pp. 99–105. Available online: http://yadda.icm.edu.pl/baztech/element/bwmeta1.element.baztech-article-BATB-0001-0012 (accessed on 1 June 2019).
15. Fan, H.; Tekeei, A.; Suppes, G.J.; Hsieh, F.-H. Rigid polyurethane foams made from high viscosity soy-polyols. *J. Appl. Polym. Sci.* **2013**, *127*, 1623–1629. [CrossRef]
16. Członka, S.; Bertino, M.F.; Strzelec, K. Rigid polyurethane foams reinforced with industrial potato protein. *Polym. Test.* **2018**, *68*, 135–145. [CrossRef]
17. Formela, K.; Hejna, A.; Zedler, Ł.; Przybysz, M.; Ryl, J.; Saeb, M.R.; Piszczyk, Ł. Structural, thermal and physico-mechanical properties of polyurethane/brewers' spent grain composite foams modified with ground tire rubber. *Ind. Crop. Prod.* **2017**, *108*, 844–852. [CrossRef]
18. Silva, M.C.; Takahashi, J.A.; Chaussy, D.; Belgacem, M.N.; Silva, G.G. Composites of rigid polyurethane foam and cellulose fiber residue. *J. Appl. Polym. Sci.* **2010**. [CrossRef]
19. Shi, X.; Yang, P.; Peng, X.; Huang, C.; Qian, Q.; Wang, B.; He, J.; Liu, X.; Li, Y.; Kuang, T. Bi-phase fire-resistant polyethylenimine/graphene oxide/melanin coatings using layer by layer assembly technique: Smoke suppression and thermal stability of flexible polyurethane foams. *Polymers* **2019**, *170*, 65–75. [CrossRef]

20. Cho, J.H.; Vasagar, V.; Shanmuganathan, K.; Jones, A.R.; Nazarenko, S.; Ellison, C.J. Bioinspired Catecholic Flame Retardant Nanocoating for Flexible Polyurethane Foams. *Chem. Mater.* **2015**, *27*, 6784–6790. [CrossRef]

21. Roberts, B.; Jones, A.; Ezekoye, O.; Ellison, C.; Webber, M.; Roberts, B. Development of kinetic parameters for polyurethane thermal degradation modeling featuring a bioinspired catecholic flame retardant. *Combust. Flame* **2017**, *177*, 184–192. [CrossRef]

22. Septevani, A.A.; Evans, D.A.; Annamalai, P.K.; Martin, D.J. The use of cellulose nanocrystals to enhance the thermal insulation properties and sustainability of rigid polyurethane foam. *Ind. Crop. Prod.* **2017**, *107*, 114–121. [CrossRef]

23. Song, Z.-L.; Ma, L.-Q.; Wu, Z.-J.; He, D.-P. Effects of viscosity on cellular structure of foamed aluminum in foaming process. *J. Mater. Sci.* **2000**, *35*, 15–20. [CrossRef]

24. Dolomanova, V.; Rauhe, J.C.M.; Jensen, L.R.; Pyrz, R.; Timmons, A.B. Mechanical properties and morphology of nano-reinforced rigid PU foam. *J. Cell. Plast.* **2011**, *47*, 81–93. [CrossRef]

25. Niyogi, D.; Kumar, R.; Gandhi, K.S. Water blown free rise polyurethane foams. *Polym. Eng. Sci.* **1999**, *39*, 199–209. [CrossRef]

26. Sung, G.; Kim, J.H. Influence of filler surface characteristics on morphological, physical, acoustic properties of polyurethane composite foams filled with inorganic fillers. *Compos. Sci. Technol.* **2017**, *146*, 147–154. [CrossRef]

27. Gu, R.; Khazabi, M.; Sain, M. Fiber reinforced soy-based polyurethane spray foam insulation. Part 2: Thermal and mechanical properties. *BioResources* **2011**, *6*, 3775–3790.

28. Luo, X.; Mohanty, A.; Misra, M. Lignin as a reactive reinforcing filler for water-blown rigid biofoam composites from soy oil-based polyurethane. *Ind. Crop. Prod.* **2013**, *47*, 13–19. [CrossRef]

29. Michałowski, S.; Hebda, E.; Pielichowski, K. Thermal stability and flammability of polyurethane foams chemically reinforced with POSS. *J. Therm. Anal. Calorim.* **2017**, *130*, 155–163. [CrossRef]

30. Michałowski, S.; Pielichowski, K. 1,2-Propanediolizobutyl POSS as a co-flame retardant for rigid polyurethane foams. *J. Therm. Anal. Calorim.* **2018**, *134*, 1351–1358. [CrossRef]

31. Hebda, E.; Ozimek, J.; Raftopoulos, K.N.; Michałowski, S.; Pielichowski, J.; Jancia, M.; Pielichowski, K. Synthesis and morphology of rigid polyurethane foams with POSS as pendant groups or chemical crosslinks. *Polym. Adv. Technol.* **2015**, *26*, 932–940. [CrossRef]

32. Cordes, D.B.; Lickiss, P.D.; Rataboul, F. Recent Developments in the Chemistry of Cubic Polyhedral Oligosilsesquioxanes. *Chem. Rev.* **2010**, *110*, 2081–2173. [CrossRef]

33. Hosaka, N.; Otsuka, H.; Hino, M.; Takahara, A. Control of Dispersion State of Silsesquioxane Nanofillers for Stabilization of Polystyrene Thin Films. *Langmuir* **2008**, *24*, 5766–5772. [CrossRef]

34. Fina, A.; Monticelli, O.; Camino, G. POSS-based hybrids by melt/reactive blending. *J. Mater. Chem.* **2010**, *20*, 9297. [CrossRef]

35. Adnan, S.; Tuan Ismail, T.N.M.; Mohd Noor, N.; Din, N.M.; Mariam, N.S.; Hanzah, N.A.; Shoot Kian, Y.; Abu Hassan, H. Development of Flexible Polyurethane Nanostructured Biocomposite Foams Derived from Palm Olein-Based Polyol. *Adv. Mater. Sci. Eng.* **2016**, *2016*, 1–12. [CrossRef]

36. Pawlak, T.; Kowalewska, A.; Zgardzińska, B.; Potrzebowski, M.J. Structure, Dynamics, and Host–Guest Interactions in POSS Functionalized Cross-Linked Nanoporous Hybrid Organic–Inorganic Polymers. *J. Phys. Chem. C* **2015**, *119*, 26575–26587. [CrossRef]

37. Blanco, I. The Rediscovery of POSS: A Molecule Rather than a Filler. *Polymers* **2018**, *10*, 904. [CrossRef] [PubMed]

38. Liu, Y.; Tseng, M.; Fangchiang, M. Polymerization and nanocomposites properties of multifunctional methylmethacrylate POSS. *J. Polym. Sci. Part A Polym. Chem.* **2008**, *46*, 5157–5166. [CrossRef]

39. Rashid, E.S.A.; Ariffin, K.; Kooi, C.C.; Akil, H.M. Preparation and properties of POSS/epoxy composites for electronic packaging applications. *Mater. Des.* **2009**, *30*, 1–8. [CrossRef]

40. Fan, X.; Cao, M.; Zhang, X.; Li, Z. Synthesis of star-like hybrid POSS-(PDMAEMA-b-PDLA)8 copolymer and its stereocomplex properties with PLLA. *Mater. Sci. Eng. C* **2017**, *76*, 211–216. [CrossRef] [PubMed]

41. Blanco, I.; Bottino, F.A.; Cicala, G.; Cozzo, G.; Latteri, A.; Recca, A. Synthesis and thermal characterization of new dumbbell shaped POSS/PS nanocomposites: Influence of the symmetrical structure of the nanoparticles on the dispersion/aggregation in the polymer matrix. *Polym. Compos.* **2015**, *36*, 1394–1400. [CrossRef]

42. Phillips, S.H.; Haddad, T.S.; Tomczak, S.J. Developments in nanoscience: Polyhedral oligomeric silsesquioxane (POSS)-polymers. *Curr. Opin. Solid State Mater. Sci.* **2004**, *8*, 21–29. [CrossRef]

43. Pielichowski, K.; Njuguna, J.; Janowski, B.; Pielichowski, J. *Polyhedral Oligomeric Silsesquioxanes (POSS)-Containing Nanohybrid Polymers*; Springer: Berlin/Heidelberg, Germany, 2006; Vol. 201, pp. 225–296.
44. Marcinkowska, A.; Przadka, D.; Dudziec, B.; Szczesniak, K.; Andrzejewska, E. Anchor Effect in Polymerization Kinetics: Case of Monofunctionalized POSS. *Polymers* **2019**, *11*, 515. [CrossRef]
45. Liu, L.; Tian, M.; Zhang, W.; Zhang, L.; Mark, J.E. Crystallization and morphology study of polyhedral oligomeric silsesquioxane (POSS)/polysiloxane elastomer composites prepared by melt blending. *Polymers* **2007**, *48*, 3201–3212. [CrossRef]
46. Liu, Y.; Huang, Y.; Liu, L. Effects of TriSilanolIsobutyl-POSS on thermal stability of methylsilicone resin. *Polym. Degrad. Stab.* **2006**, *91*, 2731–2738. [CrossRef]
47. Song, X.; Zhang, X.; Li, T.; Li, Z.; Chi, H. Mechanically Robust Hybrid POSS Thermoplastic Polyurethanes with Enhanced Surface Hydrophobicity. *Polymers* **2019**, *11*, 373. [CrossRef] [PubMed]
48. Amin, M.; Najwa, K. Cellulose Nanocrystals Reinforced Thermoplastic Polyurethane Nanocomposites. Ph.D. Thesis, The University of Queensland, Queensland, Australia, 2016. [CrossRef]
49. KAIRYTĖ, A.; Vaitkus, S.; Vėjelis, S.; Girskas, G.; Balčiūnas, G. Rapeseed-based polyols and paper production waste sludge in polyurethane foam: Physical properties and their prediction models. *Ind. Crop. Prod.* **2018**, *112*, 119–129. [CrossRef]
50. Michalowski, S.; Cabulis, U.; Kirpluks, M.; Prociak, A.; Kuranska, M. Microcellulose as a natural filler in polyurethane foams based on the biopolyol from rapeseed oil. *Polimery* **2016**, *61*, 625–632.
51. Yan, D.-X.; Xu, L.; Chen, C.; Tang, J.; Ji, X.; Li, Z. Enhanced mechanical and thermal properties of rigid polyurethane foam composites containing graphene nanosheets and carbon nanotubes. *Polym. Int.* **2012**, *61*, 1107–1114. [CrossRef]
52. Ghoreishi, R.; Al-Moameri, H.; Zhao, Y.; Suppes, G.J. Simulation Blowing Agent Performance, Cell Morphology, and Cell Pressure in Rigid Polyurethane Foams. *Ind. Eng. Chem. Res.* **2016**, *55*, 2336–2344.
53. Członka, S.; Sienkiewicz, N.; Strąkowska, A.; Strzelec, K. Keratin feathers as a filler for rigid polyurethane foams on the basis of soybean oil polyol. *Polym. Test.* **2018**, *72*, 32–45. [CrossRef]
54. Członka, S.; Bertino, M.F.; Strzelec, K.; Strąkowska, A.; Masłowski, M. Rigid polyurethane foams reinforced with solid waste generated in leather industry. *Polym. Test.* **2018**, *69*, 225–237. [CrossRef]
55. Wolska, A.; Goździkiewicz, M.; Ryszkowska, J. Thermal and mechanical behaviour of flexible polyurethane foams modified with graphite and phosphorous fillers. *J. Mater. Sci.* **2012**, *47*, 5627–5634. [CrossRef]
56. Guo, C.; Zhou, L.; Lv, J. Effects of Expandable Graphite and Modified Ammonium Polyphosphate on the Flame-Retardant and Mechanical Properties of Wood Flour-Polypropylene Composites. *Polym. Polym. Compos.* **2013**, *21*, 449–456. [CrossRef]
57. Ciecierska, E.; Jurczyk-Kowalska, M.; Bazarnik, P.; Gloc, M.; Kulesza, M.; Krauze, S.; Lewandowska, M.; Kowalski, M. Flammability, mechanical properties and structure of rigid polyurethane foams with different types of carbon reinforcing materials. *Compos. Struct.* **2016**, *140*, 67–76. [CrossRef]
58. Kim, J.M.; Han, M.S. Thermal, Morphological and Rheological Properties of Rigid Polyurethane Foams as Thermal Insulating Materials. *AIP Conference Proceedings* **2008**, *1027*, 905–907.
59. Chang, L.-C. Improving the Mechanical Performance of Wood Fiber Reinforced Bio-based Polyurethane Foam. Ph.D. Thesis, University of Toronto, Toronto, Canada, 2014.
60. Kurańska, M.; Prociak, A. Porous polyurethane composites with natural fibres. *Compos. Sci. Technol.* **2012**, *72*, 299–304. [CrossRef]
61. Kurańska, M.; Aleksander, P.; Mikelis, K.; Ugis, C. Porous polyurethane composites based on bio-components. *Compos. Sci. Technol.* **2013**, *75*, 70–76. [CrossRef]
62. Mosiewicki, M.; Dell'Arciprete, G.; Aranguren, M.; Marcovich, N. Polyurethane Foams Obtained from Castor Oil-based Polyol and Filled with Wood Flour. *J. Compos. Mater.* **2009**, *43*, 3057–3072. [CrossRef]
63. Finlay, K.A.; Gawryla, M.D.; Schiraldi, D.A. Effects of Fiber Reinforcement on Clay Aerogel Composites. *Materials* **2015**, *8*, 5440–5451. [CrossRef]
64. Hamilton, A.R.; Thomsen, O.T.; Madaleno, L.A.; Jensen, L.R.; Rauhe, J.C.M.; Pyrz, R. Evaluation of the anisotropic mechanical properties of reinforced polyurethane foams. *Compos. Sci. Technol.* **2013**, *87*, 210–217. [CrossRef]
65. Marcovich, N.E.; Kurańska, M.; Prociak, A.; Malewska, E.; Kulpa, K. Open cell semi-rigid polyurethane foams synthesized using palm oil-based bio-polyol. *Ind. Crops Prod.* **2017**, *102*, 88–96. [CrossRef]

66. Wu, L.; Van Gemert, J.; Camargo, R.E. Rheology Study in Polyurethane Rigid Foams. Auburn Hills, MI, USA, 2012. Available online: http://www.huntsman.com/polyurethanes/Media%20Library/a_MC1CD1F5AB7BB1738E040EBCD2B6B01F1/Products_MC1CD1F5AB8081738E040EBCD2B6B01F1/Construction_MC1CD1F5AEF051738E040EBCD2B6B01F1/Technical%20presentati_MC1CD1F5AF6F41738E040EBCD2B6B01F1/files/cpi_08_lifengwu_revised.pdf (accessed on 1 June 2019).

67. Hatakeyama, H.; Kosugi, R.; Hatakeyama, T. Thermal properties of lignin-and molasses-based polyurethane foams. *J. Therm. Anal. Calorim.* **2008**, *92*, 419–424. [CrossRef]

68. Ye, L.; Meng, X.-Y.; Ji, X.; Li, Z.-M.; Tang, J.-H. Synthesis and characterization of expandable graphite–poly(methyl methacrylate) composite particles and their application to flame retardation of rigid polyurethane foams. *Polym. Degrad. Stab.* **2009**, *94*, 971–979. [CrossRef]

69. Gama, N.V.; Silva, R.; Mohseni, F.; Davarpanah, A.; Amaral, V.; Ferreira, A.; Barros-Timmons, A. Enhancement of physical and reaction to fire properties of crude glycerol polyurethane foams filled with expanded graphite. *Polym. Test.* **2018**, *69*, 199–207. [CrossRef]

70. Jiao, L.; Xiao, H.; Wang, Q.; Sun, J. Thermal degradation characteristics of rigid polyurethane foam and the volatile products analysis with TG-FTIR-MS. *Polym. Degrad. Stab.* **2013**, *98*, 2687–2696. [CrossRef]

71. Levchik, S.V.; Weil, E.D. Thermal decomposition, combustion and fire-retardancy of polyurethanes—A review of the recent literature. *Polym. Int.* **2004**, *53*, 1585–1610. [CrossRef]

72. Septevani, A.A.; Evans, D.A.; Chaleat, C.; Martin, D.J.; Annamalai, P.K.; Evans, D.A.C. A systematic study substituting polyether polyol with palm kernel oil based polyester polyol in rigid polyurethane foam. *Ind. Crop. Prod.* **2015**, *66*, 16–26. [CrossRef]

73. Pagacz, J.; Hebda, E.; Michałowski, S.; Ozimek, J.; Sternik, D.; Pielichowski, K. Polyurethane foams chemically reinforced with POSS—Thermal degradation studies. *Thermochim. Acta* **2016**, *642*, 95–104. [CrossRef]

74. Chattopadhyay, D.; Webster, D.C. Thermal stability and flame retardancy of polyurethanes. *Prog. Polym. Sci.* **2009**, *34*, 1068–1133. [CrossRef]

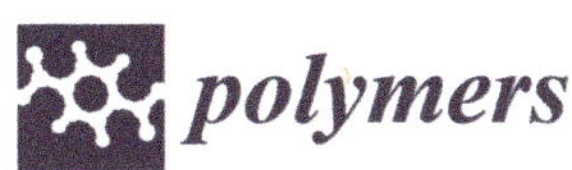 *polymers*

Article

An In Vitro Evaluation, on Polyurethane Foam Sheets, of the Insertion Torque (IT) Values, Pull-Out Torque Values, and Resonance Frequency Analysis (RFA) of NanoShort Dental Implants

Luca Comuzzi [1], Giovanna Iezzi [2], Adriano Piattelli [2,3,4] and Margherita Tumedei [2,*

[1] Private practice, via Raffaello 36/a, 31020 San Vendemiano (TV), Italy; luca.comuzzi@gmail.com
[2] Department of Medical, Oral and Biotechnological Sciences, University "G. D'Annunzio" of Chieti-Pescara, Via dei Vestini 31, 66100 Chieti, Italy; gio.iezzi@unich.it (G.I.); apiattelli@unich.it (A.P.)
[3] Biomaterials Engineering, Catholic University of San Antonio de Murcia (UCAM), Av. de los Jerónimos, 135, 30107 Guadalupe, Murcia, Spain
[4] Villaserena Foundation for Research, Via Leonardo Petruzzi 42, 65013 Città Sant'Angelo (PE), Italy
* Correspondence: margytumedei@yahoo.it; Tel.: +39-0871-355-4083

Received: 11 May 2019; Accepted: 7 June 2019; Published: 10 June 2019

Abstract: Objectives: The aim of this study was to investigate, in polyurethane foam sheets, the primary implant stability of a NanoShort implant compared to a self-condenser implant and to a standard, conventional implant. Materials and Methods: Three implant designs were evaluated in the present in vitro investigation: The Test implant (NanoShort), the Control A implant (self-condenser), and the Control B implant (standard design). The study was conducted by comparing the insertion torque values, the pull-out strength values, and the resonance frequency analysis (RFA) values of the Test and Control A and B implants inserted in polyurethane foam models of different thicknesses and densities. The foam densities were 10, 20, and 30 pounds per cubic foot (pcf). Three thicknesses of polyurethane foams (1, 2, 3 mm) were evaluated for a total of 640 experimental sites. Results: The Pearson correlation showed a moderate/strong correlation between all study groups (r > 0.3) for insertion torque and pull-out strength levels. Increased stability of the Test implants was obtained in 3 mm polyurethane sheets. The 2.5 and 3.5 mm Test implants presented good stability in 3 mm polyurethane sheets of 20–30 pcf densities. The Control implants showed better results compared to the Test implants in 1, 2, and 3 mm polyurethane sheets with densities of 10, 20, and 30 pcf. Conclusions: The NanoShort dental implant evaluated in this in vitro study showed a high level of stability in some experimental conditions, and could represent a useful tool, especially in the posterior mandible, as an alternative to vertical augmentation procedures.

Keywords: implant stability; insertion torque; pull-out strength; polyurethane foam

---

1. Introduction

During the insertion of dental implants, bone density plays a key role in determining the primary stability of the implants [1,2]. There is, then, a need to understand the relationship between bone density and primary implant stability to plan implant treatment in a proper way [1]. To get mineralized tissues at the interface with the implants, there is an absolute need to achieve primary implant stability, i.e., the initial biomechanical engagement between bone and implant, with no relative micromovements between these two structures, immediately after insertion of the implants [3–6]. Poor bone density affects primary stability in a negative way [4,5,7]. Higher bone quality is correlated with higher primary implant stability [6,8]. Therefore, the main factors influencing primary stability are the percentage of bone-to-implant contact (BIC) and the compressive stresses at the implant–bone interface [5]. Besides bone quantity and bone quality, primary stability is also related to implant geometry (length and diameter) and to the surgical technique used to prepare the insertion site of the implants [8]. A low primary stability has been reported to carry a higher risk of implant failure or loss, while with a high stability, better conditions for the formation of mineralized tissues at the implant interface are realized [8]. Moreover, it must also be underlined that a high level of primary stability is associated in a positive way with secondary implant stability [8]. It is, then, extremely important to be able to assess, in an accurate way, primary stability [8].

So far, several non-invasive and non-destructive methods have been suggested to evaluate implant stability, such as insertion torque measurement (IT) and resonance frequency analysis (RFA). IT measures the compression produced by an implant during placement into the surgical site [7] and it is closely related to the primary implant stability, which is considered the most important factor for successful implant treatment [5]. Pull-out torque measures the force needed to remove the implants.

RFA is a validated, useful, and non-invasive tool in the clinical practice [9]. This procedure is able to measure the deflection of the implant–bone complex at different time intervals [10]. Polyurethane foam has been proposed for in vitro tests because it simulates the consistency and density of bone tissue.

The intrinsic mechanical features and biocompatibility of polyurethane foam have been applied in several different medical fields, including vascular engineering and orthopaedics [11–13]. This material has been reported to show mechanical properties as described by the ASTM (American Society Testing Materials) F-1839-08 2012 standard [1,4]. Its characteristics render it a very good material to test different implant materials [1], and to standardize the procedures excluding the anatomical and structural differences present in bone [5,6]. The low-to-high densities of polyurethane foams are representative of different bone densities, according to the bone tissue classification D1–D4 proposed by Misch [14].

Short implants (less than 8 mm in length) have been proposed mainly in the clinical condition of deficient alveolar ridges in the posterior jaws [15–17]. They could avoid the use of maxillary sinus augmentation procedures or bone grafts, avoiding increased morbidity, higher costs, and higher risks of complications [18–21]. Several recent systematic reviews with meta-analyses have shown that the survivals of short and standard-length implants are similar [18–23]. In the past few years, implants with a reduced length (4, 5, and 6 mm) (Ultrashort or Extrashort implants) have been proposed [16,24]. In the market, it is also possible to find implants with a still-reduced length (2.5 or 3.5 mm) (NanoShort implants) [25].

The aim of the present investigation was to evaluate the in vitro biomechanical behavior of a NanoShort implant compared with a self-condenser implant and a standard implant.

2. Materials and Methods

2.1. Implants

NanoShort titanium dental implants (Oralplant Suisse, Mendrisio, Switzerland; Test Implants) were used for the present in vitro investigation (Figure 1). These implants had a length of 2.5 mm and a diameter of 4.5 mm (Test A), or a length of 3.5 mm and a diameter of 4.1 mm (Test B) (Figure 2A).

SinusPlant^{TM} implants (Oralplant Suisse, Mendrisio, Switzerland) were used as Control A implants, and were 4.5 mm diameter and 10 mm length.

Cylindrical screw-shaped implants were used as Control B implants, and they were of 13 mm length, had a platform diameter of 4.1 mm, and a body diameter of 3.75 mm (Restore, Keystone Dental, Burlington, MA, USA). The coronal portions of two of the experimental implants presented the same diameter, but different shapes: A troncoconical geometry for the Test implants and a cylindrical morphology for the Control B implants (Figure 2A).

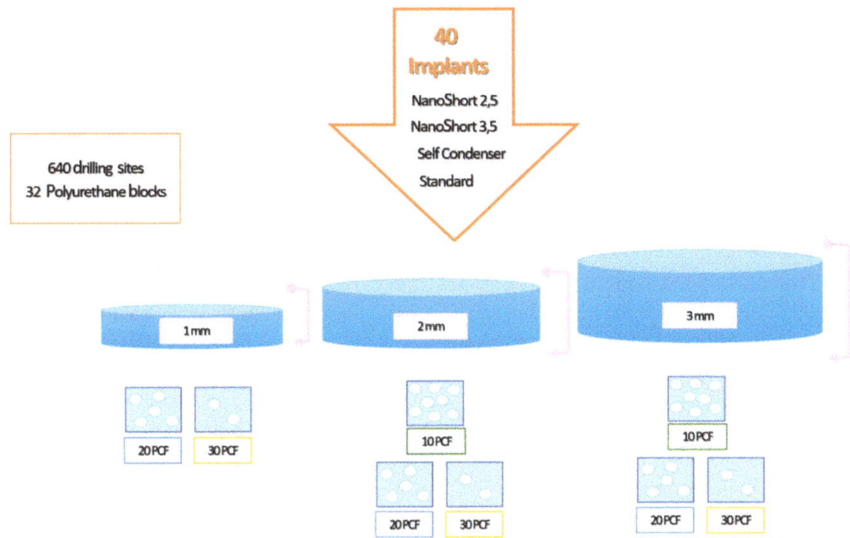

Figure 1. Summary of the model design of the study.

2.2. Polyurethane Foam Blocks

Polyurethane foam represented a useful tool for biomedical applications, including testing instruments and dental implants for the comparative testing of bone screws [16]. This material has been demonstrated to be able to eliminate the variables of human cadaver bone and of animal bone. Artificial bone blocks presented a uniform cortical bone density and depth, and were unaffected by desiccation. This material exhibited similar properties to bone, and was shown to be reliable and required no special handling or preservation. This synthetic material presented consistent mechanical characteristics.

2.3. Experimental Design of the Study

A total of 40 implants (10 Test A, 10 Test B, 10 Control A, and 10 Control B) were used in the present investigation. A total of 640 osteotomies were produced into the polyurethane blocks. Each implant was repositioned 16 times for each experiment, for a total of 10 implants for each study group. The implants were inserted following the protocol of the manufacturer, using an implant lance drill, a 2 mm drill (1600 rpm), and a 3.8 mm final drill (800 rpm) (Figure 1). The handpiece was calibrated at a speed of 70 rpm and a torque of 30 Ncm. Torque values were taken with a software (ImpDat Plus, East Lansing, MI, USA) installed on a digital card. The insertion torque (IT; Ncm) values indicated the force of the maximum clockwise movement that stripped the material. The investigation was conducted by a single operator (LC), comparing the torque insertion and pull-out strength values of the Test, Control A, and Control B implants when inserted into polyurethane foam models of different sizes and densities. Different types of solid rigid polyurethane foam (SawBones H, Pacific Research Laboratories Inc, Vashon, WA, USA) with homogeneous densities were selected for the present investigation. Solid rigid polyurethane foam provided a closed cell content range from 96.0% to 99.9%. Foam was available in a range of sizes and densities, from 0.08 to 0.80 grams per cubic centimeter (5 to 50 pounds per cubic foot). The densities of polyurethane foam used in the present study were 10 pounds per cubic foot (pcf), corresponding to a density of 0.16 g/cm3 (similar to the D3 bone type); 20 pcf, corresponding to 0.32 g/cm3 (similar to the D2 bone type); and 30 pcf, corresponding to 0.48 g/cm3 (similar to D1 bone). Furthermore three different sizes of polyurethane foams were evaluated, for a total of 32 blocks and 640 experimental sites: 13 cm × 18 cm × 1 mm (20, 30 pcf); 13 cm × 18 cm × 2 mm (10, 20, and 30 pcf); and 13 cm × 18 cm × 3 mm (10, 20, and 30 pcf).

2.4. Insertion Torque and Pull-Out Torque

The study was conducted by comparing the insertion torque and the pull-out strength values using a calibrated torque meter (UNIKA, Oralplant Suisse, Mendrisio Switzerland) with a torque range of 5–80 Ncm. The final 1 mm insertion torque of the implants into the polyurethane sheets was recorded. In the present study, mechanical torque gauges were used to assess the insertion torque and pull-out strength values. A total of 640 drilling sites (160 for each group) were produced in different sizes and densities of the polyurethane foam models. For each block, 20 drilling sites were obtained.

2.5. Implant Drill

Test A implants were inserted following a dedicated drill protocol: A surgical twist drill of 2 mm at 1500 Rpm, a pilot drill of 2–4.5 mm at 800 Rpm, and finally, burs of 4.5 mm diameter and 2.5 mm length.

Test B implants were inserted following a dedicated drill protocol: A surgical twist drill of 2 mm at 1500 Rpm, a pilot drill of 2–4.1 mm at 800 Rpm, and finally, burs of 4.1 mm diameter and 3.5 mm length.

Control A implants were inserted using a surgical twist drill of 2 mm at 1500 Rpm, a pilot drill of 2–3.8 mm at 800 Rpm, osteocondensation screw burs of 10 mm length and 4.5 mm diameter used at the top at 40 Rpm, and then the implant was placed with a predetermined maximum torque of 46 Ncm and 30 Rpm with the surgical motor.

Control B implants were inserted using a surgical twist drill of 2 mm at 1500 Rpm, a pilot drill of 2–3 mm at 800 Rpm, a final 3 mm drill, and then the implant was placed with a calibrated torque of 46 Ncm at a predetermined 30 Rpm with the surgical motor (Figure 2B–D).

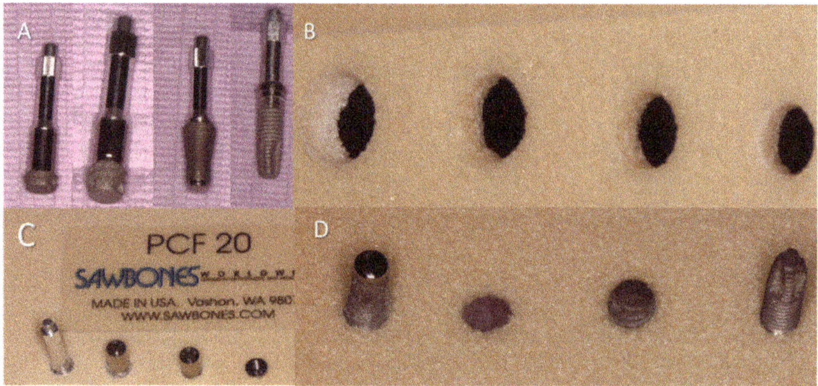

Figure 2. (**A**) From the left to the right: NanoShort 2.5, NanoShort 3.5, self-condenser implant, standard implant; (**B**) details of site preparation of the polyurethane blocks after the drillings protocols; (**C**) implants positioned into a polyurethane block; (**D**) detail of the back view of the positioned implant.

2.6. Resonance Frequency Analysis

After implant insertion, primary stability was measured using RFA values, expressed using the implant stability quotient (ISQ), with hand-screwed Smart-Pegs (type 7 for Test implants and type 3 for Control implants; Osstell Mentor Device, Integration Diagnostic AB, Savadelen, Sweden). The implant stability quotient (ISQ) ranged from 0 to 100 (measured between 3500 and 85,000 Hz), and was divided into Low (lower than 60 ISQ), Medium (60–70 ISQ), and High stability (more than 70 ISQ) [26]. For each specimen, the RFA measurement was repeated two times. Measurements were performed in two orientations separated by a 90-degree angle, and the average ISQ values were calculated.

2.7. Statistical Analysis

Differences between the values of insertion torque, pull-out strength, and RFA of the four groups were analyzed by one-way analysis of variance (ANOVA), followed by Tukey's post-hoc test. A p-value

< 0.05 was considered statistically significant. The correlation between the values of insertion torque and pull-out strength was determined by the Pearson Test. Data treatment and statistical analysis were performed by Excel origin and StatPlus6 software.

3. Results

The results were similar when comparing the 2.5 and 3.5 NanoShort implants (Test A and Test B). The Test implants had good stabilization in 3 mm polyurethane sheets of 20–30 pcf densities. In 2 mm sheets, good stability of the Test implants was only obtained in the sheets of 30 pcf density (Table 1). Control A implants showed better results when compared to the Test implants in 1, 2, and 3 mm polyurethane sheets of 10, 20, and 30 pcf densities ($p < 0.01$) (Figures 3–5). Control B implants showed good results in 1 and 2 mm polyurethane sheets. In 3 mm sheets, the values of the Control B implants were similar or slightly lower when compared to the Test A and B Implants ($p > 0.05$). The pull-out torque values of the Test implants were lower than the values of insertion torque, while in Control A implants, the values were similar, and in Control B implants, the values of the insertion torque were lower than the values of the pull-out torque (Table 1) (Figures 3–5).

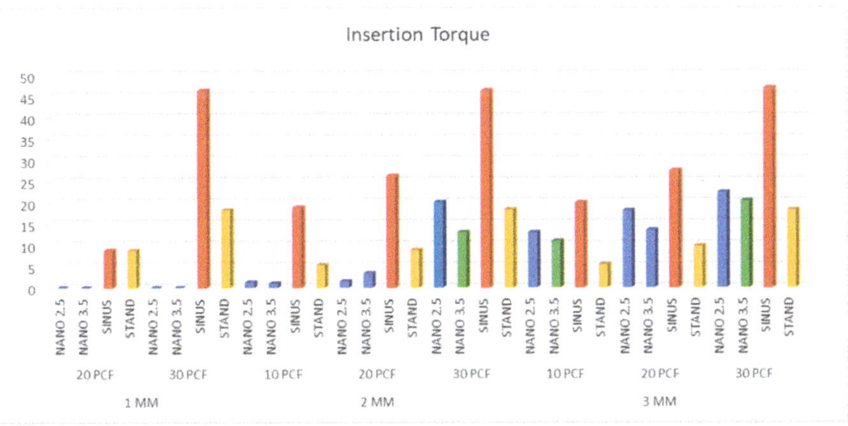

Figure 3. Insertion torque values for the four experimental groups. The self-condenser implant showed the highest ratio of stability.

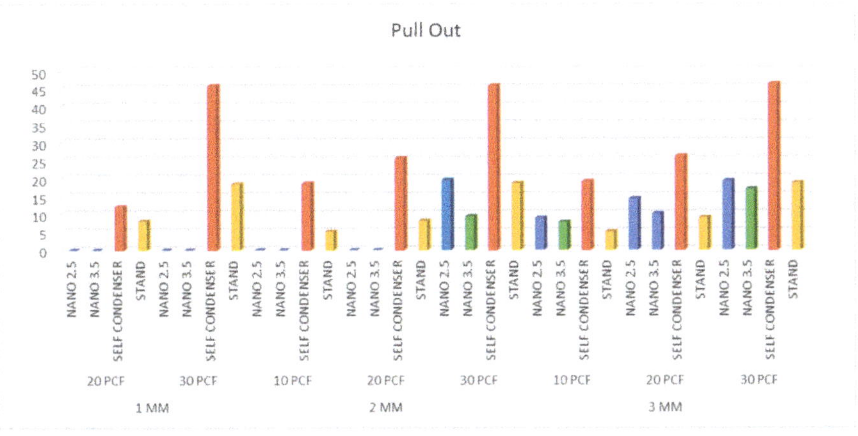

Figure 4. Results of the investigation of pull-out torque. The self-condenser implant showed the highest ratio of stability.

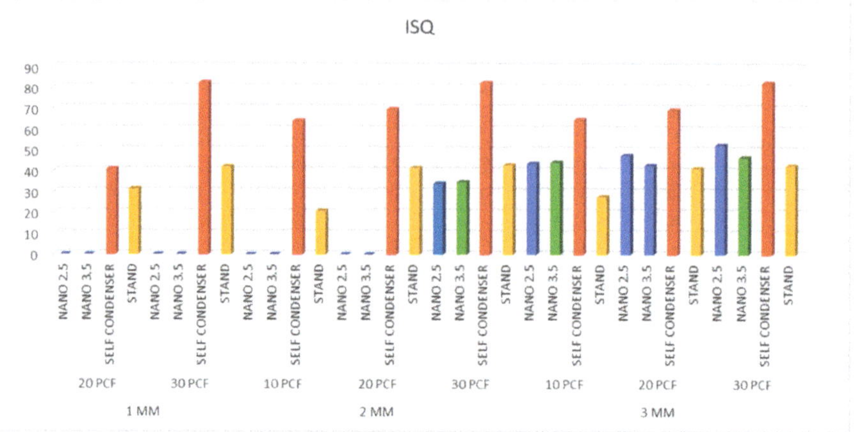

Figure 5. Resonance frequency analysis (RFA) effectiveness of the study groups.

Table 1. Values of insertion torque, pull-out strength, and RFA effectiveness of the study groups.

			INSERTION TORQUE		PULL OUT			RFA ANALYSIS	
		Groups	MEAN	SD	MEAN	SD	Pearson Correlation (r)	MEAN	SD
1 MM	20 PCF	NANO 2.5	0	0	0	0	1.000	0	0
		NANO 3.5	0	0	0	0	1.000	0	0
		SELF CONDENSER	14.14	2.049	12.1	1.723	0.69	40.9	1.107
		STAND	10.7	1.625	8.1	1.518	0.76	31.5	1.721
	30 PCF	NANO 2.5	0	0	0	0	1.000	0	0
		NANO 3.5	0	0	0	0	1.000	0	0
		SELF CONDENSER	46.2	1.704	45.8	1.508	0.56	82.43	1.173
		STAND	19.15	1.755	18.4	1.142	0.72	42.15	1.065
2 MM	10 PCF	NANO 2.5	1.25	2.221	0	0	1.000	0	0
		NANO 3.5	1	0.4588	0	0	1.000	0	0
		SELF CONDENSER	19.75	2.049	18.55	1.905	0.75	64	1.564
		STAND	5.25	0.9105	5.1	0.9119	0.34	20.93	2.38
	20 PCF	NANO 2.5	1.45	2.114	0	0	1.000	0	0
		NANO 3.5	3.35	0.7373	0	0	1.000	0	0
		SELF CONDENSER	26.2	1.963	25.55	1.317	0.48	69.58	1.633
		STAND	10.8	1.196	8.2	1.824	0.69	41.58	1.115
	30 PCF	NANO 2.5	20	1.1	19.5	1.433	0.98	34.05	2.299
		NANO 3.5	12.9	0.4224	9.4	2.529	0.61	34.85	2.529
		SELF CONDENSER	46.2	1.704	45.8	1.508	0.36	82.33	1.321
		STAND	19.25	1.803	18.6	1.667	0.80	42.93	1.29
3 MM	10 PCF	NANO 2.5	12.9	2.222	8.95	1.986	0.73	43.75	2.505
		NANO 3.5	10.9	0.2283	7.8	1.262	0.61	44.25	1.262
		SELF CONDENSER	20.9	1.651	19.1	1.518	0.56	64.7	1.712
		STAND	8.4	1.353	5.1	1.373	0.60	27.78	2.835
	20 PCF	NANO 2.5	18.05	2.012	14.35	2.498	0.60	47.65	2.72
		NANO 3.5	13.45	0.3733	10.15	1.928	0.89	42.83	1.928
		SELF CONDENSER	27.45	2.235	26.15	2.007	0.79	69.35	1.702
		STAND	10.7	1.625	9.05	1.849	0.59	41.4	1.283
	30 PCF	NANO 2.5	22.3	2.003	19.25	1.618	0.92	52.68	2.358
		NANO 3.5	20.3	0.3332	16.95	2.047	0.80	46.68	2.047
		SELF CONDENSER	46.8	2.042	46.25	1.832	0.45	82.5	1.225
		STAND	19.15	1.981	18.65	1.814	0.80	42.75	1.323

4. Discussion

Recent systematic reviews with meta-analyses agreed that short implants could be a suitable alternative to more invasive procedures when there is the need to treat alveolar atrophy of the posterior jaws [15,17–24]. Similar survival rates, when compared with standard-length implants, have been reported for short implants. A trend towards a reduction in the implant length of short implants has been observed in recent years, with the introduction in the market of Extrashort and Ultrashort implants (5–6 mm length, or even less) [15–17,24]. These implants have been shown to be able to osseointegrate and bear a functional load [24]. Furthermore, Nanoshort implants with a still-reduced implant length (2.5/3.5 mm length) are available in the market [25].

Polyurethane foam could be a useful alternative material to provide mechanical tests for human bone. The American Society for Testing and Materials (ASTM F-1839-08) has approved this material, and has recognized it as a standard for testing instruments and oral implants for the comparative testing of bone screws ("Standard specification for Rigid Polyurethane Foam for Use as a Standard Material for Testing Orthopaedic Devices for Instruments"). Given the difficulties of working with human cadaver bone and animal bone, synthetic polyurethane foams have been widely used as alternative materials in several biomechanical tests, due to the fact that these materials present a similar cellular structure and consistent mechanical characteristics. Artificial bone blocks were chosen over animal or cadaver bone, as they presented a uniform cortical bone density and depth, and were unaffected by desiccation. This material exhibited similar properties to bone, and it was reliable and required no special handling or preservation.

On the contrary, the limitation of this study is that the study design provided only an analysis of the mechanical retention of the implant, whereas in clinical situations, many biological factors affect primary stability, and the physiological and molecular events of the healing of bone tissue produce phenomena like bone resorption, neoformation, and remodeling, leading to secondary stability.

In this way, the reuse of fixture and implant drills did not produce any effect on the outcome of this research, because the study model provided an experimental design that evaluated only the macro-mechanical interactions between the surfaces in the absence of variables of biological interaction that could be conditioned in vivo.

The good primary stability obtained by the Test implants in polyurethane foams of 20 and 30 pcf densities makes them a suitable implant for use in the posterior regions of the mandible, where the bone quality is good and is similar to 30 pcf polyurethane. The polyurethane foam density values of 10–20 pcf are comparable to the D3 bone type. D3 is a bone quality typical of the anterior maxilla and the posterior regions of the jaws. The D3 bone of the anterior maxilla presents less thickness than the mandibular D3.

Yamaguchi et al. (2016) reported that the mean mineral density in the posterior maxilla is 0.31 g/cm. Test implants would probably have less utility in the posterior maxilla, where the bone quality is usually poor. In this anatomic area, Control B implants seem to have higher stability parameters in all the polyurethane heights and densities. The lack of differences between the two different types of Test implants (2.5 vs. 3.5 mm) is probably related to the fact that 2.5 mm implants had a wider diameter (4.5 vs. 4.1 mm), and the PIC (polyurethane-implant contact) was, in all probability, the same. The better results of Control B implants were probably correlated to the fact that these implants were conical, while the Test Implants were cylindrical, and they produced higher friction between the implant and the polyurethane. The similar values obtained by the 2.5 and 3.5 mm implants could probably support the hypothesis that diameter is more important than length [2,7], as the highest load is present in the cortical zone, which is more rigid than the cancellous zone [8], even though some authors have reported that implant length is of great value in obtaining primary stability, particularly in low-density bone [4]. Mohlherich et al. (2015) reported that a significant increase in implant stability was found when there was an increase in bone density [8].

The authors concluded that to achieve the highest possible primary stability, the largest possible diameter must be chosen [8]. Moreover, in cases where it could be possible to use either 2.5 or 3.5 mm Test implants, a preference for the 2.5 mm implants could be suggested due to their easier positioning. When the height of the residual alveolar bone in the posterior mandible is at least 3 mm, it could be preferable to use the Test implants instead of performing a vertical augmentation procedure that has higher morbidity, higher cost, and the need of an additional surgical procedure. The present study showed that in these cases, the Test implants could reach a good stabilization. Furthermore, the use of a higher number of implants could improve the load distribution of the prosthetic restauration. Short, Ultrashort, and NanoShort implants should have an appropriate design to improve the primary stability, even in regions with low bone density [7]. In the case of the Test implants used in the present study, a conical shape would probably have been better than a cylindrical design, and also a different pitch of the threads could have helped to better engage the material. Falco et al. (2018) demonstrated that larger implant threads with a greater pitch could contact more bone trabeculae and better compact the bone debris in a peri-implant location.

5. Conclusions

The NanoShort implants used in this study demonstrated an acceptable primary stability in solid rigid polyurethane models when compared to standard implants. The outcome of this investigation suggested a possible clinical application in cases of critical atrophy of posterior mandible, instead of using a more complicated vertical ridge augmentation procedure.

Author Contributions: A.P., conceptualization; L.C., investigation; A.P. and L.C., methodology and supervision; G.I., validation; A.P., G.I., and M.T., data curation; M.T., formal analysis; A.P. and M.T., writing—original draft; A.P. and M.T., writing—review draft.

Funding: This work was supported in part by the Ministry of University, Education, and Research (M.I.U.R.), Rome, Italy.

Acknowledgments: This work was supported in part by the Ministry of University, Education, and Research (M.I.U.R.), Rome, Italy.

Conflicts of Interest: The authors declare no conflict of interest related to this study.

References

1. Di Stefano, D.A.; Arosio, P.; Gastaldi, G.; Gherlone, E. The insertion torque-depth curve integral as a measure of implant primary stability: An in vitro study on polyurethane foam blocks. *J. Prosthet. Dent.* **2018**, *120*, 706–714. [CrossRef] [PubMed]
2. Falco, A.; Berardini, M.; Trisi, P. Correlation between implant geometry, implant surface, insertion torque, and primary stability: In vitro biomechanical analysis. *Int. J. Oral Maxillofac. Implants* **2018**, *33*, 824–830. [CrossRef] [PubMed]
3. Yamaguchi, Y.; Shiota, M.; FuJii, M.; Sekiya, M.; Ozeki, M. Development and application of a direct method to observe the implant/bone interface using simulated bone. *Springerplus* **2016**, *5*, 494. [CrossRef] [PubMed]
4. Tsolaki, I.N.; Tonsekar, P.P.; Najafi, B.; Drew, H.J.; Sullivan, A.J.; Petrov, S.D. Comparison of osteotome and conventional drilling techniques for primary implant stability: An in vitro study. *J. Oral Implantol.* **2016**, *42*, 321–325. [CrossRef] [PubMed]
5. Gehrke, S.A.; Guirado, J.L.C.; Bettach, R.; Fabbro, M.D.; Martínez, C.P.-A.; Shibli, J.A. Evaluation of the insertion torque, implant stability quotient and drilled hole quality for different drill design: An in vitro Investigation. *Clin. Oral Implants Res.* **2018**, *29*, 656–662. [CrossRef]
6. Romanos, G.E.; Delgado-Ruiz, R.A.; Sacks, D.; Calvo-Guirado, J.L. Influence of the implant diameter and bone quality on the primary stability of porous tantalum trabecular metal dental implants: An in vitro biomechanical study. *Clin. Oral Implants Res.* **2018**, *29*, 649–655. [CrossRef]
7. De Oliveira, G.J.P.L.; Barros-Filho, L.A.B.; Barros, L.A.B.; Queiroz, T.P.; Marcantonio, E. In vitro evaluation of the primary stability of short and conventional implants. *J. Oral Implantol.* **2016**, *42*, 458–463. [CrossRef]
8. Möhlhenrich, S.C.; Heussen, N.; Elvers, D.; Steiner, T.; Hölzle, F.; Modabber, A. Compensating for poor primary implant stability in different bone densities by varying implant geometry: A laboratory study. *Int. J. Oral Maxillofac. Surg.* **2015**, *44*, 1514–1520. [CrossRef]
9. Lages, F.S.; Douglas-de Oliveira, D.W.; Costa, F.O. Relationship between implant stability measurements obtained by insertion torque and resonance frequency analysis: A systematic review. *Clin. Implant. Dent. Relat. Res.* **2018**, *20*, 26–33. [CrossRef]
10. Chen, M.H.-M.; Lyons, K.M.; Tawse-Smith, A.; Ma, S. Clinical significance of the use of resonance frequency analysis in assessing implant stability: A systematic review. *Int. J. Prosthodont.* **2019**, *32*, 51–58. [CrossRef]

11. Liu, L.; Gao, Y.; Zhao, J.; Yuan, L.; Li, C.; Liu, Z.; Hou, Z. A Mild method for surface-grafting PEG onto segmented poly (ester-urethane) film with high grafting density for biomedical purpose. *Polymers* **2018**, *10*, 1125. [CrossRef] [PubMed]

12. Xu, W.; Xiao, M.; Yuan, L.; Zhang, J.; Hou, Z. Preparation, physicochemical properties and hemocompatibility of biodegradable chitooligosaccharide-based polyurethane. *Polymers* **2018**, *10*, 580. [CrossRef] [PubMed]

13. Szperlich, P.; Toroń, B. An ultrasonic fabrication method for epoxy resin/SbSI nanowire composites, and their application in nanosensors and nanogenerators. *Polymers* **2019**, *11*, 479. [CrossRef] [PubMed]

14. Misch, C.E. Bone density: A key determinant for clinical success. *Contemp. Implant. Dent.* **1999**, *8*, 109–118.

15. Deporter, D.; Ogiso, B.; Sohn, D.-S.; Ruljancich, K.; Pharoah, M. Ultrashort sintered porous-surfaced dental implants used to replace posterior teeth. *J. Periodontol.* **2008**, *79*, 1280–1286. [CrossRef]

16. Lombardo, G.; Pighi, J.; Marincola, M.; Corrocher, G.; Simancas-Pallares, M.; Nocini, P.F. Cumulative success rate of short and ultrashort implants supporting single crowns in the posterior maxilla: A 3-year retrospective study. *Int. J. Dent.* **2017**, *2017*, 8434281. [CrossRef]

17. Markose, J.; Eshwar, S.; Srinivas, S.; Jain, V. Clinical outcomes of ultrashort sloping shoulder implant design: A survival analysis. *Clin. Implant. Dent. Relat. Res.* **2018**, *20*, 646–652. [CrossRef]

18. Lemos, C.A.A.; Ferro-Alves, M.L.; Okamoto, R.; Mendonça, M.R.; Pellizzer, E.P. Short dental implants versus standard dental implants placed in the posterior jaws: A systematic review and meta-analysis. *J. Dent.* **2016**, *47*, 8–17. [CrossRef]

19. Fan, T.; Li, Y.; Deng, W.-W.; Wu, T.; Zhang, W. Short implants (5 to 8 mm) versus longer implants (>8 mm) with sinus lifting in atrophic posterior maxilla: A meta-analysis of RCTs. *Clin. Implant. Dent. Relat. Res.* **2017**, *19*, 207–215. [CrossRef]

20. Cruz, R.S.; Lemos, C.A.D.A.; Batista, V.E.D.S.; Gomes, J.M.D.L.; Pellizzer, E.P.; Verri, F.R. Short implants versus longer implants with maxillary sinus lift. A systematic review and meta-analysis. *Braz. Oral Res.* **2018**, *32*, e86. [CrossRef]

21. Nielsen, H.B.; Schou, S.; Isidor, F.; Christensen, A.-E.; Starch-Jensen, T. Short implants (≤8 mm) compared to standard length implants (>8 mm) in conjunction with maxillary sinus floor augmentation: A systematic review and meta-analysis. *Int. J. Oral Maxillofac. Surg.* **2019**, *48*, 239–249. [CrossRef] [PubMed]

22. Tolentino da Rosa de Souza, P.; Binhame Albini Martini, M.; Reis Azevedo-Alanis, L. Do short implants have similar survival rates compared to standard implants in posterior single crown? A systematic review and meta-analysis. *Clin. Implant. Dent. Relat. Res.* **2018**, *20*, 890–901. [CrossRef] [PubMed]

23. Dias, F.D.; Pecorari, V.G.; Martins, C.B.; Del Fabbro, M.; Casati, M.Z. Short implants versus bone augmentation in combination with standard-length implants in posterior atrophic partially edentulous mandibles: Systematic review and meta-analysis with the Bayesian approach. *Int. J. Oral Maxillofac. Surg.* **2019**, *48*, 90–96.

24. Urdaneta, R.A.; Daher, S.; Leary, J.; Emanuel, K.M.; Chuang, S.-K. The survival of ultrashort locking-taper implants. *Int. J. Oral Maxillofac. Implants* **2012**, *27*, 644–654. [PubMed]

25. Piattelli, A.; Balice, P.; Scarano, A.; Perrotti, V. *Short and Ultrashort Implants*; Quintessence Publishing: Berlin, Germany, 2018; pp. 59–74.

26. Sennerby, L.; Meredith, N. Implant stability measurements using resonance frequency analysis: Biological and biomechanical aspects and clinical implications. *Periodontol. 2000* **2008**, *47*, 51–66. [CrossRef] [PubMed]

Article

Degradable Poly(ether-ester-urethane)s Based on Well-Defined Aliphatic Diurethane Diisocyanate with Excellent Shape Recovery Properties at Body Temperature for Biomedical Application

Minghui Xiao [1,†]**, Na Zhang** [1,†]**, Jie Zhuang** [2]**, Yuchen Sun** [1]**, Fang Ren** [3]**, Wenwen Zhang** [3] **and Zhaosheng Hou** [1,*]

[1]   College of Chemistry, Chemical Engineering and Materials Science, Shandong Normal University, Jinan 250014, China; xiaominghui98@163.com (M.X.); sdzxnzkay@163.com (N.Z.); yuchensun12365@163.com (Y.S.)

[2]   Shandong Academy of Pharmaceutical Sciences, Shandong Provincial Key Laboratory of Biomedical Polymer, Jinan 250101, China; cpfzhuangjie@126.com

[3]   Success Bio-tech Co., Ltd., Jinan 250101, China; fangren_2008@126.com (F.R.); 15662698340@163.com (W.Z.)

*   Correspondence: houzs@sdnu.edu.cn

†   These authors contributed equally to this work.

Received: 29 April 2019; Accepted: 3 June 2019; Published: 5 June 2019

Abstract: The aim of this study is to offer a new class of degradable shape-memory poly(ether-ester-urethane)s (SMPEEUs) based on poly(ether-ester) (PECL) and well-defined aliphatic diurethane diisocyanate (HBH) for further biomedical application. The prepolymers of PECLs were synthesized through bulk ring-opening polymerization using ε-caprolactone as the monomer and poly(ethylene glycol) as the initiator. By chain extension of PECL with HBH, SMPEEUs with varying PEG content were prepared. The chemical structures of the prepolymers and products were characterized by GPC, ^{1}H NMR, and FT-IR, and the effect of PEG content on the physicochemical properties (especially the shape recovery properties) of SMPEEUs was studied. The microsphase-separated structures of the SMPEEUs were demonstrated by DSC and XRD. The SMPEEU films exhibited good tensile properties with the strain at a break of 483%–956% and an ultimate stress of 23.1–9.0 MPa. Hydrolytic degradation in vitro studies indicated that the time of the SMPEEU films becoming fragments was 4–12 weeks and the introduction of PEG facilitates the degradation rate of the films. The shape memory properties studies found that SMPEEU films with a PEG content of 23.4 wt % displayed excellent recovery properties with a recovery ratio of 99.8% and a recovery time of 3.9 s at body temperature. In addition, the relative growth rates of the SMPEEU films were greater than 75% after incubation for 72 h, indicating good cytocompatibility in vitro. The SMPEEUs, which possess not only satisfactory tensile properties, degradability, nontoxic degradation products, and cytocompatibility, but also excellent shape recovery properties at body temperature, promised to be an excellent candidate for medical device applications.

Keywords: poly(ether-ester-urethane)s; poly(ethylene glycol); well-defined hard segments; degradability; shape memory behavior

---

1. Introduction

Shape-memory polymers (SMPs), as a kind of smart polymeric material, have the ability to remember and recover their permanent shape upon application of an external stimulus, such as heat, humidity, pH, light, electromagnetic induction, or a solvent [1–6]. Compared with shape-memory alloys (SMAs), SMPs have a lot of advantages, such as flexible transition temperatures, high recoverable strains, low density, low manufacturing cost, and easy processing [2]. Nowadays, SMPs have been proposed for application in several kinds of medical devices [7–9]. As an example, SMPs were used to fabricate medical bandages by the Luo group, and they found that SMPs could produce gradient pressures acting on an ulcer of the leg to accelerate blood circulation and the healing process [10]. Although SMPs are very useful, their applications in medical implant materials are limited because of their nonbiodegradability and low biocompatibility. For medical implantations, SMPs are required to be degradable, possess high biocompatibility and have a recovery temperature near the human body temperature. Therefore, it is urgently required to develop biocompatible and degradable SMPs with adequate tensile properties to fabricate novel kinds of medical devices [11–13]. In 2017, the Becker group [14] provided an overview of SMPs being used in medical applications, and after that, a large number of works related to biocompatible SMPs have been published [15–17].

Segmented polyurethanes (PUs), as one of the important types of shape-memory materials, have received attention recently because of their unique properties, such as high shape recoverability, a wide range of shape recovery temperatures, good tensile properties, and adequate biocompatibility [18,19]. The unique properties of segmented PUs are attributed to microphase separation, and the primary driving forces behind microphase separation are both the immiscibility of the soft and hard segments and the strong intermolecular interaction of H-bonding among the hard segments, with hydrogen bonding energies of 12~36 kJ/mol [20,21].

As implant medical devices, many commercial medical PUs, including Elasthane^{TM}, Biomer$^{®}$, Biospan^{TM}, and ChronoFlex$^{®}$ AR, are quintessentially synthesized based on 4,4′-diphenylmethane diisocyanates (MDI) [22,23]. Aromatic diamines are carcinogenic degradation products that are produced and released during the degradation process [24]. Clearly, PUs prepared from aliphatic diisocyanate can conquer these shortcomings; however, compared with MDI-based Pus, these types of PUs possess weak tensile properties because they lack a significant length of hard segments [25]. Long well-defined hard segments are significant for good tensile properties [26]. In our previous report [27], a kind of degradable medical PU containing well-defined hard segments was prepared by using the chain extender of aliphatic diurethane diisocyanate. The well-defined chemical structure of the hard segments enhances the microphase separation degree of the hard and soft segments, and in addition, the existence of multiple H-bonds between the urethane units gives a compact network structure physical-linked by H-bonds, which leads to comparative or even better tensile properties than MDI-based PUs.

On the other hand, most of the commercially available shape-memory PUs (SMPUs) are thermo-responsive. They have a higher recovery temperature (50–90 °C) than human body temperature, which makes them not particularly suitable for medical device applications as they would need high activation temperatures [28]. Therefore, it is desirable to design degradable SMPUs with a near-body recovery temperature for medical applications. Poly(ethylene glycol) (PEG) not only possesses good hydrophilicity, biocompatibility, and nontoxicity, but also exhibits excellent flexibility and a low glass-transition temperature [29,30]. The introduction of a PEG chain into hydrophobic polyester chains as the soft segment can produce a low transition temperature. Thus, the corresponding SMPUs based on the poly(ether-ester) have both biodegradability and a near-body recovery temperature. Besides, the flexible PEG chain can serve as a plasticizer to enhance the distance between the soft segments, and thus enhances the shape memory properties of the SMPUs, as other papers have reported [10,31,32].

This work describes the preparation and properties of a new class of shape-memory poly(ether-ester-urethane)s (SMPEEUs) based on poly(ether-ester) and well-defined aliphatic diurethane diisocyanate. The SMPEEUs are expected to have not only excellent shape recovery properties at body temperature, but also acceptable physicochemical properties, degradability, biocompatibility, and nontoxic degradation products for further biomedical application. First, the poly(ether-ester) of triblock poly(ε-caprolactone)-poly(ethylene glycol)-poly(ε-caprolactone) (PECL) was synthesized by bulk ring-opening polymerization with ε-caprolactone (ε-CL) as the monomer and

PEG as the initiator. Then, the SMPEEUs with well-defined hard segments were obtained by one-step chain extension of PECL with diurethane diisocyanate. The chemical structures of the PECLs and SMPEEUs were characterized, and the influences of PEG content on the tensile properties, in vitro degradability, and shape-memory properties of the SMPEEU films were researched. In addition, the biocompatibility (cytocompatibility) of the SMPEEU films was evaluated by a cytotoxicity test.

2. Materials and Methods

2.1. Materials

ε-CL was purchased from Acros Chimica, Geel, Belgium and distilled from CaH_2 under reduced pressure. Dibutyltin dilaurate (DBTDL) and 1,6-hexanediisocyanate (HDI) were supplied by Sigma-Aldrich Chemical Co., St. Louis, MO, USA and used without further purification. PEG (M_n = 400, 600, 1000, 2000 g/mol) and 1,4-butanediol (BDO) (Shanghai Aladdin Reagent Co., Shanghai, China) were dried at 110 °C under reduced pressure for about 4 h prior to use. Diurethane diisocyanate of 1,6-hexanediisocyanate-1,4-butanediol-1,6-hexanediisocyanate (HBH) was synthesized in our lab, as shown in Figure 1a according to our previous report [33] and HR-MS and NMR were adopted to confirm its chemical structure. *N,N*-dimethylformamide (DMF, Beijing Chemical Reagent Co., Ltd, Beijing, China) was dried with phosphorus pentoxide for 8 h and then distilled under vacuum before use. Other reagents (AR grade) were purified by standard methods.

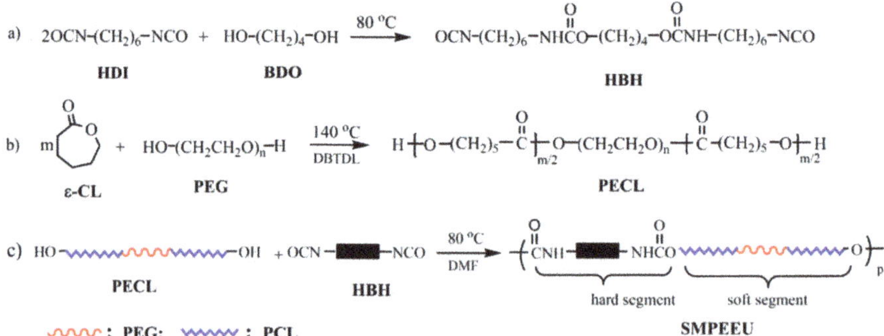

Figure 1. Synthetic pathways of (**a**) diurethane diisocyanate (HBH), (**b**) PECL, and (**c**) SMPEEU.

2.2. Polyurethane Synthesis

2.2.1. Synthesis of PECLs

OH-terminated PECLs were prepared by bulk ring-opening polymerization with ε-CL as the monomer and PEG as the initiator [34]. Briefly, the predetermined amount of ε-CL, PEG and catalyst DBTDL (0.25 wt % of ε-CL) were mixed in a vacuum flask at room temperature, and the system was deoxygenated with dry argon three times. After the flask was sealed, the reaction was performed at 140 °C under vacuum conditions (~50 Pa) with an oil bath for 24 h. A small amount of chloroform was added to the system to dissolve the raw product, and the solution was precipitated with cold diethyl ether to obtain the PECLs, which were then dried thoroughly under reduced pressure (~150 Pa) at room temperature. The PECLs based on PEG-400, PEG-600, PEG-1000, and PEG-2000, are named PECL-I, PECL-II, PECL-III, and PECL-IV, respectively. The reaction scheme is shown in Figure 1b, and the feed ratios and molecular weights of the PECLs are listed in Table 1.

Table 1. Molecular weight and feed ratios of the poly(ether-ester) (PECL).

PECLs	ε-CL/g	PEG/g				M_{theo}	M_n	$Đ_M$	M_{NMR}
		−400	−600	−1000	−2000				
-I	36	4.0	-	-	-	4000	3890	1.11	3990
-II	34	-	6.0	-	-	4000	3920	1.08	4020
-III	30	-	-	10.0	-	4000	3880	1.10	3970
-IV	20	-	-	-	20.0	4000	3940	1.12	4040

Note: $M_{theo.}$, theoretical molecular weight calculated from the molar ratio of CL/PEG; M_n and $Đ_M$, number average molecular weight and molecular weight dispersity obtained from GPC; M_{NMR}, molecular weight calculated from the peak integration in ^{1}H NMR.

2.2.2. Preparation of SMPEEUs and SMPEEU Films

A representative process was as follows: A certain amount of PECL was added to a two-necked flask and the system was heated to 80 °C with an oil bath under dried argon. Then the DMF solution of HBH (~0.25 g/mL) was added dropwise into the system with mechanical stirring, and the ratio of –NCO/–OH was 1.02 (mol:mol). When the reaction became too viscous to be stirred, a small amount of DMF could be added to keep the reaction homogeneous. After that, the reaction mixture proceeded at the same temperature with dried argon until the characteristic peak of –NCO at 2250~2280 cm^{-1} disappeared completely in FT-IR (approximately 3.5 h), and then was diluted to ~0.045 g/mL with DMF. The diluted solution was degassed under reduced pressure and then gently poured into a polytetrafluoroethylene mold, in which most of the DMF was volatilized at 40 °C for five days. The last trace of DMF was removed under vacuum for two days to give semitransparent films with a thickness of 0.3 ± 0.02 mm. The SMPEEU films, based on PECL-I, PECL-II, PECL-III, and PECL-IV are named SMPEEU-I, SMPEEU-II, SMPEEU-III, and SMPEEU-IV, respectively. The reaction scheme is displayed in Figure 1c, and the molecular weight and feed ratios of the SMPEEUs are shown in Table 2.

Table 2. Molecular weights and feed ratios of the shape-memory poly(ether-ester-urethane)s (SMPEEUs).

SMPEEUs	PECL/g				HBH/g	PEG Content/wt %	M_n (kDa)	$Đ_M$
	-I	-II	-III	-IV				
-I	19.5	-	-	-	2.17	9.25	118	1.38
-II	-	19.6	-	-	2.17	13.8	113	1.41
-III	-	-	19.4	-	2.17	23.2	115	1.43
-IV	-	-	-	19.7	2.17	45.7	108	1.37

Note: PEG content, PEG content in SMPEEUs; M_n and $Đ_M$, number average molecular weight and molecular weight dispersity obtained by gel permeation chromatography (GPC).

2.3. Instruments and Characterization

Characterization: Nuclear magnetic resonance (NMR) spectra were obtained by employing a Bruker 400 MHz Avance II spectrometer (Rheinstetten, Germany) at room temperature. DMSO-d^6 and CDCl$_3$ were used as the solvents for HBH and polymer analysis, respectively. Fourier transform infrared (FT-IR) spectra were conducted on a Bruker Alpha spectrometer (Rheinstetten, Germany) in the range of 4000–400 cm^{-1}, with a resolution of 4 cm^{-1}. The number average molecular weight (M_n) and molecular weight dispersity ($Đ_M$) were determined using Viscotec TriSEC302 (Kennesaw, GA, USA) gel permeation chromatography (GPC) at 35 °C, with tetrahydrofuran as the continuous phase (flow rate: 1.0 mL/min) and monodisperse polystyrene as the calibration standard.

Thermal transition: Thermal transition behaviors were researched with a Q20 differential scanning calorimeter (DSC) (TA Instrument, New Castle, DE, USA) under nitrogen (flow rate: 30 mL/min). Samples, which were encapsulated in standard aluminum pans with lids, were first heated up to 150 °C with a heating rate of 10 °C/min to relieve the thermal history. The samples were then cooled to −75 °C at 5 °C/min, and finally scanned from −75 to 150 °C at 10 °C/min. The reported thermal transition values were collected from the second heating cycle, and the value of the glass transition temperature (T_g) was obtained from the center of the change of slope.

Crystallization behaviors: X-ray powder diffraction (XRD) was adopted to investigate the crystallization behaviors of SMPEEU films. The data were collected on a Max 2200PC power X-ray diffractometer (Rigaku, Tokyo, Japan) with Cu K_α radiation (wavelength: 1.54051 Å) at 40 kV and 20 mA. The sample holder containing film samples was scanned from 5° to 55° with a step size of $2\theta = 0.02°$.

Tensile properties: The tensile properties were tested with a single-column tensile test machine (Model HY939C, Dongguan Hengyu Instruments, Ltd., Dongguan, China) according to national standard GB/T1040.2-2006. The film samples were cut into dumb-bell shapes with a neck length and width of 4.0 and 30 mm, respectively. The tests were performed at room temperature with a cross-head speed of 50 mm/min. The ultimate stress, strain at break, and initial modulus were extracted from the stress-strain curves. Property values obtained here were the average of at least three tensile samples.

Bulk hydrophilicity: The water absorption was adopted to measure the bulk hydrophilicity of the SMPEEU films. A film disc with a 10.0 mm diameter was immersed in distilled water (10 mL) at 37 ± 0.1 °C until water absorption equilibrium was reached. The water absorption was obtained from the formula (1):

$$\text{Waterabsorption} (\%) = \frac{W_s - W_o}{W_o} \times 100 \tag{1}$$

where W_s and W_o are the weights of the swollen and original samples, respectively. The results were the average of three samples.

Surface hydrophilicity: The film surface hydrophilicity was evaluated by sessile static water contact angle measurement, which was executed on an optical goniometer (CAM200, KSV Instruments, Helsinki, Finland) with a sessile drop method. Ultrapure water (~0.02 mL) was used as a liquid probe, and the measurement of each contact angle were finished within 5 s after steady state for the angle was reached. All the measurements were carried out at room temperature and the results were the average of at least six replicate measurements on each sample.

In vitro degradation: In vitro hydrolytic degradation tests of swollen films were carried out in phosphate-buffered saline (PBS) with a pH value of 7.4, employing the mass loss. The swollen film discs (diameter: 10.0 mm) were put into individual sealed vial containing PBS (10 mL), and incubated in a biochemical incubator at a temperature of 37 ± 0.1 °C. At given time intervals, the discs were taken out, wiped with filter paper, and weighed to determine the weight loss. The measurements were performed until the discs lost their tensile properties and became fragments. The weight loss was calculated according to the formula (2):

$$\text{Weightloss} (\%) = \frac{m_s - m_d}{m_s} \times 100 \tag{2}$$

where the m_s is the mass of the swollen sample after immersing it in water for ~2 hours, and m_d is the mass of the sample after hydrolysis at the given time. The measurements were performed with three independent discs and the average results were reported.

Cytotoxicity tests: According to the reported method [35], mouse fibroblast cells (L929) were adopted as the test model to evaluate the cytotoxicity of possible substance that could leach from SMPEEU films. Before the cytotoxicity tests, SMPEEU films in the form of 10 ± 0.02 mm discs were rinsed with ethanol to eliminate the effects of impurity on the cells, and then were sterilized with UV radiation for 30 min on each side of the discs. The sterilized film discs were incubated in 2.0 mL culture medium (Dulbecco's modified Eagle's medium with 10% (*v/v*) fetal bovine serum) overnight at 37 ± 0.5 °C. The extract solution was filtrated with a filter membrane (0.22 μm pore size) to avoid the possible presence of solid particles released from the samples, and the filtrate was diluted with equivoluminal culture medium. L929 cells with a density of 5.0×10^4 cells/mL were seeded into a

96-well tissue culture plate containing 100 μL of the respective film extracted dilutions, and wells which only contained the cells and culture medium served as controls. After incubation at 37 °C with 5% CO_2 for 72 h, cell viability was measured with the thiazolyl blue tetrazolium bromide (MTT) assay method. The optical density (OD) was obtained with a multiwell microplate reader (Multiskan Mk3-Thermolabsystems, Thermo Fisher Scientific, Inc., Waltham, MA, USA) at 570 nm, and values relative to the control were reported.

Shape memory behaviors: The shape memory behaviors of SMPEEU films were evaluated with the "fold-deploy shape memory test" method [36,37] using rectangular strips with dimensions of 40 mm × 10 mm × 0.30 mm. First, the sample was immersed in a water bath with temperature of ~55 °C (25 °C above the T_g) for 0.5 min, and then was folded in half under the aid of external force. Second, the folded sample was quickly quenched by immersion in an ice water bath (0 °C) under the existence of constant external force to fix the shape. A few minutes later (typically 1.5 min), the applied external force was removed and a marginal recovery occurred. The blending angle was recorded as θ_f. Finally, the sample with the fixed shape was immersed in a water bath at body temperature (37.5 °C) and reheated to recover its original shape. The corresponding bending angle after recovery was recorded as θ_r, and the time of the bending angle changing from θ_f to θ_r was recorded as the recovery time (T_r). In order to quantitatively describe the shape memory properties of the films, the fixity ratio (R_f) and recovery ratio (R_r) were roughly defined as the following formulas (3) and (4), respectively:

$$R_f\ (\%) = \frac{\theta_f}{180} \times 100 \tag{3}$$

$$R_r\ (\%) = \frac{180 - \theta_r}{180} \times 100 \tag{4}$$

3. Results and Discussion

3.1. Characterization

^{1}H NMR and FT-IR measurements were adopted to confirm the structure of the chain extender, prepolymer, and polyurethane. Representative ^{1}H NMR and FT-IR spectra of HBH, PECL and SMPEEU are displayed in Figures 2 and 3, respectively.

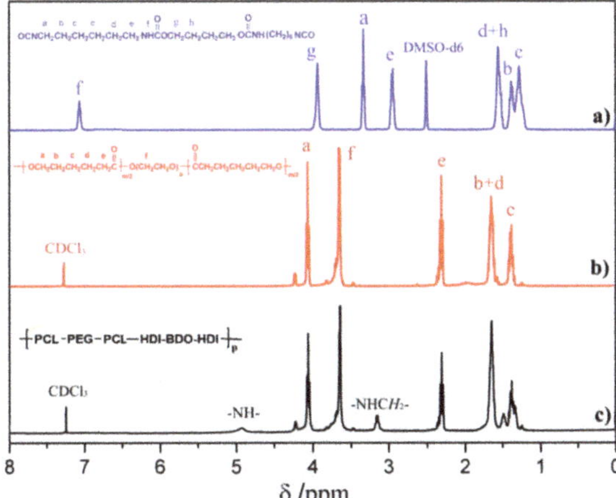

Figure 2. Representative ^{1}H NMR spectra of (**a**) HBH, (**b**) PECL-III, and (**c**) SMPEEU-III.

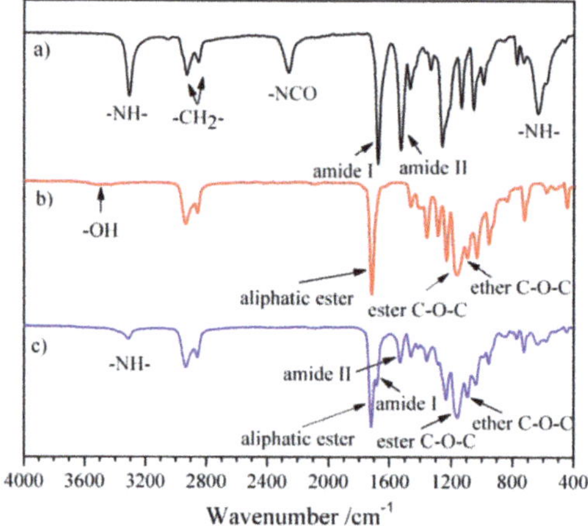

Figure 3. Representative FT-IR spectra of (**a**) HBH, (**b**) PECL-III, and (**c**) SMPEEU-III.

The chain extender of diurethane diisocyanate (HBH) was obtained through the reaction of eight-fold excess of HDI with BDO without using a catalyst. The data obtained from the ^{1}H NMR spectrum (Figure 2a) were consistent with the results of our previous paper [33] and matched the chemical structure of HBH (Figure 1a), indicating no longer segments existing in the products. The peaks at 3319, 2262, 1680, and 1521 cm^{-1} in the FT-IR spectrum (Figure 3a) were attributed to the characteristic absorptions of –NH–, –NCO, amide I, and amide II, respectively. The FT-IR results provide additional support for the formation of the desired diurethane diisocyanate structure.

Typical signals appeared at δ 4.06, δ 2.29, and δ 3.64 ppm in the ^{1}H NMR spectrum of PECL (Figure 2b), and were attributed to proton signals of ω-CH$_2$, α-CH$_2$ from [CL] units and repeat units of –O–CH$_2$–CH$_2$– ([EO]) from PEG segments, respectively [38]. According to the molecular weight of the PEG macroinitiator, the number of average repeat units of [CL] in the prepolymer could be obtained by calculating the peak integration at δ 4.06 and/or δ 2.29 with δ 3.64 ppm. The average repeat units of [CL] ($\overline{DP}_{CL}$) in the prepolymer and the molecular weight (M_{NMR}) of the prepolymer were obtained according to Equations (5) and (6), respectively [39]:

$$\overline{DP}_{CL} = \frac{M_{PEG}}{44} \times \frac{S_{\alpha,\omega-CH_2}}{S_{EO}} \tag{5}$$

$$M_{NMR} = M_{PEG} + \overline{DP}_{CL} \times 114 \tag{6}$$

where $S_{\alpha,\omega-CH_2}$ and S_{EO} are the peak integration of α,ω-CH$_2$ proton signals from [CL] units and [EO] proton signals from PEG segments, and 44 and 144 are the molecular weight of [EO] and [CL] repeat units. The values of M_{NMR} matched with those of $M_{theo.}$ calculated from the stoichiometry of CL/PEG (mol:mol) and M_n obtained from GPC (Table 1), indicating a complete ring-opening reaction and the absence of residual ε-CL in the PECL. In the FT-IR spectrum (Figure 3b), the stretching vibration of cyclic ester C=O (~1760 cm^{-1}) disappeared completely, and another strong absorption band at ~1723 cm^{-1} appeared which should be attributed to the aliphatic ester C=O stretching vibration. In addition, the absorption bands at 3510 and 1092 cm^{-1} belonged to ether C–O–C stretching frequencies and terminal hydroxyl groups. This was further support that the ring-opening reaction indeed occurred.

SMPEEUs were prepared via chain extension of PECLs with HBH at 80 °C. The transesterification reaction, which had been reported to be severe when using diol as a chain extender at 80 °C [25], could be limited by using HBH diisocynantes as the chain extender, resulting in low $Đ_M$ (Table 2), The chemical structures of the SMPEEUs were characterized by ^{1}H NMR (Figure 2c) and FT-IR (Figure 3c). The proton signal of –CH$_2$– next to –NCO (δ 3.33 ppm, 'a' in Figure 1a) and the absorption band of –NCO (~2262 cm^{-1}) of HBH and the terminal –OH (~3319 cm^{-1}) of PECL disappeared completely, alongside the appearance of the proton signals of –NH– in urethane groups and –CH$_2$– next to urethane groups at δ 4.90 and 3.15 ppm, respectively (the signals appeared at δ 7.07 and 2.94 ppm of HBH in DMSO-d^6, Figure 2a). In addition, the absorption peaks at 1093, 1168, 1535, 1675 and 1723 cm^{-1} in the FT-IR spectrum (Figure 3c) belonged to the stretching vibration of ether C–O–C, ester C–O–C, amide II, amide I, and aliphatic ester C=O, respectively. All the results represented a thorough chain extension reaction.

3.2. Thermal Transition

The typical DSC curves of the SMPEEU films with varying PEG content are presented in Figure 4 and the corresponding transition temperatures obtained from the thermograms are listed in Table 3. Two glass transition temperatures (T_{g1}, T_{g2}) were founded in their curves. The first glass transition points of T_{g1} −41.9 to −48.5 °C belonged to the soft domains, which was consistent with the result of polyurethanes-based poly(3-hydroxybutyrate) and PCL-PEG-PCL in a previous study [40]. Only one T_g appeared at a low temperature, indicating the high miscibility of PEG and PCL segments. The second glass transition points (T_{g2}) appeared at higher temperature of ~32 °C, belonging to the well-defined hard domains. Because the polarity of urethane groups in hard segments is much higher than that of ester and ester groups in soft segments, they are hardly miscible, leading to microphase separation structure and two T_g for soft and hard domains [41]. The T_{g1} decreased slightly with the increment of PEG content in SMPEEU, which meant that the introduction of PEG could improve the flexibility of the SMPEEU films and act as a plasticizer [42]. Moreover, one broad endothermic peak (T_m) at about 50~80 °C was observed, which should belong to the crystalline melting transition of PCL segments and hard domains. The ΔH_f decreased from 56.7 to 19.6 J/g (Table 3) with the PEG content increasing from 9.25 to 47.5 wt % in SMPEEU (Table 2), which signified that the PEG segments introduced to the main chain of the SMPEEU destroyed the crystallization of PCL and/or hard segments.

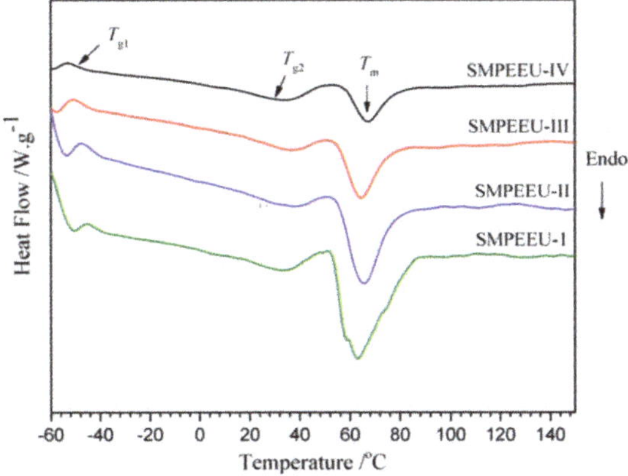

Figure 4. The typical differential scanning colorimetry (DSC) curves of the SMPEEU films in the second heating cycle.

Table 3. The transition temperatures of SMPEEU films.

SMPEEUs	-I	-II	-III	-IV
T_{g1} (°C)	−41.9	−43.7	−46.1	−48.5
T_{g2} (°C)	30.4	32.8	33.4	31.6
T_m (°C)	52–81	53–80	53–77	55–78
ΔH_f (J/g)	56.7	43.8	29.3	19.6

3.3. Crystallization Behaviors

The crystallization behaviors of the SMPEEU films were studied by means of XRD, and the results are presented in Figure 5. The SMPEEU-I film exhibited a broad diffraction zone with two clear diffraction peaks (2θ: 21.4°, 23.6°) in the scattering pattern, suggesting some degree of crystallinity. This was assigned to the crystalline soft domains (mainly PCL segments) and regular hard regions formed by the well-defined structure [43,44]. The intensity of the broad peaks decreased with the increasing PEG content in the SMPEEUs (SMPEEU-I~SMPEEU-IV), which indicated that the degree of crystallinity decreased gradually. It was supported that the introduction of PEG could affect the crystallinity of the SMPEEUs, and the results were consistent with DSC analysis.

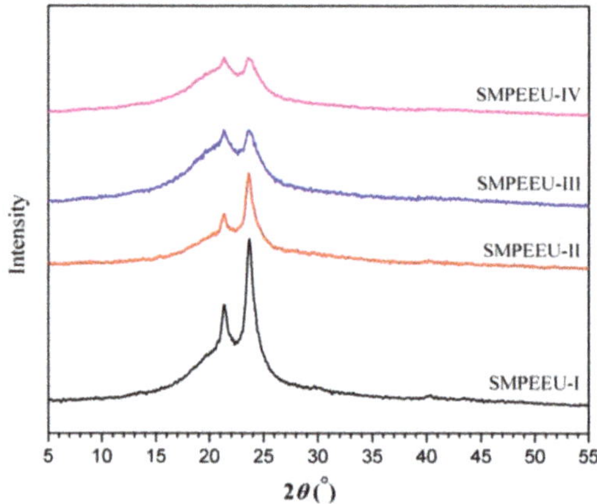

Figure 5. X-ray diffraction (XRD) patterns of SMPEEU films.

3.4. Tensile Properties

Quintessential stress–strain curves of SMPEEU films with varying PEG content are displayed in Figure 6, and the tensile properties obtained from the curves are listed in Table 4. All the films first presented an elastic deformation, and then necking was observed clearly with a yield point, which indicated a smooth transition from elastic to plastic deformation. With PEG content increasing from 9.25 to 47.5 wt %, the strain at break increased from 483% to 956%, while the ultimate stress and initial modulus decreased from 23.1 to 9.0 MPa and 48 to 8.7 MPa, respectively (Table 4). The trend appears to be correlated with the PEG chain in the polymers which is much more flexible than a PCL chain of comparable molecular weight. The introduction of a flexible PEG chain hinders the crystallization of PCL soft segments and/or well-defined hard segments, and weakens the average crystallite size as shown by XRD analysis [45]. Consequently, the tensile properties can be tailored by adjustment of the PEG content to meet the requirements of medical materials.

Table 4. Tensile properties of SMPEEU films.

Films	Strain at Break (%)	Ultimate Stress (MPa)	Yield Stress (MPa)	Yield Strain (%)	Initial Modulus (MPa)
SMPEEU-I	483 ± 12	23.1 ± 1.7	9.7 ± 0.76	20.2 ± 1.2	48.0
SMPEEU-II	589 ± 15	18.5 ± 1.5	9.5 ± 0.69	24.0 ± 1.4	39.6
SMPEEU-III	675 ± 18	14.8 ± 1.3	5.8 ± 0.52	34.1 ± 1.6	17.0
SMPEEU-IV	956 ± 23	9.0 ± 0.9	4.7 ± 0.27	54.1 ± 1.9	8.7

Note: "±" represents standard error of mean; n = 3.

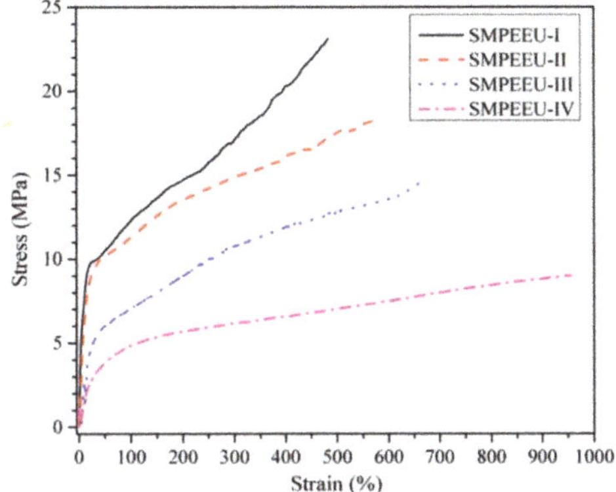

Figure 6. Stress-strain behaviors of SMPEEU films.

3.5. Bulk and Surface Hydrophilicity

The bulk and surface hydrophilicity, which can be reflected directly by the measurements of water absorption and water surface contact angle, are very significant for biomaterials because the wettability can affect the hydrolytic degradation rate and the interaction of cells with the materials [46]. The water contact angle and water absorption of SMPEEU films with varying PEG contents are shown in Figures 7 and 8, respectively. With the increment of PEG content in SMPEEUs, the water contact angle gradually decreased from 72.8° to 41.7°, indicating an improving surface hydrophilicity. The decrease in water contact angles should be ascribed to the excellent hydrophilicity of PEG chains exposed on the surfaces, and the better surface hydrophilicity means higher biocompatibility, such as reduced tissue adhesion [42]. The water absorption of the films occurred with a rapid water penetration at early immersion and reached the equilibrium after ~90 min (Figure 8), displaying a Fickian diffusion behavior [47,48]. The equilibrium water absorption increased sharply from 3.1 to 23.6 wt % with the increasing amount of PEG incorporated into the polymer structure, which is also attributed to the unique hydration effect of the PEG. The intermolecular forces between water and the PEG chains caused the water to actively penetrate into the films. From the results, the bulk and surface hydrophilic ability—important roles in the hydrolytic stability—are mainly influenced by the hydrophilicity of the components [49]. Therefore, it is deduced that the degradation rate of the SMPEEU films can be affected by the content of hydrophilic PEG.

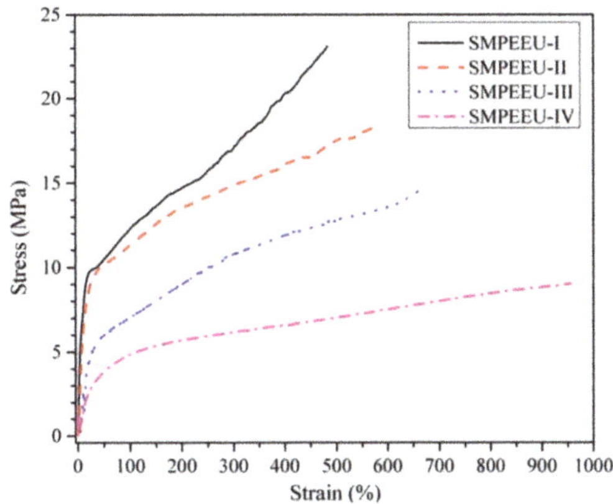

Figure 7. Water surface contact angle of SMPEEU films with varying PEG contents. (Error bars represent standard error of mean; $n = 6$).

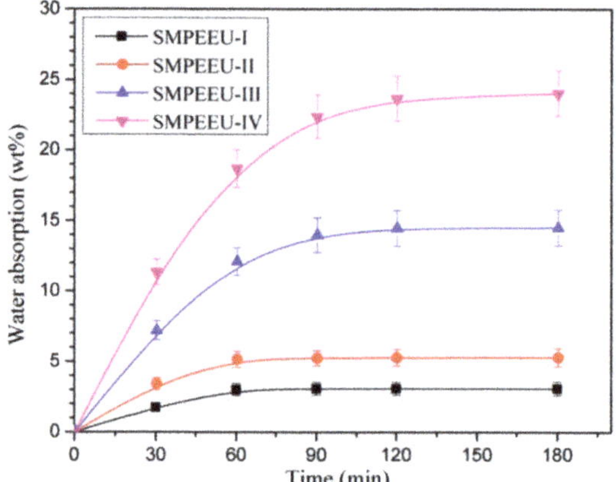

Figure 8. Water absorption of SMPEEU films with varying PEG contents. (Error bars represent standard error of mean; $n = 3$).

3.6. In Vitro Hydrolytic Degradation

In vitro hydrolytic degradation tests of the SMPEEU films were carried out at 37 °C in PBS solution with a pH value of 7.4. The time dependence of the percentage weight loss is shown in Figure 9. The degradation process included two stages: the samples presented a slight weight loss of less than 10 wt % in the first two weeks, followed by a sharp increase in the weight loss rate up to the end of the test (degradation criterion is the film becoming fragments). The swelling and hydration of the films are contained in the incipient stage, which leads to a slow weight loss rate in appearance. The substantial weight loss in the second stage should be due to the loss of water-soluble molecules formed during the hydrolysis of the ester groups. Obviously, introduction of hydrophilic PEG could enhance the degradation rate of the polymers, and the time of the SMPEEU-I, -II, -III, and -IV films becoming fragments was 12, 9, 6, and 4 weeks, respectively. The trend can be explained by two factors: PEG content and crystallinity. As in the description of the measurement of water absorption, high hydrophilic PEG content can bind more water molecules. The introduction of PEG can slightly destroy the order of the polymer structure and decrease the crystallinity of the polymer, as verified in XRD, which allows the water molecules to easily pass through the film. Thus, the ester groups approached the water molecules easily, generating an enhanced hydrolytic degradation rate.

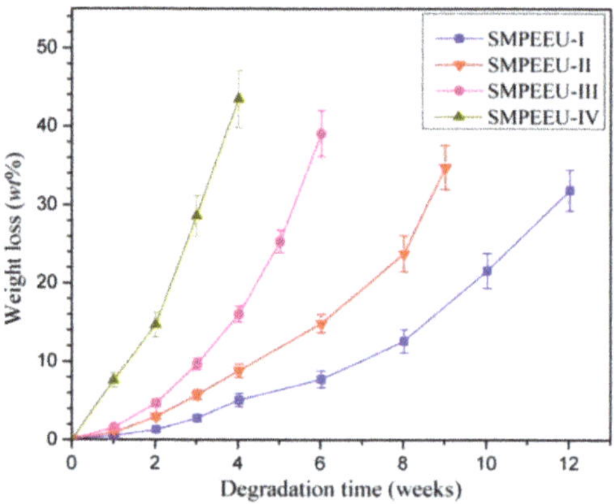

Figure 9. Degradation behaviors of SMPEEU films with varying PEG content. (Error bars represent standard error of mean; $n = 5$).

3.7. Shape Memory Properties

The shape memory properties are the most significant characteristics by which to evaluate the quality of SMPs. The purpose of this paper is to offer a new shape-memory biomaterial for medical device applications, so the recovery process is carried out at body temperature—37.5 °C.

The shape memory properties, including the fixity ratio (R_f), recovery ratio (R_r), and recovery time (T_r), are quantitatively measured to assess the shape memory behaviors of the developed SMPEEU films, and the results are displayed in Table 5. In the process for the "fold-deploy shape memory test", the fixed angle θ_f was almost equal to 180° (less than 1°), and R_f is more than 99.5%, so that the films had high R_f and could be fixed to the desired shape completely. After the shape-fixed films were immerged in a 37.5 °C water bath, all the samples needed only several seconds to recover their original shape, indicative of an excellent shape recovery performance. As previous papers have reported [20,21,50,51], the hydrogen bonds play an important role in the structure change in the deformation and shape memory processes, so it can be assumed that the excellent shape memory properties are related to the denser hydrogen bonds in the SMPEEU structure. Four urethane groups are contained in each hard segment; therefore, multiple hydrogen bonds can exist not only among the hard segments but also between the hard and soft segments. The influence mechanism of denser hydrogen bonds on the shape memory performance needs further research. In addition, the crystalline PCL chain segments in the soft phase contributed to the shape memory performance [52,53]. The SMPEEU with a PEG content of 23.4 wt % (SMPEEU-III) manifested more excellent recovery properties with an R_r of 99.8% and T_r of 3.9 s, meaning that introduction of flexible PEG chain into the SMPEEU structure improved the recoverability at body temperature and the optimum PEG content in SMPEEUs was approximately 23.4 wt %. The representative recovery process of the an SMPEEU-III film is visualized in Figure 10. In order to more directly exhibit the shape memory performances, the SMPEEU-III film was designed as "A" shape and dyed, and the shape fixity and recovery procedure was demonstrated in Movie S1 (Supplementary Materials) in the ESI[†]. The deformed shape of the random clew needs only 3.5 s to recover its original "A" shape.

Table 5. Shape memory properties of SMPEEU films.

Films	Thickness (mm)	R_f (%)	R_r (%)	T_r (s)
SMPEEU-I	0.30	99.5 ± 0.02	81.1 ± 1.4	12.2 ± 0.5
SMPEEU-II	0.31	99.6 ± 0.02	90.7 ± 1.2	7.5 ± 0.4
SMPEEU-III	0.30	99.7 ± 0.01	99.8 ± 0.06	3.9 ± 0.2
SMPEEU-IV	0.29	99.8 ± 0.01	92.8 ± 0.7	8.4 ± 0.5

Note: "±" represents standard error of mean; n = 3.

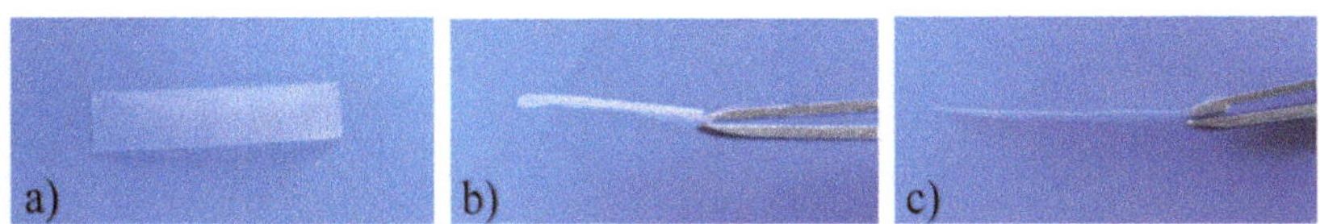

Figure 10. Representative shape memory process of an SMPEEU-III film. (**a**) original shape; (**b**) fixed shape; (**c**) recovered shape.

Repeatability of shape memory recovery, which can characterize the fatigue resistance of materials, is an important performance parameter for SMPs. The influence of repeated fold-deploy cycles on the recovery properties of R_r and T_r was studied, and the results are shown in Figure 11. The SMPEEU-III film could almost recover its original shape rapidly over four cycles. The slight decrease of T_r may be due to the adjustment of multiple hydrogen bonds existing in the SMPEEU structures. The friction among molecules was reduced with frequent bending, and thereafter, the molecules became more submissive, resulting in a reduced T_r [36,54]. During the subsequent cycles, the R_r and recovery rate decreased dramatically, which could be caused by material fatigue, demonstrating that the multiple hydrogen bonds were partly destroyed so that the material gradually lost its shape memory property.

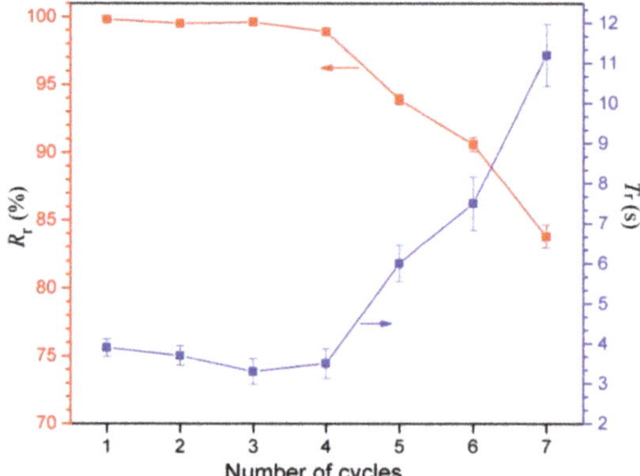

Figure 11. Shape memory properties of an SMPEEU-III film after several fold-deploy cycles. (Error bars represent standard error of mean; $n = 3$).

3.8. Cytotoxicity Test

The first step to evaluate the biocompatibility of biomedical materials is the cytotoxicity test. L929 cells were cultured with SMPEEU film extracts to measure whether the film extracts were toxic to cells. From the average OD values calculated with Equation (7), the relative growth rate (RGR) of the test biomaterial can be obtained. An RGR ≥ 75% implies low or no cytotoxicity, and the material can be utilized in medical applications [35]. The results of an MTT assay of L929 cells in SMPEEU film extracts after 72 h are presented in Figure 12. Although a slight reduction of RGR was observed compared with the control, all the RGRs were larger than 75%, indicating that the SMPEEU films had low toxicity to cells in vitro and met the requirement of medical materials. The RGR increased gradually with the increase of PEG content, which can be attributed to the excellent cytocompatibility of PEG.

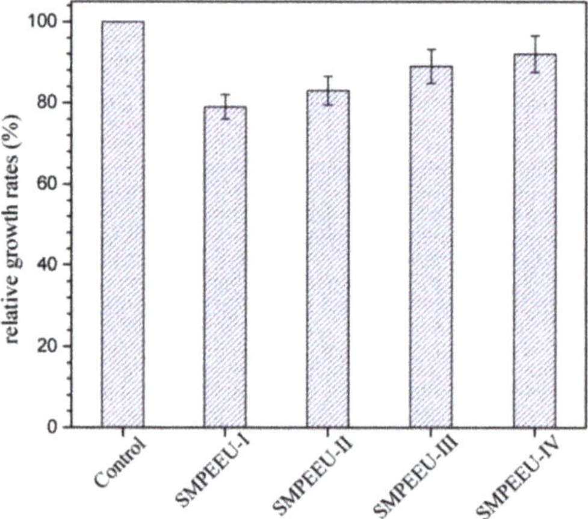

Figure 12. MTT assay of L929 cells on SMPEEU film extracts. (Error bars represent standard error of mean; $n = 3$).

$$\mathrm{RGR}\,(\%) = \frac{\text{OD value of sample suspension}}{\text{OD value of negative control suspension}} \times 100 \tag{7}$$

4. Conclusions

In this paper, a new class of degradable SMPEEUs with well-defined hard segments was prepared by one-step chain extension. The chemical structures of PECLs and SMPEEUs were confirmed by ^{1}H NMR, FT-IR, and GPC. The effect of PEG content on the physicochemical properties, including the thermal transition, crystallization behavior, bulk and surface hydrophilicity, tensile properties, and in vitro degradability, was extensively studied. DSC and XRD research demonstrated that the SMPEEUs had a microphase-separated structure, and the introduction of PEG segments could lead to a low transition temperature and decrease the crystallinity of the polymers. The surface and bulk hydrophilicity of SMPEEU films were found to be closely related to the hydrophilic PEG content. With the PEG content increasing from 9.25 to 47.5 wt %, the strain at break increased from 483 to 956%, and the ultimate stress decreased from 23.1 to 9.0 MPa. Hydrolytic degradation in vitro studies indicated that the time of the SMPEEU films becoming fragments was 4–12 weeks, and the introduction of PEG could facilitate the degradation rate of the films. The shape memory properties were evaluated by the fold-deploy test. The SMPEEU-film with a PEG content of 23.4 wt % displayed excellent recovery properties with a recovery ratio of 99.8% and recovery time of 3.9 s at body temperature, and the film could almost recover its original shape rapidly after four fold-deploy cycles. In addition, a cytotoxicity test of the film extracts was conducted using L929 cell, and the RGR was larger than 75% after incubation for 72 h, indicating good cytocompatibility in vitro. The SMPEEUs possess not only satisfactory tensile properties, biodegradability, nontoxic degradation products, and good cytocompatibility, but also excellent recovery properties at body temperature, suggesting their potential for application as biomedical devices.

Supplementary Materials: The supplementary materials are available online at http://www.mdpi.com/2073-4360/11/6/1002/s1.

Author Contributions: Z.H., M.X. and N.Z. conceived and designed the experiments; M.X., J.Z., Y.S., F.R. and W.Z. performed the experiments; Z.H. and M.X. analyzed the data and wrote the paper.

Funding: This research was funded by Shandong Provincial Natural Science Foundation, China (Project No. ZR2018MEM024) and National Undergraduate Training Programs for Innovation and Entrepreneurship, China (Project No. 201810445155).

Conflicts of Interest: The authors declare no conflict of interest.

References

1. Huang, W.M.; Yang, B.; Zhao, Y.; Ding, Z. Thermo-moisture responsive polyurethane shape-memory polymer and composites: A review. *J. Mater. Chem.* **2010**, *20*, 3367–3381. [CrossRef]
2. Ratna, D.; Karger-Kocsis, J. Recent advances in shape memory polymers and composites: A review. *J. Mater. Sci.* **2008**, *43*, 254–269. [CrossRef]
3. Liu, Y.; Lv, H.; Lan, X.; Leng, J.; Du, S. Review of electro-active shape-memory polymer composite. *Compos. Sci. Technol.* **2009**, *69*, 2064–2068. [CrossRef]
4. Huang, W.M.; Zhao, Y.; Wang, C.C.; Ding, Z.; Purnawali, H.; Tang, C.; Zhang, J.L. Thermo/chemo- responsive shape memory effect in polymers: A sketch of working mechanisms, fundamentals and optimization. *J. Polym. Res.* **2012**, *19*, 9952. [CrossRef]
5. Hardy, J.G.; Palma, M.; Wind, S.J.; Biggs, M.J. Responsive biomaterials: Advances in materials based on shape-memory polymers. *Adv. Mater.* **2016**, *28*, 5717–5724. [CrossRef] [PubMed]
6. Lu, H.; Liu, Y.; Leng, J.; Du, S. Qualitative separation of the effect of the solubility parameter on the recovery behavior of shape-memory polymer. *Smart Mater. Struct.* **2009**, *18*, 085003. [CrossRef]
7. Lendlein, A.; Langer, R. Biodegradable, elastic shape-memory polymers for potential biomedical applications. *Science* **2002**, *296*, 1673–1676. [CrossRef] [PubMed]

8.   Wang, K.; Strandman, S.; Zhu, X.X. A mini review: Shape memory polymers for biomedical applications. *Front. Chem. Sci. Eng.* **2017**, *11*, 143–153. [CrossRef]

9.   Govindarajan, T.; Shandas, R. Shape memory polymers containing higher acrylate content display increased endothelial cell attachment. *Polymers* **2017**, *9*, 572. [CrossRef]

10. Ahmad, M.; Luo, J.; Xu, B.; Purnawali, H.; King, P.J.; Chalker, P.R.; Fu, Y.Q.; Huang, W.M.; Miraftab, M. Synthesis and characterization of polyurethane-based shape-memory polymers for tailored T_g around body temperature for medical applications. *Macromol. Chem. Phys.* **2011**, *212*, 592–602. [CrossRef]

11. Hearon, K.; Wierzbicki, M.A.; Nash, L.D.; Landsman, T.L.; Laramy, C.; Lonnecker, A.T.; Gibbons, M.C.; Ur, S.; Cardinal, K.O.; Wilson, T.S.; et al. A processable shape memory polymer system for biomedical applications. *Adv. Healthc. Mater.* **2015**, *4*, 1386–1398. [CrossRef] [PubMed]

12. Zhan, M.Q.; Yang, K.K.; Wang, Y.Z. Shape-memory poly(p-dioxanone)-poly(ε-caprolactone)/sepiolite nanocomposites with enhanced recovery stress. *Chin. Chem. Lett.* **2015**, *26*, 1221–1224. [CrossRef]

13. Balk, M.; Behl, M.; Wischke, C.; Zotzmann, J.; Lendlein, A. Recent advances in degradable lactide-based shape-memory polymers. *Adv. Drug Delivery Rev.* **2016**, *107*, 136–152. [CrossRef] [PubMed]

14. Peterson, G.I.; Dobrynin, A.V.; Becker, M.L. Biodegradable shape memory polymers in medicine. *Adv. Healthcare Mater.* **2017**, *6*, 1700694. [CrossRef] [PubMed]

15. Kunkel, R.; Laurence, D.; Wang, J.; Robinson, D.; Scherer, J.; Wu, Y.; Bohnstedt, B.; Chien, A.; Liu, Y.; Lee, C.H. Synthesis and characterization of bio-compatible shape memory polymers with potential applications to endovascular embolization of intracranial aneurysms. *J. Mech. Behav. Biomed.* **2018**, *88*, 422–430. [CrossRef] [PubMed]

16. Wang, J.; Kunkel, R.; Luo, J.; Li, Y.; Liu, H.; Bohnstedt, B.N.; Liu, Y.; Lee, C.H. Shape memory polyurethane with porous architectures for potential applications in intracranial aneurysm treatment. *Polymers* **2019**, *11*, 631. [CrossRef] [PubMed]

17. Wang, J.; Luo, J.; Kunkel, R.; Saha, M.; Bohnstedt, B.N.; Lee, C.H. Development of shape memory polymer nanocomposite foam for treatment of intracranial aneurysms. *Mater. Lett.* **2019**, *250*, 38–41. [CrossRef]

18. Sokolowski, W.; Metcalfe, A.; Hayashi, S.; Yahia, L.; Raymond, J. Medical applications of shape memory polymers. *Biomed. Mater.* **2007**, *2*, S23–27. [CrossRef]

19. Sobczak, M. Biodegradable polyurethane elastomers for biomedical application-synthesis methods and properties. *Polym-Plast. Technol.* **2015**, *54*, 155–172. [CrossRef]

20. Zhang, C.; Hu, J.; Wu, Y. Theoretical studies on hydrogen-bonding interactions in hard segments of shape memory polyurethane-III: isophorone diisocyanate. *J. Mol. Struct.* **2014**, *1072*, 13–19. [CrossRef]

21. Zhang, C.; Hu, J.; Li, X.; Wu, Y.; Han, J. Hydrogen-bonding interactions in hard segments of shape memory polyurethane: toluene diisocyanates and 1,6-hexamethylene diisocyanate. A theoretical and comparative study. *J. Phys. Chem. A* **2014**, *118*, 12241–12255. [CrossRef] [PubMed]

22. Reed, A.M.; Potter, J.; Szycher, M. A solution grade biostable polyurethane elastomer: ChronoFlex® AR. *J. Biomater. Appl.* **1994**, *8*, 210–236. [CrossRef] [PubMed]

23. Yeh, J.; Gordon, B.; Rosenberg, G. Moisture diffusivity of Biomer® versus Biomer®-coated polyisobutylene polyurethane urea (PIB-PUU): A potential blood sac material for the artificial heart. *J. Mater. Sci. Lett.* **1994**, *13*, 1390–1391. [CrossRef]

24. Szycher, M. Biostability of polyurethane elastomers: acritical review. *J. Biomater. Appl.* **1998**, *3*, 297–402. [CrossRef]

25. Groot, J.H.D.; Spaans, C.J.; Dekens, F.G.; Pennings, A.J. On the role of aminolysis and transesterification in the synthesis of ε-caprolactone and L-lactide based polyurethanes. *Polym. Bull.* **1998**, *41*, 299–306. [CrossRef]

26. Spaans, C.J.; Groot, J.H.D.; Belgraver, V.W.; Penning, A.J. A new biomedical polyurethane with a high modulus based on 1,4-butanediisocyanate and ε-caprolactone. *J. Mater. Sci. Mater. Med.* **1998**, *9*, 675–678. [CrossRef] [PubMed]

27. Jia, Q.; Xia, Y.; Yin, S.; Hou, Z.; Wu, R. Influence of well-defined hard segment length on the properties of medical segmented polyesterurethanes based on poly(ε-caprolactone-*co*-L-lactide) and aliphatic urethane diisocyanates. *Int. J. Polym. Mater. Po.* **2017**, *66*, 388–397. [CrossRef]

28. Schmidt, C.; Neuking, K.; Eggeler, G. Functional fatigue of shape memory polymers. *Adv. Eng. Mater.* **2008**, *10*, 922–927. [CrossRef]

29. Herold, D.A.; Keil, K.; Bruns, D.E. Oxidation of polyethylene glycols by alcohol dehydrogenase. *Biochem. Pharmacol.* **1989**, *38*, 73–76. [CrossRef]

30. Milton, H.J. Introduction to biotechnical and biomedical applications of poly(ethylene glycol). In *Poly(ethylene glycol) chemistry: Biotechnical and biomedical applications*; Plenum: New York, NY, USA, 1992; pp. 1–14.

31. Ding, X.M.; Hu, J.L.; Tao, X.M.; Hu, C.P. Preparation of temperature-sensitive polyurethanes for smart textiles. *Text. Res. J.* **2006**, *76*, 406–413. [CrossRef]

32. Chun, B.C.; Cho, T.K.; Chang, Y.C. Enhanced mechanical and shape memory properties of polyurethane block copolymers chain-extended by ethylene diamine. *Eur. Polym. J.* **2006**, *42*, 3367–3373. [CrossRef]

33. Qu, W.Q.; Xia, Y.R.; Jiang, L.J.; Zhang, L.W.; Hou, Z.S. Synthesis and characterization of a new biodegradable polyurethanes with good mechanical properties. *Chin. Chem. Lett.* **2016**, *27*, 135–138. [CrossRef]

34. Zhen, W.; Zhu, Y.; Wang, W.; Hou, Z. Synthesis and properties of amphipathic poly(D,L-lactideco-glycolide)-polyethylene glycol-poly(D,L-lactide-coglycolide) triblock copolymers. *Aust. J. Chem.* **2015**, *68*, 1593–1598. [CrossRef]

35. Shi, R.; Zhu, A.; Chen, D.; Jiang, X.; Xu, X.; Zhang, L.; Tian, W. In vitro degradation of starch/PVA films and biocompatibility evaluation. *J. Appl. Polym. Sci.* **2010**, *115*, 346–357. [CrossRef]

36. Liu, Y.; Han, C.; Tan, H.; Du, X. Thermal, mechanical and shape memory properties of shape memory epoxy resin. *Mat. Sci. Eng. A* **2010**, *527*, 2510–2514. [CrossRef]

37. Zhang, Z.; Liao, F.; He, Z.; Yang, J.; Huang, T.; Zhang, N.; Wang, Y.; Gao, X. Tunable shape memory behaviors of poly(ethylene vinyl acetate) achieved by adding poly(L-lactide). *Smart Mater. Struct.* **2015**, *24*, 125002. [CrossRef]

38. Reddy, T.T.; Kano, A.; Maruyama, A.; Takahara, A. Synthesis, characterization and drug release of biocompatible/biodegradable nontoxic poly(urethane urea)s based on poly(ε-caprolactone)s and lysine-based diisocyanate. *J. Biomater. Sci. Polym. Ed.* **2010**, *21*, 1483–1502. [CrossRef] [PubMed]

39. Huang, M.; Li, S.; Coudane, J.; Vert, M. Synthesis and characterization of block copolymers of ε-caprolactone and DL-lactide initiated by ethylene glycol or poly(ethylene glycol). *Macromol. Chem. Phys.* **2003**, *204*, 1994–2001. [CrossRef]

40. Naguib, H.F.; Aziz, M.S.A.; Sherif, S.M.; Saad, G.R. Synthesis and thermal characterization of poly(ester-ether urethane)s based on PHB and PCL-PEG-PCL blocks. *J. Polym. Res.* **2011**, *18*, 1217–1227. [CrossRef]

41. Krol, P. Synthesis methods, chemical structures and phase structures of linear polyurethanes. Properties and applications of linear polyurethanes in polyurethane elastomers, copolymers and ionomers. *Prog. Mater. Sci.* **2007**, *52*, 915–1015. [CrossRef]

42. Lee, J.H.; Go, A.K.; Oh, S.H.; Lee, K.E.; Yuk, S.H. Tissue anti-adhesion potential of ibuprofen-loaded PLLA-PEG diblock copolymer films. *Biomaterials* **2005**, *26*, 671–678. [CrossRef] [PubMed]

43. Xu, W.; Xiao, M.; Yuan, L.; Zhang, J.; Hou, Z. Preparation, physicochemical properties and hemocompatibility of biodegradable chitooligosaccharide-based polyurethane. *Polymers* **2018**, *10*, 580. [CrossRef] [PubMed]

44. Liu, X.; Xia, Y.; Liu, L.; Zhang, D.; Hou, Z. Synthesis of a novel biomedical poly(ester urethane) based on aliphatic uniform-size diisocyanate and the blood compatibility of PEG-grafted surfaces. *J. Biomate. Appl.* **2018**, *32*, 1329–1342. [CrossRef] [PubMed]

45. Gu, X.; Mather, P.T. Entanglement-based shape memory polyurethanes: synthesis and characterization. *Polymer* **2012**, *53*, 5924–5934. [CrossRef]

46. Barrioni, B.R.; Carvalho, D.S.M.; Rrefice, R.L.; Oliveira, A.A.R.; Pereira, M.D.M. Synthesis and characterization of biodegradable polyurethane films based on HDI with hydrolyzable crosslinked bonds and a homogeneous structure for biomedical applications. *Mat. Sci. Eng. C* **2015**, *52*, 22–30. [CrossRef] [PubMed]

47. Nosbi, N.; Akil, H.M.; Ishak, Z.A.M.; Bakar, A.A. Degradation of compressive properties of pultruded kenaf fiber reinforced composites after immersion in various solutions. *Mater. Des.* **2010**, *31*, 4960–4964. [CrossRef]

48. Zuraida, A.; Humairah, A.R.N.; Suraya, M.R.N.; Nazariah, M. Evaluation of kenaf fibres reinforced starch based biocomposite film through water absorption and biodegradation properties. *J. Eng. Sci.* **2014**, *10*, 31–39.

49. Zhang, N.; Yin, S.N.; Hou, Z.S.; Xu, W.W.; Zhang, J.; Xiao, M.H.; Zhang, Q.K. Preparation, physicochemical properties and biocompatibility of biodegradable poly(ether-ester-urethane) and chitosan oligosaccharide composites. *J. Polym. Res.* **2018**, *25*, 212. [CrossRef]

50. Peterson, G.I.; Dobrynin, A.V.; Becker, M.L. α-Amino acid-based poly(ester urea)s as multishape memory polymers for biomedical applications. *ACS Macro Lett.* **2016**, *5*, 1176–1179. [CrossRef]

51. Ban, J.; Mu, L.; Yang, J.; Chen, S.; Zhuo, H. New stimulus-responsive shape-memory polyurethanes capable of UV light-triggered deformation, hydrogen bond-mediated fixation, and thermal-induced recovery. *J. Mater. Chem. A* **2017**, *5*, 14514–14518. [CrossRef]

52. Ping, P.; Wang, W.; Chen, X.; Jing, X. Poly(ε-caprolactone) polyurethane and its shape-memory property. *Biomacromolecules* **2005**, *6*, 587–592. [CrossRef] [PubMed]

53. Jing, X.; Mi, H.Y.; Huang, H.X.; Turng, L.S. Shape memory thermoplastic polyurethane (TPU)/poly(ε-caprolactone) (PCL) blends as self-knotting sutures. *J. Mech. Behav. Biomed.* **2016**, *64*, 94–103. [CrossRef] [PubMed]

54. Liu, Y.; Sun, H.; Tan, H.; Du, X. Modified shape memory epoxy resin composites by blending activity polyurethane. *J. Appl. Polym. Sci.* **2013**, *127*, 3152–3158. [CrossRef]

Article

Polyhexamethylene Biguanide:Polyurethane Blend Nanofibrous Membranes for Wound Infection Control

Anna Worsley [1,2,*], **Kristin Vassileva** [1,2], **Janice Tsui** [2], **Wenhui Song** [2] **and Liam Good** [1]

[1]   Royal Veterinary College, Department of Pathobiology and Population Sciences, 4 Royal College Street,
      London NW1 0TU, UK; Kristin.vassileva.16@ucl.ac.uk (K.V.); lgood@rvc.ac.uk (L.G.)
[2]   University College London, Centre for Biomaterials in Surgical Reconstruction and Regeneration,
      Division of Surgery & Interventional Science, 9th floor, Royal Free Hospital, Pond Street,
      London NW3 2QG, UK; janice.tsui@ucl.ac.uk (J.T.); w.song@ucl.ac.uk (W.S.)
*   Correspondence: aworsley@rvc.ac.uk

Received: 20 March 2019; Accepted: 20 May 2019; Published: 22 May 2019

Abstract: Polyhexamethylene biguanide (PHMB) is a broad-spectrum antiseptic which avoids many efficacy and toxicity problems associated with antimicrobials, in particular, it has a low risk of loss of susceptibility due to acquired antimicrobial resistance. Despite such advantages, PHMB is not widely used in wound care, suggesting more research is required to take full advantage of PHMB's properties. We hypothesised that a nanofibre morphology would provide a gradual release of PHMB, prolonging the antimicrobial effects within the therapeutic window. PHMB:polyurethane (PU) electrospun nanofibre membranes were prepared with increasing PHMB concentrations, and the effects on antimicrobial activities, mechanical properties and host cell toxicity were compared. Overall, PHMB:PU membranes displayed a burst release of PHMB during the first hour following PBS immersion (50.5–95.9% of total released), followed by a gradual release over 120 h ($\leq$25 wt % PHMB). The membranes were hydrophilic (83.7–53.3°), gradually gaining hydrophobicity as PHMB was released. They displayed superior antimicrobial activity, which extended past the initial release period, retained PU hyperelasticity regardless of PHMB concentration (collective tensile modulus of 5–35% PHMB:PU membranes, 3.56 ± 0.97 MPa; ultimate strain, >200%) and displayed minimal human cell toxicity (<25 wt % PHMB). With further development, PHMB:PU electrospun membranes may provide improved wound dressings.

Keywords: PHMB; polyhexanide; wound dressing; polyhexamethylene biguanide; polyurethane; electrospinning; nanofibres; antimicrobial; antiseptic

---

1. Introduction

Preventing and stabilising infection is a global challenge that is growing within healthcare systems. In addition to being a major cause of death, slow healing increases costs and perpetuates patient suffering. For example, one of the most common complications of diabetes mellitus is a chronic diabetic foot ulcer [1]; infection of these ulcers can contribute to serious negative outcomes including limb loss and sepsis [2]. The challenge of infection control is becoming more difficult with rising rates of acquired antimicrobial resistance. This is a worldwide concern that should be taken into consideration in wound infection control. With many populations still overusing antibiotics [3], research into non-antibiotic alternatives is essential.

Topical antiseptic treatments are commonly used to prevent infection in wounds; they are effective against a wide range of different types of yeasts, fungi and bacteria and result in relatively low levels of antimicrobial resistance [4,5]. There are many different antiseptic treatments currently used in healthcare [5], and many different delivery systems are being investigated. For example, silver nanoparticles have shown promise in hydrogel [6] and microemulsion [7] delivery systems, chlorhexidine in gel form has displayed strong antimicrobial properties [8], and a povidone-iodine foam dressing fully prevented infection in a prospective phase 4 study [9]. Despite comprehensive research on antiseptics for wound infection control, many have undesired side effects or have poor delivery systems: Silver leads to skin sensitisation and has limited tissue penetration; chlorhexidine causes skin sensitisation; iodine-based antiseptics stain the skin and some of the delivery systems for iodine can reduce wound healing or lead to skin sensitisation [4,5,10]. Therefore, there is a requirement for more research on alternative antimicrobials and delivery systems that may provide reduced side effects.

Polyhexamethylene biguanide (PHMB), also known as polyhexanide, is a polymerised biguanide compound used as a broad-spectrum antiseptic [4,11], disinfectant [4], and preservative [12]; see Scheme 1. It has proven to be effective against a wide range of pathogens, including strains of *Escherichia coli* [11,13], *Staphylococcus epidermidis* [13,14], and even the Protista *Acanthamoeba castellanii* [14]. It has previously been thought to work primarily through microbial membrane disruption [11,13,15,16], however, more recently it was reported to also selectively bind and condense bacterial DNA, arresting bacterial cell division. This mechanical antimicrobial mechanism of action may help explain why PHMB has a low risk for antimicrobial resistance, of which, none has been recorded despite extensive testing since its first synthesis [11]. Mammalian cells are comparatively unaffected by the polymer as it is segregated into endosomes, seemingly protecting the nuclei from its harmful effects [11], leading to faster wound closures compared to other antimicrobials [17].

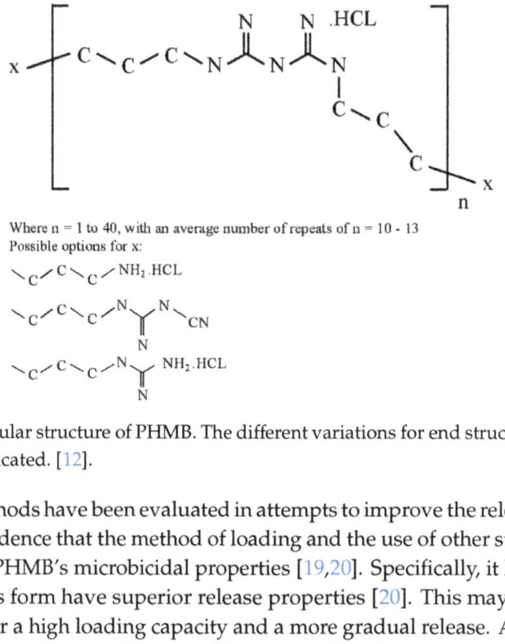

Where n = 1 to 40, with an average number of repeats of n = 10 - 13
Possible options for x:

Scheme 1. The molecular structure of PHMB. The different variations for end structures (x) and number of repeats (n) are indicated. [12].

Many different methods have been evaluated in attempts to improve the release of PHMB [16,18,19]. For example, there is evidence that the method of loading and the use of other substances are important factors for maintaining PHMB's microbicidal properties [19,20]. Specifically, it has been suggested that antimicrobials in fibrous form have superior release properties [20]. This may be due to an increased surface area, allowing for a high loading capacity and a more gradual release. As a strategy to improve release properties, electrospinning offers a relatively simple and versatile technique to manufacture a range of nanostructured drug delivery systems, from monolithic nanofibres to various multiple drug composition systems [21]. Llorens et al. used electrospinning to fabricate polylactide:PHMB nanofibres [22]. Interestingly, PHMB was only released in the first 1–4 h with a substantial amount not released at all, meaning only higher concentrations of spun PHMB gave sufficient antimicrobial properties when compared to controls [22]. These features indicate a possible route to producing a tunable release of PHMB, where optimisation of the electrospinning process and alternative backbone polymers may improve performance and duration of antimicrobial activity. Despite this, PLA:PHMB electrospun membranes allowed for MDCK cell attachment and were hydrophobic in nature [22].

This may cause cell surface damage when removing the wound dressing [23], as well as not allowing for wound exudate movement and therefore full PHMB release. Hydrophilic polymer matrices, such as polyurethane, may give more beneficial results by avoiding these problems, while matching the mechanical properties of the skin. Polyurethanes have favourable hyperelasticity, strength, wettability and biocompatibility for biomaterial applications [24–27]. They have been used to incorporate a wide range of compounds, including antimicrobial membranes for wound infection control, and have been intensively characterised [24–31].

This project developed hyperelastic nanofibrous membranes of PHMB and polyurethane (PU) blend through the technique of electrospinning, with the aim of producing an improved antimicrobial wound dressing with a gradual release of PHMB. As the electrospinning of PU with PHMB is unexplored, antimicrobial activity, human cell toxicity and release profile investigations were required. The hypothesis is that electrospinning could be used to engineer the nanostructure of PU-based wound dressings to have high compliance and a sustainable PHMB release overtime to improve antimicrobial effects, with minimal host cell damage.

2. Materials and Methods

2.1. Materials

PHMB with an average molecular weight of 3000 g/mol was obtained from Arch Chemicals (Castleford, UK) and Tecrea Ltd., (London, UK). Selectophore™ thermoplastic PU (a medical-grade aliphatic poly(ether-urethane) [25]), 2,2,2-trifluoroethanol, Dulbecco's Modified Eagle Medium (DMEM), fetal bovine serum (FBS), penicillin/streptomycin, glutaraldehyde, hexamethyldisilazane and ethanol were purchased from Sigma-Aldrich (Dorset, UK). Phosphate buffered saline (PBS) with a pH of 7.4 was purchased from ThermoFisher Scientific (London, UK). Mueller Hinton broth and agar were purchased from Merck (Watford, UK). Alamar blue was purchased from Invitrogen (London, UK). Surgical glue Med1-4013 was obtained from Polymer Systems Technology Ltd. (High Wycombe, UK). Actisorb Silver 220 (Systa genix) was donated by a diabetic foot clinic at the Royal Free hospital, London, UK. HaCaT keratinocytes were obtained from Dr. Amir Sharili, Queen Mary University of London, London, UK.

2.2. Preparation of PHMB:PU Nanofibrous Membranes via Electrospinning

2.2.1. Solution Preparation and Electrospinning Optimisation

Eight percent PU (wt %) in 2,2,2-trifluoroethanol was prepared overnight on a magnetic spinner at a temperature of 40 °C. PHMB was then added as a weight percentage (wt %) of PU and stirred at 40 °C for between 30 and 60 min until fully dissolved. A typical vertical electrospinning setup was custom built in house, as described by Wang et al. [25]. For all characterisation experiments, 5, 15, 25 and 35 (wt %) of PHMB were used (5-35PHMB:PU membranes); these were compared to a PU only electrospun control group (0PHMB:PU membranes). High concentrations of PHMB were chosen to optimise the percentage of PHMB released from the wound dressing. Previous research into electrospinning showed that a large quantity of PHMB is trapped within the fibres and is not released [22], suggesting higher concentrations should be included to maintain strong, clinically relevant antimicrobial properties overtime. To find the optimal electrospinning parameters for nanofibrous PHMB:PU membranes, 25PHMB:PU membranes were spun with varying parameters onto glass slides for between 0.5 and 4 min depending on the speed set. These fibres were analysed with a light microscope and tested parameters were chosen in response to previous research papers on electrospinning [32] and initial results. A distance of 17 cm from the collector, a voltage of 25 kV and a flow rate of 2 mL/hour were chosen for all future experiments.

2.2.2. Sample Preparation

For all experiments, excluding tensile strength tests, electrospinning was carried out for 90 min onto collection platforms (smooth aluminum foil). For tensile strength analyses, samples were spun over 180 min. The edges of all electrospun membranes were then taped down to prevent the material from contracting and creasing while excess solvent evaporated. For PHMB release and biological characterisation studies, the electrospun membranes were cut with a laser cutter (Speedy 100R, trotec, Wells, Austria) into disks with 13-mm diameter. For all other experiments, samples were cut with a scalpel into rectangular shapes. Electrospun samples were not used immediately and contracted and shrunk; disk samples shrunk to an average diameter of 6.84 mm by the time they were tested. These disk samples, used for all release and biological characterisations, weighed an average of 2.30 ± 0.65 mg and contained approximately 0.12, 0.35, 0.58 and 0.81 mg of PHMB in 5-, 15-, 25- and 35PHMB:PU membranes respectively. All samples were stored in normal room conditions.

2.3. Structural, Physical and Mechanical Property Characterisation

2.3.1. Morphology and Fibre Diameter of PHMB:PU Membranes via SEM

A scanning electron microscope (SEM, Carl Ziess EVO HD LS15, Ziess, Oberkochen, Germany) was used with the associated SmartSEM 5.07 software to visualise the morphology of the membranes. Average fibre diameter was then calculated by measuring 30–50 fibre diameters from three different areas of each electrospun membrane. Samples tested included untreated membranes and membranes that had undergone a 24 h immersion in PBS to mimic use in vivo. The latter were washed in distilled water to remove released PHMB that may still reside on the sample surface and air dried before being measured.

2.3.2. Surface Chemistry of PHMB:PU Membranes via ATR-FTIR

Attenuated total reflectance Fourier-transform infrared spectroscopy (ATR-FTIR) was used to assess PHMB levels on the surface of untreated PHMB:PU electrospun membranes and on membranes that had been immersed in PBS for 1 h, 4 h and then every 24 h up to 120 h to mimic use in vivo. These soaked membranes were washed in distilled water and air dried before being measured. Three different material samples were measured for each membrane type. The Jasco ATR-FTIR machine (Jasco, Dunmow, UK) was used for all measurements at room temperature.

2.3.3. Static Tensile Mechanical Properties of PHMB:PU Membranes

Static tensile strength was measured using an Instron 5565 testing system and Instron Bluehill 3 software (Instron, High Wycombe, UK). The samples were all spun for 180 min to have a similar thickness of 0.17 ± 0.07 mm. Samples were cut with a scalpel to a width of 10 mm and length of 8 mm. Calculations of tensile modulus at 5–10 mm extension, strain at break-up, strength of the membranes and result compilation was done by the software. Tensile toughness was calculated by finding the area under the stress-strain curve. Measurements from three different material samples were taken for each membrane type.

2.3.4. Pore Size and Distribution of PHMB:PU Membranes

The membranes pore size and distribution were characterised by the gas-liquid displacement method using Porolux 1000 (Porometer nv, Nazareth, Belgium). The membranes were cut to have a measurable area of 0.785 cm^2 and wetted with the specific wetting liquid PorefilL (Porometer nv, Nazareth, Belgium; surface tension of 16 mN·m^{-1}). The pressure of the testing gas (N$_2$) was increased from 0 to 34.5 bar in a step-by-step manner to replace the wetting liquid inside the pores of each sample. The pressure and flow were stabilised within ±1% for 2 s at each step before the data were recorded. The relevant pore size corresponding to each operating pressure was calculated using the

Young-Laplace equation. As well as untreated samples, membranes that had been immersed in PBS for 24 h were tested to mimic use in vivo. These samples were washed in distilled water and air dried before being measured.

2.3.5. Surface Wettability of PHMB:PU Membranes

The contact angle was measured using sessile drop analysis with the DSA 100 instrument and drop shape analysis software (KRUSS, Hamburg, Germany). A flat tip needle with a 0.5 mm diameter was used as part of the syringe pump to dispense the 3 µL volume sessile drop of deionised water. All experiments were undertaken at room temperature. As well as untreated samples, membranes that had been immersed in PBS for 1 h, 4 h and then every 24 h up to 120 h were tested to mimic use in vivo. These samples were washed in distilled water and air dried before being measured. Three different material samples were measured for each membrane type.

2.3.6. PHMB Release Kinetics from PHMB:PU Membranes

The release of PHMB from electrospun membranes was measured using optical density (OD; 230–249 nm [33,34]) readings using the nanoDrop ND1000 spectrophotometer with associated software (version 3.8.1; ThermoFisher Scientific, London, UK). All samples were immersed in PBS in a 96-well plate and left at room temperature. OD readings were taken after 1 h, 4 h and then every 24 h up to 120 h. Before readings were taken, the wells were mixed with a pipette to make sure the released PHMB was evenly distributed for accurate measurements. Concentrations of PHMB released into the liquid were calculated using a standard curve produced using known PHMB concentrations (See Appendix A, Figure A1). Percentage release was then calculated using the known quantity of PHMB within each sample. As well as 5-35PHMB:PU membranes, a non-electrospun PHMB-only control group (ctrl PHMB) was tested. These ctrl PHMB samples were prepared by injecting 20 µL of a 1 mg/mL PHMB/water solution into paper disks with a pipette. These disks were then air dried, forming non-electrospun samples carrying 0.02 mg of PHMB (the equivalent of approximately 1% PHMB when compared to electrospun samples). The ctrl PHMB disks had a similar diameter to the electrospun samples. Three different material samples were measured for each sample type with three measurements taken per sample.

2.4. Biological Interaction Characterisations

2.4.1. Antimicrobial Properties of PHMB:PU Membranes

Disk diffusion and liquid bacterial inoculation assays were carried out using *Staphylococcus aureus* RN4220. An overnight liquid bacterial culture was prepared using MH broth, incubated at 37 °C. Samples included the 0-35PHMB:PU electrospun membranes; the non-electrospun PHMB-only controls (ctrl PHMB) as described above in Section 2.3.6; and the commercial wound dressing Actisorb Silver 220 (Systa genix, Skipton, UK), a silver and charcoal woven antimicrobial dressing used in diabetic foot clinics. Three different material samples were used for each sample type. For disk diffusion assays, agar plates were air dried of condensation and evenly inoculated with 1,000,000 colony forming units (CFU) in 100 µL of MH broth from an overnight culture. These were semi-air-dried before the samples were added and then incubated overnight at 37 °C. The zone of growth inhibition that formed overnight was measured and results were normalised by subtracting the disk diameter. For liquid bacterial inoculation assays, samples were incubated in 96-well plates overnight at 37 °C with 280 µL of phenol-free high glucose DMEM containing 1.5×10^8 CFU per mL of *Staphylococcus aureus* RN4220 from an overnight culture. Untreated samples and samples that had been immersed in PBS overnight were tested. After the overnight incubation, 3×10 µL diluted samples from each well were seeded onto MH agar plates for analysis. These were incubated for 24 h at 37 °C before the total CFU count for each well was measured.

2.4.2. Human Cell Viability and Attachment on PHMB:PU Membranes

HaCaT keratinocytes were cultured in low glucose DMEM with phenol red, supplemented with 10% FBS and 1% penicillin/streptomycin in a 37 °C, 5% CO_2 humidified atmosphere. Samples tested were the same as in the antimicrobial studies undertaken, with tissue culture plate as an extra control group. For SEM analysis, glass controls were also used. Three different material samples were used for each sample type. For toxicity assays, cells were seeded in 12-well plates, with 150,000 cells per well. The first Alamar blue assay was carried out 24 h after seeding. Samples were then added and further Alamar blue assays were carried out after another 24 h and 48 h to assess the toxicity of released PHMB. For each assay, 10% Alamar blue with cell culture media was added to wells and incubated (37 °C, 5% CO_2 humidified atmosphere) for 4 h before readings were taken. Four × 100 μL samples of the Alamar blue/media were taken per well and placed into 96-well plates. Relative fluorescence readings were taken at 530 nm emission and 620 nm excitation.

For cell attachment analysis, material was electrospun as described before, but instead of being laser cut, glass coverslips were glued on around its circumference with surgical glue. These were then cut out with a scalpel and sterilised under UV light for 30 min. Cell culture was as described above, but with 1,000,000 cells per well being seeded. After 24 h of incubation, samples were moved to new 12-well plates and incubated in 10% Alamar blue for 4 h before analysis. This experiment was then repeated with samples that had been immersed in cell culture media overnight before cell seeding. For these samples, Alamar blue assays were undertaken 24 h and 48 h after cell seeding. The cells attached to these soaked samples were then fixed for SEM analysis. To fix these samples, membranes were incubated in 2.5% *w/v* glutaraldehyde/PBS for 1 h. Samples were then washed in distilled water and dehydrated in a series of ethanol (20%, 40%, 80% and 100% for 10 min each) at room temperature. Samples were then incubated in hexamethyldisilazane at room temperature for 15 min and air dried. Dried samples were sputter coated with 20 nm of gold using a Quorum Q150RS instrument (Laughton, UK).

2.5. Statistical Analysis

Statistical significance was assessed using GraphPad Prism 7 software (San Diego, CA, USA). A statistically significant difference was set at $p < 0.05$ (*), $p < 0.001$ (**) and $p < 0.0001$ (***). All error bars are a standard deviation from the mean. One-way Analysis of Variance (ANOVA) tests were used for single time point experiments and two-way ANOVA for multiple time point experiments.

3. Results

3.1. Initial Optimisations

As the combination of PU with PHMB had not been evaluated previously, electrospinning parameters were first optimised. Different parameters were compared and the optimal distance between the needle and collector was found to be 17 cm with a voltage of 25 kV and a rate of 2 mL/hour for the most uniform and straight fibres. These parameters were therefore used for all subsequent material production.

3.2. Structure, Physical and Mechanical Properties

3.2.1. Morphology and Fibre Diameter of PHMB:PU Membranes via SEM

SEM was used to analyse the fibre morphology of untreated electrospun membranes and membranes after a 24 h immersion in PBS to mimic 24 h of use. Fibres of electrospun PHMB:PU membranes before treatment were long, straight and uniform. However, with increasing amounts of PHMB they became thicker and more sinuous (Figures 1 and 2, 0 h). There was a gradual increase in fibre diameter as PHMB concentration increased, however there were no immediate significant differences between PHMB:PU membranes, see Figures 1 and 2, 0 h. All membranes also had thinner web-like fibres in-between the main fibres, these were found to be much thinner in diameter (52 ± 21 nm in all membrane samples).

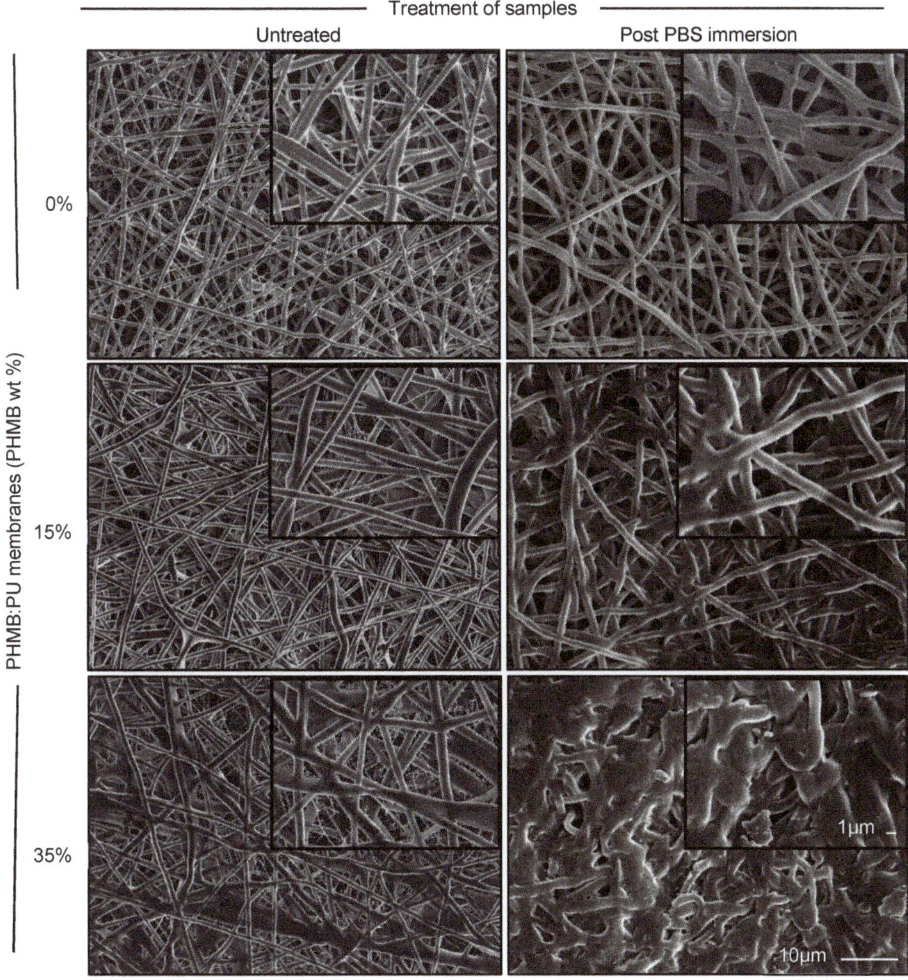

Figure 1. Representative scanning electron microscopy images of 0-, 15- and 35PHMB:PU electrospun membranes. Images are representative of untreated membranes and membranes after immersion in PBS for 24 h. Scale bars are for both untreated and treated samples.

After a 24-h immersion in PBS, fibres in all membrane samples became less uniform and distinct. This change was more prominent with higher PHMB contents (Figure 1, 24 h). Fibres with a lower concentration of PHMB (<25%) become more sinuous after immersion, but retained their fibrous structure. In 35PHMB:PU membranes, the fibres were severely deformed and adhered together, with the apparent closure of pores. Immersion in PBS also caused fibres to swell, with fibre diameters increasing by 51%, 18.4%, 59%, 18.1% and 40% in 0-35PHMB:PU membranes respectively after 24 h (Figure 2, $p < 0.0001$ for 0-, 15- and 35PHMB:PU membranes, $p < 0.05$ for 25PHMB:PU membranes). The relatively large standard deviation for all sample measurements reflects a relatively wide distribution of the fibre diameters in both untreated and treated membranes (Figure 2).

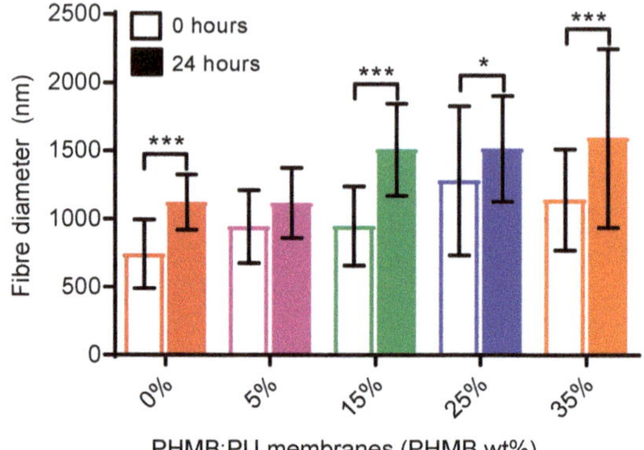

Figure 2. Fibre diameters (nm) of electrospun PHMB:PU membranes. Samples included untreated membranes (0 h) and membranes after immersion in PBS for 24 h. N = 20–50, taken from three different areas of material imaged with SEM. Error bars describe the standard deviation from the mean. * signifies a significant percentage change after immersing samples in PBS (* for $p < 0.05$; *** for $p < 0.0001$).

3.2.2. Surface Chemistry of PHMB:PU Membranes via ATR-FTIR

The FTIR spectra for electrospun PHMB:PU membranes had characteristic absorption peaks for PU (functional groups of N–H, C–H, aromatic C=C and ester groups C=O and C–O at wavelengths of 3320, 2928 and 2852, 1530, 1696, 1230 and 1100 cm^{-1} respectively [25]) and PHMB (functional groups of C=N, N–H and C–H with wavelengths of 1600 cm^{-1} [stretch], 3310 cm^{-1}, 2928 and 2852 cm^{-1} respectively [35]), see Figure 3a–c. Percentage transmission at these wavelengths exhibited a trend of PHMB influence on PU peaks as PHMB levels increased (Figure 3). In particular, the stretch corresponding to PHMB's imine group became more pronounced as PHMB concentration increased (Figure 3a–c). The amine peak (N–H) characteristic of PU and PHMB also had a dampened percentage transmission with the addition of PHMB (Figure 3a–c), significantly decreasing for 25- and 35PHMB:PU membranes before PBS immersion (Figure 3d, 0 h). This is likely related to a secondary peak forming at around 3150 cm^{-1} with higher PHMB concentrations, corresponding to PHMB's primary amine groups as opposed to PU which only has secondary amine groups.

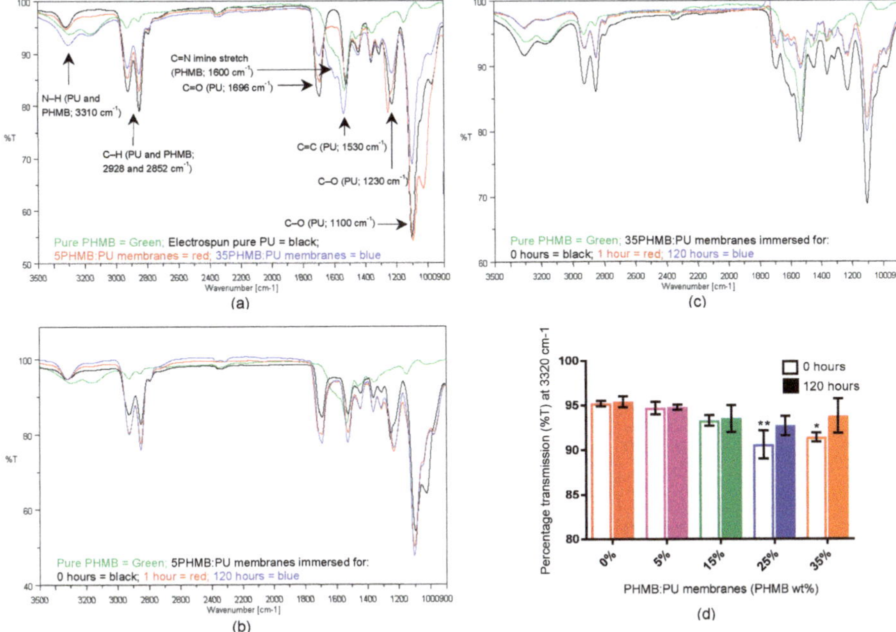

Figure 3. FTIR measurements of PHMB:PU electrospun membranes. (**a**) Pure PHMB (green), untreated membranes: 0-, 5- and 35PHMB:PU electrospun membranes (black, red and blue respectively). (**b**,**c**), 5- and 35PHMB:PU membranes respectively that were immersed in PBS for 0, 1 and 120 h (black, red and blue respectively). These were compared to pure PHMB (green). (**d**) Percentage transmission for the 3320 cm^{-1} peak (N–H) before and after membranes were immersed in PBS for 0 and 120 h. * indicates a significant difference (* for $p < 0.05$; ** for $p < 0.001$).

3.2.3. Tensile Mechanical Properties of PHMB:PU Membranes

The overall tensile mechanical properties of PHMB:PU membranes, in terms of the strain at break, elastic modulus, ultimate strength and toughness, were reduced in the presence of PHMB, compared to pure PU electrospun membranes (0PHMB:PU), as shown in Figure 4. In particular, the ultimate strength and tensile toughness of PHMB:PU electrospun membranes were seen to decrease significantly by an average of 58.65% and 70.95% respectively when compared to electrospun PU only membranes (Figure 4c,d, $p < 0.05$ and $p < 0.001$ respectively). However, this trend did not reflect the increasing load of PHMB to electrospun PU membranes as no significance was recorded between 5-35PHMB:PU membranes.

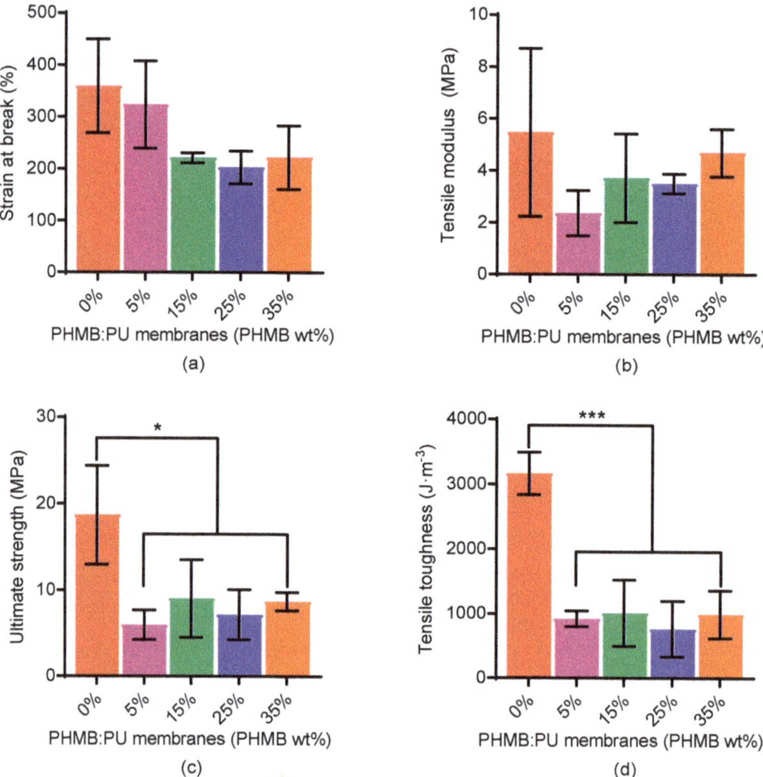

Figure 4. Tensile mechanical properties of electrospun PHMB:PU membranes. (**a**) Strain at break; (**b**) Tensile modulus at 5–10 mm extension; (**c**) Ultimate strength; (**d**) Tensile toughness. Samples tested were the 0-35PHMB:PU electrospun membranes. Error bars describe the standard deviation from the mean, with $n = 3$. * signifies a significant difference between samples (* for $p < 0.05$; *** for $p < 0.0001$).

3.2.4. Pore Size and Distribution of PHMB:PU Membranes

Pore size and distribution were evaluated to access the breathability and permeability of membranes. As the content of PHMB increased, the pore size of the membranes reduced drastically, but were not fully exhausted (significant differences recorded between 0-, 5- and 15-35PHMB:PU membranes, Figure 5). After the PHMB:PU membranes were immersed for 24 h in PBS, pore size decreased further with 35PHMB:PU membranes becoming impermeable.

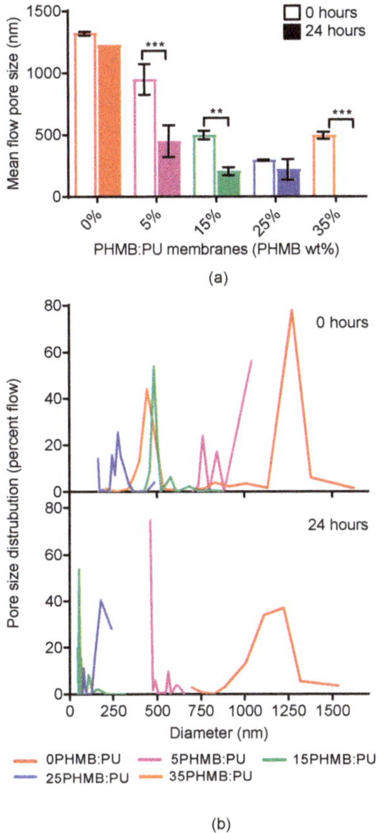

Figure 5. The effect of increasing PHMB concentration on pore size in PHMB:PU electrospun membranes. (**a**) Mean flow pore size (nm). Error bars describe the standard deviation from the mean, with n = 20–114. * signifies a significant percentage change after immersing samples in PBS (** for $p < 0.001$; *** for $p < 0.0001$). (**b**) Representative pore size distributions (percentage flow of N2) after a 0-h and 24-h immersion in PBS. 35PHMB:PU membranes were not included in (**b**), 24 h, as they were recorded as impermeable.

3.2.5. PHMB Release Kinetics of PHMB:PU Membranes

To prove the hypothesis that an electrospun nanofibrous structure would allow for a gradual delivery of PHMB, the PHMB:PU membranes were immersed in PBS and the PHMB release was measured over time. All membranes showed a significant burst release of PHMB within the first hour; see Figure 6 ($p < 0.05$ for 5PHMB:PU membranes, $p < 0.001$ for 15PHMB:PU and $p < 0.0001$ for 25-35PHMB:PU membranes). The percentage of the total released PHMB released in this initial burst increased as PHMB concentration increased from a 15% content (after 1 h, 5-35PHMB:PU membranes released 50.5%, 50.4%, 71.9% and 95.9% of their total PHMB content released respectively). The initial burst release was followed by a more gradual release of PHMB in 5-25PHMB:PU electrospun membranes (no significant differences between each immediately succeeding time point, but significant differences measured between 1 h and 120 h time points, $p < 0.0001$). In contrast, 35PHMB:PU electrospun membranes showed no further significant release after the first hour. All electrospun membranes released a significantly lower percentage of their total PHMB content compared to non-electrospun PHMB samples, reflecting work done by Llorens et al. who electrospun PHMB with polylactide [22],

($p < 0.0001$, Figure 6; percentage released at 120 h: 16.4% ± 0.69% from 5PHMB:PU membranes, equating to an average of 0.02 mg PHMB; 18.8% ± 1.2% from 15PHMB:PU membranes, equating to an average of 0.07 mg PHMB; 36.3% ± 2.7% from 25PHMB:PU membranes, equating to an average of 0.2 mg PHMB; 23.9% ± 2.7% from 35PHMB:PU membranes, equating to an average of 0.2 mg PHMB).

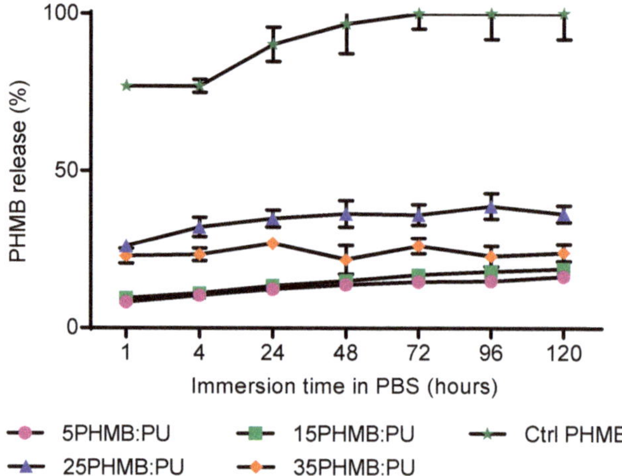

Figure 6. Average percentage release of PHMB from electrospun PHMB:PU membranes after immersion in PBS. Error bars describe the standard deviation from the mean, with $n = 3$. Paper disks freely infused with PHMB (Ctrl PHMB) were used as controls, as described in the methods section.

3.2.6. Surface Wettability of PHMB:PU Membranes

As PHMB content increased in the PHMB:PU membranes, surface hydrophilicity increased (as seen in Figure 7, 0 h measurements; the contact angles for 0-35PHMB:PU electrospun membranes were measured at 93° ± 2°, 84° ± 1°, 66° ± 1°, 53° ± 1° and 55° ± 1° respectively, significant differences between all values excluding between 0- and 5PHMB:PU membranes and between 25- and 35PHMB:PU membranes, $p < 0.05$). 35PHMB:PU membranes at 0 h were measured to have a contact angle similar to that seen for 25PHMB:PU membrane, however, the droplets were unstable and dissipated after a few minutes, indicating higher levers of hydrophilicity than measured.

The contact angle of electrospun PHMB:PU membranes was recorded after membranes had been immersed in PBS, revealing the change in wettability as PHMB was released. 35PHMB:PU membranes had a significant decrease in wettability in the first hour, which then stabilised and remained constant over the next 119 h analysed. 15- and 25PHMB:PU electrospun membranes also showed this initial significant decrease in wettability (the surface contact angle of 15-, 25- and 35PHMB:PU membranes increased by 27.39%, 51.89% and 34.21% respectively after 1 h of PBS immersion, $p < 0.0001$, Figure 7). However, after this first hour, 5-25PHMB:PU electrospun membranes had a more gradual decrease in wettability up to the 120 h time point (significant increases in contact angle recorded between 1 and 120 h but not between each immediately succeeding time point; the overall increase of surface contact angle for 5-, 15- and 25- PHMB:PU membranes was 26.73%, 30.54% and 21.98% respectively, $p < 0.001$). After 120 h, wettability of 5-25PHMB:PU electrospun membranes decreased to values similar to PU only membranes (0PHMB:PU, Figure 7).

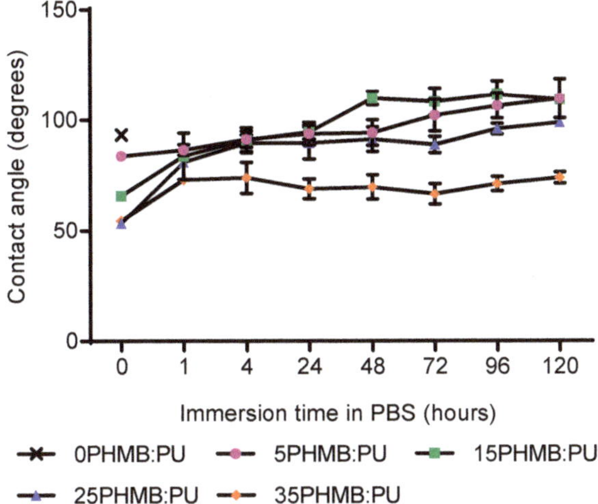

Figure 7. The contact angle change of electrospun PHMB:PU membranes after immersion in PBS. Changes in contact angle inversely indicate changes in wettability as PHMB was released. Error bars describe the standard deviation from the mean, with $n = 3$. 0 h measurements for 35PHMB:PU membranes were taken but were found to give unstable droplet formation, indicating higher hydrophilicity than reported.

3.3. Effects on Pathogen and Host Cells

3.3.1. Antibacterial Activities of PHMB:PU Membranes

The antimicrobial activity of the PHMB:PU membranes was first examined using disk diffusion assays, where sample disks were placed on lawns of bacteria and the diameter of growth inhibition was measured after 24 h. Antimicrobial activity increased as PHMB content increased, with all PHMB:PU membranes having a significantly higher antimicrobial activity than PU only membranes, as shown in Figure 8b. 5PHMB:PU membranes showed similar antimicrobial activity compared to control PHMB paper disks despite containing more PHMB, indicating a gradual release of PHMB from 5PHMB:PU membranes (Figure 8b).

Measurements from liquid inoculation incubations, where samples were suspended in infected liquid broth, also displayed impressive antimicrobial activity (Figure 8a); 0 CFU/mL survived after 24 h of incubation with all PHMB:PU electrospun membranes. After 24 h of immersion in PBS, electrospun 15-35PHMB:PU membranes showed a continuation of their antimicrobial activity strength. 5PHMB:PU electrospun membranes lost some of their antimicrobial activity after the 24 h of immersion, although this was not statistically significant; the level of antimicrobial activity was still significantly lower than 0PHMB:PU groups ($p < 0.05$) and was not significantly different to 15-35PHMB:PU membranes (Figure 8a). Control PHMB samples still showed antimicrobial activity after 24 h of immersion, however this was seen as a false positive result. Unlike the PHMB:PU membranes, the paper control disks could not be efficiently washed to remove excess PHMB collected on the surface because it had partially deteriorated with immersion, allowing for released PHMB to collect and remain on the sample surface after removal from the PBS.

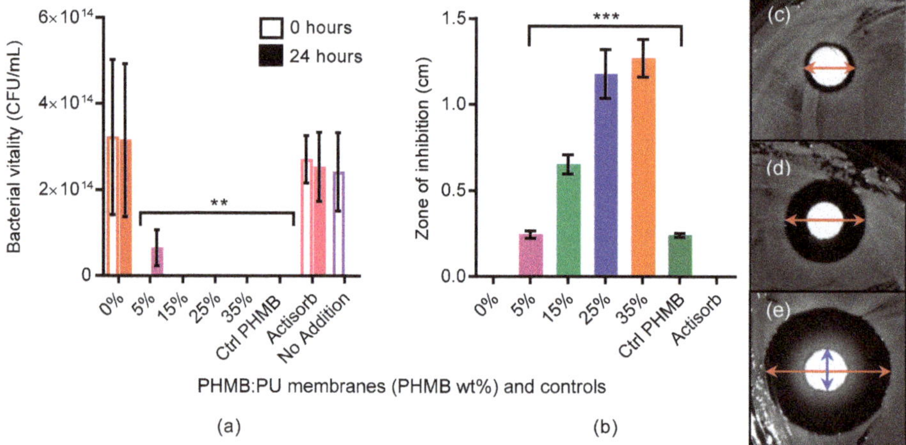

Figure 8. Antibacterial activities of PHMB:PU membranes. (**a**) Antimicrobial activity in liquid *S.aureus* RN4220 cultures. Surviving bacterial colony forming units (CFU/mL) after overnight incubation with samples. Samples included PHMB:PU electrospun membranes and controls before (0 h) and after a 24 h immersion in PBS. (**b**) Zones of bacterial (*S.aureus* RN4220) growth inhibition using disk diffusion assays. Error bars describe the standard deviation from the mean, with $n = 3$. * indicates a significant difference to 0% electrospun membrane samples (** for $p < 0.001$; *** for $p < 0.0001$). Controls used were PU only (0PHMB:PU membranes), PHMB freely infused in paper (Ctrl PHMB), the commercial product Actisorb Silver 220 (Actisorb), and for (**a**), no addition of material or reagent to bacterial liquid cultures. Images (**c–e**) are representative of the zones of bacterial growth inhibition measured as shown in (**b**), (5-, 15- and 35PHMB:PU electrospun membranes respectively). The red arrows indicate the inhibition zone diameter measured, this was normalised by subtracting the sample diameter (**e**, blue arrow).

3.3.2. HaCaT Cell Responses

Human Cell Toxicity of PHMB:PU Membranes

Alamar blue assays were used to assess the toxicity of PHMB:PU electrospun membranes against cultured human keratinocytes via metabolic activity. 5PHMB:PU electrospun membranes showed no cell toxicity at all time points (Figure 9). The same was seen for pure PU only electrospun membranes, displaying the non-toxic nature of PU. After 24 h of incubation with samples, the cell viability levels of 15PHMB:PU membranes were not significantly different to the commercial product Actisorb Silver 220 or the non-electrospun PHMB control groups, all of which showed a significant amount of cell toxicity (Figure 9, $p < 0.05$). However, after 48 h, Actisorb Silver 220 showed slight but significant improvements in cell viability, while 15PHMB:PU membranes did not. Finally, 25- and 35PHMB:PU electrospun membranes showed high levels of toxicity that were not recoverable during the monitoring period.

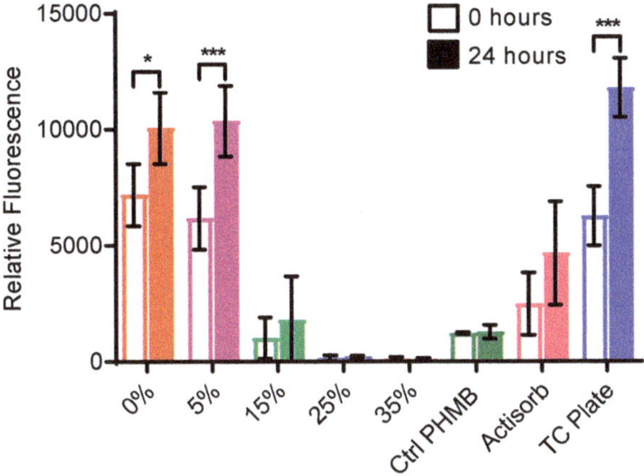

Figure 9. HaCaT cell toxicity. Alamar blue results taken 24 and 48 h after sample addition. Controls included TC plate (tissue culture plate), paper disks freely infused with PHMB (Ctrl PHMB) and the commercial product Actisorb Silver 220 (Actisorb). Error bars are the standard deviation from the mean, with $n = 3$. * indicates a significant difference between time points within samples groups (* for $p < 0.05$ and *** for $p < 0.0001$ respectively).

Human Cell Attachment on PHMB:PU Membranes

All electrospun PHMB:PU membranes showed no successful cell attachment after 24 h of incubation. Cells were then seeded onto PHMB:PU electrospun membranes that had been immersed in cell culture media for 24 h to allow for more protein attachment and to mimic use overtime. Some attachment was seen with these samples (Figure 10). However, as seen in the SEM images, these cells were rounded in morphology showing unfavorable attachment conditions (Figure 11). 25- and 35PHMB:PU electrospun membranes showed minimal cell attachment but showed high levels of toxicity in the previous section; therefore, they could not be imaged.

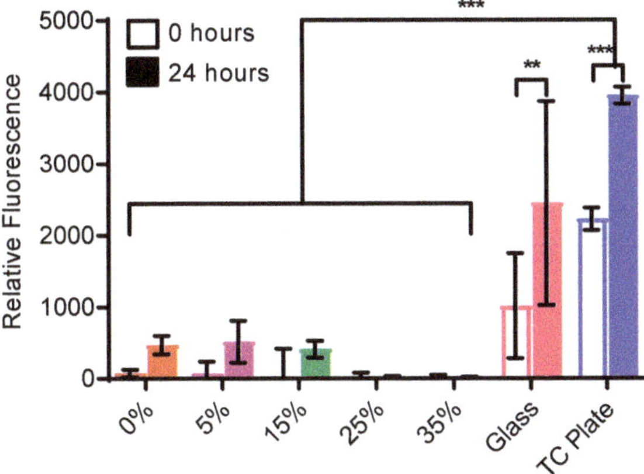

Figure 10. HaCaT cell metabolic activity on PHMB:PU membranes. Cell metabolic activity after 24 and 48 h using the Alamar blue assay on membranes which had been immersed in cell culture media overnight before cell seeding to mimic attachment after use in vivo. Tissue culture plate (TC plate) and glass were used as controls. Error bars describe the standard deviation from the mean, with $n = 3$. * indicates a significant difference (** for $p < 0.001$; *** for $p < 0.0001$).

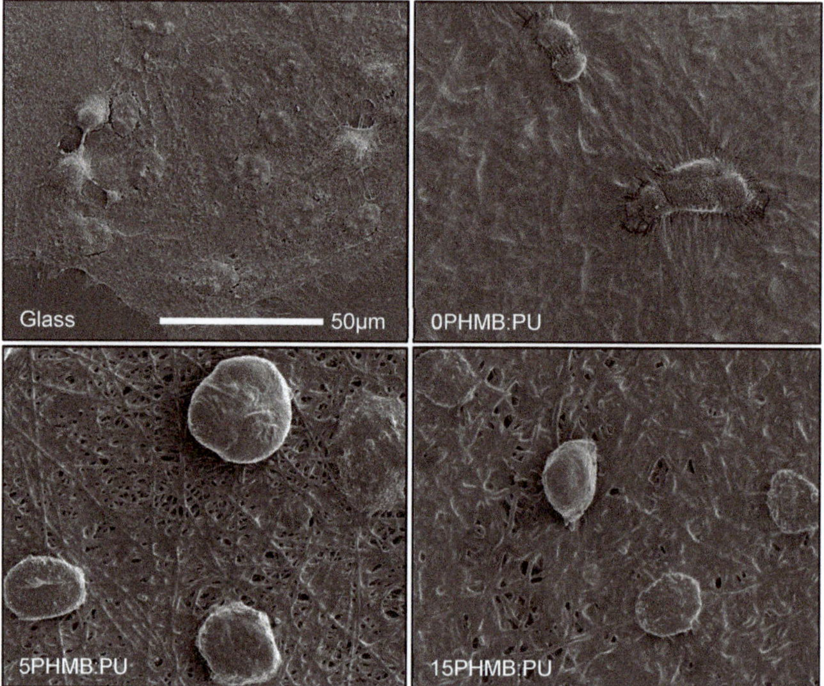

Figure 11. HaCaT cell morphology on PHMB:PU membranes. SEM images showing cell morphology after 48 h on membranes which had been immersed in cell culture media overnight before cell seeding to mimic attachment after use in vivo.

4. Discussion

PHMB was successfully incorporated within PU nanofibre membranes at concentrations ranging from 5% to 35% (Figure 3). The aim of this project was to develop a wound dressing which would allow PHMB to be released in a gradual manner, allowing for the optimisation of PHMB's antimicrobial properties. This was accomplished using 5-25PHMB:PU membranes; when immersed in PBS, these membranes had an initial burst release of PHMB followed by a more gradual release over the 120 h measured (Figure 6). This was reflected in the change of wettability also recorded with PBS immersion overtime (Figure 7). The initial burst release measured within the first hour mirrors similar results recorded by Llorens et al. [22] and could be beneficial for heavily infected wounds. The gradual release after the initial burst would prevent bacteria growth recovery and colonisation overtime, providing ideal conditions to reduce bacterial load while minimising host cell toxicity. The frequency of wound dressing changes may also be reduced as the PHMB:PU membranes retain their antimicrobial activity overtime [36].

5-25PHMB:PU membranes were found to be porous even after use (Figure 5), allowing for breathability and permeability for wound exudate passage. Fibres became swollen and more sinuous when immersed in PBS (Figure 1), causing pores to shrink but not fully close up. This change in fibre diameter and morphology may be due to the hydrophilic nature of PHMB, which could promote some liquid absorption prior to PHMB release and is consistent with the observed increase in wettability (Figure 7). Along with the hydrophilic nature of the PHMB:PU membranes (Figure 7), the continued permeability would aid continued PHMB release and may allow for a moist wound bed that does not become overly wet, aiding wound healing further [36]. The decrease in pore size with increasing PHMB content also displays a possible route for tunability.

The addition of PHMB to PU electrospun membranes slightly reduced their elasticity and tensile strength and significantly decreased their ultimate strength and tensile toughness (Figure 4). This suggests that PHMB:PU electrospun membranes are elastic but are quicker to tear with deformity than with pure PU electrospun membranes. Despite this, the overall tensile strength, modulus and ultimate strain of the blend membranes remained favorable for a wound dressing as they still retained the hyperelastic and durable properties of polyurethane that are relevant to skin [37].

As 5PHMB:PU electrospun membranes showed no cell toxicity and 15PHMB:PU membranes had comparable toxicity levels to non-electrospun PHMB samples, despite a higher PHMB content (Figure 9), these samples are the most promising for further development. Cell death in response to 25- and 35PHMB:PU membranes was possibly because too much PHMB released in the first burst; therefore, the amount of PHMB entering the cells was too high for them to cope leaving some to enter the nuclei and condense chromosomes, as seen in bacterial cells [11]. These higher concentration membranes are unlikely to be used in future work unless the release is slowed in further material optimisations.

Cell attachment and cell morphology analysis indicated that PHMB:PU membrane surfaces were not ideal for cell attachment (Figures 10 and 11 respectively). This may be related to the high positive charge of PHMB causing some abnormal protein attachment; when released onto the membrane surface, proteins may bind too strongly, resulting in abnormal conformation and protein denaturation. It is important to clarify that low cell attachment is in fact beneficial for wound dressings as it would prevent healthy tissue removal on dressing changes [23]. Therefore, our results suggest that PHMB:PU electrospun membranes could avoid associated complications with healthy cell removal, improving the wound healing process.

5. Conclusions

Electrospun PHMB:PU nanofibrous membranes offer an exciting alternative strategy for providing a gradual release of PHMB in wound infection prevention and control, and show promise for the development of a future antimicrobial wound dressing. Their elastic and hydrophilic nature could aid patient comfort and help prevent host cell damage. The strong antimicrobial properties sustained overtime may provide an ideal dressing for clearing bacterial load while enhancing healing. Sustained PHMB release may also mean fewer wound dressing changes, saving time, waste products and money. For future work, the tunability of PHMB:PU membranes will be further investigated to produce an optimised wound dressing material.

Author Contributions: Conceptualization, L.G., W.S., and J.T.; methodology, A.W., L.G., W.S., K.V.; validation, A.W., L.G., W.S., J.T.; formal analysis, A.W., L.G., W.S.; investigation, A.W.; resources, A.W., L.G., W.S.; data curation, A.W.; writing—original draft preparation, A.W.; writing—review and editing, A.W., L.G., W.S., J.T.; visualization, A.W., L.G., W.S., J.T.; supervision, L.G., W.S., J.T.; project administration, A.W., L.G., W.S.; funding acquisition, L.G., W.S.

Funding: This research was funded by the BBSRC-LIDO DTP studentship (reference 1764829) and Engineering and Physical Sciences Research Council (United Kingdom, EPSRC grants Nos. EP/L020904/1, EP/M026884/1 and EP/R02961X/1).

Acknowledgments: We would like to thank all colleagues and collaborators who helped and support this work, including Thomas Maltby and Antonio for building the electrospinning system (London South Bank University, London, UK), as well as Bo Wang and Kang Li, Imperial Collage (London, UK) for their kind assistance in analysing membrane pore size.

Conflicts of Interest: The authors declare no conflict of interest.

Appendix A

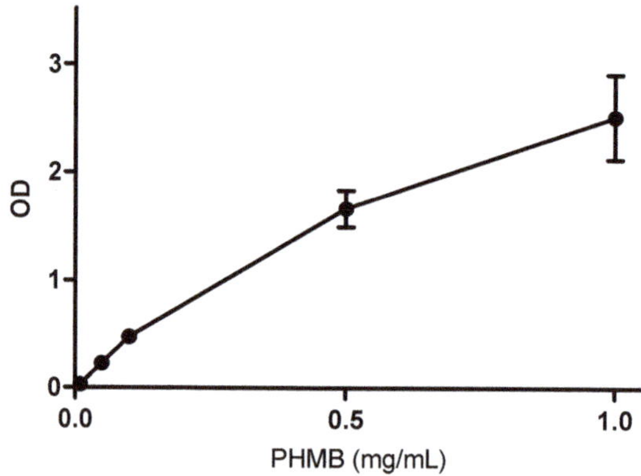

Figure A1. PHMB standard curve in PBS.

References

1. Ndosi, M.; Wright-Hughes, A.; Brown, S.; Backhouse, M.; Lipsky, B.A.; Bhogal, M.; Reynolds, C.; Vowden, P.; Jude, E.B.; Nixon, J.; et al. Prognosis of the infected diabetic foot ulcer: A 12-month prospective observational study. *Diabetic Med.* **2018**, *35*, 78–88. [CrossRef]
2. Noor, S.; Zubair, M.; Ahmad, J. Diabetic foot ulcer-a review on pathophysiology, classification and microbial etiology. *Diabetes Metab. Syndr. Clin. Res. Rev.* **2015**, *9*, 192–199. [CrossRef]
3. Antibiotic Resistance—A Threat to Global Health Security and the Case for Action. Available online: https://www.gov.uk/government/publications/antibiotic-resistance-a-threat-to-global-health-security-and-the-case-for-action (accessed on 21 May 2019).
4. Mulder, G.D.; Cavorsi, J.P.; Lee, D.K. Polyhexamethylene biguanide (phmb): An addendum to current topical antimicrobials. *Wounds* **2007**, *19*, 173–182. [PubMed]
5. Landis, S.J. Chronic wound infection and antimicrobial use. *Adv. Skin Wound Care* **2008**, *21*, 541–542. [CrossRef]
6. Verma, J.; Kanoujia, J.; Parashar, P.; Tripathi, C.B.; Saraf, S.A. Wound healing applications of sericin/chitosan-capped silver nanoparticles incorporated hydrogel. *Drug Delivery Transl. Res.* **2017**, *7*, 77–88. [CrossRef]
7. Chhibber, S.; Gondil, V.S.; Singla, L.; Kumar, M.; Chhibber, T.; Sharma, G.; Sharma, R.K.; Wangoo, N.; Katare, O.P. Effective topical delivery of h-agnps for eradication of klebsiella pneumoniae-induced burn wound infection. *AAPS Pharmscitech* **2019**, *20*, 169. [CrossRef] [PubMed]
8. Ferreira, M.O.C.; Lima, I.S.D.; Morais, A.I.S.; Silva, S.O.; Carvalho, R.D.F.D.; Ribeiro, A.B.; Osajima, J.A.; Silva, E.C. Chitosan associated with chlorhexidine in gel form: Synthesis, characterization and healing wounds applications. *J. Drug Delivery Sci. Technol.* **2019**, *49*, 375–382. [CrossRef]
9. Pak, C.S.; Park, D.H.; Oh, T.S.; Lee, W.J.; Jun, Y.J.; Lee, K.A.; Oh, K.S.; Kwak, K.H.; Rhie, J.W. Comparison of the efficacy and safety of povidone-iodine foam dressing (betafoam), hydrocellular foam dressing (allevyn), and petrolatum gauze for split-thickness skin graft donor site dressing. *Int. Wound J.* **2019**, *16*, 379–386. [CrossRef] [PubMed]
10. Punjataewakupt, A.; Napavichayanun, S.; Aramwit, P. The downside of antimicrobial agents for wound healing. *Eur. J. Clin. Microbiol. Infect. Dis.* **2019**, *38*, 39–54. [CrossRef]
11. Chindera, K.; Mahato, M.; Sharma, A.K.; Horsley, H.; Kloc-Muniak, K.; Kamaruzzaman, N.; Kumar, S.; McFarlane, A.; Stach, J.; Bentin, T.; et al. The antimicrobial polymer phmb enters cells and selectively condenses bacterial chromosomes. *Sci. Rep.* **2016**, *6*, 23121. [CrossRef]

12. Scientific Committee on Consumer Safety (European Commission). Opinion on the Safety of Poly(Hexamethylene) Biguanide Hydrochloride (PHMB). Available online: http://ec.europa.eu/health/ /sites/health/files/scientific_committees/consumer_safety/docs/sccs_o_157.pdf (accessed on 21 May 2019).

13. Gilbert, P.; Pemberton, D.; Wilkinson, D. Synergism within polyhexamethylene biguanide biocide formulations. *J. Appl. Bacteriol.* **1990**, *69*, 593–598. [CrossRef]

14. Gilbert, P.; Das, J.; Jones, M.; Allison, D. Assessment of resistance towards biocides following the attachment of micro-organisms to, and growth on, surfaces. *J. Appl. Microbiol.* **2001**, *91*, 248–254. [CrossRef] [PubMed]

15. Mafra, C.; Carrijo-Carvalho, L.; Chudzinski-Tavassi, A.; Taguchi, F.; Foronda, A.; Carvalho, F.; Freitas, D.D. Antimicrobial action of biguanides on the viability of acanthamoeba cysts and assessment of cell toxicity. *Invest. Ophthalmol. Visual Sci.* **2013**, *54*, 6363. [CrossRef] [PubMed]

16. Forstner, C.; Leitgeb, J.; Schuster, R.; Dosch, V.; Kramer, A.; Cutting, K.; Leaper, D.; Assadian, O. Bacterial growth kinetics under a novel flexible methacrylate dressing serving as a drug delivery vehicle for antiseptics. *Int. J. Mol. Sci.* **2013**, *14*, 10582–10590. [CrossRef]

17. Kramer, A.; Roth, B.; Müller, G.; Rudolph, P.; Klöcker, N. Influence of the antiseptic agents polyhexanide and octenidine on fl cells and on healing of experimental superficial aseptic wounds in piglets. *Skin Pharmacol. Physiol.* **2004**, *17*, 141–146. [CrossRef]

18. Eberlein, T.; Haemmerle, G.; Signer, M.; Gruber-Moesenbacher, U.; Traber, T.; Mittlboeck, M.; Abel, M.; Strohal, R. Comparison of phmb-containing dressing and silver dressings in patients with critically colonised or locally infected wounds. *J. Wound Care* **2012**, *21*, 12–20. [CrossRef] [PubMed]

19. Napavichayanun, S.; Yamdech, R.; Aramwit, P. The safety and efficacy of bacterial nanocellulose wound dressing incorporating sericin and polyhexamethylene biguanide: In vitro, in vivo and clinical studies. *Arch. Dermatol. Res.* **2016**, *308*, 123–132. [CrossRef] [PubMed]

20. Siadat, S.; Mokhtari, J. Fabrication of novel antimicrobial bio-fibres using silk wastage, study of poly (hexamethylene) biguanide, and silver nanoparticles interaction. *J. Nat. Fibres* **2017**, *14*, 1–11. [CrossRef]

21. Song, W.; Mitchell, G.; Burugapalli, K. Electrospinning for biomedical applications. In *Electrospinning: Principles, Practics and Possibilities*; Royal Society of Chemistry: London, UK, 2015.

22. Llorens, E.; Calderón, S.; Valle, L.D.; Puiggalí, J. Polybiguanide (phmb) loaded in pla scaffolds displaying high hydrophobic, biocompatibility and antibacterial properties. *Mater. Sci. Eng. C* **2015**, *50*, 74–84. [CrossRef]

23. Waring, M.; Bielfldt, S.; Mätzold, K.; Wilhelm, K.; Butcher, M. An evaluation of the skin stripping of wound dressing adhesives. *J. Wound Care* **2011**, *20*, 412, 414, 416–422. [CrossRef]

24. Wang, N.; Burugapalli, K.; Song, W.; Zheng, Y.; Ma, Y.; Wu, Z.; Li, K. Electrospun co-axial polyurethane-gelatin nano-fibrous coating for implantable glucose biosensor. *Biofabrication* **2014**, *6*, 015002. [CrossRef] [PubMed]

25. Wang, N.; Burugapalli, K.; Song, W.; Hall, J.; Moussy, F.; Wu, Z.; Li, K. Tailored fibro-porous structure of electrospun polyurethane membranes, their size-dependent properties and trans-membrane glucose diffusion. *J. Membr. Sci.* **2013**, *247*, 207–217. [CrossRef]

26. Jiang, L.; Jiang, J.; Stiadle, J.; Wang, X.; Wang, L.; Li, Q.; Shen, C.; Thibeault, S.L.; Turng, L. Electrospun nanofibrous thermoplastic polyurethane/poly(glycerol sebacate) hybrid scaffolds for vocal fold tissue engineering applications. *Mater. Sci. Eng. C* **2019**, *94*, 740–749. [CrossRef] [PubMed]

27. Akduman, C.; Özgüney, I.; Kumbasar, E.P.A. Preparation and characterization of naproxen-loaded electrospun thermoplastic polyurethane nanofibers as a drug delivery system. *Mater. Sci. Eng. C* **2016**, *64*, 383–390. [CrossRef]

28. Aydogdu, M.O.; Oprea, A.E.; Trusca, R.; Surdu, A.V.; Ficai, A.; Holban, A.M.; Lordache, F.; Paduraru, A.V.; Filip, D.G.; Altun, E.; et al. Production and characterization of antimicrobial electrospun nanofibers containing polyurethane, zirconium oxide and zeolite. *Bionanoscience* **2018**, *8*, 154–165. [CrossRef]

29. Huang, C.C.; Rwei, S.P.; Jang, S.C.; Tsen, W.C.; Chuang, F.S.; Ku, T.H.; Chow, J.D.; Chen, C.C.; Shu, Y.C. Antibacterial of silver-containing polydimethylsiloxane urethane nanofibrous, hollow fibrous, using the electrospinning process. *J. Nanosci. Nanotechnol.* **2017**, *17*, 1975–1982. [CrossRef]

30. Jaganathan, S.K.; Mani, M.P. Electrospun polyurethane nanofibrous composite impregnated with metallic copper for wound-healing application. *3 Biotech* **2018**, *8*, 327. [CrossRef]

31. Shababdoust, A.; Ehsani, M.; Shokrollahi, P.; Zandi, M. Fabrication of curcumin-loaded electrospun nanofiberous polyurethanes with anti-bacterial activity. *Prog. Biomater.* **2018**, *7*, 23–33. [CrossRef] [PubMed]

32. Pham, Q.; Sharma, U.; Mikos, A. Electrospinning of polymeric nanofibres for tissue engineering applications: A review. *Tissue Eng.* **2006**, *12*, 1197–1211. [CrossRef] [PubMed]

33. De Paula, G.F.; Netto, G.I.; Mattoso, L.H.C. Physical and chemical characterization of poly(hexamethylene biguanide) hydrochloride. *Polymers* **2011**, *3*, 928–941. [CrossRef]

34. Küstersa, M.; Beyera, S.; Kutschera, S.; Schlesingera, H.; Gerhartz, M. Rapid, simple and stability-indicating determination of polyhexamethylene biguanide in liquid and gel-like dosage forms by liquid chromatography with diode-array detection. *J. Pharm. Anal.* **2013**, *3*, 408–414. [CrossRef]

35. Ashraf, S.; Akhtar, N.; Ghauri, M.A.; Rajoka, M.I.; Khalid, Z.M.; Hussain, I. Polyhexamethylene biguanide functionalized cationic silver nanoparticles for enhanced antimicrobial activity. *Nanoscale Res. Lett.* **2012**, *7*, 267. [CrossRef] [PubMed]

36. Sood, A.; Granick, M.S.; Tomaselli, N.L. Wound dressings and comparative effectiveness data. *Adv. Wound Care (New Rochelle)* **2014**, *3*, 511–529. [CrossRef]

37. Agache, P.; Monneur, C.; Leveque, J.; Rigal, J.D. Mechanical properties and young's modulus of human skin in vivo. *Arch. Dermatol. Res.* **1980**, *269*, 221–232. [CrossRef] [PubMed]

Article

Preparation, Physicochemical Properties, and Hemocompatibility of the Composites Based on Biodegradable Poly(Ether-Ester-Urethane) and Phosphorylcholine-Containing Copolymer

Jun Zhang [1], Bing Yang [2], Qi Jia [3], Minghui Xiao [1] and Zhaosheng Hou [1,*]

[1] College of Chemistry, Chemical Engineering and Materials Science, Shandong Normal University, Jinan 250014, China; zj971127@163.com (J.Z.); xiaominghui98@163.com (M.X.)
[2] Key Laboratory of Public Security Management Technology in Universities of Shandong, Shandong Management University, Jinan 250357, China; yb197325@163.com
[3] Qilu Pharmaceutical Co. Ltd., Jinan 250104, China; haperwork@163.com
* Correspondence: houzs@sdnu.edu.cn; Tel.: +86-13791076340

Received: 4 April 2019; Accepted: 9 May 2019; Published: 11 May 2019

Abstract: To improve the hemocompatibility of the biodegradable medical poly(ether-ester-urethane) (PEEU), containing uniform-size aliphatic hard segments that was prepared in our lab, a copolymer containing phosphorylcholine (PC) groups was blended with the PEEU. The PC-copolymer of poly(MPC-co-EHMA) (PMEH) was first obtained by copolymerization of 2-methacryloyloxyethyl phosphorylcholine (MPC) and 2-ethylhexyl methacrylate (EHMA), and then dissolved in mixed solvent of ethanol/chloroform to obtain a homogeneous solution. The composite films (PMPU) with varying PMEH content were prepared by solvent evaporation method. The physicochemical properties of the composite films with varying PMEH content were researched. The PMPU films exhibited higher thermal stability than that of the pure PEEU film. With the PMEH content increasing from 5 to 20 wt%, the PMPU films also possessed satisfied tensile properties with ultimate stress of 22.9–15.8 MPa and strain at break of 925–820%. The surface and bulk hydrophilicity of the films were improved after incorporation of PMEH. In vitro degradation studies indicated that the degradation rate increased with PMEH content, and it took 12–24 days for composite films to become fragments. The protein adsorption and platelet-rich plasma contact tests were adapted to evaluate the surface hemocompatibility of the composite films. It was found that the amount of adsorbed protein and adherent platelet on the surface decreased significantly, and almost no activated platelets were observed when PMEH content was above 5 wt%, which manifested good surface hemocompatibility. Due to the biodegradability, acceptable tensile properties and good surface hemocompatibility, the composites can be expected to be applied in blood-contacting implant materials.

Keywords: poly(ether-ester-urethane); MPC copolymers; composites; physicochemical properties; hemocompatibility

---

1. Introduction

Polyurethanes (PUs) are a kind of polymer with carbamate groups (–NHCOO–) on the backbones. Compared with other biomedical materials, PUs possess a multitude of advantages, including high tenacity, chemical resistance, and adjustable mechanical flexibility [1–3]. Due to their excellent physical-mechanical properties and adequate biocompatibility, PUs have been widely used in the medical field for almost half a century as heart valves, pacemaker wires, vascular grafts, cardioids, artificial skin joints, and catheters [4–7].

Biodegradable PUs are designed to undergo hydrolytic degradation to produce noncytotoxic products [8]. Most PUs based on aliphatic diisocyanate are prepared for biomedical application because they have lower toxicity compared to aromatic ones [9,10]. However, because of the absence of hard segments of significant length, these kinds of PUs exhibit dissatisfactory tensile properties like low tensile strength [11]. Penning and coworkers found that PUs containing long uniform-size hard segments possess excellent tensile properties [12]. In our previous reports [13,14], several kinds of biodegradable medical PUs based on aliphatic diurethane diisocyanate were prepared. The uniform chemical structure of hard segments improves the microphase separation degree of soft and hard segments, and, at the same time, the denser hydrogen bonds among carbamate units give a more compact physical-linking network structure, leading to comparative or even better tensile properties than aromatic PUs.

Biocompatibility, especially hemocompatibility, is another important requirement for the clinical implant or/and blood-contacting materials [15–18]. When PUs are used as long-term blood-contacting materials, proteins can accumulate rapidly on the material surface, subsequently platelets are activated, and then blood coagulation and thrombus occur [19,20]. Therefore, many strategies have been developed in order to achieve improved hemocompatibility of PUs. The most widely used method is chemical surface modification [21–23], which mainly includes introduction a high activation barrier to repel proteins by grafting hydrophilic polymers or biomimicking materials on the surface. However, the surface modification approaches inevitably deteriorate the bulk properties, especially the tensile properties of the substrates, which is difficult to overcome.

Recently, much attention has been concentrated on synthetic/natural polymer composites because it is simple and efficient to obtain new materials from mixing two different polymer materials. In the previous articles [24–26], the composites of PU and bioactive polymers were prepared and their fundamental properties were evaluated. The hemocompatibility of the composites, with attention to protein adsorption and platelet adhesion, was much better than that of original PU material. In addition, the tensile properties had no obvious change after addition a small quantity of bioactive polymer to PU. It provides a novel technique for exploitation of PU in biomedical application.

Phosphorylcholine (PC) is a hydrophilic zwitterionic head group of cell membrane phosphatidylcholine that endows the cell membrane with ideal biocompatibility, especially in resistance to protein adsorption and platelet adhesion [27,28]. In order to exploit new biocompatible materials, the methacrylate monomer-bearing PC group 2-methacryloyloxyethyl phosphorylcholine (MPC) was synthesized by Nakabayashi [29], and later Chapman [30]. Research on surface modification of biomaterials with PC functionality has been developed rapidly after publication of the MPC [31,32]. Because of the polymerizable methacrylate moiety, MPC can easily be copolymerized with other monomers to enable the design of numerous materials with various molecular architectures. The modification of biomedical PU with the MPC copolymer by blending to improve the biocompatibility has been studied by many researchers for nearly thirty years [33–35]. For example, Ishihara groups [36,37] prepared the composite films of MPC polymer and commercial PU (Tecoflex® and Pellethane® 2363-90), and the film materials exhibited excellent nonthrombogenicity in contact with human whole blood. The PC groups could be concentrated effectively near the surface in the plasma resulting in the formation of a 'self-assembled biomimetic membrane bilayers', which can form a thermodynamic hydration barrier over the surface and suppress any unfavorable interaction with blood cells and proteins. Therefore, it can be hypothesized that incorporation of MPC polymers will improve the biocompatibility (especially hemocompatibility) of new kinds of medical PU with uniform-size hard segments prepared in our lab.

In this paper, the composites of the PU–PC copolymer were prepared by simple physical blending to improve the hemocompatibility of PU materials. The biodegradable poly(ether-ester-urethane) (PEEU), which contains uniform-size hard segments, was obtained in our lab and used as model PU. Based on the fundamental properties of PC-containing polymers, the MPC copolymer was used as a polymeric additive, which could blend and interact with PEEU. The effect of the MPC copolymer introduced in the PEEU films on the physicochemical properties of PEEU was investigated. Furthermore, the surface hemocompatibility of the composite films was evaluated by protein adsorption and platelet adhesion tests.

2. Materials and Methods

2.1. Materials

MPC (>98%) were purchased from J&K Scientific Ltd. (Beijing, China) and used without further purification. 2-Ethylhexyl methacrylate (EHMA, Shanghai Macklin Biochemical Co., Ltd., Shanghai, China) was dried with anhydrous magnesium sulfate and distilled under reduced pressure, then saved at -20 °C before use. 2,2′-Azobisisobutyronitrile (AIBN) was obtained from Shanghai Macklin Biochemical Co., Ltd. (Shanghai, China) and recrystallized several times from ethanol. Poly (ethylene glycol) (PEG, M_n = 600, Aladdin Reagent Co., Ltd., China) was dehydrated at 110 °C for ~4 h under vacuum. L-lactide (L-LA, J&K Scientific Co., Ltd.) was recrystallized several times times from dry ethyl acetate. ε-Caprolactone (ε-CL, Sigma-Aldrich, city, country) was distilled from CaH_2 under reduced pressure. *N,N*-dimethylformamide (DMF, Beijing Chemical Reagent Co., Ltd., Beijing, China) was refluxed with phosphorus pentoxide for about 5 h and then distilled under reduced pressure. Diurethane diisocyanate (hexanediisocyanate-1,4-butanediol-hexanediisocyanate, HBH) was synthesized in our lab according to our previous paper [13]; NMR and HRMS analyses were used to confirme its chemical structure. Phosphate buffer saline (PBS, pH = 7.4) was supplied by Beijing Chemical Reagent Co., Ltd. and used as received. Other regents were AR grade and purified by standard methods.

2.2. Preparation of PEEU

The PEEU was prepared referring to our published literature according to Figure S1 [13]. In brief, ε-CL (0.15 mol), L-LA (0.19 mol), and PEG600 (0.05 mol) were mixed in a vacuum flask under dried argon atmosphere. After three rounds of deoxygenation, catalyst stannous octoate (0.1 wt% of monomers) was added and the reaction was carried out at 140 °C for 36 h under vacuum to obtain the prepolymer (Figure S1a). Then, the DMF solution of HBH (25 wt%) was added dropwise into the prepolymer at 80 °C under dried argon atmosphere (molar ratio of –NCO/–OH was controlled at 1.05). After that, the reaction was allowed to proceed at the same temperature for about 3.5 h until the NCO peak (~2270 cm^{-1}) in the FT-IR spectrum disappeared completely. Subsequently, the solution was diluted to about 5 wt% and precipitated in cold diethyl ether. The product was dried to a constant mass at 40 °C under reduced pressure to obtain the white filiform PEEU.

The chemical structure of PEEU was characterized by ^{1}H NMR (Figure S2), FT-IR (Figure S3) and GPC. ^{1}H NMR (400 MHz, CDCl$_3$, ppm): δ 5.10–5.31 (CH$_3$–C$\underline{H}$–), 4.80 (–N$\underline{H}$CO–), 4.06–4.23 (–COOC$\underline{H}_2$–), 3.65–3.70 (–C$\underline{H}_2$–CH$_2$O– of PEG), 3.15 (–NHC$\underline{H}_2$–), 2.29–2.41 (–C$\underline{H}_2$CO–), 1.67 (CH$_3$–), 1.46–1.52 (–OCH$_2$ (C$\underline{H}_2$)$_3$CH$_2$CO, –OCH$_2$(C$\underline{H}_2$)$_2$CH$_2$O–), 1.34 (NHCH$_2$(C$\underline{H}_2$)$_4$CH$_2$NH). FT-IR (ATR, cm^{-1}): 3321 (N–H), 2937, 2864 (–CH$_2$–), 1731 (C=O), 1687 (amide I), 1533 (amide II), 1091 (C–O–C of ester). GPC (THF): M_w-GPC = 141,200, M_n = 101,300, M_w/M_n = 1.39.

2.3. Preparation of Poly(MPC-co-EHMA) (PMEH)

Monomers of EHMA (0.1 mol, 13.8 g) and MPC (0.025 mol, 7.4 g) were placed in a vacuum flask and the mixture was dissolved with ethanol (10 mL). After dried argon was bubbled into the solution to remove oxygen, AIBN (1 wt% of monomers) was added and the vacuum flask was sealed. The polymerization was carried out under vacuum at 60 °C for 12 h. The reaction mixture was cooled to room temperature, and then precipitated with a large amount of hexane. The precipitate was filtered off and dried under reduced pressure at room temperature. The copolymer was purified by Soxhlet extraction, first with water and then with ether, to remove the unreacted monomers. The reaction scheme is displayed in Figure S4.

The chemical structure of PMEH was characterized by ^{1}H NMR (Figure S5), FT-IR (Figure S6), GPC and elemental analysis. ^{1}H NMR (400 MHz, CDCl$_3$, ppm): δ 4.52 (–COC$\underline{H}_2$C$\underline{H}_2$P–, –N$^+$CH$_2$C$\underline{H}_2$OP, –COC$\underline{H}_2$C$\underline{H}$–), 3.65–3.83 (–N$^+$C$\underline{H}_2$–, –N$^+$C$\underline{H}_3$), 3.28 (–C$\underline{H}_2$–C–), 1.83 (–CH$_2$C$\underline{H}$CH$_2$–), 1.28 (–CHC$\underline{H}_2$CH$_3$,

–CH(C$\underline{H}_2$)$_3$CH$_3$), 0.84 (C$\underline{H}_3$CH$_2$–, C$\underline{H}_3$C–). FT-IR (ATR, cm^{-1}): 3380 (–OH of H$_2$O), 2957, 2928, 2863 (–CH$_2$–), 1724 (C=O), 1238 (P=O), 1090 (C–O–C of ester), 1067 (P–O), 953 (–N$^+$(CH$_3$)$_3$). GPC (THF): M_w = 27,000, M_n = 23,100, M_w/M_n = 1.17. Anal. Calcd (%): C, 65.09; H, 10.21; O, 20.58; N, 1.29. Found: C, 65.12; H, 10.25; O, 20.52; N, 1.21.

2.4. Preparation of PEEU/PMEH Composite Films

The composite films were prepared by a solvent evaporation technique solvent evaporation technique according to Figure S7. Briefly, four weight percent solutions of both PEEU and PMEH solutions were prepared separately using ethanol/chloroform mixture (1/1 by volume) as solvent. The homogeneous solution containing the predetermined amounts of PEEU and PMEH (4.0 g/100 mL) was poured on to a Teflon mold, and the solvent was evaporated at room temperature for 72 h. Subsequently, the formed film was dried under reduced pressure for 24 h to eliminate the last traces of solvent. The PEEU/PMEH (PMPU) composite films were obtained with 0.25 ± 0.02 mm thickness. The chemical composition of PMPU films is listed in Table 1, and the PMPUs are described as PMPU-X (X: the weight percent of PMEH in the composites).

Table 1. The chemical composition of PEEU/PMEH (PMPU) films.

Films / Components	PMPU-0	PMPU-5	PMPU-10	PMPU-20
PEEU/g	4.0	3.8	3.6	3.2
PMEH/g	0	0.2	0.4	0.8
PMEH content/wt%	0	5	10	20

The chemical structure of PMPUs was characterized by FT-IR and the representative spectrum is shown in Figure S8. FT-IR (ATR, cm^{-1}): 3324 (N–H), 2932, 2869 ((–CH$_2$–), 1730 (C=O), 1683 (amide I), 1535 (amide II), 1240 (P=O), 1086 (C–O–C of ester), 1059 (P–O), 957 (–N$^+$(CH$_3$)$_3$).

2.5. Characterization and Instruments

Characterization: ^{1}H NMR spectra were recorded on a 400 MHz Avance II spectrometer (Bruker, Rheinstetten, Germany) using CDCl$_3$ as solvent. FT-IR spectra were recorded between 4000 and 400 cm^{-1} on a Bruker Alpha infrared spectrometer (Bruker, Rheinstetten, Germany) equipped with a Bruker platinum ATR accessory. The weight average molecular weight (M_w), number average molecular weight (M_n) and polydispersity index (M_w/M_n) of polymers were measured by gel permeation chromatography (GPC) on a Water Alillance GPC 2000 system (Waters, Milford, MA, USA) with tetrahydrofuran as the eluting solvent. Elemental analysis (C, H, N, and O) was performed on Elementar Vario E1 III analyzer (Elementar, Langenselbold, German).

Thermogravimetric analysis (TGA): TGA tests were conducted using a TGA 2050 analyzer (Universal, New Brunswick, NJ, USA). The mass loss of the dried samples was performed under nitrogen inert atmosphere (N$_2$, 40 mL/ min) from 50 to 600 °C at a heating rate of 15 °C/min.

Tensile properties: According to the national standard GB/T1040-2006, tensile stress–strain tests were measured using a single-column tensile test machine (Model HY939C, producer, Dongguan, China) at room temperature with a cross-head speed of 50 mm/min. The films were cut in a dumbbell shape with neck width of 4.0 mm and length of 30 mm, respectively. For each data, the result was the average value of five parallel measurements.

Water absorption: The amount of absorbed water by the film is the parameter to evaluate the water-swelling property of the films. The preweighed dry film disks (m_o) with ~10 mm diameter were immersed in deionized water at 37 ± 0.1 °C and equilibrated for about 48 h. The swollen samples were blotted with laboratory tissue to remove the surplus absorbed water, and weighed immediately (m_t). The water-swelling property of the film was expressed as the water absorption calculated from the following formula. Water absorption (%) = (m_t − m_o)/m_o × 100. The results reported were the average values for at least five replicated samples.

Water contact angle: The surface hydrophilicity was assessed by measuring the water contact angles formed between the water drops and the film surface. The sessile static water contact angles against the film surface were measured by a contact angle setup of KSV CAM 200 (KSV Instruments, Helsinki, Finland) using a sessile drop method at room temperature. Before measurement, the film specimens were immerged in distilled water for 2 h and then dried under vacuum. To ensure that the droplets did not penetrate the compact material, the test was carried out within 5 s. The test was performed on six samples, and three drops were applied on each sample.

In vitro degradation: Film degradation was quantified by the weight loss in PBS (pH = 7.4). The film in a disk shape (diameter: 10 mm) was placed into a sealed bottle which containing 10 mL PBS solution, and incubated at the temperature of 37 ± 0.1 °C. At given time intervals, the sample was removed from the solution, washed with distilled water and dried in a vacuum oven at 25 °C until constant weight. The weight loss was calculated using the following equation to evaluate the films degradation. Weight loss (%) = $(W_o - W_r)/W_o \times 100$, where W_r is the rest weight of the sample after degradation for a predetermined time and W_o is the weight of the dry sample. The tests were carried out until the films lost tensile properties and became fragments. Each test was repeated at least three times and the results were the average.

Surface morphologies: The samples after in vitro degradation for a fixed time period were collected to observe the surface morphologies. The dried samples were coated with gold for the morphological observation by using a SU8010 FE-SEM (Hitachi, Tokyo, Japan).

Protein adsorption: The Bradford protein determining method was used to determine the amount of adsorbed protein onto the film surface when using bovine serum albumin (BSA, Shanghai Aladdin Reagent Co., Ltd., Shanghai, China) [38,39] and human plasma fibrinogen (HPF, Shanghai Macklin Biochemical Co., Ltd., Shanghai, China) [40,41] as model proteins. The film discs (~10 mm diameter) were equilibrated with PBS (pH = 7.4) for ~12 h to achieve complete hydration, and then immersed in a solution of 1.0 mL protein solution (BSA: 45 µg/mL; HPF: 30 µg/mL) for 3 h at the temperature of 37 ± 0.5 °C. The discs were gently taken off and rinsed sufficiency with PBS to remove the unbound BSA. After sonication in sodium dodecylsulfonate aqueous solution (1 wt%) for 30 min to detach the adsorbed protein on the surface, a micro-Bradford protein analysis kit (Sangon Biotech Co., Ltd., Shanghai, China) with a multiwall microplate reader (Multiskan Mk3-Thermolabsystems, Thermo Fisher Scientific, Inc., USA) was used to determine the concentration of the adsorbed BSA and HPF in the solutions at 595 nm and 562 nm, respectively. The amount of proteins adsorbed on the surface could be calculated from the protein concentration in the solution. At least three replicate samples were tested to ensure reproducibility of the measurements, and values relative to the controls (PBS) were collected.

Platelet adhesion: To evaluate the interactions between blood and films, the platelet adhesion tests were performed in this study. Platelet-rich plasma (PRP) was obtained from fresh rabbit blood (Shandong Success Biotechnology Co., Ltd., Jinan, China) by centrifugation of blood in a sodium citrate buffer at 2000 rpm for 20 min at 4 °C. The disk-shaped samples (diameter: 10 mm) were contacted with PBS (pH = 7.4) for 2 h to equilibrate the surface, and then removed from the solution and incubated with 1.0 mL PRP at 37 °C for 1 h. The samples were rinsed thoroughly with fresh PBS to remove nonadherent platelets. The platelets adhering to the surface were fixed with 2.5% glutaraldehyde for 30 min at 40 °C. Then, the discs were dehydrated by treating with gradual ethanol/water solutions (60, 70, 80, 90, 100% (*v/v*)) for 30 min in each step and allowed to dry at room temperature on a clean bench. Finally, the platelet-attached surfaces were coated with gold prior to observation by FE-SEM on different fields of surfaces.

3. Results and Discussion

3.1. Thermal Stability

TGA analysis is often used to evaluate the thermal stability of materials; Figure 1 shows the TGA and differential thermal gravimetric analysis (DTGA) curves of the composite films with varying PMEH content. There were two clear consecutive weight losses observed in curve of pure PEEU (PMPU-0). The first weight loss occurred at ~247–353 °C with the maximum decomposition temperature (T_{max}) of 320 °C and the weight loss was 68 wt%, which was attributed to the decomposition of carbamate and ester bonds. Another weight loss occurred at a higher temperature of 354–438 °C; T_{max} of 385 °C was assigned to the decomposition of ether bonds. The remaining weight was lower than 2 wt%, indicating that the PMPU-0 decomposed almost completely. While the PMEH exhibited obvious three-step weight loss, which was due to the complex structure. The initial decomposition temperature (~200 °C) was lower that of pure PEEU, but the PMEH had higher thermal stability than pure PEEU at high-temperature region (above 400 °C). The residue weight was more than 20 wt%, which should be ascribed to the nitrogen and phosphonium salt formed after the decomposition of PMEH segments. Compared with PEEU, the thermal stability of composite films (PMPU-5~PMPU-20) increased obviously and the T_{max} was ~20 °C higher than that of pure PEEU, which could be due to the interaction (maybe H bonds) among the chains of PMEH and PEEU. Obviously, the increasing residue weight should be attributed to the increase of PC content in composite films. No other weight loss steps found in the curves of PMPU films indicated that the PMEH component was compatible with PEEU component.

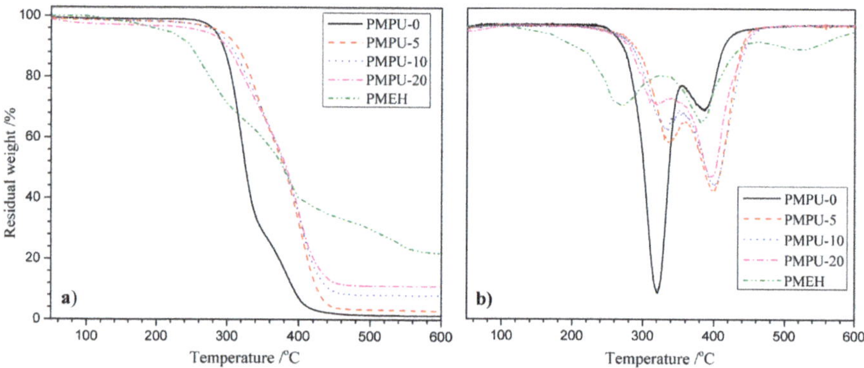

Figure 1. (**a**) Thermogravimetric analysis (TGA) and (**b**) differential gravimetric thermal analysis (DTGA) curves of composite films with varying PMEH content.

3.2. Tensile Properties

Tensile property was an important quality for long-term implant biomaterials. The typical stress–strain curves of the composite films with varying PMEH content are shown in Figure 2, and the corresponding characteristic values obtained from the curves are listed in Table 2. From the curves, it could be found that the composite films with PMEH content increasing from 0 to 20 wt% (PMPU-0~PMPU-20) behaved as soft elastic materials, displaying a smooth transition from the elastic to plastic deformation regions [42]. The pure PEEU (PMPU-0) containing uniform-size hard segments exhibited good tensile properties with ultimate stress of 20.8 MPa, strain at break of 930% and initial modulus of 19.5 MPa, which was due to the compact physical-linking network structure formed by denser H bonds existing not only among carbamate groups but between carbamate and ether/ester groups [43]. When the PMEH was blended in PEEU, the strain at break decreased gradually. However, with the PMEH content increasing from 5 to 20 wt%, the ultimate stress of the composites films first expressed no obvious change and then decreased gradually. When the PMEH content in composites is as low as 5 wt% (PMPU-5), the PMEH can be homo-dispersed in PEEU and acts as a filler; additionally, the PMEH content maybe too low to be reflected, the two reasons result in almost unchanged strain at break and ultimate stress. When the PMEH content is higher than 5 wt% (PMPU-10 and PMPU-20), the residual brittle PMEH can be aggregated as a stress concentration point and destroy the tensile properties, thus the strain at break, ultimate stress and initial modulus of the composite films decrease [44]. However, the composite film with high PMEH content up to 20 wt% (PMPU-20) exhibited good tensile properties with strain at break of 820% and ultimate stress of 15.8 MPa, which could also meet the clinical requirements of long-term implant medical biomaterials, such as cartilage.

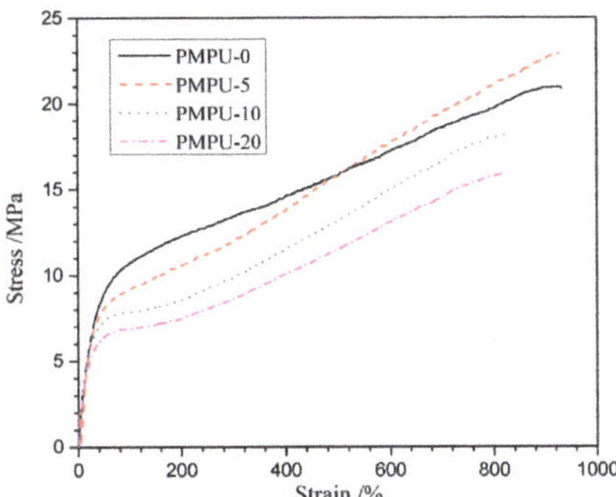

Figure 2. Stress–strain behaviors of PMPU films with varying PMEH content.

Table 2. Tensile properties of PMPU films with varying PMEH content.

Films	Strain at Break (%)	Ultimate Stress (MPa)	Yield Stress (MPa)	Yield Strain (%)	Initial Modulus (MPa)
PMPU-0	932 ± 41	20.8 ± 2.4	9.97 ± 1.2	51.1 ± 4.2	19.5
PMPU-5	925 ± 38	22.9 ± 2.3	8.15 ± 1.05	43.3 ± 3.3	18.8
PMPU-10	825 ± 32	18.2 ± 1.9	7.26 ± 0.92	41.3 ± 3.4	17.5
PMPU-20	820 ± 34	15.8 ± 1.4	6.20 ± 0.76	39.8 ± 3.0	15.6

3.3. Surface and Bulk Hydrophilicity

The surface and bulk hydrophilicity of the composite films with varying PMEH content were characterized by measuring the surface water contact angle and water absorption, and the results are displayed in Figure 3. The pure PEEU film exhibited a hydrophobic surface with the water contact angle of 87.4°, and the water contact angle showed a trend to decline dramatically after PMEH was introduced. With the PMEH content varying from 5 to 20 wt% (PMPU-5~PMPU-20) in composite films, the water contact angle decreased dramatically from 64.5° to 29.8°, presenting an increasingly hydrophilic surface. After water immersion of the composite films, the hydrophilic PC units can rearrange to the water-side and the concentration of the PC units on the surface increases. The zwitterionic PC groups interact with water molecules by hydrogen bonding, which forms a hydration layer on the film surface and improves the surface hydrophilicity [45]. The equilibrium water absorption of the composite films, which was reached after immersion in water for 48 h, increased gradually from 8.4 to 31.5 wt% with the PMEH content varying from 0 to 20 wt% (Figure 3). It is also ascribed to the hydrophilicity of PC units which can bind much water into film. The results manifested that the surface and bulk hydrophilicity, an important role in the hydrolytic degradation and surface hemocompatibility, could be affected by the content of hydrophilic PC units in composites.

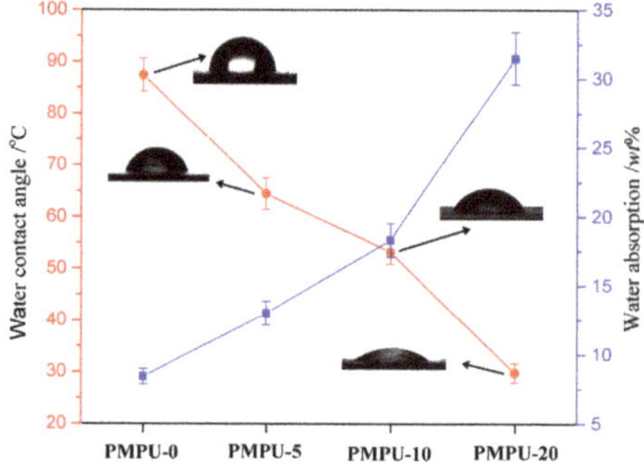

Figure 3. Water contact angle and water absorption of composite films with varying PMEH content.

3.4. In Vitro Degradation

In vitro hydrolytic degradation was examined by the weight loss in PBS solution at 37 °C, and the degradation behaviors of composite films are shown in Figure 4. The degradation process included three stages. In the first several days, all films exhibited a similar degree of degradation with slight weight loss less than 10 wt%, which should be attributed to the hydration and swelling of the films in the incipient stage. After that, the weight loss rate increased sharply, mainly because the films were cleaved to water-soluble molecules that were soluble in the media. Finally, the films become fragments and lost the tensile properties. In addition, the degradation rate of composite films increased with the increase of PMEH content, and the time taken for PMPU-0, -5, -10, and -20 films to become fragments was ~24, 18, 14, and 12 days, respectively. As the description in test of water absorption, high hydrophilic PC units can bind more water molecules, which make the ester groups in PEEU chains easily expose to water molecules. Thus, the chain scission occurs easily through hydrolysis of ester bonds, which leads to an increased degradation rate. From the results, it can be found that the degradation rate of composite films can be adjusted by changing the PMEH content.

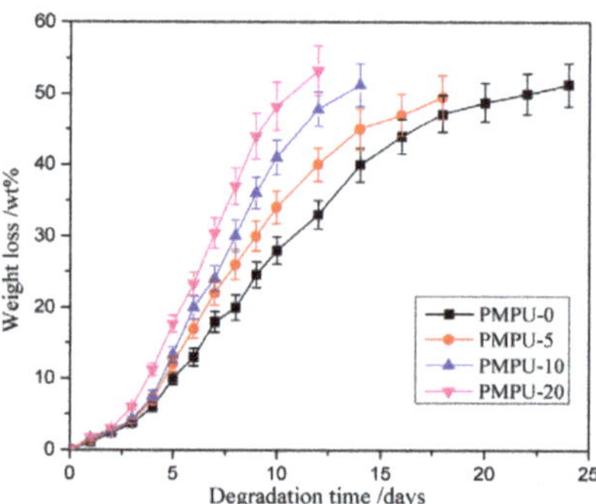

Figure 4. Weight loss curves of composite films with varying PMEH content in PBS (pH = 7.4) at 37 ± 0.1 °C.

The degradation process can be directly reflected by morphological changes in film surface. The typical surface morphologies of PMPU-10 after different degradation periods were presented in Figure 5. The nondegraded film (Figure 5a) was semitransparent with a relatively smooth surface. A rough surface was observed when the film was degraded for 3 days (Figure 5b), and more and more irregular cavities appeared on the film surface with the further degradation (Figure 5c,d). A large number of holes were observed after 10 and 12 days of degradation (Figure 5e,f), which indicates the film gradually losing its tensile properties.

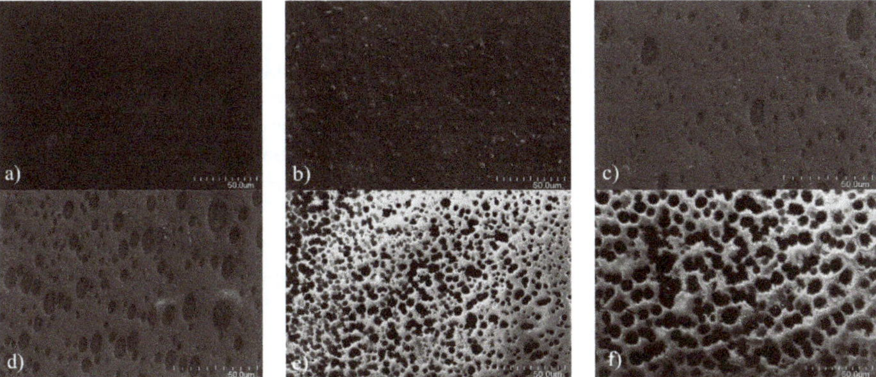

Figure 5. Surface morphologies of PMPU-10 film in PBS (pH = 7.4) at 37 ± 0.1 °C after (**a**) 0, (**b**) 3, (**c**) 5, (**d**) 7, (**e**) 10, and (**f**) 12 days of degradation.

3.5. Protein Adsorption

The protein adsorption on biomaterials' surfaces has been considered the first step to test many undesired biological reactions [46], and the formation of thrombus at the interface is dependent on the number and state of the protein adsorption layer [47]. The adsorption behaviors of protein (BSA and HPF) on the composite film surface are exhibited in Figure 6. As shown in Figure 6, the amount of adsorbed protein on pure PEEU (PMPU-0) film surface was 1.99 µg/cm^2 for BSA and 2.52 µg/cm^2 for HPF, both decreased significantly after the introduction of PMEH into the films. With the PMEH content increasing from 5 to 20 wt%, the absorbed amount of BSA and HPF on the composite film surface decreased gradually from 1.65 to 0.45 µg/cm^2 and from 1.92 to 0.53 µg/cm^2, respectively. It may be ascribed to the presence of a hydrated layer around the zwitterionic PC groups on the film surface which reduces molecular interactions with protein [48]. In addition, the proteins are very difficult to replace the water molecule because of the strong interaction between water molecules and PC groups [27]. The lower protein adsorption capacity of composite films means better surface hemocompatibility.

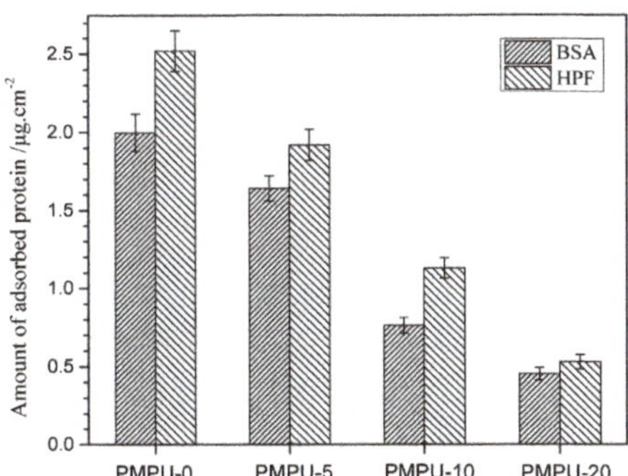

Figure 6. The amount of adsorbed protein on the surface of composite films with varying PMEH content at 37 ± 0.5 °C.

3.6. Platelet Adhesion

The platelet adhesion on materials surface is an effective and frequently used measurement to evaluate the hemocompatibility of blood-contacting biomaterials [49]. The platelet adhesion on the composite films after contacting with rabbit PRP was assessed by SEM observation, and the representative micrographs are given in Figure 7. Massive platelets adhered on the surface of pure PEEU film (PMPU-0, Figure 7a). Some of the platelets aggregated to some extent and some exhibited deformation and formed pseudopods, presenting the highly activated state. The platelet adhesion on the surface was effectively suppressed after PMEH was introduced to the PEEU, and moreover, deformation of adherent platelets was also hindered on the composite film surface. When the composition of PMEH in the composites was 20 wt% (PMPU-20, Figure 7d), very few adherent platelets were observed, which proved an excellent antiplatelet adhesion surface. It has been reported that PC groups on the surface can migrate and reorientate onto the interface between the film surface and water, which forms a mimetic structure of cell outer membrane and possesses outstanding antiplatelet adhesion property [50,51]. Although the results of platelet adhesion test demonstrated that the films surface hemocompatibility was improved dramatically by adding a small amount of the additional of PMEH, the surface hemocompatibility need further evaluations, such as using human whole blood and subcutaneous implant tests.

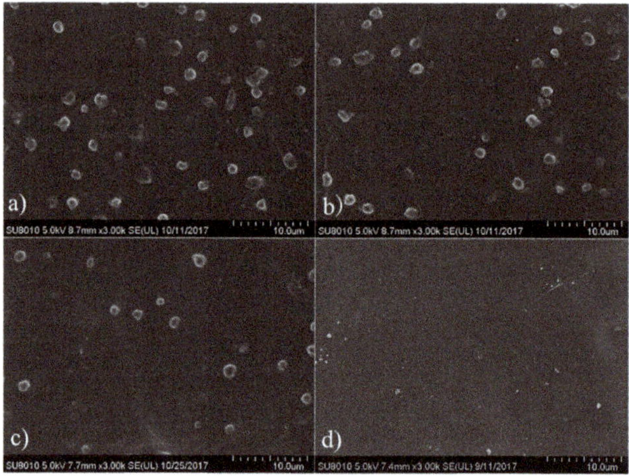

Figure 7. Representative SEM images of platelet adhesion on the surface of (**a**) PMPU-0, (**b**) PMPU-5, (**c**) PMPU-10, and (**d**) PMPU-20 films.

4. Conclusions

In this paper, the novel composites of the biodegradable PEEU and blood-compatible PC copolymer (PMEH) were prepared by blending from a homogeneous solution, and the corresponding films were obtained by a solvent evaporation method. The physicochemical properties of the composite films with varying PMEH content were studied. The composite films (PMPU) exhibited higher thermal stability than that of original PEEU film. With PMEH content increasing from 5 to 20 wt%, the PMPU films also possessed satisfied tensile properties with ultimate stress of 22.9–15.8 MPa and strain at break of 925–820%. The surface and bulk hydrophilicity of PMPU films were closely related to the content of hydrophilic PC groups. In vitro degradation studies indicated that the degradation rate increased with the increment of PMEH content, and the time of the composite films becoming fragments was 12–24 days. The reduction of protein adsorption and platelet adhesion on the composite films surface manifested good surface hemocompatibility. Due to the biodegradability, satisfied tensile properties and good surface hemocompatibility, the composites can be expected to be applied in blood-contacting implant materials.

Supplementary Materials: The supplementary materials are available online at http://www.mdpi.com/2073-4360/11/5/860/s1.

Author Contributions: Z.H. and J.Z. conceived and designed the experiments; B.Y., Q.J., and M.X. performed the experiments; Z.H. and J.Z. analyzed the data and wrote the paper.

Funding: This research was funded by the Shandong Provincial Natural Science Foundation, China (Project No. ZR2018MEM024) and the Scientific Research Fund Project of Undergraduates, Shandong Normal University, China (Project No. 2018BKSKYJJ53).

Conflicts of Interest: The authors declare no conflicts of interest.

References

1. Heath, D.E.; Cooper, S.L. *Polyurethanes*; Elsevier: New York, NY, USA, 2013.
2. Han, J.; Chen, B.; Ye, L.; Zhang, A.; Zhang, J.; Feng, Z. Synthesis and characterization of biodegradable polyurethane based on poly(ε-caprolactone) and L-lysine ethyl ester diisocyanate. *Front. Mater. Sci.* **2009**, *3*, 25–32. [CrossRef]
3. Xu, W.; Xiao, M.; Yuan, L.; Zhang, J.; Hou, Z. Preparation, physicochemical properties and hemo-compatibility of biodegradable chitooligosaccharide-based polyurethane. *Polymers* **2018**, *10*, 580. [CrossRef]
4. He, W.; Hu, Z.; Xu, A.; Liu, R.; Yin, H.; Wang, J.; Wang, S. The preparation and performance of a new polyurethane vascular prosthesis. *Cell. Biochem. Biophys.* **2013**, *66*, 855–866. [CrossRef]
5. Shen, Z.; Kang, C.; Chen, J.; Ye, D.; Qiu, S.; Guo, S.; Zhu, Y. Surface modification of polyurethane towards promoting the ex vivo cytocompatibility and in vivo biocompatibility for hypopharyngeal tissue engineering. *J. Biomater. Appl.* **2012**, *28*, 607–616. [CrossRef]
6. Thomas, V.; Kumari, T.V.; Jayabalan, M. In vitro studies on the effect of physical cross-linking on the biological performance of aliphatic poly(urethane urea) for blood contact applications. *Biomacromolecules* **2001**, *2*, 588–596. [CrossRef]
7. Silvestri, A.; Sartori, S.; Boffito, M.; Mattu, C.; Rienzo, A.M.D.; Boccafoschi, F.; Ciardelli, G. Biomimetic myocardial patches fabricated with poly(ε-caprolactone) and polyethylene glycol-based polyurethanes. *J. Biomed. Mater. Res. B* **2014**, *102*, 1002–1013. [CrossRef]
8. Guelcher, S.A. Biodegradable polyurethanes: Synthesis and applications in regenerative medicine. *Tissue Eng. B* **2008**, *14*, 3–17. [CrossRef]
9. Kobayashi, H.; Hyon, S.; Ikada, Y. Water-curable and biodegradable prepolymers. *J. Biomed. Mater. Res.* **1991**, *25*, 1481–1494. [CrossRef]
10. Guan, J.; Sacks, M.S.; Beckman, E.J.; Wagner, W.R. Biodegradable poly(ether ester urethane) urea elastomers based on poly(ether ester) triblock copolymers and putrescine: synthesis, characterization and cytocompatibility. *Biomaterials* **2004**, *25*, 85–96. [CrossRef]
11. Groot, J.H.D.; Spaans, C.J.; Dekens, F.G.; Pennings, A.J. On the role of aminolysis and transesterification in the synthesis of ε-caprolactone and L-lactide based polyurethanes. *Polym. Bull.* **1998**, *41*, 299–306. [CrossRef]
12. Spaans, C.J.; Groot, J.H.D.; Dekens, F.G.; Pennings, A.J. High molecular weight polyurethanes and a polyurethane urea based on 1,4-butanediisocyanate. *Polym. Bull.* **1998**, *41*, 131–138. [CrossRef]
13. Qu, W.; Xia, Y.; Jiang, L.; Zhang, L.; Hou, Z. Synthesis and characterization of a new biodegradable polyurethanes with good mechanical properties. *Chin. Chem. Lett.* **2016**, *27*, 135–138. [CrossRef]
14. Jia, Q.; Xia, Y.; Yin, S.; Hou, Z.; Wu, R. Influence of well-defined hard segment length on the properties of medical segmented polyesterurethanes based on poly(ε-caprolactone-co-L-lactide) and aliphatic urethane diisocyanates. *Int. J. Polym. Mater. Po.* **2017**, *66*, 388–397. [CrossRef]
15. Chen, J.; Guo, B.; Eyster, T.W.; Ma, P.X. Super stretchable electroactive elastomer formation driven by aniline trimer self-assembly. *Chem. Mater.* **2015**, *27*, 5668–5677. [CrossRef]
16. Zhao, X.; Guo, B.; Wu, H.; Liang, Y.; Ma, P.X. Injectable antibacterial conductive nanocomposite cryogels with rapid shape recovery for noncompressible hemorrhage and wound healing. *Nat. Commun.* **2018**, *9*, 2784. [CrossRef]
17. Liang, Y.; Zhao, X.; Hu, T.; Chen, B.; Yin, Z.; Ma, P.X.; Guo, B. Adhesive hemostatic conducting injectable composite hydrogels with sustained drug release and photothermal antibacterial activity to promote full-thickness skin regeneration during wound healing. *Small* **2019**, *15*, 1900046. [CrossRef]
18. Qu, J.; Zhao, X.; Liang, Y.; Zhang, T.; Ma, P.X.; Guo, B. Antibacterial adhesive injectable hydrogels with rapid self-healing, extensibility and compressibility as wound dressing for joints skin wound healing. *Biomaterials* **2018**, *183*, 185–199. [CrossRef]

19. Li, D.; Chen, H.; McClung, W.G.; Brash, J.L. Lysine-PEG-modified polyurethane as a fibrinolytic surface: Effect of PEG chain length on protein interactions, platelet interactions and clot lysis. *Acta Biomater.* **2009**, *5*, 1864–1871. [CrossRef]

20. Bochynska, A.I.; Hannink, G.; Grijpma, D.W.; Buma, P. Tissue adhesives for meniscus tear repair: An overview of current advances and prospects for future clinical solutions. *J. Mater. Sci. Mater. M.* **2016**, *27*, 85–102. [CrossRef]

21. Liu, X.; Xia, Y.; Liu, L.; Zhang, D.; Hou, Z. Synthesis of a novel biomedical poly(ester urethane) based on aliphatic uniform-size diisocyanate and the blood compatibility of PEG-grafted surfaces. *J. Biomater. Appl.* **2018**, *32*, 1329–1342. [CrossRef]

22. Liu, L.; Gao, Y.; Zhao, J.; Yuan, L.; Li, C.; Liu, Z.; Hou, Z. A mild method for surface-grafting PEG onto segmented poly(ester-urethane) film with high grafting density for biomedical purpose. *Polymers* **2018**, *10*, 1125. [CrossRef]

23. Gao, B.; Feng, Y.; Lu, J.; Zhang, L.; Zhao, M.; Shi, C.; Khan, M.; Guo, J. Grafting of phosphorylcholine functional groups on polycarbonate urethane surface for resisting platelet adhesion. *Mat. Sci. Eng. C* **2013**, *33*, 2871–2878. [CrossRef] [PubMed]

24. Zeng, M.; Zhang, L.; Zhou, Y. Effects of solid substrate on structure and properties of casting waterborne polyurethane/carboxymethylchitin films. *Polymer* **2004**, *45*, 3535–3545. [CrossRef]

25. Mohamed, N.; Salama, H.; Sabaa, M.; Saad, G. Synthesis and characterization of biodegradable copoly(ether-ester-urethane)s and their chitin whisker nanocomposites. *J. Therm. Anal. Calorim.* **2016**, *125*, 163–173. [CrossRef]

26. Zhang, N.; Yin, S.; Hou, Z.; Xu, W.; Zhang, J.; Xiao, M.; Zhang, Q. Preparation, physicochemical properties and biocompatibility of biodegradable poly(ether-ester-urethane) and chitosan oligosaccharide composites. *J. Polym. Res.* **2018**, *25*, 212. [CrossRef]

27. Zhang, S.; Wang, L.; Yang, S.; Gong, Y. Improved biocompatibility of phosphorylcholine end-capped poly(butylene succinate). *Sci. China Chem.* **2013**, *56*, 174–180. [CrossRef]

28. Hao, N.; Wang, Y.; Zhang, S.; Shi, S.; Nakashima, K.; Gong, Y. Surface reconstruction and hemo-compatibility improvement of a phosphorylcholine end-capped poly(butylene succinate) coating. *J. Biomed. Mater. Res. A* **2014**, *102*, 2972–2981. [CrossRef]

29. Ishihara, K.; Ueda, T.; Nakabayashi, N. Preparation of phospholipid polylners and their properties as polymer hydrogel membranes. *Polym. J.* **1990**, *22*, 355–360. [CrossRef]

30. Hayward, J.A.; Chapman, D. Biomembrane surfaces as models for polymer design: The potential for haemocompatibility. *Biomaterials* **1984**, *5*, 135–142. [CrossRef]

31. Korematsu, A.; Takemoto, Y.; Nakaya, T.; Inoue, H. Synthesis, characterization and platelet adhesion of segmented polyurethanes grafted phospholipid analogous vinyl monomer on surface. *Biomaterials* **2002**, *23*, 263–271. [CrossRef]

32. Liu, Y.; Inoue, Y.; Mahara, A.; Kakinoki, S.; Yamaoka, T.; Ishihara, K. Durable modification of segmented polyurethane for elastic blood-contacting devices by graft-type 2-methacryloyloxyethyl phosphorylcholine copolymer. *J. Biomater. Sci. Polym. E* **2014**, *25*, 1514–1529. [CrossRef]

33. Ishihara, K.; Iwasaki, Y. Biocompatible elastomers composed of segmented polyurethane and 2-methacryloyloxyethyl phosphorylcholine polymer. *Polym. Adv. Technol.* **2000**, *11*, 626–634. [CrossRef]

34. Hong, Y.; Ye, S.H.; Nieponice, A.; Soletti, L.; Wagner, W.R. A small diameter, fibrous vascular conduit generated from a poly (ester urethane) urea and phospholipid polymer blend. *Biomaterials* **2009**, *30*, 2457–2467. [CrossRef]

35. Ho, S.P.; Nakabayashi, N.; Iwasaki, Y.; Boland, T.; Laberge, M. Frictional properties of poly(MPC-co-BMA) phospholipid polymer for catheter applications. *Biomaterials* **2003**, *24*, 5121–5129. [CrossRef]

36. Ishihara, K.; Tanaka, S.; Furukawa, N.; Kurita, K.; Nakabayashi, N. Improved blood compatibility of segmented polyurethanes by polymeric additives having phospholipid polar groups. I. Molecular design of polymeric additives and their functions. *J. Biomed. Mater. Res.* **1996**, *32*, 391–399. [CrossRef]

37. Ishihara, K.; Hanyuda, H.; Nakabayashi, N. Synthesis of phospholipid polymers having a urethane bond in the side chain as coating material on segmented polyurethane and their platelet adhesion-resistant properties. *Biomaterials* **1995**, *16*, 873–879. [CrossRef]

38. Sheikh, Z.; Khan, A.; Roohpour, N.; Glogauer, M.; Rehman, I. Protein adsorption capability on polyurethane and modified-polyurethane membrane for periodontal guided tissue regeneration applications. *Mat. Sci. Eng. C* **2016**, *68*, 267–275. [CrossRef] [PubMed]

39. Roohpour, N.; Wasikiewicz, J.M.; Moshaverinia, A.; Paul, D.; Grahn, M.F.; Rehman, I.U.; Vadgama, P. Polyurethane membranes modified with isopropyl myristate as a potential candidate for encapsulating electronic implants: A study of biocompatibility and water permeability. *Polymers* **2010**, *2*, 102–119. [CrossRef]

40. Kung, F.C.; Yang, M.C. Effect of conjugated linoleic acid immobilization on the hemocompatibility of cellulose acetate membrane. *Colloid. Surface. B* **2006**, *47*, 36–42. [CrossRef]

41. Huang, M.H.; Yang, M.C. Swelling and biocompatibility of sodium alginate/poly(γ-glutamic acid) hydrogels. *Polym. Adv. Technol.* **2010**, *21*, 561–567. [CrossRef]

42. Caracciolo, P.; Queiroz, A.; Higa, O.; Buffa, F.; Abraham, G. Segmented poly(esterurethane urea)s from novel urea-diol chain extenders: Synthesis, characterization and in vitro biological properties. *Acta Biomater.* **2008**, *4*, 976–988. [CrossRef] [PubMed]

43. Yin, S.; Xia, Y.; Jia, Q.; Hou, Z.; Zhang, N. Preparation and properties of biomedical segmented polyurethanes based on poly(ether ester) and uniform-size diurethane diisocyanates. *J. Biomat. Sci. Polym. E* **2007**, *28*, 119–138. [CrossRef]

44. Zuo, D.; Wang, Y.; Xu, W.; Liu, H. Effects of polyvinylpyrrolidone on structure and performance of composite scaffold of chitosan superfine powder and polyurethane. *Adv. Polym. Tech.* **2012**, *31*, 310–318. [CrossRef]

45. Chi, C.; Sun, B.; Zhou, N.; Zhang, M.; Chu, X.; Yuan, P.; Shen, J. Anticoagulant polyurethane substrates modified with poly(2-methacryloyloxyethyl phosphorylcholine) via SI-RATRP. *Colloid. Surface. B* **2018**, *163*, 301–308. [CrossRef]

46. Kwon, M.; Bae, J.; Kim, J.; Na, K.; Lee, E. Long acting porous microparticle for pulmonary protein delivery. *Int. J. Pharmaceut.* **2007**, *333*, 5–9. [CrossRef]

47. Liu, Y.; Inoue, Y.; Sakata, S.; Kakinoki, S.; Yamaoka, T.; Ishihara, K. Effects of molecular architecture of phospholipid polymers on surface modification of segmented polyurethanes. *J. Biomat. Sci. Polym. E* **2014**, *25*, 474–486. [CrossRef] [PubMed]

48. Kyomoto, M.; Ishihara, K. Self-initiated surface graft polymerization of 2-methacryloyloxyrthyl phosphorylcholine on poly(ether ether ketone) by photoirradiation. *Acs. Appl. Mater. Inter.* **2009**, *1*, 537–542. [CrossRef] [PubMed]

49. Brockman, K.; Kizhakkedathu, J.; Santerre, J. Hemocompatibility studies on a degradable polar hydrophobic ionic polyurethane (D-PHI). *Acta Biomater.* **2017**, *48*, 368–377. [CrossRef]

50. Yang, S.; Zhang, S.; Winnik, F. Group reorientation and migration of amphiphilic polymer bearing phosphorylcholine functionalities on surface of cellular membrane mimicking coating. *J. Biomed. Mater. Res. A* **2008**, *84*, 837–841. [CrossRef]

51. Nederberg, F.; Bowden, T.; Nilsson, B.; Hong, J.; Hilborn, J. Phosphoryl choline introduces dual activity in biomimetic ionomers. *J. Am. Chem. Soc.* **2004**, *126*, 15350–15351. [CrossRef] [PubMed]

Article

Designed Polyurethanes for Potential Biomedical and Pharmaceutical Applications: Novel Synthetic Strategy for Preparing Sucrose Containing Biocompatible and Biodegradable Polyurethane Networks

Lajos Nagy [1], Miklós Nagy [1], Bence Vadkerti [1], Lajos Daróczi [2], György Deák [1], Miklós Zsuga [1] and Sándor Kéki [1,*]

[1] Department of Applied Chemistry, Faculty of Science and Technology, University of Debrecen, Egyetem tér 1, H-4032 Debrecen, Hungary; nagy.lajos@science.unideb.hu (L.N.); miklos.nagy@science.unideb.hu (M.N.); bencevadkerti94@gmail.com (B.V.); deak.gyorgy@science.unideb.hu (G.D.); zsuga.miklos@science.unideb.hu (M.Z.)

[2] Department of Solid State Physics, Faculty of Science and Technology, University of Debrecen, Bem tér 18/b, H-4026 Debrecen, Hungary; daroczi.lajos@science.unideb.hu

* Correspondence: keki.sandor@science.unideb.hu; Tel: +36-52-512-900 (ext. 22455)

Received: 16 April 2019; Accepted: 2 May 2019; Published: 7 May 2019

Abstract: In this paper the preparation and detailed characterization of designed polyurethanes (SPURs) are reported for potential biological, biomedical and/or pharmaceutical applications. Importantly, in order to fulfill these goals all reactants and solvents used were selected according to the proposal of EUR-8 Pharmacopoeia. For the synthesis, a novel strategy was introduced and elaborated. A series of SPUR samples was prepared from poly(ε-caprolactone)-diol, 1,6-hexamethylene diisocyanate and sucrose as a chain extender/crosslinking agent to obtain sucrose containing polyurethanes. In addition, the mol ratios of the sucrose were varied within an order of magnitude. The prepolymers and the products of the syntheses were investigated by matrix-assisted laser desorption/ionization time-of-flight mass spectrometry (MALDI-TOF MS) and infrared spectroscopy (IR), respectively. It was found that the reactivity of the eight free hydroxyl groups of sucrose are different, and after curing the SPUR samples at 60 °C no free isocyanate groups can be observed. Furthermore, swelling experiments performed with various solvents of different polarities revealed that the highest degree of swelling took place in dimethyl-sulfoxide. However, low degrees of swelling were recognized in water and hexane. It is important to note that the gel contents were around 90% in all cases, which demonstrate that the crosslinking was almost complete. In addition, the kinetics of swelling were also evaluated and successfully modeled. The crosslink densities were calculated from the data of the swelling experiments by means of the Flory-Rehner equation. Unexpectedly, it was found that the crosslink density decreased with the increasing sucrose content also in line with the results obtained by relaxation modulus experiments and dynamic mechanical analysis (DMA). The Tg and Tm of SPUR samples, determined from DSC and DMA measurements, were around −57 °C and 27 °C, respectively. According to the mechanical tests the SPUR samples showed high elongation at break values, i.e., high flexibilities. Furthermore, the stress-strain curves were also modeled and discussed.

Keywords: poly(ε-caprolactone), 1,6-hexamethylene-diisocyanate; sucrose; polyurethane, swelling; mechanical testing

---

1. Introduction

The last few decades have witnessed an unprecedented increase in human life expectancy and life quality. The key to keep up with these improvements can be offered by tissue engineering and regeneration, which involve the use of a tissue scaffold for the formation of new viable tissue for a medical purpose [1]. The certain mechanical and structural properties the tissues require for proper functioning are to be provided by biocompatible synthetic polymers, such as poly(ε-caprolactone), polyurethanes and polylactic acid, since through macromolecular engineering their properties can be tailored and fine-tuned more precisely than those of natural polymers [1].

Poly(ε-caprolactones) (PCLs) are useful biomaterials whose utilities span from as materials for biomedical devices to food packaging applications. Importantly, the Food and Drug Administration (FDA) has approved PCLs for specific human applications. For example, FDA allowed their use in long term implantable devices, absorbable sutures and in certain drug delivery systems. However, there are some shortcomings with PCLs, such as their low degradation rates and poor mechanical properties, which are not beneficial for their special applications, e.g., for tissue engineering [2–4]. One potential solution for improving the physical, chemical and biological properties of PCLs, among others, is to prepare polyurethane derivatives involving PCL blocks.

Polyurethanes (PURs) play an important and unavoidable role in everyday life [5]. They are frequently used as flexible foams, rigid foams, elastomers and so on [5]. The physical, chemical and biological properties of PURs are fundamentally determined by their chemical structures. A general synthetic approach for producing of PURs involves polyaddition reaction between isocyanates and polyols. If the target PURs are intended to be thermosets, the use of crosslinking agent is necessary [5]. The step-growth polyaddition reaction carries a great variability in the compositions including the types and functionality of both the isocyanates and polyols as well as of other additives [5–9]. Furthermore, both isocyanates and polyols can originate from natural resources, such as polyols, e.g., from carbohydrates [6]. Thus, one can vary the chemical structure of PURs in a wide range to tailor their physical, chemical and biological properties for specific applications. The use of naturally occurring carbohydrates for the synthesis of PURs is well-documented [6]. Garcon et al. synthesized PURs from protected sugar derivatives and 1,6-hexamethylene diisocyanate (HDI) [7]. M. Barikani et al reacted corn starch with a prepolymer obtained from the reaction of HDI and poly(ε-caprolactone)-diol (PCLD) to prepare hydrophobic PUR-copolymers [8]. Kizuka and Inonue reported the preparation of PURs from aromatic 4,4′-diphenylmethane diisocyanate (MDI), and different polyols and sucrose as a cross linker were reported by putting all reactants in one pot to obtain copolymers [9]. Since our primary purpose was to synthesize flexible, biocompatible and biodegradable PUR-s for potential biological use in general, and pharmaceutical and/or biomedical applications in particular, we encountered some serious limitations. Importantly, we should exclude all aromatic isocyanates and various unsupported solvents from the synthesis. Thus, the solvents should be selected according to the European Pharmacopoeia (Ph. Eur. 8th edition) requirements, i.e., the solvents should be of class 2 and class 3 [10]. Under these constraints, we elaborated a novel synthetic method to prepare PURs with well-defined structures and networks as well as to control their swellability and flexibility. Our novel concept is based on the followings: (i) first, a prepolymer is synthesized in melt from a biodegradable, biocompatible polyester polyol, i.e., from poly(ε-caprolactone) diol (PCLD) to obtain prepolymer diisocyanate; (ii) in the second step a multifunctional chain extender and/or crosslinking agent dissolved in DMSO is added to the melted prepolymer to obtain the main polymer chain. It is important to note that the substituents of the chain extender/crosslinking agent should have different reactivity and furthermore, the chain extender/crosslinking agent should have two highly reactive functional groups to yield the main polymer backbone chain. Then the reaction proceeds between the excess of free isocyanate groups of the prepolymer and the lesser reactive functional groups in the chain extender/crosslinking agent to form the final polymer network. On the other hand, upon variation of the M_n of the polyol moiety in the prepolymer one can vary the distances between the main polymer chains. The crosslink density can thus be regulated deliberately simply by adjusting the molar ratio of the prepolymer and the chain extender/crosslinking agent.

We have recently reported that the reactivity of the primary OH groups towards isocyanates is higher than that of the secondary ones [11]. As it is well known, the sucrose has eight free OH-groups, from which 3 of them are primary OH groups (one of them is sterically hindered) and five of them are secondary OH groups. Taking into consideration this fact, the sucrose fulfills the requirement to be an appropriate chain extender/crosslinking agent, therefore, we selected sucrose as a chain extender/crosslinking agent to synthesize the target PURs.

Our purpose was to solve the challenge of the synthesis to fulfill the strict requirements of the different biological applications. Based on the strong, well established literature background, we accept that all ingredient used in this work are biocompatible and biodegradable. The biocompatibility and the biodegradation behaviors of both poly(ε-caprolactone) and certain poly(ε-caprolactone) containing polyurethanes are also extensively investigated and proved therefore we concentrated our efforts to the chemical solutions [12,13].

In this paper, we report a one pot synthesis and detailed characterization of biocompatible and completely biodegradable PUR networks synthesized from PCLD/HDI/sucrose systems.

2. Materials and Methods

2.1. Materials

Poly(ε-caprolactone) diol (PCLD, Mn = 2000 g/mol), dimethyl-sulfoxide (99.9 %, DMSO, stored on molecular sieve), 1,6-hexamethylene diisocyanate (HDI, reagent grade), Tin(II) 2-ethylhexanoate (reagent grade) were purchased from Sigma-Aldrich (Darmstadt, Germany) and were used as received. D(+)-sucrose (puriss, Ph. Eur. 6.) from Reanal (Budapest, Hungary) was powdered and dried in a vacuum oven at 40 °C for overnight before use. Toluene (analytical grade), acetone (99.9%, HPLC grade) from Sigma-Aldrich (Darmstadt, Germany), methanol (HPLC grade) from Merck (Darmstadt, Germany) and hexane (HPLC grade) from VWR (Debrecen, Hungary) were used without any purification.

2.2. Synthesis of the Prepolymer

The synthesis of prepolymer was basically carried out according to ref. [8] in melt. The modified method applied was the following: A 100 mL round-bottom, three-necked flask equipped with a mechanical stirrer, nitrogen inlet, reflux condenser and a nitrogen outlet was used. The temperature was kept at 90 °C by means of a silicon oil bath. Before assembly, all the glassware were dried at 140 °C for a day. The flask was charged with 20 g (10 mmol) of PCLD under continuous nitrogen flow. After complete melting of the PCLD the mechanical stirrer was turned on and 3.5 mL (22 mmol) HDI was introduced to the reactor. The homogeneous mixture was stirred at 80 °C for four hours. During the prepolymer synthesis only a negligible amount of gas formation may be observed. When the mixing time expired a sample (around 0.2 mL) was taken out from the reaction mixture for further analysis.

2.3. Synthesis of Sucrose Containing (SPUR) Polyurethane Networks

Predetermined amount of vacuum dried sucrose (9.5–0.95 mmol) was dissolved in 50 mL of anhydrous DMSO at 60 °C. After complete dissolution of sucrose, the solution was added in one portion to the melt prepolymer (synthesized according to Section 2.2.) during vigorous stirring (0.08 g, 0.2 mmol). Tin(II) 2-ethylhexanoate catalyst was also introduced to the reaction flask. The temperature was kept at 80 °C. The opaque mixture was then stirred until the mixture became homogenous again. It usually took place within a minute. To avoid the fast gelation the resulting viscous mixture was poured into Teflon coated pans within 2 min and cured at 60 °C for 24 h. The obtained SPUR sheets were dried in a vacuum oven at 40 °C for several (2–3) days for constant weight. During the synthesis no weight loss was observed.

2.4. Characterization

2.4.1. Scanning Electron Microscopy (SEM)

To visualize the morphology of the SPUR samples scanning electron microscopy (SEM) was used. A secondary electron image was taken from the cut edge of SPUR samples after covering them by 30 nm conductive gold layer by a Hitachi S-4300 scanning electron microscope (Tokyo, Japan).

2.4.2. Attenuated Total Reflectance Fourier-Transform Infrared (ATR-FTIR)

Attenuated total reflectance Fourier-transform infrared (ATR-FTIR) spectra were recorded on a PerkinElmer Spectrum Two Instrument equipped with Ultra ATR Two Sampling Accessory having a diamond-zinc-selenium composite prism (Waltham, MA, USA). The penetration depth of the IR beam into the SPUR sample is 6 μm. Thickness of the specimens were ca. 0.5 mm. Sixteen scans were taken. The spectra were evaluated by Spectrum ES 5.0 program.

2.4.3. Swelling Experiments

SPUR samples (having initial measures: $10 \times 15 \times 0.2\text{--}0.4$ mm, and initial weight: m_o) were placed in 20 mL of different solvents in Erlenmeyer flasks of 50 mL at room temperature. The following solvents were used for swelling: acetone, DMSO, hexane, methanol, toluene and water. The swelled SPUR samples were taken out and the immersion fluid was wiped off from the surfaces with a paper towel. The weight gain was followed by an analytical balance in every hour. The weight gain was monitored until equilibrium was reached. It usually took around two days. The degree of swelling, i.e., the swelling ratio (%) (Q) is defined by Equation (1) as:

$$Q = \frac{m - m_o}{m_o} \times 100 \tag{1}$$

where m and m_0 are the masses of the swollen and the initial (dry) sample, respectively.

After completing the swelling test, the absorbed solvent was removed in a vacuum oven at 40 °C until constant weight (m_1). The gel content ($G\%$) of the SPUR samples were calculated according to the following Equation (2):

$$G(\%) = \frac{m_1}{m_o} \times 100 \tag{2}$$

The crosslink density of SPUR samples was calculated according to the Flory-Rehner equation [14] as given by Equation (3):

$$\nu_e = \frac{-\ln\left[(1 - Y_p) + Y_p + \chi Y_p^2\right]}{V_s\left(Y_p^{1/3} - \frac{Y_p}{2}\right)} = \frac{\rho_p}{M_c} \tag{3}$$

where ν_e = crosslink density, Y_p = the volume fraction of the polymer, V_s is the molar volume of the solvent, χ is the polymer-solvent interaction parameter, ρ_p is the density of the polymer and M_c is the number average molecular weight between crosslink points. The polymer-solvent interaction parameter for the SPUR networks was determined as detailed in the results and discussion.

2.4.4. Matrix-Assisted Laser Desorption/Ionization Time-of-Flight Mass Spectrometry (MALDI-TOF MS)

The Matrix-Assisted Laser Desorption/Ionization Time-of-Flight Mass Spectrometry (MALDI-TOF MS) measurements were carried out with a Bruker Autoflex Speed mass spectrometer equipped with a time-of-flight/time-of-flight (TOF/TOF) mass analyser. In all cases 19 kV acceleration voltage was used in the positive ion mode. To obtain appreciable resolution and mass accuracy ions were detected in the reflectron mode and 21 kV and 9.55 kV were applied as reflector voltage 1 and voltage 2, respectively. A solid phase laser (355 nm, ≥100 μJ/pulse) operating at 500 Hz was applied to produce laser desorption and 5000 shots were summed. The MALDI-TOF MS spectra were externally calibrated with poly(ethylene glycol) standard (M_n = 1540 g/mol).

Samples were withdrawn from the reaction mixture and dissolved in a mixture of THF and methanol (50/50 *v/v*) at a concentration of 10 mg/mL and allowed the free isocyanate groups to react with the methanol for 2 days. Samples for MALDI-TOF MS were prepared with 2,5-dihydroxy benzoic acid (DHB) matrix. The matrix was dissolved in a mixture of THF and methanol (50/50 *v/v*) at a concentration of 20 mg/mL. The matrix solution, the sample solution and sodium trifluoroacetate solution (5 mg/mL in THF/methanol (50/50 *v/v*)), used as the cationization agent to promote the ionization, were mixed in a 10:2:1 (*v/v*) ratio (matrix/analyte/cationization agent). A volume of 0.5 µL of the solution was deposited onto a metal sample plate and allowed to air-dry.

2.4.5. Mechanical Tests

Instron 4302 type mechanical testing machine (Instron, Norwood, MA, USA), equipped with a 1 kN load cell, was used for tensile testing of the SPUR series. Four dumbbell specimens were cut (clamped length 60 mm) from the SPUR samples and the tensile was loaded at a strain rate 50 mm/min. For the data evaluation Instron series 9 Automated Materials Tester—version 8.30.00 software was used.

For the stress relaxation experiments an Instron 3366 (Instron, Norwood, MA, USA) type mechanical testing machine was used. The specimens were subjected to 100% elongation and the decay of the load at this strain was measured and evaluated under the supervision of the Instron Bluehill Universal V 4.05 (2017) software.

2.4.6. Differential Scanning Calorimetry (DSC)

The thermal properties of the samples were evaluated by Differential Scanning Calorimetry (DSC) applying a DSC Q2000 power compensation equipment (TA Instruments, New Castle, DE, USA) operating at 10 °C/min heating rate. Nitrogen was used as protective atmosphere.

2.4.7. Dynamic Mechanical Analysis (DMA)

The dynamic mechanical properties of the samples were carried out using Dynamic Mechanical Analysis (DMA) testing with a DMA Q800 device (TA Instruments, New Castle, DE, USA). The DMA traces were recorded in tension mode (dimension of the specimens: length: 25 mm, clamped length: 12 mm, width: 7 mm, thickness: ca. 0.5 mm) at an oscillation amplitude of 0.2% with a frequency of 1 Hz and applying a static load of 1 N. The temperature was varied between −100 °C and 200 °C with a heating rate of 3 °C/min.

3. Results and Discussion

Our purpose was to elaborate a one pot synthesis for the preparation of PCLD and sucrose containing polyurethane networks with regular structures and to improve their physical and chemical properties, such as swelling and flexibility for biological applications.

For the synthesis we worked out a novel synthetic strategy: first, a macro diisocyanates (prepolymer) was prepared from PCLD and HDI in melt. To obtain the main polymer chains of the network, sucrose as a multifunctional, small molecular weight chain extender/crosslinking agent dissolved in DMSO was used. The main polymer chains were thus formed by the polyaddition reaction between the prepolymer and the 6 and 6′ primary hydroxyl groups of sucrose (see Scheme 1).

The crosslinking reaction took place by the reaction of the excess of the free isocyanate groups in the prepolymer with the residual primary hydroxyl group, most probably, in the 1′position of sucrose. Thus, by varying the Mn-s of PCLD and the molar ratios of the prepolymer to sucrose, the distances between the main polymer chains and the crosslink density can also be altered, respectively (Scheme 1)

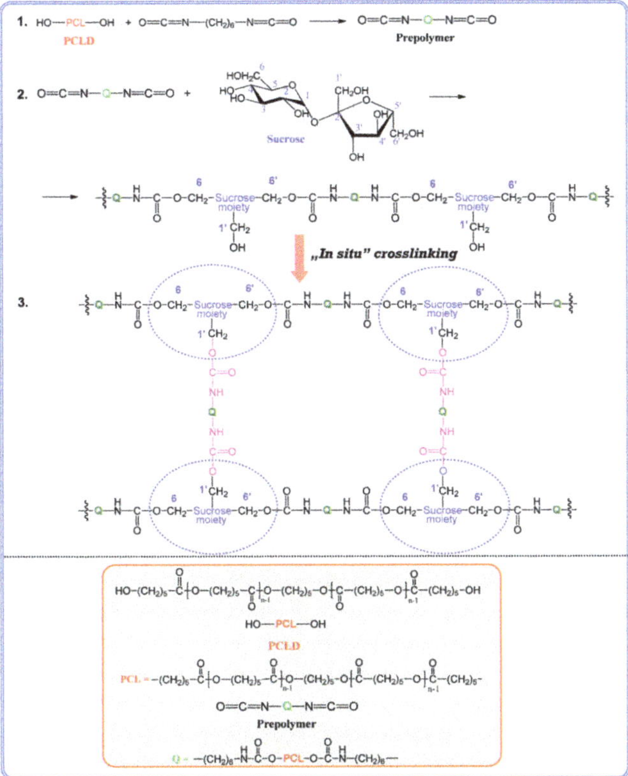

Scheme 1. Synthesis path for the preparation of SPURs.

Based on the synthesis route presented in Scheme 1, a series of SPUR samples was prepared. The compositions, densities and Shore A hardness of SPUR samples are summarized in Table 1.

Table 1. Sample name, composition, density and Shore A hardness of sucrose containing PUR polymers (SPUR samples).

Sample Name	Composition (mol Ratio) PCLD/HDI/Sucrose	Sucrose % (m/m)	Density (g/cm^3)	Hardness Shore A Scale (%)
SPUR-1	1/2.2/0.95	12.1	1.136	67
SPUR-2	1/2.2/0.71	9.3	1.133	62
SPUR-3	1/2.2/0.475	6.4	1.109	51
SPUR-4	1/2.2/0.238	3.3	1.124	52
SPUR-5	1/2.2/0.095	1.4	1.102	66

As can be seen in Table 1 the mol ratio of prepolymer and the sucrose was varied within one order of magnitude. The prepolymer concentration was kept constant while the concentration of sucrose in DMSO was changed gradually as shown in Table 1.

During the synthesis there was no weight loss, indicating that all reactants were built in the SPUR networks formed. This finding agrees well with the results from the swelling experiments (see later).

3.1. Infrared Spectroscopy of the SPUR Samples

In order to get information on the molecular structures of the SPUR networks formed Attenuated Total Reflectance Fourier-Transform Infrared (ATR-FTIR) spectra were recorded (Figure 1).

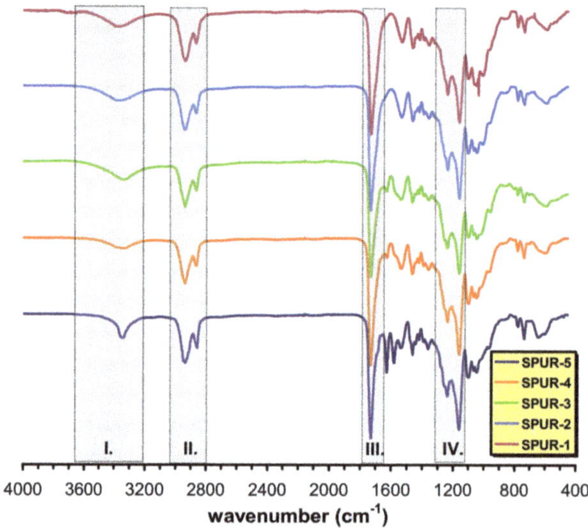

Figure 1. ATR-FTIR spectra of the SPUR samples.

As seen in Figure 1, the intermolecularly bonded OH stretching, which might overlap with the NH stretching vibrations, especially at lower prepolymer/sucrose molar ratios, occur in the range of 3358–3334 cm^{-1} (shaded area I). A double band at 2937 and 2863 cm^{-1} (shaded area II) and at 1361 cm^{-1} belonging to CH$_2$ stretching and bending vibrations, respectively can also be observed and a strong absorption band of C=O stretching is visible at 1727 cm^{-1} (shaded area III). An absorption band at 1537 cm^{-1} is attributed to the C-N stretching. The presence of this band in the ATR-FTIR spectrum confirms the formation of polyurethane linkages. The =C-O/-C-O-C- vibrations are assigned to appear in the range of 1150–1230 cm^{-1} (shaded area IV). In addition, no absorption can be seen at around 2275–2230 cm^{-1} (i.e., no absorption in the 2750–1850 cm^{-1} region), indicating no residual isocyanate groups present after 24 h of curing at 60 °C.

3.2. MALDI-TOF MS Investigations

The MALDI-TOF MS spectrum obtained on the prepolymer poly(ε-caprolactone)-diol (PCLD) with two HDI end-groups is shown in Figure 2. Prior to recording the MALDI-TOF mass spectrum the free isocyanate groups in the prepolymer were reacted with methanol in order to prevent further reaction of the free isocyanate groups with the matrix molecules. Furthermore, to enhance the cationization and to obtain mainly sodiated oligomers, sodium trifluoroacetate was added to the reaction mixture. As seen in Figure 2, the mass difference between the neighboring peaks is 114 Da that corresponds to the mass of a caprolactone repeat unit.

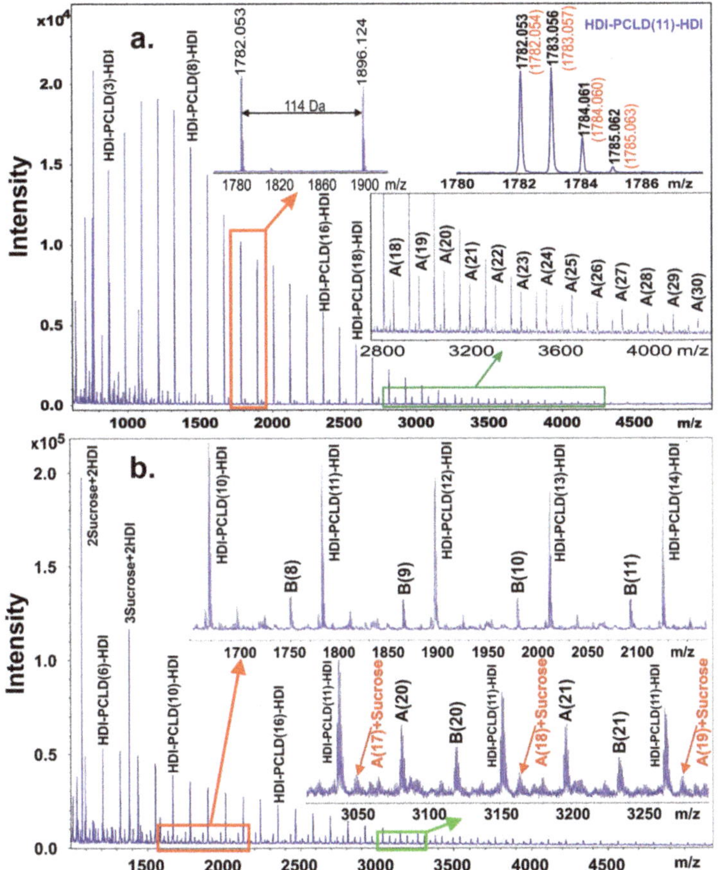

Figure 2. MALDI-TOF mass spectra of the samples obtained from the PCLD-HDI reaction mixture (Reaction conditions: temperature = 90 °C, reaction time: 4 h) (**a**) and from the HDI-PCLD prepolymer sucrose reaction (Reaction conditions: temperature = 90 °C, reaction time: 3 min, the molar ratio for PCLD/HDI/sucrose is 1/2.2/0.95, sample SPUR-1) (**b**). The numbers in the brackets represents the number of the caprolactone repeating unit. The isotopic distribution, the measured and the calculated (red in brackets) *m/z* values are shown in the inset (top right corner of Figure 2a).

Furthermore, the measured masses match those of the PCLD with two HDI and two methanol units as expected for a prepolymer HDI-PCLD-HDI. For example, as illustrated in the top right corner of Figure 2a. inset, the measured monoisotopic mass at *m/z* 1782.053 is in line with that found for the composition $[C_{89}H_{154}N_4O_{30}+Na]^+$ (*m/z* 1782.054). On the other hand, as seen in the zoomed MALDI-TOF MS spectrum (Figure 2a., in the range of *m/z* 2800–4000) in addition to the main series, series A corresponding to a HDI-PCLD-HDI-PCLD-HDI oligomer series also appeared.

In the next synthetic step the resulting prepolymer was further reacted with sucrose in DMSO. After a short reaction time, in order to avoid considerable cross-linking, a sample was withdrawn from the reaction mixture and analyzed by MALDI-TOF MS to prove that successful coupling reactions have been taken place between the prepolymer and sucrose. Indeed, as it turns out from Figure 2b, series B appeared in the MALDI-TOF MS spectrum, which was identified as HDI-PCLD-HDI-Sucrose oligomers. Furthermore, at higher masses, series A reacted with sucrose was also detected as demonstrated in the lower inset of Figure 2b. In the low mass region of the MALDI-TOF MS spectrum products resulting from the reaction of sucrose with HDI, i.e., sucrose + HDI containing no PCLD oligomer were also detected. This latter finding is very intriguing, since two sucrose + 2HDI and three sucrose + 2 HDI were also formed as seen in Figure 2b. Thus, the presence of these compounds in the reaction mixture indicates that albeit sucrose has eight OH groups to be reacted with the isocyanate, they are not equally reactive. Furthermore, it seems unambiguous from these findings that two of these OH groups are more reactive than the others thus yielding linear versions of sucrose + HDI oligomers at the early stage of the reaction. This finding has also consequence on the cross-linking process with sucrose that will be discussed later.

3.3. Swelling Experiments

In order to gain insight into the crosslink densities and the nature of the network formed by the reaction of the prepolymer HDI-PCLD-HDI with sucrose, the resulting crosslinked polymers were swollen with solvents (such as n-hexane, toluene, acetone, methanol, DMSO and water) of different polarities and solubility parameters. The swelling of the SPUR samples was followed in time by measuring the weight of the swollen samples with respect to those of the dry ones and the swelling ratios (%) as a function of the time were plotted (Figure 3). To describe the swelling ratio (%) (Q) versus time curves various models were tried to fit to the experimental data. The investigated models included the single exponential (SE, Equation (4)), the integrated form of the pseudo second order rate equation (PSOE, Equation (5)) [15], the power law equation (PLE, Equation (6)) [16] and the stretched exponential function of Kolmogorov, Erofeev, Kozeeva, Avrami, Mampel (KEKAM, Equation (7)) [17].

$$Q = A(1 - e^{-kt}) \tag{4}$$

$$Q = \frac{kt}{A + kt} \tag{5}$$

$$Q = kt^{\gamma} \tag{6}$$

$$Q = A[1 - e^{-(kt)^{\gamma}}] \tag{7}$$

where k and A are the rate coefficients and the equilibrium value of Q respectively, while A and γ stand for the constant of the stretched exponential function.

Figure 3a illustrates the results of fitting of Equations (4)–(7) to the experimental swelling ratios (%) versus time data obtained for the sample SPUR-4 and Figure 3a demonstrates the differences (residuals) between the fitted and the experimental data.

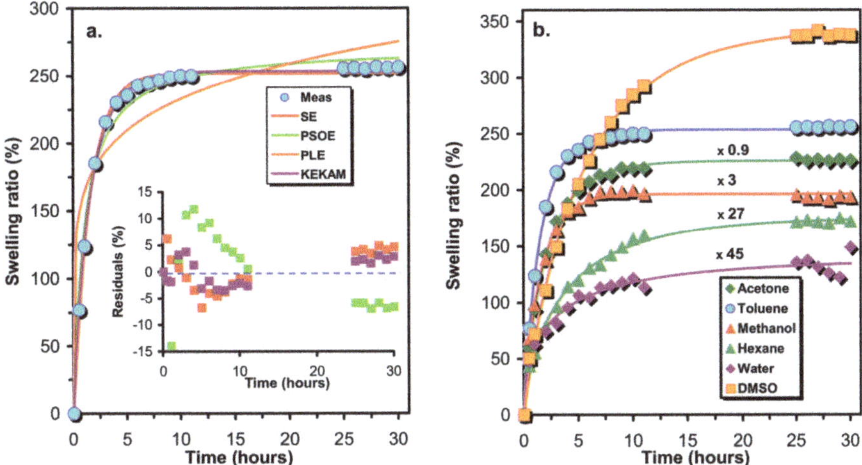

Figure 3. Variation of the swelling ratios with time for the sample SPUR-4 and fitting of the different models to the experimental data (solid lines) (a) and the dependence of the swelling ratio on the time together with the curves fitted by the streched exponential function (solid lines) (b). The inset in Figure 4a. shows the residuals as the difference between the experimental and the fitted data by the various models. The fitted parameters can be found in the Supplementary Materials (Table S1).

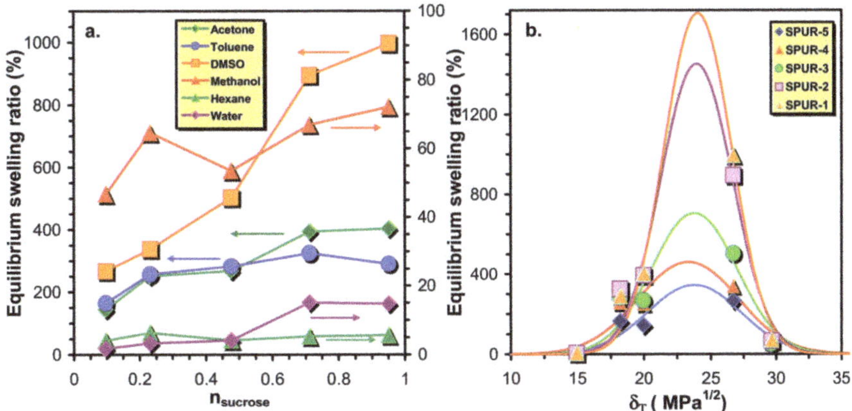

Figure 4. Variations of the swelling ratios with the molar ratios of sucrose (a) and the dependences of the swelling ratios on the values of the Hildebrand solubility parameter (δ_T) (b). The solid lines in Figure 4b represent the fitted curves by Equation (8).

As seen in Figure 3a, the PLE model (Equation (6)) is unable to render the main characteristics of the experimental swelling ratio (%)-time dependences, whereas the other three models may give acceptable fitting. However, as it turns out from Figure 3a, KEKAM model (Equation (7)) provides the best fitting and this finding is also true for the rest of the SPUR samples investigated. Thus, the KEKAM model was applied for the description of the swelling properties of these polymer samples. As seen in Figure 3b, Equation (7) is indeed capable of the description of the variation of the swelling ratio with the time for various solvents. The parameters of the KEKAM model obtained by fitting it to the experimental data are compiled in Supplementary Table S1. According to the data of Table S1, the value of the parameter γ is around 0.8 independently of the solvent, while values of k are spannig from 0.16 to 1.65, however, no rigorous relationships can be established between these values and the degrees of crosslinks.

In Figure 4a the variations of the equilibrium swelling ratios (Q_e) with the molar ratios of sucrose to that of prepolymer HDI-PCLD-HDI are plotted. As a general trend, it was found that the values of Q_e decrease with the sucrose molar ratio from which it can be surmised that the crosslink densities also decrease with the increasing sucrose content. An explanation for this unusual finding will be given later. Furthermore, it is also evident from Figure 4a that the Q_e −s at higher sugar content decrease in the order of DMSO > acetone > toluene > methanol > water > n-hexane, whereas at low sucrose content no considerable differences can be found in the Q_e values of acetone and toluene.

To evaluate the dependence of the equilibrium swelling ratio on the solvent, the experimental equilibrium swelling ratios were plotted as a function of the solvent solubility parameters (δ_T) and Equation (8) [18] were fitted to these data to determine the solubility parameters of the crosslinked polymer networks (δ_N).

$$Q = Q_e e^{-\omega(\delta_T - \delta_N)^2} \tag{8}$$

where Q_e is the equilibrium swelling ratio, and $\omega = \frac{\kappa \overline{V}_1}{RT}$, in which κ, $\overline{V}_1$, R and T are the empirical parameter, the molar volume of the solvent, the gas-constant and the temperature, respectively.

In Equation (8), parameters Q_e at $\delta = \delta_T$, δ_N and ω were determined and the fitted curves by Equation (5) together with the corresponding experimental data are shown in Figure 4b and the fitted values of the parameters are compiled in Table 2.

Table 2. The values of parameters Q_e, ω and δ_N obtained by fitting of Equation (8) to the corresponding experimental data.

Sample	Q_e (%)	ω (MPa^{-1})	δ_N (MPa$^{1/2}$)
SPUR-1	1705 ± 540	0.078 ± 0.026	24.0 ± 0.3
SPUR-2	1452 ± 496	0.069 ± 0.027	23.9 ± 0.4
SPUR-3	705 ± 253	0.049 ± 0.024	23.7 ± 0.5
SPUR-4	460 ± 126	0.036 ± 0.016	23.3 ± 0.5
SPUR-5	344 ± 113	0.041 ± 0.019	23.7 ± 0.6

It was found that the values of δ_N fall into a narrow range with values of δ_N = 23.3–24.0 (MPa)$^{1/2}$, i.e., crosslinking in the applied extent does not alter significantly the solubility of the polymer network formed. In addition, we have also investigated the effect of the components of the Hansen solubility parameters [19], i.e., the dispersion δ_D, the dipole δ_P and hydrogen-bonded δ_H solubility parameters on the swelling ratios. However, no significant correlation between the swelling ratios and the separated Hansen solubility parameters were found indicating that all of these three solubility parameters have an important role in determining the swelling properties of these polymer networks. It is to be noted that δ_N values in Table 3. were used for the determination of crosslink densities. The crosslink densities were calculated using the Flory-Rehner equation (as indicated in the Materials and Methods). The results are compiled in Table 3.

Table 3. The gel content ($G(\%)$), crosslink densities (ν_e) and the average molecular weights of segment between crosslinks (M_c) with different sucrose content determined in DMSO.

Sample	$G(\%)$	$\nu_e \times 10^4$ (mol/cm^3)	$M_c \times 10^3$ (g/mol)
SPUR-1	91.0	0.89	12.6
SPUR-2	81.8	0.91	12.4
SPUR-3	90.6	2.5	4.4
SPUR-4	98.3	3.7	3.1
SPUR-5	95.1	7.3	1.5

The graphical presentations of the data in Table 3. are shown in Figure 5.

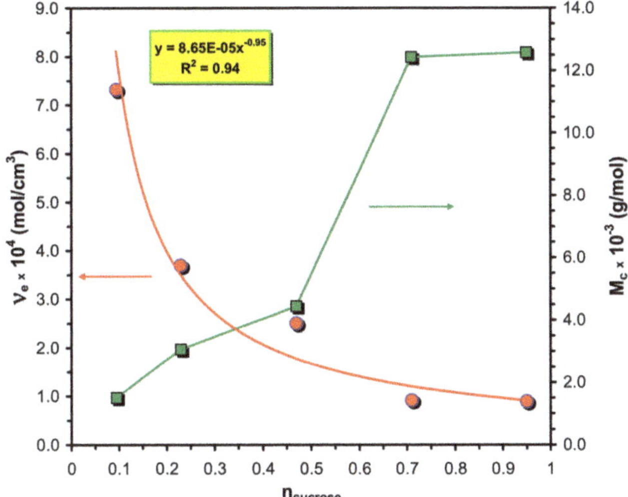

Figure 5. Dependence of the crosslink densities (v_e) and the average molecular weights of the segment between crosslinks (M_c) on the sucrose content.

According to the data of Table 3 the crosslink densities unexpectedly increase from 8.5×10^{-5} mol/cm^3 to 7.3×10^{-4} mol/cm^3 with the decreasing sucrose content and the dependence of v_e on the sucrose content can be described as $v_e = 8.65 \times 10^{-5} n_{\text{sucrose}}^{-0.95}$ (see Figure 5), i.e., v_e is approximately inversely proportional to n$_{\text{sucrose}}$. Furthermore, it also turns out from the M$_c$ data in Table 3 that at the highest crosslink density M$_c$ closely reflects to the M$_n$ of the PCLD segment. To interpret this unique finding one should consider that there is a competitive reaction between the isocyanate groups of the prepolymer and the OH groups of sucrose. The sucrose bears eight free OH-groups, from which three are primary OH groups (one of them is sterically hindered) and five of them are secondary OH groups. The main polymer chains of the SPURs are formed by the reaction of the prepolymer and the primary OH groups at the 6 and 6' positions of sucrose. The crosslinking reaction between the main polymer chains may take place by the reaction of the free isocyanate groups of the prepolymer with the less reactive primary OH group at the 1' position of sucrose to form the major skeleton of the SPUR networks (see Scheme 1). Moreover, upon depleting the primary OH groups, i.e., if the prepolymer concentration is significantly higher than that of the sucrose, which is the case for the samples SPUR-3, SPUR-4 and SPUR-5, the excess of prepolymer can react with the secondary OH groups of sucrose resulting in increasing crosslink densities.

3.4. Morphology of SPURs by SEM

The SEM images of SPUR samples are shown in Figure 6.

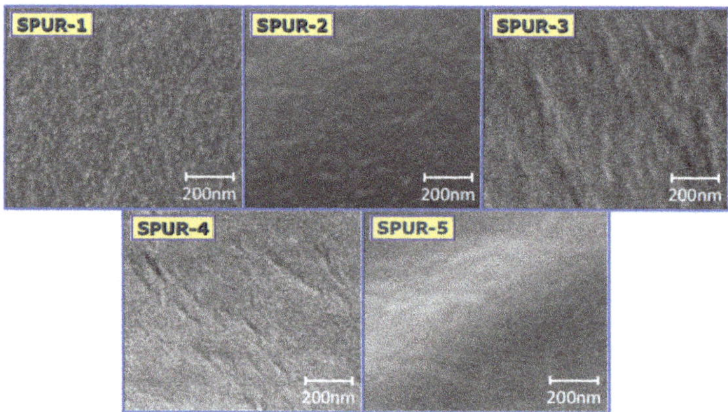

Figure 6. SEM images of SPUR samples.

As can be seen in Figure 6, similar, but not identical microstructures were found. There are no deep inclusions and no significant phase separation occur. These facts suggest that the polymer networks obtained have regular structure.

3.5. Mechanical Properties of SPURs

The data in Table 4 shows the results of the uniaxial tensile measurements. All the polymers were very flexible and transparent before the tests.

Table 4. Uniaxial tensile mechanical properties of SPURs.

Sample	Elastic Modulus (MPa)	Ultimate Elongation (%)	Stress at Break (MPa)
SPUR-1	5.0 ± 0.4	810 ± 110	22.6 ± 2.2
SPUR-2	3.3 ± 0.8	800 ± 40	14.8 ± 6.4
SPUR-3	2.1 ± 0.3	940 ± 330	15.8 ± 1.0
SPUR-4	2.8 ± 0.3	945 ± 110	24.3 ± 4.7
SPUR-5	5.3 ± 0.6	1100 ± 80	26.3 ± 1.3

As seen in Table 4, all the SPUR samples show low elastic modulus and high ultimate elongation values. Elastic (Young) moduli vary in the range of 2.1–5.3 MPa according to a minimum curve depending on the composition, i.e., the crosslink density.

However, the ultimate elongation data reveals an increasing trend with the decreasing amount of sucrose (i.e., with increasing crosslink density, see Table 3). On the other hand, more crosslink gives higher stress at break (sample SPUR-1 is an exception to that) and increasing ultimate elongation at the same time.

An attempt was made to describe the resulting stress-strain ($\sigma-\varepsilon$) curves at low and high strains. The description of the $\sigma-\varepsilon$ relationship was based on the standard linear solid (SLS) viscoelastic model. As it was shown earlier in uniaxial tensile mode and at constant strain rate (dε/dt) "the equation of motion" can be given by Equation (9) [20,21].

$$\sigma + a_1 \left(\frac{d\varepsilon}{dt} \right) \frac{d\sigma}{d\varepsilon} = a_2 \varepsilon + a_3 \frac{d\varepsilon}{dt} \tag{9}$$

where a_1, a_2 and a_3 are the parameters that contain the moduli of springs and the viscosity of the dashpot in the constitutive SLS model.

Equation can be solved analytically to obtain Equation (10).

$$\sigma = b_1[\varepsilon + b_2(1 - e^{-b_3\varepsilon})] \tag{10}$$

where $b_1 = a_2$, $b_2 = (d\varepsilon/dt)(a_3/a_2 - a_1)$ and $b_3 = [(d\varepsilon/dt)a_1]^{-1}$

The stress-strain ($\sigma - \varepsilon$) curves for the samples SPUR-5 and SPUR-2 are shown in Figure 7. and in the inset of Figure 7. As it turns out from the inset of Figure 7, Equation (9) is capable of rendering the experimental σ-ε curves up to moderate strains ($\varepsilon < 2.2$). However, it fails to describe the whole (σ-ε) curves especially at higher strains where marked deviations from the calculated ones have been observed. These deviations can be attributed to the upward curvature of the σ-ε plot, i.e., σ increases more rapidly with the increasing ε above ca. 2.2. These upward curvatures were supposed to be the effect of strain hardening, i.e., crystallization of the polymer segments can take place upon increasing strain, yielding an increase in the apparent modulus [20]. It is to be noted that a similar effect was observed and described for the epoxy-polyurethane (EPU) shape memory polymers [21]. The model that takes into account the strain hardening above a critical value of ε ($\varepsilon > \varepsilon_L$) is similar to that reported for the EPUs, with an extension that the present model provides a more general description of strain-hardening by introducing a new variable (β) as seen in Equation (11).

$$a_2(\varepsilon) = a_2 + \alpha(\varepsilon - \varepsilon_L)^\beta \ (\text{if } \varepsilon > \varepsilon_L) \tag{11}$$

$$d\sigma/d\varepsilon = [(d\varepsilon/dt)a_1]^{-1}[a_2(\varepsilon)\varepsilon + a_3(d\varepsilon/dt) - \sigma] \tag{12}$$

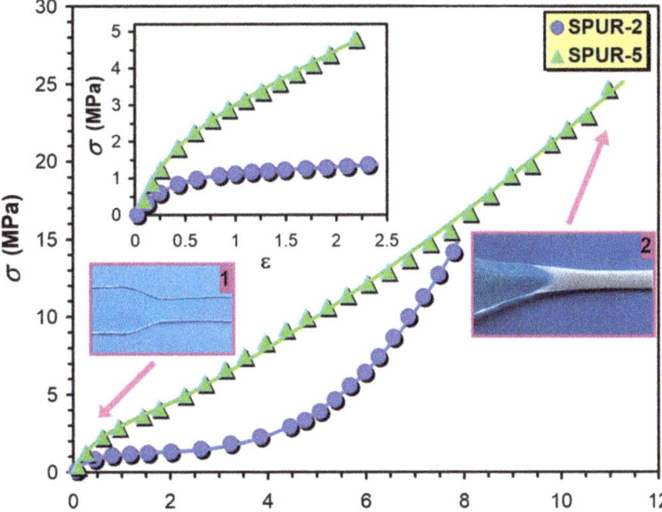

Figure 7. The stress-strain (σ-ε) curves for the samples SPUR-5 and SPUR-2. The symbols and the solid lines stand for the experimental data and the fitted curves, respectively. The inset figure reveals the lower part of the σ-ε curves up to $\varepsilon = 2.2$. The pictures illustrate the dumbbell from sample SPUR-5 at low (1) and at high strain (2). The fitted parameters for the sample SPUR-5 are $b_1 = 1.41$ MPa, $b_2 = 1.18$ and b3 = 3.24 (i.e., $[(d\varepsilon/dt)a_1]^{-1} = 3.24$, $a_2 = 1.41$ MPa, $(d\varepsilon/dt)a_3 = 2.10$ MPa and $\alpha = 0.136$ MPa and $\beta = 0.75$), while for the sample SPUR-2 these parameters are $b_1 = 0.12$ MPa, $b_2 = 8.80$ and $b_3 = 3.15$ i.e., $[(d\varepsilon/dt)a_1]^{-1} = 3.15$, $a_2 = 0.12$ MPa, $(d\varepsilon/dt)a_3 = 1.09$ MPa and $\alpha = 0.084$ MPa and $\beta = 1.75$). (The stress-strain (σ-ε) curves for the samples SPUR-1, SPUR-3 and SPUR-4 are shown in Figures S1–S3, respectively in the Supplementary Materials).

Integrating Equation (12) numerically (after the substitution of Equation (11) into Equation (12) and the fitted values obtained by Equation (10)) and then fitting it to the experimental data over the whole ε range, the values of α and β can now be determined. One can also realize that the fitted (solid lines) and the experimental (symbols) values are in good agreement (see Figure 7).

It is interesting to note that σ increases more rapidly with ε in the case of sample SPUR-2 than it does for the sample SPUR-5 and this finding is also reflected in the values of β which are 0.75 and 1.75 for the sample SPUR-5 and SPUR-2, respectively. Furthermore, as seen in Figure 7, the stress at break for the sample SPUR-2 is lower than for the sample SPUR-5. These findings are in line with the crosslink densities that are 9.1×10^{-5} and 7.3×10^{-4} mol/cm^3 for sample SPUR-2 and SPUR-5, respectively. At lower crosslink densities there are larger free spaces available for the polymer chains to crystallize giving rise to a steeper increase in σ with ε, and providing lower stress at break at the same time.

In addition, as illustrated by Figure 7, at low strains the sample is transparent (1), whereas at high strain it becomes opaque (white) due to the crystallization of the polymer segments upon the effect of strain (2), which also supports the validity of our approach.

In addition to these mechanical investigations, stress relaxation experiments were also performed to get deeper insight into the relaxation processes of these SPUR samples and in parallel, to determine the crosslink densities. The stress and the relative stress relaxation curves are demonstrated in Figure 8.

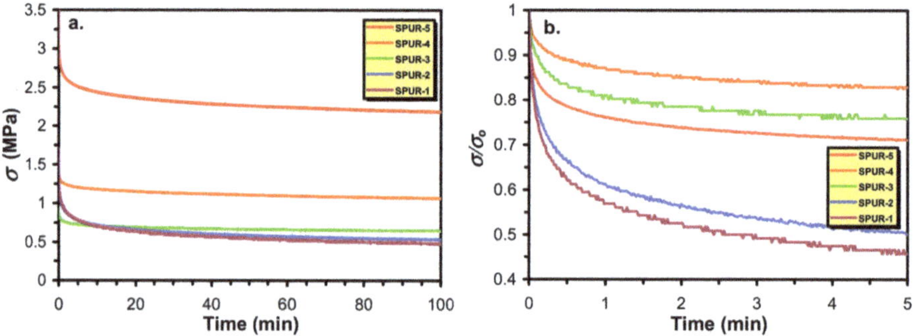

Figure 8. Stress (σ) (**a**) and relative stress (σ/σ_o) (**b**) relaxation curves for the SPUR samples. The relative stress is defined as σ/σ_o, where σ_o and σ are the initial and the instant stress, respectively.

As it turns out from Figure 8a there is a relative fast relaxation process at the beginning of the relaxation curves that is followed by a slower relaxation period and eventually reaching a plateau value of σ that corresponds to the equilibrium relaxation modulus. Furthermore, as it is also seen from Figure 8b, the extent of decrease in the values of σ/σ_o with time and to the equilibrium value follows an order of SPUR-4 > SPUR-3 > SPUR-5 > SPUR-2 > SPUR-1. This trend is almost in line with the decreasing crosslink densities for these SPURs except sample SPUR-5. This exception may highlight the fact that although the relative "viscous" part (relaxation to the equilibrium state) is mainly determined by the crosslink densities other factors can also affect the relaxation process. Moreover, the equilibrium stress value called as equilibrium modulus (σ_{eq}) is directly related to the crosslink density of elastomers (ν_r) according to theory of rubbers [22] as shown by Equation (13).

$$\nu_r = \frac{\sigma_{eq}}{RT(\lambda - 1/\lambda^2)} \tag{13}$$

where λ, R and T are the extension ratio (i.e. the ratio of the extended and the initial length of the sample), the universal gas-constant and the temperature in Kelvin, respectively.

The corresponding crosslink densities calculated by Equation (13) are compiled in Table 5.

Table 5. The crosslink densities calculated by Equation (13). for the SPUR samples.

Sample	$v_r \times 10^4$ (mol/cm^3)
SPUR-1	1.1
SPUR-2	1.2
SPUR-3	1.5
SPUR-4	2.4
SPUR-5	5.0

Comparing the data of Table 5 with that of Table 3, one can realize that the crosslink densities determined by the two methods (swelling and stress relaxation) agree relatively well. Hence, the decreasing crosslink densities with the increasing sucrose content is again supported by an independent method.

3.6. Thermal and Thermomechanical Properties of PURs

The thermal properties of the SPUR samples were investigated by differential scanning calorimetry (DSC) to determine their thermal transitions such as the glass transition temperature (T_g) and the melting temperature (T_m). A typical DSC trace for the sample SPUR-3 along with the characteristic thermal transitions are shown in Figure 9.

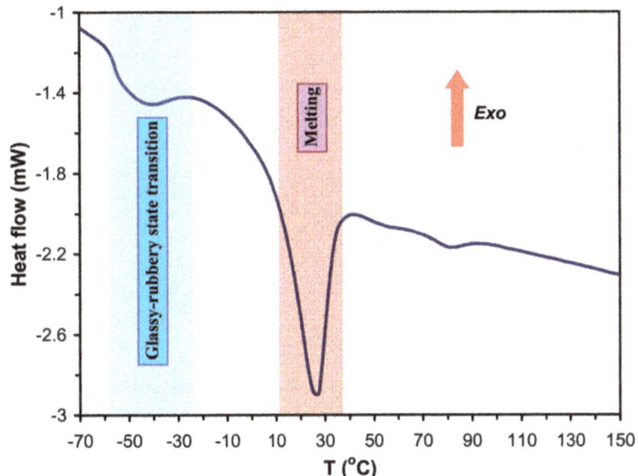

Figure 9. Representative DSC trace for the sample SPUR-3 ($T_g = -56$ °C, and $T_m = 26$ °C) (The DSC traces for the samples SPUR-1, SPUR-2, SPUR-4 and SPUR-5 can be seen in Figures S4–S7, respectively in the Supplementary Materials).

As seen in Figure 9 the T_g occurs between -50°C and -60 °C, while T_m are present at 20–30 °C. The values of T_g and T_m of SPUR samples determined from DSC measurements are presented in Table 6.

Table 6. The glass-transition (T_g) and the melting temperature (T_g) determined from DSC measurements.

Sample	T_g (°C)	T_m (°C)
SPUR-1	−58	27
SPUR-2	−58	27
SPUR-3	−56	26
SPUR-4	−52	30
SPUR-5	−51	23

As it turns out from Table 6, the T_g-s of the SPUR samples vary from -58 °C to -51 °C with the decreasing sucrose content, i.e., T_g increase with the increasing crosslink density. This finding is in line with the fact that chain mobility is gradually reduced with the increasing crosslink densities resulting in an increasing trend for T_g values. Furthermore, it can also be expected that at higher crosslink densities the crystallization is more hindered giving rise to lower T_m values. On the other hand, the T_m values for the PCLD segments in the SPUR samples are close to the room temperature, and thus are much lower than that of the high molecular weight poly(ε-caprolactone) (PCL), which melts at ca. 60 °C. That means the PCLD segments in the SPUR networks are in amorphous rubbery physical state. This fact is highly favorable for improving the chemical and/or biodegradation process in biological environments.

The dynamical mechanical properties of SPUR samples were also characterized by DMA. The dependences of the storage moduli (E′) on the temperature for the samples SPUR-5, SPUR-4 and SPUR-3 are shown in Figure 10.

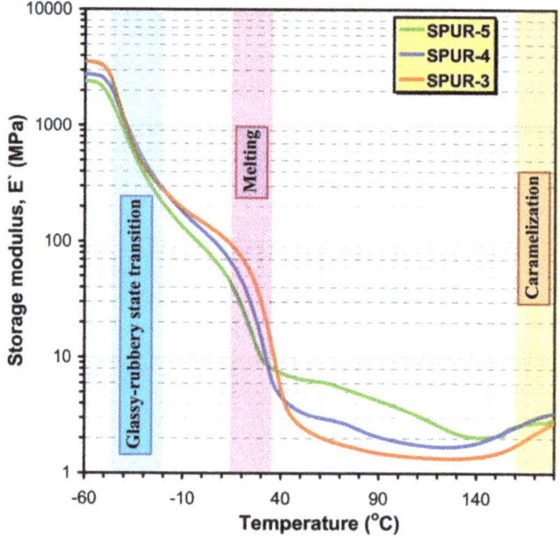

Figure 10. Variation of the storage modulus (E′) with the temperature for the samples SPUR-5, SPUR-4 and SPUR-3. (The DMA traces for the samples SPUR-1 and SPUR-2 are shown in Figures S8 and S9, respectively in the Supplementary Materials).

As seen in Figure 10 different regions can be distinguished on the DMA curves: at low temperature (at about -50 °C) the considerable decrease in E′ can be attributed to the transition from the glassy to the rubbery state and a further, marked decrease at around 20–30 °C is due to the melting of the crystalline segments. The temperatures at which these processes occur are in good agreement with the values of T_g and T_m obtained from DSC measurements (see Table 6). At higher temperatures, however, the melting is followed by a gradual loosening of the physical network, but owing to the chemical crosslinks, DMA curves exhibit rubbery like plateaus in the temperature range of 100–150 °C. Typically, above 140 °C, an increase in the storage modulus with the temperature can be observed. This finding is most probably due to the caramelization of the sucrose and thus this process leads to the formation of tighter crosslinks between these sucrose moieties. Furthermore, it can also be surmised that higher crosslink densities yield lower T_m-s owing to the reasons outlined above. Indeed, as seen in Figure 10. the melting of the crystalline phase occurs in the order of T_m: SPUR-5 < SPUR-4 < SPUR-3 and this order is in line with the crosslink densities determined from swelling and the stress relaxation experiments (see Tables 3 and 5) as well as with the data of Table 6. In addition, the rubbery-like plateau provides another way to estimate the crosslink density (v_d) from DMA results according to Equation (14).

$$v_d = \frac{E\prime}{3RT} \tag{14}$$

where E' is the value of the storage modulus at the plateau value of T and R is the universal gas-constant, and T is the temperature in Kelvin.

The crosslink densities from DMA data for samples SPUR-5, SPUR-4, SPUR-3, SPUR-2 and SPUR-1 were determined to be 2.1×10^{-4} mol/cm^3, 1.7×10^{-4} mol/cm^3, 1.3×10^{-4} mol/cm^3, 8.8×10^{-5} mol/cm^3 and 8.6×10^{-5} mol/cm^3, respectively. Although these values are lower than those determined by swelling and stress relaxation experiments the order of v_d agrees well.

4. Conclusions

For improving the mechanical, chemical and biological properties of poly(ε-caprolactone) for potential biomedical and pharmaceutical applications novel polyurethanes were synthesized involving sucrose as chain extender/crosslinking agent and HDI. For the selection of the reactants and solvents the proposals of European Pharmacopoeia-8 were considered, i.e., no aromatic diisocyanates were used and the solvents applied were selected from class 2 and class 3 groups.

For the synthesis, a novel synthetic strategy was elaborated and used. First a prepolymer was synthesized in melt from PCLD and HDI. The main polymer chains of the networks were prepared "in situ" by reacting the prepolymer with the two most reactive primary hydroxyl groups at the 6 and 6' position of the sucrose molecule. The networks were formed after reacting the sucrose less reactive primary OH (-OH group at 1' position) with the excess of the prepolymer. This way, the distances between the main polymer chains can be changed deliberately by varying the M_n-s of the prepolymer.

The crosslink density can simply be adjusted with the molar ratio of the prepolymer and the sucrose. The prepared rubbery sheets showed highly improved swelling properties. Furthermore, the mechanical properties such as the tensile behaviors and flexibility of SPURs can be chemically tailored.

Supplementary Materials: The following are available online at http://www.mdpi.com/2073-4360/11/5/825/s1, Table S1. The A, k, and γ parameters obtained by fitting of the KEKAM model to the experimental data. Figure S1: The stress-strain ($\sigma-\varepsilon$) curve for the SPUR-1 sample. Figure S2: The stress-strain ($\sigma-\varepsilon$) curve for the SPUR-3 sample. Figure S3: The stress-strain ($\sigma-\varepsilon$) curve for the SPUR-4 sample. Figure S4: DSC trace for the SPUR-1 sample (Tg = -58 °C, and Tm = 27 °C). Figure S5: DSC trace for the SPUR-2 sample (Tg = -58 °C, and Tm = 27 °C). Figure S6: DSC trace for the SPUR-4 sample (Tg = -52 °C, and Tm = 30 °C). Figure S7: DSC trace for the SPUR-5 sample (Tg = -51 °C, and Tm = 23 °C). Figure S8: DMA trace (variation of the storage modulus (E′) with the temperature in the range of -60–$+180$ °C) for the SPUR-1 sample. Figure S9: DMA trace (variation of the storage modulus (E′) with the temperature in the range of -60–$+180$ °C) for the SPUR-2 sample.

Author Contributions: L.N, G.D., M.N., M.Z. and S.K. conceived and designed the experiments; M.N., B.V., G.D. performed the experiments; L.D. performed the SEM investigations; L.N., L.D., B.V., G.D., M.Z. and S.K. analyzed the data; and L.N., M.N., M.Z. and S.K. wrote the paper.

Funding: This work was supported by the Higher Education Institutional Excellence Programme of the Ministry of Human Capacities in Hungary, within the framework of the Biotechnology thematic programme of the University of Debrecen, the GINOP-2.3.3-15-2016-00021 project, the grant No. FK-128783 from National Research, Development and Innovation Office (NKFI). Furthermore, this paper was also supported by the János Bolyai Research Scholarship of the Hungarian Academy of Sciences (M.N.).

Acknowledgments: The authors would like to express their thanks to Tamás Bárány (Head of the Department) for providing possibility for the DSC and DMA measurements and to Balázs Pinke for his assistance in these measurements (Department of Polymer Engineering, Budapest University of Technology and Economics).

Conflicts of Interest: The authors declare no conflict of interest.

References

1. Dhandayuthapani, B.; Yoshida, Y.; Maekawa, T.; Kumar, D.S. Polymeric Scaffolds in Tissue Engineering Application: A Review. *Int. J. Polym. Sci.* **2011**, *2011*, 290602. [CrossRef]
2. Azimi, B.; Nourpanah, P.; Rabiee, M.; Arbab, S. Poly (ε caprolactone) fiber: An overview. *J. Engin. Fib. Fab.* **2014**, *9*, 74–90.

3. Hajiali, F.; Tajbakhsh, S.; Shojaei, A. Fabrication and properties of polycaprolactone composites calcium phosphate-based ceramics and bioactive glasses in bone tissue engineering: A review. *Polym. Rev.* **2018**, *58*, 164–207. [CrossRef]

4. Moers-Carpi, M.M.; Sherwood, S. Polycaprolactone for the correction of nasolabial folds: A 24-month, prospective, randomized, controlled clinical trial. *Am. Soc. Dermatol. Surg.* **2013**, *39*, 457–463. [CrossRef] [PubMed]

5. Odian, G. *Principles of Polymerization*, 3rd ed.; John Wiley & Sons Inc.: New York, NY, USA, 1991; pp. 136–138.

6. Galbis, J.A.; de Gracia García-Martín, M.; de Paz, M.V.; Galbis, E. Bio-based polyurethanes from carbohydrate monomers. In *Aspects of Polyurethanes*; Chapter 7; InTech: London, UK, 2017; pp. 155–192.

7. Garcon, R.; Clerk, C.; Gesson, J.-P.; Bordado, J.; Nunes, T.; Caroco, S.; Gomes, P.T.; Minas da Piedade, M.E.; Rauter, A.P. Synthesis of novel polyurethanes from sugars and 1,6-hexamethylene diisocyanate. *Carbohydr. Polym.* **2001**, *45*, 123–127. [CrossRef]

8. Barikani, M.; Mohammadi, M. Synthesis and characterization of starch-modified polyurethane. *Carbohydr. Polym.* **2007**, *68*, 773–780. [CrossRef]

9. Kizuka, K.; Inoue, S.-I. Synthesis and properties of polyuerthane elastomers containing sucrose as a cross-linker. *Open J. Org. Polym. Mater.* **2015**, *5*, 103–112. [CrossRef]

10. *European Pharmacopoeia 8*, 8th ed. European Directorate for the Quality of Medicines & Health Care: Strasbourg, France, 2013; Volume 1, 639–644.

11. Nagy, L.; Nagy, T.; Kuki, A.; Purgel, M.; Zsuga, M.; Keki, S. Kinetics of uncatalyzed reactions of 2,4′- and 4,4′-diphenylmethane-diisocyanate with primary and secondary alcohols. *Int. J. Chem. Kinet.* **2017**, *49*, 643–655. [CrossRef]

12. Mondal, S.; Martin, D. Hydrolytic degradation of segmented polyurethane copolymers for biomedical applications. *Polym. Degrad. Stab.* **2012**, *97*, 1553–1561. [CrossRef]

13. Peng, Z.; Zhou, P.; Zhang, F.; Peng, X. Preparation and Properties of Polyurethane Hydrogels Based on Hexamethylene Diisocyanate/Polycaprolactone-Polyethylene Glycol. *J. Macromol. Sci. B* **2018**, *57*, 187–195. [CrossRef]

14. Flory, P.J. *Principles of Polymer Chemistry*; Cornell University Press: Ithaca, NY, USA, 1953; 579p.

15. Marecka, A.; Mianowski, A. Kinetics of CO_2 and CH_4 sorption on high rank coal at ambient temperatures. *Fuel* **1993**, *77*, 1691–1696. [CrossRef]

16. Malana, M.A.; Zafar, Z.I.; Zuhra, R. Effect of Cross Linker Concentration on Swelling Kinetics of a Synthesized Ternary Co-Polymer System. *J. Chem. Soc. Pak.* **2012**, *34*, 793–801.

17. Brouers, F.; Sotolongo-Costa, O. Generalized fractal kinetics in complex systems (application to biophysics and biotechnology). *Phys. A Stat. Mech. Appl.* **2006**, *368*, 165–175. [CrossRef]

18. Chee, K.K. Solubility Parameters of Polymers from Swelling Measurements at 60 °C. *J. App. Polym. Sci.* **1995**, *58*, 2057–2062. [CrossRef]

19. Hansen, C.M. 50 Years with solubility parameters—Past and future. *Prog. Org. Coat.* **2004**, *51*, 77–84. [CrossRef]

20. Karger-Kocsis, J.; Kéki, S. Review of Progress in Shape Memory Epoxies and Their Composites. *Polymers* **2018**, *10*, 34. [CrossRef] [PubMed]

21. Czifrák, K.; Lakatos, C.; Karger-Kocsis, J.; Daróczi, L.; Zsuga, M.; Kéki, S. One-Pot Synthesis and Characterization of Novel Shape-Memory Poly(ε-Caprolactone) Based Polyurethane-Epoxy Co-networks with Diels-Alder Couplings. *Polymers* **2018**, *10*, 504. [CrossRef] [PubMed]

22. Sekkar, V.; Gopalakrishnan, S.; Ambika Devi, K. Studies on allophanate–urethane networks based on hydroxyl terminated polybutadiene: Effect of isocyanate type on the network characteristics. *Eur. Polym. J.* **2003**, *39*, 1281–1990. [CrossRef]

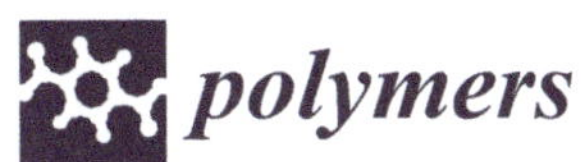 *polymers*

Article

An Ab Initio Investigation of the 4,4′-Methlylene Diphenyl Diamine (4,4′-MDA) Formation from the Reaction of Aniline with Formaldehyde

R. Zsanett Boros [1,2], László Farkas [1], Károly Nehéz [3], Béla Viskolcz [2,*] and Milán Szőri [2,*]

[1] Wanhua-BorsodChem Zrt, Bolyai tér 1., H-3700 Kazincbarcika, Hungary;
renata.boros@borsodchem.eu (R.Z.B.); laszlo.farkas@borsodchem.eu (L.F.)
[2] Institute of Chemistry, University of Miskolc, Miskolc-Egyetemváros A/2, H-3515 Miskolc, Hungary
[3] Department of Information Engineering, University of Miskolc, Miskolc-Egyetemváros Informatics Building,
H-3515 Miskolc, Hungary; aitnehez@uni-miskolc.hu
* Correspondence: bela.viskolcz@uni-miskolc.hu (B.V.); milan.szori@uni-miskolc.hu (M.S.);
Tel.: +36-46-565-373 (B.V.); +36-46-565-111/1337 (M.S.)

Received: 30 January 2019; Accepted: 27 February 2019; Published: 1 March 2019

Abstract: The most commonly applied industrial synthesis of 4,4′-methylene diphenyl diamine (4,4′-MDA), an important polyurethane intermediate, is the reaction of aniline and formaldehyde. Molecular understanding of the 4,4′-MDA formation can provide strategy to prevent from side reactions. In this work, a molecular mechanism consisted of eight consecutive, elementary reaction steps from anilines and formaldehyde to the formation of 4,4′-MDA in acidic media is proposed using accurate G3MP2B3 composite quantum chemical method. Then G3MP2B3-SMD results in aqueous and aniline solutions were compared to the gas phase mechanism. Based on the gas phase calculations standard enthalpy of formation, entropy and heat capacity values were evaluated using G3MP2B3 results for intermediates The proposed mechanism was critically evaluated and important side reactions are considered: the competition of formation of protonated p-aminobenzylaniline (PABAH$^+$), protonated aminal (AMH$^+$) and o-aminobenzylaniline (OABAH$^+$). Competing reactions of the 4,4′-MDA formation is also thermodynamically analyzed such as the formation of 2,4-MDAH$^+$, 3,4-MDAH$^+$. AMH$^+$ can be formed through loose transition state, but it becomes kinetic dead-end, while formation of significant amount of 2,4-MDA is plausible through low-lying transition state. The acid strength of the key intermediates such as N-methylenebenzeneanilium, PABAH$^+$, 4-methylidenecyclohexa-2,5-diene-1-iminium, and AMH$^+$ was estimated by relative pK_a calculation.

Keywords: bis(4-aminophenyl)methane; MDA; PABA; aniline; water; reaction mechanism; ab initio; G3MP2B3; transition state; pK_a; standard enthalpy of formation

---

1. Introduction

Polyurethane industry requires large amount of methylene diphenyl diisocyanate (MDI) as raw material and global market of MDI reached 6 Mt in 2016 [1]. The MDI production is mostly based on methylene diphenyl diamine (methylenedianiline, MDA). MDA can be also applied as ingredient of epoxy resins, intermediate for pigments, organic dyes, coatings, plastic fibers, and insulation materials [2] making it an important intermediate for chemical industry.

The most commonly applied industrial MDA synthesis is the reaction of aniline with formalin in the presence of hydrochloric acid under mild (60–110 °C) reaction conditions [3]. To avoid the use of corrosive hydrochloric acid and the formation of large amount of salt solution as waste, several attempts [4–11] had been made to replace the current technology with catalytic process using solid acids, zeolites, delaminated materials, ionic liquids, or ion exchange resins as catalysts. However, neither of the catalytic MDA productions have gone beyond the laboratory stage. Although, proposed reaction mechanism for current MDA synthesis is presented in Ref. [12] (see left side of Figure 1), it has not been clarified entirely. According to this mechanism, aniline (A) is reacted with formaldehyde (F) producing N-hydroxymethyl aniline as the initial reaction step. In acidic medium this product loses water rapidly to form N-methylidene anilinium which reacts with aniline to form N-(p-aminobenzyl)aniline (PABA). PABA is then decomposed to 4-aminobenzylium and aniline. In the last step of this rearrangement 4,4′-methylene diphenyl diamine (4,4′-MDA) is formed as final product. Beside missing elementary steps, this early mechanism also cannot clarify the detected side products of MDA.

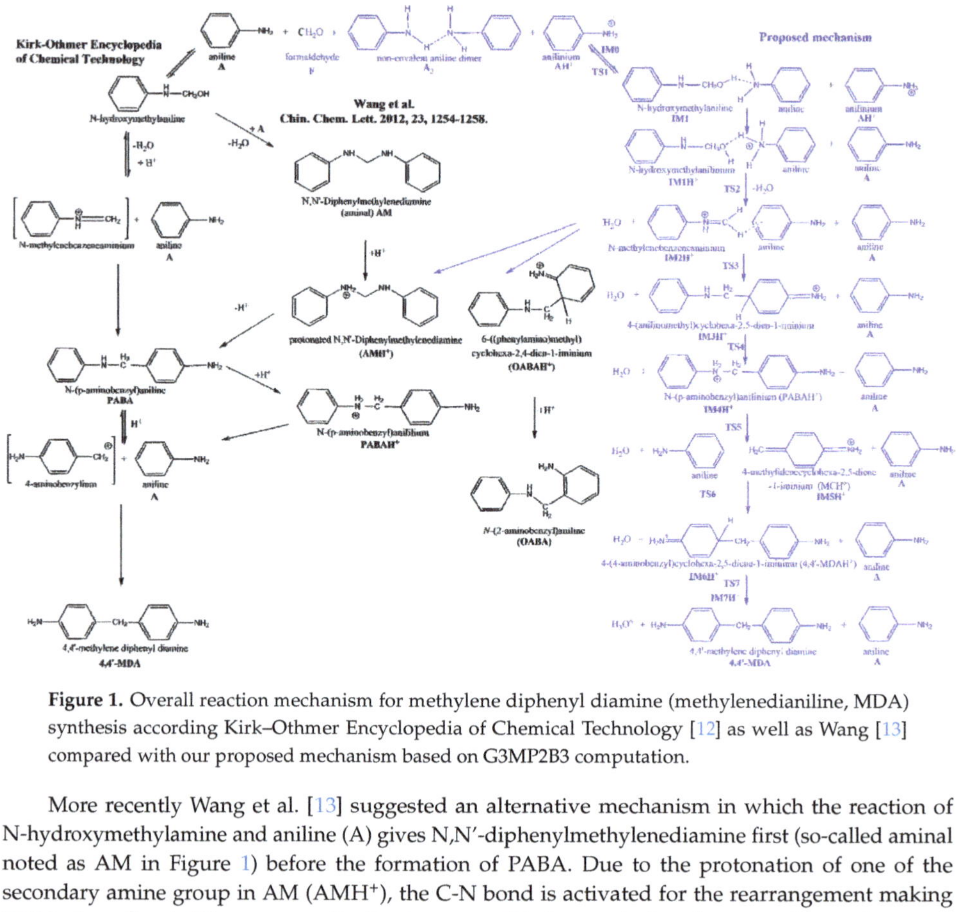

Figure 1. Overall reaction mechanism for methylene diphenyl diamine (methylenedianiline, MDA) synthesis according Kirk–Othmer Encyclopedia of Chemical Technology [12] as well as Wang [13] compared with our proposed mechanism based on G3MP2B3 computation.

More recently Wang et al. [13] suggested an alternative mechanism in which the reaction of N-hydroxymethylamine and aniline (A) gives N,N′-diphenylmethylenediamine first (so-called aminal noted as AM in Figure 1) before the formation of PABA. Due to the protonation of one of the secondary amine group in AM (AMH$^+$), the C-N bond is activated for the rearrangement making PABA. According to Wang, same activation is supposed to happen to the other secondary amine group to form MDA: PABA is protonated (PABAH$^+$) which then initiates aniline rearrangement. Wang et al. found some evidences to support the latter mechanism by isolation and identification of aminal using the combination of isotope labeling and HPLC-MS [13]. Albeit other intermediates has not been characterized at all. In the same study by-products including oligomers of MDA (e.g., 3- and 4-ring MDA) were also suggested. As a continuation of this work stabilities, potential protonation sites and structural characterization of these MDA oligomers were determined using ion mobility-mass spectrometry (IM-MS) and tandem mass spectrometry (MS/MS) techniques [14]. This resulted in two partially overlapping, but still gappy mechanism for MDA production as seen from the left side and the center of Figure 1. Furthermore, initial step of both mechanisms requires close contact of aniline and aqueous formaldehyde. Although aniline and water have similar density [15], but they are only slightly soluble in each other and aniline has large viscosity (3.770 cP) due to strong hydrogen bond between amine groups, [16] therefore, solvent effect can be crucial in the MDA production for these solvents.

Although, the MDA synthesis is essential to MDI production, the thermochemical properties of the participated species are also poorly characterized [17]. To the best of our knowledge, only the standard reaction enthalpy of formation for MDA, PABA, and AM is estimated using Benson group additivity method [18]. Only the heat of formation of 4,4′-MDA had been reported very recently using G3MP2B3 quantum chemistry protocol and group additivity increments for OCN and NHCOCl groups has been recommended [19] based on protocols mentioned in Ref. [20,21].

The aim of this study is to establish an ab initio-based molecular mechanism consisting of consecutive elementary reaction steps from anilines and formaldehyde to the formation of 4,4′-MDA. These elementary reactions can include short living species with low steady state concentrations which makes them inaccessible for most of the current detection techniques. Furthermore, we characterized thermodynamically the detailed reaction mechanism of the current industrial MDA production using quantum chemical calculations. The effects of the surrounding media as well as the protonation state of the intermediates is also characterized. Few important competing reaction channels are also included to provide theoretically established evidence for by-products.

2. Methods

G3MP2B3 [22] composite method was applied for obtaining thermodynamic properties such as zero-point corrected relative energy (ΔE_0), relative enthalpy (ΔH), standard enthalpy of formation ($\Delta_{f,298.15K}H(g)$), relative molar Gibbs free energy (ΔG), standard molar entropy (S), and heat capacity (C_V) for the species involved in the reaction mechanism of the formation of MDA from the reaction of aniline molecules with formaldehyde in acidic medium. As part of G3MP2B3 protocol B3LYP [23] functional was applied in combination with the 6-31G(d) basis set for geometry optimizations and frequency calculations, except for solvated TS2 structures (Since B3LYP/6-31G(d) was unable to localize these transition states (TS), it was replaced by BH&HLYP/6-31G(d) level of theory). Normal mode analysis was performed on the optimized structures at the same level of theory to characterize their identities on the potential energy surface (PES). TS structures were also checked by visual inspection of the intramolecular motions corresponding to the imaginary wavenumber using GaussView05 [24] as well as confirmed by intrinsic reaction coordinate (IRC) calculations [25] for mapping out the minimal energy pathways (MEP).

Although accuracy of G3MP2B3 is proven in our previous work for similar reaction system [19], gas phase standard heat of formation values obtained at G3MP2B3 and CBS-QB3 [26] computations were compared for further verification purposes. The gas phase standard heat of formation values at T = 298.15K, $\Delta_{f,298.15K}H(g)$, were achieved using an atomization scheme (AS) [27] and isodesmic reaction (IR). Accurate literature data necessary for AS calculation was collected from Ruscic's Active Thermochemistry Tables [28].

SMD polarizable continuum model developed by Truhlar and co-workers [29] was used to estimate the effect of the surrounding solvent (water and aniline). The acid dissociation constant ($pK_{a,aq}$) in aqueous solution of the intermediates were also derived from G3MP2B3 calculations. The $pK_{a,aq}$ values can be calculated using Gibbs free energy of the gas phase species in neutral and cationic form (noted as $G(A_g)$ and $G(AH^+_g)$, respectively) which corrected by the solvation Gibbs free energies (ΔG_{aq}) of the species involved [30]:

$$pK_{a,aq} = [G(A_g) - G(AH^+_g) + \Delta G_{aq}(A) - \Delta G_{aq}(AH^+) + G_g(H^+) + G_{aq}(H^+) + RT\ln(V_m)]/RT\ln10, \quad (1)$$

where gas phase Gibbs free energy for proton $G_g(H^+)$ comes from Sackur-Tetrode equation (26.3 kJ/mol [31]) and solvation Gibbs free energy for proton, $\Delta G_{aq}(H^+)$, is −1107.1 kJ/mol [32].

Molar volume (V_m) at reference state is 24.46 dm^3. Although, such direct pK_a calculation via thermodynamics cycle usually suffers from larger error, therefore proton exchange scheme was applied to improve the accuracy of a pK_a calculation (pK_a of anilinium, AH$^+$, used as reference). All quantum chemical calculations were performed by Gaussian09 [33] software package.

3. Results and Discussion

3.1. Main Reaction Mechanism in Gas Phase

Similar to our previous study [19] gas phase mechanism is used as reference to measure the solvent effect as the presence of solvent can make a significant difference on the reaction energy profile.

According to the X-ray scattering experiment of aniline [16], closest intermolecular distance between nitrogen atoms in a pair of aniline (A) molecules is 3.31 Å, which is close to the distance (3.159 Å) found in the V-shape non-covalent bonded aniline dimer (A$_2$) computed at G3MP2B3 in gas phase. This short distance and orientational preference are due to the strong hydrogen bond between two amine groups explaining the large viscosity of liquid aniline [16], which is in line with our G3MP2B3 results (the formation of the gas phase non-covalent aniline dimer is -19.5 kJ/mol in term of zero-point corrected energy). Therefore, in our calculations, A$_2$ structure had been selected as initial structure (see right side of the Figure 1). The energy reference used in Figure 2 (also given in Table 1) corresponds to the sum of non-covalent A$_2$ dimer, formaldehyde (F) and protonated aniline (AH$^+$). The latter species supposes to mimic the acidic environment used by chemical industry.

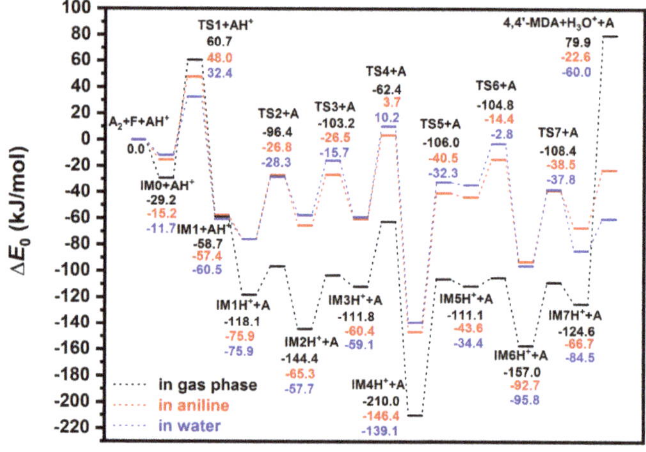

Figure 2. G3MPB3 energy profile (zero-point corrected) for MDA synthesis in gas phase (black), in aniline (red) and in water (blue).

The protonated aniline dimer (A$_2$) approaches the formaldehyde (F) to form the pre-reactive complex (IM0, see Figure 3), in which formaldehyde is strongly hydrogen bonded to one of the anilines (r_{O-H} = 2.108 Å) and its carbon atom approaches the nitrogen atom of the other amine (r_{C-N} = 2.671 Å). This structure is 29.2 kJ/mol lower in energy compared to the separated formaldehyde, protonated aniline and aniline dimer (reference state). As seen in Figure 3, the transition state for formaldehyde addition to the aniline (TS1) is a six-membered ring structure, where the carbon of F got closer to the nitrogen of the amine by roughly 1 Å (r_{C-N} = 1.644 Å) compared to the IM0. Simultaneously, attacked nitrogen releases a hydrogen which is transferred to the oxygen of formaldehyde through the other aniline amine group. The corresponding energy is 60.7 kJ/mol higher than that of the reactants making this structure as the highest energy TS along the entire reaction mechanism studied here.

As next step, the resulted N-hydroxymethylaniline (IM1) forms molecular complex with protonated aniline (AH$^+$), noted as IM1H$^+$, in which the protonated amine group is in vicinity of the OH group of N-hydroxymethylaniline (r_{OH} = 1.599 Å). This protonated complex has significantly lower in energy (−118.1 kJ/mol) compared to the previous neutral complex (−58.7 kJ/mol).

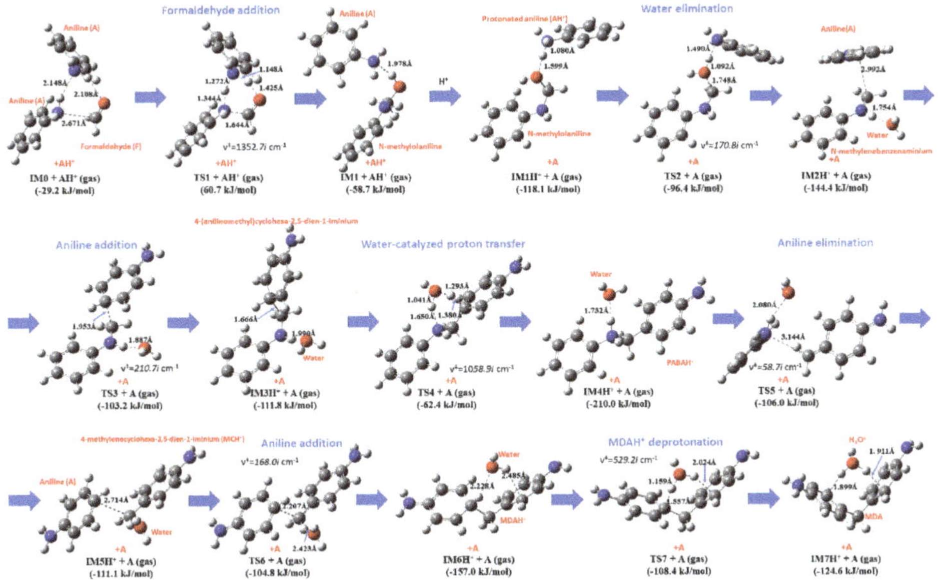

Figure 3. Transition state structures (obtained at B3LYP/6-31G(d) level of theory) for MDA synthesis in gas phase. The G3MP2B3 relative energies are also given.

Transition state of the water elimination (TS2) is structurally similar to the previous IM1H$^+$, although the O-H being formed become significantly shorter (r_{O-H} = 1.092 Å), while the C-O bond expanded to 1.748 Å. TS2 is 21.7 kJ/mol higher than IM1H$^+$. The product of this exothermic reaction (IM2H$^+$) is a trimolecular complex (aniline-water and N-methylenebenzeneaminium) having −144.4 kJ/mol of relative energy. In this structure, the carbon of the aniline at para position is just 2.992 Å far from the methylene group of N-methylenebenzenaminium and the water oxygen is hydrogen bonded to the hydrogen atom of the secondary amine group. The critical C-C distance is 1.953 Å in the transition state structure of the aniline addition (TS3) and water is only slightly rotated around the elongated hydrogen bond (1.887 Å). The activation energy of TS3 is 41.2 kJ/mol, while its relative energy become −103.2 kJ/mol. By linking these two aromatic ring structures, water complex of 4-(anilinomethyl)cyclohexa-2,5-dien-1-iminium is formed (IM3H$^+$) in a slightly endothermic reaction ($\Delta_r H$ = 29.5 kJ/mol obtained from data in Table 1). Despite the partial loss of the aromatic nature of the aniline, the IM3H$^+$ structure is −111.8 kJ/mol lower than the reference energy. In the next step of the proposed mechanism, the water reoriented to initiate the transfer of the positive charge to the amine nitrogen in such a way that the second aromatic ring can also be formed (TS4). This six-centered transition state with the activation energy of 49.4 kJ/mol resulted in the product (noted as IM4H$^+$) with the lowest relative energy (ΔE_0(IM4H$^+$) = −210.0 kJ/mol) structures in the entire mechanism, the hydrogen bonded complex of the N-(p-aminobenzyl)anilinium (PABAH$^+$) and water. Due to the proton of the amine, PABAH$^+$ is activated to dissociate to aniline (A) and 4-methylidenecyclohexa-2,5-diene-1-iminium (MCH$^+$) noted as IM5H$^+$ via C-C bond scission (TS5). The activation energy of this reaction step is 104.0 kJ/mol. To make rearrangement of the released aniline happen, the formed PABAH$^+$ should be chemically activated and stabilization of PABAH$^+$

should be avoided (e.g., through deprotonation of PABAH$^+$). Firstly, we considered the aniline addition occurs at para position to the MCH$^+$ to form protonated 4,4'-methylenedianiline (4,4'-MDAH$^+$) noted as IM6H$^+$. After the formation of the loosely bounded complex of MCH$^+$ and aniline (IM5H$^+$), these two species can form TS6 structure in which C-C bonds being formed is 2.207 Å. Only small rotational motion of the water molecule contributes to the reaction coordinate ($\nu^{\ddagger}$ = 168.0i cm^{-1}) in this case. The energy level of TS6 is −104.8 kJ/mol compared to the entrance level.

Table 1. G3MP2B3 thermochemical properties calculated in gas phase, in aniline and in water including zero-point corrected relative energies (ΔE_0), relative enthalpies (ΔH(T)) and relative Gibbs free energies (ΔG(T,P)) at T = 273.15 K, and P = 1 atm.

Species	ΔE_0			ΔH(T)			ΔG(T,P)		
	Gas	Aniline	Water	Gas	Aniline	Water	Gas	Aniline	Water
A$_2$ + F + AH$^+$	0.0	0.0	0.0	0.0	0.0	0.0	0.0	0.0	0.0
IM0 + AH$^+$	−29.2	−15.2	−11.7	−30.0	−15.2	−11.7	17.2	26.3	31.3
TS1 + AH$^+$	60.7	48.0	32.4	52.4	40.2	24.6	121.7	106.3	92.5
IM1 + AH$^+$	−58.7	−57.4	−60.5	−64.0	−61.9	−65.6	−6.2	−10.1	−8.3
IM1H$^+$ + A	−118.1	−75.9	−75.9	−123.3	−80.3	−81.0	−68.6	−26.4	−23.8
TS2 + A	−96.4	−26.8	−28.3	−121.6	−53.0	−53.8	38.2	106.2	102.3
IM2H$^+$ + A	−144.4	−65.3	−57.7	−162.4	−83.2	−75.2	−18.5	58.0	64.0
TS3 + A	−103.2	−26.4	−15.7	−124.4	−46.3	−35.1	30.9	104.1	114.2
IM3H$^+$ + A	−111.8	−60.4	−59.1	−132.9	−80.5	−79.6	23.8	70.0	72.7
TS4 + A	−62.4	3.7	10.2	−88.3	−20.5	−13.6	81.1	141.0	148.1
IM4H$^+$ + A	−210.0	−146.4	−139.1	−231.8	−167.5	−160.5	−72.9	−14.6	−6.9
TS5 + A	−106.0	−40.5	−32.3	−125.0	−57.7	−49.9	21.0	79.9	87.8
IM5H$^+$ + A	−111.1	−43.6	−34.4	−127.3	−58.9	−49.3	15.1	72.3	83.0
TS6 + A	−104.8	−14.4	−2.8	−123.0	−29.9	−20.7	28.0	108.8	128.2
IM6H$^+$ + A	−157.0	−92.7	−95.8	−177.3	−110.7	−114.3	−20.0	34.5	33.4
TS7 + A	−108.4	−38.5	−37.8	−132.3	−60.5	−60.1	35.7	98.4	101.2
IM7H$^+$ + A	−124.6	−66.7	−84.5	−147.3	−88.1	−06.3	17.0	69.0	53.1
4,4'-MDA + H$_3$O$^+$ + A	79.9	−22.6	−60.0	60.0	−41.5	−79.3	176.7	67.6	32.8

In IM6H$^+$, water binds to the MDAH$^+$ by hydrogen bond and interaction occurs of its other hydrogen and the aromatic ring. This structure shows similarity for the deprotonation transition state, that is TS7, in which the distance of C-H being broken is significantly elongated (1.557 Å) and the critical distance of H-O bond is short (1.159 Å). Moving along the reaction coordinate the water molecule reoriented again and its one of the alone pairs is now pointed toward the extra proton of IM6H$^+$ making the post-reaction complex, IM7H$^+$ after a slight reorientation of the H$_3$O$^+$ cation. This structural change resulted in an energy decrease by 16.2 kJ/mol (ΔE_0(IM7H$^+$) = −124.6 kJ/mol). The relative energy of last three transition states is also energetically close to each other (ΔE_0(TS5) = −106.0 kJ/mol, ΔE_0(TS6) = −104.8 kJ/mol, and ΔE_0(TS7) = −108.4 kJ/mol). Finally, the post-reaction complex converted into the final product that is the 4,4'-MDA (ΔE_0(4,4'-MDA) = 79.9 kJ/mol) which has the highest relative energy considering the whole reaction coordinate in gas phase. Basically, this consistent mechanism can be considered as an extension of the mechanism found in Kirk–Othmer Encyclopedia of Chemical Technology [12] by three new intermediate elementary steps (formation of IM3H$^+$, IM4H$^+$, and IM6H$^+$). No experimental evidence found for 4-(aminomethyl)cyclohexa-2,5-dien-1-iminium (part of IM3H$^+$ complex) as well as protonated 4,4'-MDA (part of IM6H$^+$ complex). One plausible reason for that is their destroyed aromatic nature making them short living and strong acid. Although, amongst these new species, IM4H$^+$ differs only in the protonation state from the PABA intermediate suggested in Ref. [12]. On the other hand, the one step formation of aminal (AM) suggested by Wang [13] is unlikely, since the corresponding hypothetical transition state would be crowded around the methylene carbon, therefore reaction of N-methylenebenzeneaminium and aniline is considered as a formation of protonated aminal (AMH$^+$) instead (for its detailed discussion see Important side reactions section). Before the solvent effect is discussed the gas phase thermochemistry of the species playing role is evaluated.

Polymers **2019**, *11*, 398

3.2. Gas Phase Thermodynamic Properties of Reactants, Intermediates, and Products

As shown in Table 2, the calculated standard enthalpies of formation for aniline, 4,4' MDA and all intermediates are endothermic, while the formation of formaldehyde (-111.5 kJ/mol) and N-hydroxymethylaniline (-71.8 kJ/mol) are exothermic.

Highly accurate standard enthalpy of formation ($\Delta_{f,298.15K}H^0$) value is only reported for aniline and formaldehyde in the literature [28,34]. The largest error for their G3MP2B3 computation is 2.3 kJ/mol, which is significantly smaller than those for CBS-QB3, therefore only the G3MP2B3 results are discussed latter. While the G3MP2B3 value for 4,4'-MDA (171.2 kJ/mol) is also consistent with the estimated values based on group additivity rules by Benson [18] and Benson and Stein [36–38] (165.6 kJ/mol [11] and 172.0 kJ/mol, respectively), only CBS-QB3 estimates more endothermic enthalpy of formation for 4,4'-MDA. Interestingly, ortho and meta isomers of MDA (2,4-MDA, 2',4-MDA, and 3,4-MDA) have less endothermic formation than 4,4'-MDA, while among their protonated forms 2,4-MDA and 4,4'-MDA found to be less endothermic. For these species, only modest difference in molar entropy had been found. Furthermore, G3MP2B3 and the group additivity values are also consistent with each other in the case of PABA. To the best of our knowledge, no literature $\Delta_{f,298.15K}H^0$ was found for other intermediates presented here (Table 2). The computed standard molar entropy ($S^\circ(g)$) and molar heat capacity ($C_V(g)$) values are also tabulated in Table 2, and their deviation from accurate literature values [34,35] is less than 8.6 J/molK, while larger deviation is observed from the results obtained form group additivity (22.6 J/molK for 4,4'-MDA) which is probably due to missing correction terms in the group additivity.

Table 2. Gas phase thermochemical properties for reactants, products and all the intermediates MDA synthesis as well as MDA and MDAH$^+$ isomers. Standard enthalpy of formation ($\Delta_{f,298.15K}H^\circ(g)$) is calculated from G3MP2B3 and CBS-QB3 enthalpies by means of atomization scheme (AS) at 1 atm pressure at 298.15 K. Absolute deviation is given in parenthesis.

Species	$\Delta_{f,298.15K}H^0$ (g) kJ/mol	Method	Ref.	S_0(g) J/molK	C_V(g) J/molK	Ref.
aniline (A)	86.5 (0.5)	AS(G3MP2B3)	1	319.0	96.6	1
	96.0 (9.0)	AS(CBS-QB3)	1	317.3	97.4	
	87.0 ± 0.88	Burcat	[34]	311.6	104.5	[34]
non-covalent aniline dimer (A$_2$)	156.2	AS(G3MP2B3)	311.7	529.6	214.3	1
formaldehyde (F)	−111.5 (2.3)	AS(G3MP2B3)	1	224.4	26.8	1
	−113.3 (4.1)	AS(CBS-QB3)	1	224.3	26.8	
	−109.2 ± 0.11	Ruscic ATcT	[28]	218.8	35.4	[35]
4,4'-methylene diphenyl diamine (4,4'-MDA)	171.2	AS(G3MP2B3)	[19]	500.1	221.4	1
	191.5	AS(CBS-QB3)		503.6	223.3	
	165.6	additivity rule	[18]	522.7	n.a.	[18]
	172	additivity rule [36,37]	NIST [37,38]	511.6	234.7	[38]
2,4-MDA	159.4	AS(G3MP2B3)	1	490.4	220.6	1
2',4-MDA	168	AS(G3MP2B3)	1	484.0	219.8	1
3,4-MDA	168.3	AS(G3MP2B3)	1	499.4	221.3	1
N-(p-aminobenzyl)aniline (PABA)	202.4	AS(G3MP2B3)	1	496.3	216.5	1
	201.3	additivity rule	[18]	514.4	n.a.	[18]
N-hydroxymethylaniline	−71.8	AS(G3MP2B3)	1	379.3	129.3	1
protonated aniline (AH+)	739.4	AS(G3MP2B3)	1	339.9	97.2	1
N-methylenebenzeneaminium	828.6	AS(G3MP2B3)	1	346.2	107.9	1
4-(anilinomethyl)cyclo-hexa-2,5-dien-1-iminium	858.5	AS(G3MP2B3)	1	491.2	220.0	1
p-aminobenzylaniline (PABAH+)	785.1	AS(G3MP2B3)	1	501.2	219.4	1
4-methylidenecyclohexa-2,5-diene-1-iminium (MCH+)	802.8	AS(G3MP2B3)	1	340.9	113.7	1
4,4'-MDAH$^+$	814.3	AS(G3MP2B3)	1	494.8	225.1	1
2,4-MDAH$^+$	812.2	AS(G3MP2B3)	1	494.7	226.4	1
2',4-MDAH$^+$	830.9	AS(G3MP2B3)	1	487.1	224.4	1
3,4-MDAH$^+$	858.6	AS(G3MP2B3)	1	534.2	236.5	1

^{1} this work.

3.3. Solvent Effect

The presence of solvents (aniline and water) changes significantly energy profile of the reaction as shown in Figure 2, the solvent effect is very similar regardless which solvent is considered. The most dramatic change is the stabilization of the final product (see 4,4′-MDA+H_3O^++A in Table 1) by 102.5 kJ/mol for aniline solution and by 139.9 kJ/mol for aqueous solution which is mainly due to the solvation energy difference between protonated aniline (AH^+) and hydronium cation (H_3O^+). The highest lying barrier in gas phase, that is TS1, is also decreased due to solvation by 12.7 kJ/mol (in aniline) and 28.3 kJ/mol (in water). While its pre-reaction (IM0 + AH^+) and post-reaction (IM1 + AH^+) complexes are destabilized slightly (14.0 kJ/mol and 1.3 kJ/mol in aniline, respectively), large destabilization effect can be observed for the intermediates and transition states after the protonation of IM1 intermediates (IMx and TSx, where $2 \leq x$), its magnitude is in the range of 42.2–90.5 kJ/mol for the aniline solution, while it is 40.1–102.1 kJ/mol for the aqueous solution (about 60 kJ/mol in average). The largest increase in relative energy belongs to the transition state of the second aniline addition (TS6) for both solutions, this energy shift was 90.5 kJ/mol and 102.1 kJ/mol for aniline and for aqueous solution, respectively. Similarly, the transition state for first aniline addition (TS3) and its pre-complex (IM2H^+) are also significantly destabilized by solvation compared to the other TSs and intermediates (in the range of 76.8–87.5 kJ/mol). Structural changes according to solvent effect are shown in the Supplementary Materials (see Figures S1 and S2 in for aqueous and aniline solution, respectively).

As Table 1 shows, relative enthalpies (ΔH^0) show same trend as ΔE_0, ΔH^0 values tend to be larger, but not more than 26.6 kJ/mol. Most of the cases, difference in relative enthalpies ($\Delta \Delta H^0_{an \rightarrow aq}$) by comparing aniline to aqueous phase is less than 9.2 kJ/mol. Larger difference in ΔH^0 values found in those cases, where large solvation effect for ΔE_0 had already been observed, such as TS1, TS3, TS6, IM7H^+ and 4,4′-MDA ($\Delta \Delta H^0_{an \rightarrow aq}$(TS1) = −15.6 kJ/mol, $\Delta \Delta H^0_{an \rightarrow aq}$(TS3) = 10.7 kJ/mol, $\Delta \Delta H^0_{an \rightarrow aq}$(TS6) = 11.6 kJ/mol, $\Delta \Delta H^0_{an \rightarrow aq}$(IM7$H^+$) = −17.8 kJ/mol, and $\Delta \Delta H^0_{an \rightarrow aq}$(4,4′MDA) = −37.3 kJ/mol). Analysis of the relative Gibbs free energies (ΔG^0) in Table 1 show also that solvation Gibbs free energies (ΔG_{aq}^0) and $\Delta E_{0,aq}$ are in linear relationship for both surrounding media.

3.4. Proton Dissociation Constants (pK_a) of Intermediates

The species in proposed mechanism are proton activated, their possible deprotonation can lead to side reactions and appearance of deactivated intermediates in the industrial process. Therefore, the affinity of these species for deprotonation is important and it can be described by site-specific acid dissociation constant (pK_a) [39,40]. Ghalami-Choobar et al. performed calculation for aqueous pK_b values of aniline and its substituted derivatives with good accuracy [41]. Behjatmanesh-Ardakani [42] and Lu [43] also reported calculated pK_a for some aniline derivatives. Although, direct pK_a calculation via thermodynamics cycle usually suffers from larger error than the relative method. In this study, the G3MP2B3-SMD based absolute pK_a value of anilinium (AH^+) was found to be 2.86 in aqueous phase using the direct pK_a estimation which is smaller by 1.74 pK_a units than the reference value ($pK_{a,aq}$ = 4.60 in Ref. [44], so we have used the deprotonation half-reaction of anilinium as reference for the relative pK_a calculation shown in Table 3.

N-methylenebenzeneanilium, N-(p-aminobenzyl)anilinium and PABAH$^+$ have similar pK_a values which corresponds to weak acid and they can lose their potential to turn into MDA by deprotonation, while acid strength of MCH$^+$ is far less therefore it likes to be protonated. Indeed, instead of PABAH$^+$, deprotonated PABAH$^+$ (PABA) has been detected from aniline-formaldehyde condensation mixture [45].

Table 3. Protonation dissociation constants ($pK_{a,aq}$) for the protonated intermediates in aqueous solution. The dissociative proton presented in red.

Species		$pK_{a,aq}$	$pK_{a,aq}$
AH$^+$		4.6 [1]	4.60 [44]
N-methylenebenzeneanilium	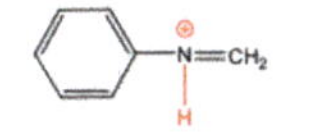 N-methylenebenzeneaminium	4.2	
PABAH$^+$	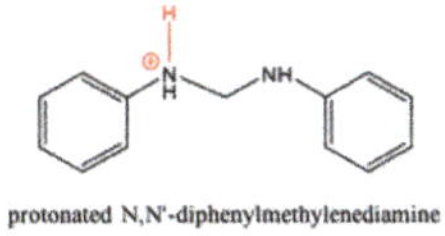 protonated N,N'-diphenylmethylenediamine	6.7	
MCH$^+$	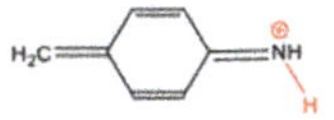 4-methylidenecyclohexa-2,5-diene-1-iminium	11.4	
AMH$^+$	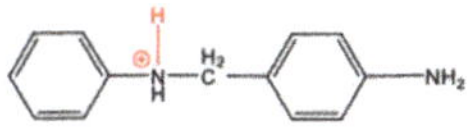 N-(p-aminobenzyl)anilinium	5.1	

[1] used as reference.

3.5. Important Side Reactions

Alternative aniline addition reactions can also be proposed. One of these side reactions can be the protonated aminal formation (see AMH$^+$ in Figure 1), which is essentially an alternative to the formation of IM3H$^+$ from aniline to the N-methylenebenzeneaminium ('the first' aniline addition). However, the aniline addition occurs through the amine group instead of the aromatic carbon at para position. Interestingly, loose transition state had been found for this reaction in both condensed media, which was also proven by scan of the potential energy surface via the C-N bond stretching of the protonated aminal at B3LYP/6-31G(d) level of theory. The B3LYP curve was also reproduced by BHandHLYP and MP2 methods. In contrast, the formation of ortho adduct (o-aminobenzylaniline, OABAH$^+$) undergoes tight submerged transition state ($\Delta^{\ddagger}E_0 = -9.4$ kJ/mol in aniline phase, $\Delta^{\ddagger}E_0 = -5.0$ kJ/mol in aqueous phase).

As seen from Table 4, aminal (AMH$^+$) formation reaction is exothermic, but it has only moderate exergonicity in each media studied in contrast to the competitive reaction step, IM3H$^+$ formation. However, the thermochemical and kinetic favor of formation of AMH$^+$ over IM3H$^+$ is obvious, AMH$^+$ is without relevant new exit channel. Similarly, formation of OABAH$^+$ is more exothermic than that of IM3H$^+$, although it is more endergonic by 32 kJ/mol than in the case of IM3H$^+$.

Table 4. Thermochemical properties for some side reactions of the formation of 4,4'-MDAH$^+$.

Reaction	Aminal (AMH$^+$) Formation			IM3H$^+$ Formation			OABA$^+$ Formation		
	Gas	Aniline	Water	Gas	Aniline	Water	Gas	Aniline	Water
$\Delta_r E_0$(kJ/mol)	−90.7	−75.2	−77.2	32.7	4.9	−1.4	−61.1	−24.9	−26.8
$\Delta_r H^0$(kJ/mol)	−92.2	−76.6	−78.8	29.6	2.7	−4.4	−62.4	−25.8	−28.0
$\Delta_r G^0$(kJ/mol)	-41.1	−26.5	−28.6	42.2	12.0	8.7	−7.7	27.1	26.0

Beside the formation of 4,4'-MDAH$^+$, the aniline addition to MCH$^+$ (second aniline addition) can also result in the formation of ortho and meta MDAH$^+$ isomers (2,4-MDAH$^+$, 2',4-MDAH$^+$ and 3,4-MDAH$^+$). The thermodynamic properties of these competitive reactions are collected in Table 5.

Table 5. Energetic description for aniline addition to 4-methylidene-cyclohexa-2,5-diene-1-iminium (MCH^+) reactions.

Reaction	$\Delta_r E_0$ (kJ/mol)		$\Delta^{\ddagger} E_0$ (kJ/mol)	
	Water	Aniline	Water	Aniline
$IM5H^+ + A \rightarrow TS6 \rightarrow IM6H^+ + A$ [1]	−49.2	−61.4	29.2	31.6
$A+MCH^+ \rightarrow TS6 \rightarrow 4,4'\text{-MDAH}^+$	−62.2	−64.4	17.8	22.5
$A+MCH^+ \rightarrow TS6^* \rightarrow 2,4\,MDAH^+$	−45.1	−49.0	15.5	18.9
$A+MCH^+ \rightarrow TS6^* \rightarrow 2',4\text{-MDAH}^+$	−52.7	−53.6	−1.9	2.3
$A+MCH^+ \rightarrow TS6^* \rightarrow 3,4\text{-MDAH}^+$	48.4	45.2	54.8	57.4

[1] $IM5H^+$ represents complex of MCH^+, A and water while $IM6H^+$ stands for water complex of $4,4'\text{-MDAH}^+$ (see Figure 1).

From both thermodynamic and kinetic points of view, aniline addition to ortho position (2,4) is at least as preferred as the para position (4,4) due to the low activation energy and energy release. In contrast, the meta addition (3,4) is endothermic with high activation barrier in both condensed phases. More interestingly, one of the ortho addition steps has submerged transition state (−1.9 kJ/mol in aqueous phase) due to the strong ion-dipole interaction manifested in the vicinity (2.2 Å) of amine group of A and benzene ring of MCH^+ as shown in Figure 4.

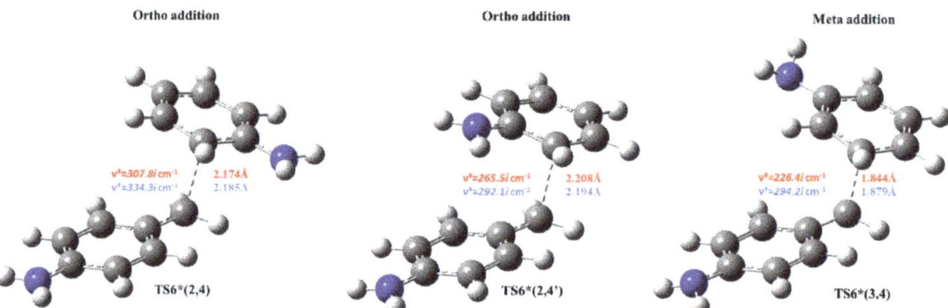

Figure 4. Transition state structures (obtained at B3LYP/6-31G(d) level of theory) for $4,4'$-MDA isomer formation (Critical distance are also given for both condensed phases).

Based on these results, significant amount of 2,4-MDA should be formed as a product of the title reaction which is in clear contradiction with laboratory observations (more than 92 *w/w%* of the product is $4,4'$-MDA while 7 *w/w%* is 2,4'-MDA [46]). This contradiction might be caused by incomplete quantum chemical description of the solvent effect and/or neglecting the role of the counter ion in the mechanism. On the other hand, 33.5 *w/w%* of the product is oligomeric and polymeric MDA [46]. Another hypothesis might be that the 2,4'-MDA is more reactive towards further aniline addition making oligomeric structures (e.g., 3-ring) overrepresented in these forms. To bear in mind that the above mentioned experiments are on the several minutes time scale, that makes reactive interferences between the intermediates and products possible (the w% of $4,4'$-MDA became saturated at 20% after ca. 40 min in Knjasev experiment at T = 70 °C, where the initial molar ratio of the mixture was A:F:HCl:H_2O = 4:2:1:13 [45]). An example for such interference is the highly preferred addition of the N-methylenebenzeneaminium (the '3-ring' structure) in the para position which is also supported by the ^{2}H- and ^{13}C-NMR-based observation by Knjasev [45]. This step can be followed by proton transfer which might make the dissociation of the adduct favorable to MCH^+ and PABA as it shown in Figure 5.

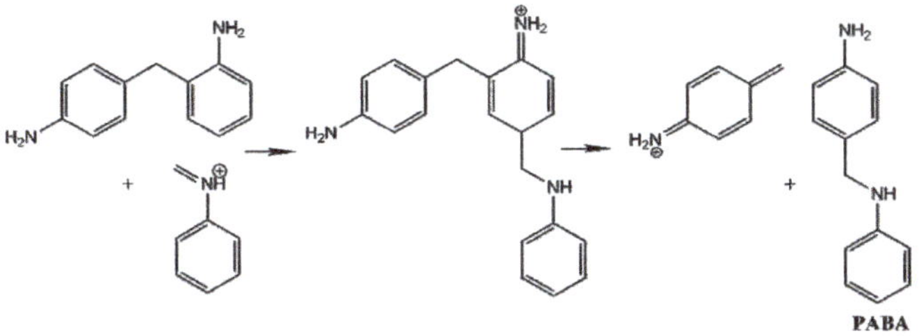

Figure 5. Schematic reaction mechanism of the formation of the 3-ring adduct and its dissociation to N-methylenebenzeneaminium and PABA.

4. Conclusions

Reaction mechanism of the MDA formation from aniline and formaldehyde was determined using G3MP2B3 quantum chemical method in gas phase and in industrially relevant solvents such as aniline and water. The non-covalent aniline dimer approaches formaldehyde to form the prereactive complex from which eight elementary reaction steps resulted in 4,4′-MDA as the final product. Our important findings for the whole reaction mechanism are:

1. The highest lying transition state (60.7 kJ/mol) corresponds to formaldehyde addition to aniline (TS1) leading to N-hydroxymethylaniline formation and its barrier heights significantly decreased by solvation. This step can be the main kinetic bottleneck for 4,4′-MDA production. After exothermic water elimination, aniline addition to N-methylenebenzenaminium took place through either tight transition state (resulted in 4-methylidenecyclohexa-2,5-diene-1-iminium) or loose transition state to form protonated aminal. However, aminal formation is both thermodynamically and kinetically preferred, there is no exit channel belong to it. Afterwards, proton shift can undergo in 4-methylidenecyclohexa-2,5-diene-1-iminium to produce protonated PABA, PABAH$^+$, being the global minimum at this reactive potential energy surface, which can be also along with the line with experimental observation of PABA during MDA production. Two steps rearrangement of aniline resulted the protonated MDA isomers (MDAH$^+$). It was found that aniline addition in ortho position is competitive with that of in the para position from both kinetic and thermodynamic points of view. The deprotonation of MDAH$^+$ is thermodynamically more favorable in water phase.

2. The species in proposed mechanism are proton activated, their possible deprotonation can lead to side reactions and appearance of deactivated intermediates in the industrial process. Therefore, the acid strength of four important intermediates such as N-methylenebenzeneanilium (4.2), PABAH$^+$ (6.7), MCH$^+$ (11.4), and AMH$^+$ (5.1) was estimated using relative pK_a calculation. Although, most of them found to be weak acid in aqueous solution, but they got more acidic in aniline (basic) environment which can then deactivate the intermediates.

3. Aniline addition-type side reactions had been also investigated and it was found that aminal formation is both thermodynamically and kinetically preferable, but it is a kinetic dead-end. Both 2,4- and 4,4′-MDAH$^+$ formation from MCH$^+$ and A has low-lying transition state and TS is submerged in the case of 2,4-MDAH$^+$ making likely the formation of 2,4-MDAH$^+$ beside 4,4-MDAH$^+$.

4. Gas phase thermodynamic properties for the reactants, products and intermediates were determined and carefully compared to the literature. Based on our G3MP2B3 and CBS-QB3 calculations, accurate standard enthalpy of formation is recommended for the intermediates.

Supplementary Materials: The following are available online at http://www.mdpi.com/2073-4360/11/3/398/s1, Figure S1: Transition state structures (obtained at B3LYP/6-31G(d) level of theory) for MDA synthesis in aqueous phase. The G3MP2B3 relative energies are also given, Figure S2: Transition state structures (obtained at B3LYP/6-31G(d) level of theory) for MDA synthesis in aniline. The G3MP2B3 relative energies are also given.

Author Contributions: Conceptualization, M.S. and L.F.; methodology, M.S.; software, K.N.; validation, M.S., B.V.; formal analysis, R.Z.B.; investigation, R.Z.B., M.S.; resources, L.F., K.N and B.V..; data curation, R.Z.B.; writing—original draft preparation, R.Z.B.; writing—review and editing, B.V. and L.F.; visualization, R.Z.B.; supervision, M.S.; project administration, B.V.; funding acquisition, B.V.

Funding: This research was supported by the European Union and the Hungarian State, co-financed by the European Regional Development Fund in the framework of the GINOP-2.3.4-15-2016-00004 project, aimed to promote the cooperation between the higher education and the industry. Milán Szőri gratefully acknowledges the financial support by the János Bolyai Research Scholarship of the Hungarian Academy of Sciences (BO/00113/15/7) and the additional financial support (Bolyai+) from the New National Excellence Program of the Ministry of Human Capacities (ÚNKP-18-4-ME/4).

Acknowledgments: Authors are thankful to opportunity provided by Wanhua-BorsodChem to conduct this study and to Tamás Purzsa, Vice President of BorsodChem and Zhao Nan Director of Production Technology and Process Optimization of BorsodChem for his support. The GITDA (Governmental Information-Technology Development Agency, Hungary) is also gratefully acknowledged for allocating computing resources used in this work.

Conflicts of Interest: The authors declare no conflict of interest. The funders had no role in the design of the study; in the collection, analyses, or interpretation of data; in the writing of the manuscript, or in the decision to publish the results.

References

1. Sonnenschein, M.F. *Polyurethanes: Science, Technology, Markets, and Trends*; Dow Chemical Company: Midland, MI, USA, 2015; p. 73.

2. Mustata, F.R.; Tudorachi, N.; Bicu, I. Epoxy Resins Cross-Linked with Bisphenol A/Methylenedianiline Novolac Resin Type: Curing and Thermal Behavior Study. *Ind. Eng. Chem. Res.* **2012**, *51*, 8415–8424. [CrossRef]

3. Perkins, G.T. Process for the Preparation of 4,4′-methylenedianiline. U.S. Patent 3,367,969, 6 February 1968.

4. Botella, P.; Corma, A.; Carr, R.H.; Mitchell, C.J. Towards an industrial synthesis of diamino diphenyl methane (DADPM) using novel delaminated materials: A breakthrough step in the production of isocyanates for polyurethanes. *Appl. Catal. A Gen.* **2011**, *398*, 143–149. [CrossRef]

5. de Angelis, A.; Ingallina, P.; Perego, C. Solid Acid Catalysts for Industrial Condensations of Ketones and Aldehydes with Aromatics. *Ind. Eng. Chem. Res.* **2004**, *43*, 1169–1178. [CrossRef]

6. Nafziger, J.L.; Rader, L.A.; Seward, I.J. Process for preparing polyamines with ion exchange resin catalysts. U.S. Patent 4,554,378, 19 November 1985.

7. Corma, A.; Botella, P.; Mitchell, C. Replacing HCl by solid acids in industrial processes: Synthesis of diamino diphenyl methane (DADPM) for producing polyurethanes. *Chem. Commun.* **2004**, *10*, 2008–2010. [CrossRef] [PubMed]

8. Salzinger, M.; Fichtl, M.B.; Lercher, J.A. On the influence of pore geometry and acidity on the activity of parent and modified zeolites in the synthesis of methylenedianiline. *Appl. Catal. A Gen.* **2011**, *393*, 189–194. [CrossRef]

9. Keller, T.C.; Arras, J.; Wershofen, S.; Perez-Ramirez, J. Design of hierarchical zeolite catalysts for the manufacture of polyurethane intermediates. *ACS Catal.* **2015**, *5*, 734–743. [CrossRef]

10. Tian, J.; An, H.; Cheng, X.; Zhao, X.; Wang, Y. Synthesis of 4,4′-methylenedianiline catalyzed by SO_3H-functionalized ionic liquids. *Ind. Eng. Chem. Res.* **2015**, *54*, 7571–7579. [CrossRef]

11. Wegener, G.; Brandt, M.; Duda, L.; Hofmann, J.; Klesczewski, B.; Koch, D.; Kumpf, R.J.; Orzesek, H.; Pirkl, H.G.; Six, C.; et al. Trends in industrial catalysis in the polyurethane industry. *Appl. Catal. A Gen.* **2001**, *221*, 303–335. [CrossRef]

12. Moore, W.M. Amines, Aromatic, Methylenedianiline. In *Kirk-Othmer Encyclopedia of Chemical Technology*; Grayson, M., Ed.; John Wiley and Sons: Hoboken, NJ, USA, 2007; pp. 338–348.

13. Wang, C.Y.; Li, H.Q.; Wang, L.G.; Cao, Y.; Liu, H.T.; Zhang, Y. Insights on the mechanism for synthesis of methylenedianiline from aniline and formaldehyde through HPLC-MS and isotope tracer studies. *Chin. Chem. Lett.* **2012**, *23*, 1254–1258. [CrossRef]

14. Stow, S.M.; Onifer, T.M.; Forsythe, J.G.; Nefzger, H.; Kwiecien, N.W.; May, J.C.; McLean, J.A.; Hercules, D.M. Structural characterization of methylenedianiline regioisomers by ion mobility-mass spectrometry, tandem mass spectrometry, and computational strategies. 2. Electrospray spectra of 3-ring and 4-ring isomers. *Anal. Chem.* **2015**, *87*, 6288–6296. [CrossRef] [PubMed]

15. Griswold, J. Analysis of aniline-water solutions. *Ind. Eng. Chem. Anal. Ed.* **1940**, *12*, 89–90. [CrossRef]

16. Katayama, M.; Ashiki, S.; Amakasu, T.; Ozutsumi, K. Liquid structure of benzene and its derivatives as studied by means of X-ray scattering. *Phys. Chem. Liq.* **2010**, *48*, 797–809. [CrossRef]

17. Kishore, K.; Santhanalakshmi, K.N. A thermochemical study on the reactions of aniline with formaldehyde in the presence of acid medium. *Thermochim. Acta* **1983**, *68*, 59–74. [CrossRef]

18. Wang, C.; Li, H.; Cao, Y.; Liu, H.; Hou, X.; Zhang, Y. Thermodynamic analysis on synthesis of 4,4-methylenedianiline by reaction of aniline and formaldehyde. *CIESC J.* **2012**, *63*, 2348–2355.

19. Boros, R.Z.; Koós, T.; Wafaa, C.; Nehéz, K.; Farkas, L.; Viskolcz, B.; Szőri, M. A theoretical study on the phosgenation of methylene diphenyl diamine (MDA). *Chem. Phys. Lett.* **2018**, *706*, 568–576. [CrossRef]

20. Marsi, I.; Viskolcz, B.; Seres, L. Application of the group additivity method to alkyl radicals: An ab initio study. *J. Phys. Chem. A* **2000**, *104*, 4497–4504. [CrossRef]

21. Viskolcz, B.; Berces, T. Enthalpy of formation of selected carbonyl radicals from theory and comparison with experiment. *Phys. Chem. Chem. Phys.* **2002**, *2*, 5430–5436. [CrossRef]

22. Curtiss, L.A.; Redfern, P.C.; Raghavachari, K.; Rassolov, V.; Pople, J.A. Gaussian-3 theory using reduced Møller-Plesset order. *J. Chem. Phys.* **1999**, *110*, 4703–4709. [CrossRef]

23. Becke, A.D. Density-functional thermochemistry. III. The role of exact exchange. *J. Chem. Phys.* **1993**, *98*, 5648–5652. [CrossRef]

24. Dennington, R.D.; Keith, T.A.; Millam, J.M. *GaussView05*; Semichem Inc.: Shawnee Mission, KS, USA, 2009.

25. Gonzalez, C.; Schlegel, H.B. An improved algorithm for reaction path following. *J. Chem. Phys.* **1989**, *90*, 2154–2161. [CrossRef]

26. Montgomery, J.A.; Frisch, M.J.; Ochterski, J.W.; Petersson, G.A. A complete basis set model chemistry. VI. Use of density functional geometries and frequencies. *J. Chem. Phys.* **1999**, *110*, 2822–2827. [CrossRef]

27. Nicolaides, A.; Rauk, A.; Glukhovtsev, M.N.; Radom, L. Heats of formation from G2, G2(MP2), and G2(MP2, SVP) total energies. *J. Phys. Chem.* **1996**, *100*, 17460–17464. [CrossRef]

28. Active Thermochemical Tables (ATcT) Values Based on Ver. 1.122 of the Thermochemical Network. 2016. Available online: https://atct.anl.gov/ (accessed on 17 December 2017).

29. Marenich, A.V.; Cramer, C.J.; Truhlar, D.G. Universal solvation model based on the generalized born approximation with asymmetric descreening. *J. Chem. Theory Comput.* **2009**, *5*, 2447–2464. [CrossRef] [PubMed]

30. Liptak, M.D.; Gross, K.C.; Seybold, P.G.; Feldgus, S.; Shields, G.C. Absolute pK_a Determinations for Substituted Phenols. *J. Am. Chem. Soc.* **2002**, *124*, 6421–6427. [CrossRef] [PubMed]

31. McQuarrie, D.A. *Statistical Mechanics*; Harper & Row: New York, NY, USA, 1970.

32. Liptak, M.D.; Shields, G.C. Accurate pKa calculations for carboxylic acids using complete basis set and Gaussian-n models combined with CPCM continuum solvation methods. *J. Am. Chem. Soc.* **2001**, *123*, 7314–7319. [CrossRef] [PubMed]

33. Frisch, M.J.; Trucks, G.W.; Schlegel, H.B.; Scuseria, G.E.; Robb, M.A.; Cheeseman, J.R.; Scalmani, G.; Barone, V.; Mennucci, B.; Petersson, G.A.; et al. *Gaussian 09, Revision E. 01*; Gaussian, Inc.: Wallingford, CT, USA, 2009.

34. Extended Third Millennium Ideal Gas and Condensed Phase Thermochemical Database for Combustion with Updates from Active Thermochemical Tables. 2005. Available online: http://garfield.chem.elte.hu/Burcat/hf.doc (accessed on 24 November 2017).

35. NIST Computational Chemistry Comparison and Benchmark Database. NIST Standard Reference Database No. 101. 2006. Available online: https://cccbdb.nist.gov/exp1x.asp (accessed on 24 November 2017).

36. Benson, S.W. *Thermochemical Kinetics: Methods for the Estimation of Thermochemical Data and Rate Parameters*, 2nd ed.; John Wiley: New York, NY, USA, 1976.

37. Heller, S.R. NIST Structures and Properties Database and Estimation Program. *J. Chem. Inf. Comput. Sci.* **1991**, *31*, 432–434. [CrossRef]

38. NIST Chemistry WebBook, SRD 69. Group Additivity Based Estimates. Available online: http://webbook.nist.gov/chemistry/grp-add/ (accessed on 24 November 2017).

39. Zhang, J.; Sun, Y.; Mao, C.; Gao, H.; Zhou, W.; Zhou, Z. Theoretical study of pK_a for perchloric acid. *J. Mol. Struct. THEOCHEM* **2009**, *906*, 46–49. [CrossRef]

40. Muckerman, J.T.; Skone, J.H.; Ning, M.; Wasada-Tsutsui, Y. Toward the accurate calculation of pK_a values in water and acetonitrile. *Biochim. Biophys. Acta* **2013**, *1827*, 882–891. [CrossRef] [PubMed]

41. Ghalami-Choobar, B.; Ghiami-Shomami, A.; Nikparsa, P. Theoretical Calculation of pK_b Values for Anilines and Sulfonamide Drugs in Aquous Solution. *J. Theor. Comput. Chem.* **2012**, *11*, 283–295. [CrossRef]

42. Behjatmanesh-Ardakani, R.; Safaeian, N. pKa predictions of some aniline derivatives by ab initio calculations. *Iran. Chem. Commun.* **2014**, *2*, 147–156.

43. Lu, H.; Chen, X.; Zhan, C.G. First-principles calculation of pK_a for cocaine, nicotine, neurotransmitters, and anilines in aqueous solution. *J. Phys. Chem. B* **2007**, *111*, 10599–10605. [CrossRef] [PubMed]

44. Perrin, D.D. *Dissociation Constants of Organic Bases in Aqueous Solution*; Butterworths: London, UK, 1972.

45. Knjasev, V. Beiträge zur reaktionskinetischen Untersuchung der säureinduzierten Anilin-Formaldehyd-Kondensation. Ph.D. Thesis, Universität Stuttgart, Stuttgart, Germany, 2007.

46. Wanhua-BorsodChem Zrt; Bolyai tér 1., H-3700 Kazincbarcika, Hungary. Personal Communication, 2019.

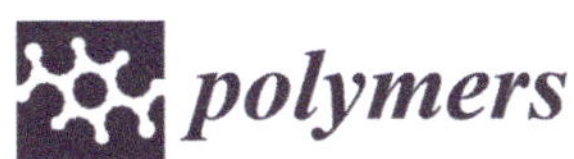 

Article

Comparison of Adhesive Properties of Polyurethane Adhesive System and Wood-plastic Composites with Different Polymers after Mechanical, Chemical and Physical Surface Treatment

Barbora Nečasová [1,*], **Pavel Liška** [1], **Jakub Kelar** [2] **and Jiří Šlanhof** [1]

[1] Faculty of Civil Engineering, Brno University of Technology, Veveří 331/95, 602 00 Brno, Czech Republic; liska.p@fce.vutbr.cz (P.L.); slanhof.j@fce.vutbr.cz (J.S.)

[2] Faculty of Science, Masaryk University, Kotlářská 2, 611 37 Brno, Czech Republic; jakub.kelar@mail.muni.cz

* Correspondence: necasova.b@fce.vutbr.cz; Tel.: +420-541-147-991

Received: 20 January 2019; Accepted: 20 February 2019; Published: 1 March 2019

Abstract: The cost of most primary materials is increasing, therefore, finding innovative solutions for the re-use of residual waste has become a topic discussed more intensely in recent years. WPCs certainly meet some of these demands. The presented study is focused on an experimental analysis of the effect of surface treatment on the adhesive properties of selected WPCs. Bonding of polymer-based materials is a rather complicated phenomenon and modification of the bonded area in order to improve the adhesive properties is required. Two traditional types of surface treatments and one entirely new approach have been used: mechanical with sandpaper, chemical with 10 wt % NaOH solution and physical modification of the surface by means of a MHSDBD plasma source. For comparison purposes, two high-density polyethylene based products and one polyvinyl-chloride based product with different component ratios were tested. A bonded joint was made using a moisture-curing permanently elastic one-component polyurethane pre-polymer adhesive. Standardized tensile and shear test methods were performed after surface treatment. All tested surface treatments resulted in an improvement of adhesive properties and an increase in bond strength, however, the MHSDBD plasma treatment was proven to be a more suitable surface modification for all selected WPCs.

Keywords: adhesion; adhesive; bond; cohesion; composite; joint; multi-hollow surface dielectric barrier discharge plasma source (MHSDBD); sandpaper; wood-plastic composite (WPC)

---

1. Introduction

Wood-plastic composites (known as WPCs) belong to the category of fiber-reinforced composite materials [1]. They combine the stability of wood fibers with the durability of synthetic thermoplastic polymers, particularly PE, PP and PVC [2–6]. This combination allows a wide range of applications, while also offering the option of using the waste products of the forestry and wood industries as well as some types of recycled plastic waste [3–8]. Even though the technology appeared almost 100 years ago [5,9], the greatest success of production and demand for these products has only been seen in recent years [5,10–12]. The statistics show that the major sector in which WPCs are applied is the construction industry, with a 76% share [8,11]. WPCs find uses primarily as flooring, fencing and façade cladding [11–14].

The presented study is focused on an experimental analysis of the effects of surface treatment on the adhesive properties of selected WPC façade cladding. Currently, established traditional joining methods, e.g., rivets or screws, are commonly used, however, as some recent studies indicate [12,15–17] these methods cause higher failure rates. The main reason can be seen as the high stress concentration at the joint, which results in the occurrence of fatigue-caused cracks [16,17]. Welded or adhesive-bonded joints have proven to be more suitable and durable alternatives [4,10,15,18–21]. They ensure better stress distribution in the joint that consequently results in greater construction stiffness and greater stress resistance [22–24]. Unfortunately, the bonding of WPCs is more complicated than the bonding of traditional materials [2,4,18]. This statement was also confirmed by authors in previous research, where the adhesive properties of solid timber, cement-bonded particle board and WPC were studied [25]. The bondability of WPCs is highly affected by the thermoplastic component of the product, i.e., its thermoplastic matrix, which has a very low surface energy and bad wettability properties [1,2,11,26,27]. The major prerequisite for perfect adhesion is the creation of interaction forces [23,24,26,27]. It has been settled that the higher the value of surface stress the higher the polarity of the given surface. Therefore, it is advisable to use an adhesive with lower polarity than that of the adherend. This allows the wetting of the adherend contact surface. However, in façade systems, this step is greatly complicated by the limited range and number of products. At present, less than eight certified adhesive systems are available on the market. Moreover, there are only polyurethane or modified polymer based adhesives. Both types have rather poor wetting characteristics in combination with plastics. The wrong combination of materials of a façade system may result in considerably shorter service life and it can significantly affect the bondability and adhesive properties. As a result, it is advisable to increase surface polarity by modifying the bonded areas.

As proven by selected case studies, the bonding of WPCs is almost impossible without prior surface modification [11,17–21,26,28] and also without the application of primer or any promoting agent. Using primer puts greater demands on the cleanness of the work environment and prolongs the installation time, therefore, a surface treatment which would allow for the promoting agent to be eliminated is desired. According to some authors [11,22], WPC surface modifications can be divided into three basic categories: mechanical modification (i.e., sandblasting and roughening), chemical modification (i.e., application of acid or alkaline solutions as, for example, chromic acid, sodium hydroxide or fluorine) and physical modification (i.e., LP/AP plasma, corona, flame or laser). It is believed that the selection of the appropriate method depends on the matrix material and the used WPC formulation [11,18,19]. For this reason, three different WPC façade claddings with dissimilar formulations, polymer matrix and wood flour were selected in this study to verify the effect and versatility of surface modifications.

In this paper the effect of three different surface treatment methods on the improvement of the adhesive properties of WPCs was studied: Mechanical with sandpaper, chemical with a 10 wt % NaOH solution and physical modification of the surface by means of a Multi-Hollow Surface Dielectric Barrier Discharge plasma source (MHSDBD). The first selected surface treatment is the mechanical modification using P40 grid sandpaper. According to Kraus et al. [11] and Oporto et al. [18] joining of WPC products is usually performed after brushing or sawing. It is a very common, economical and undemanding preparation process that allows for the more prominent appearance of wood fibers on the surface. This modification regularly shows very good wetting results and strength increases. The sandpaper coarseness was determined based on previous experience modifying cement-based composites and WPC [25,29,30]. P80 and P240 sandpaper was used in previous research cases, both types are less abrasive and leaves a finer surface. Even though the final results were very promising and a 100% increase in shear strength was monitored, a visual inspection showed that the surface was not sufficiently modified, therefore, a rougher coarseness was tested here. The chemical alkali treatment is the next of the selected surface treatments. Application of a 10 wt % sodium hydroxide was chosen due to its common availability, low cost and simple application process. Moreover, the concentration used has been shown to work well with different types of materials, especially with wood fibers [11,31,32]. Agarwal et al. [33] monitored a 120% increase in impact strength, if wood fibers in WPC were treated with a 1 wt % and 3 wt % NaOH solution. The last selected surface treatment was physical modification performed with a Multi-Hollow Surface Dielectric Barrier Discharge plasma source (MHSDBD). Plasma treatment is one of the most versatile surface treatment techniques, it is considered the most effective, sustainable and low-cost method for surface treatment of polymer-based materials compared with traditional treatments [27]. The positive effect of plasma treatment has already been proven in combination with different formulations of WPCs. Kraus et al. [11] investigated and compared LP and AP plasma surface treatments of PP-based WPC, Wolkenhauer et al. [26] studied the potential of dielectric barrier discharge (DBD) at atmospheric pressure and ambient air in combination with PP- and PE-based WPCs. Surfaces of WPCs formulated with HDPE and PP were treated and the tensile bond strength increased 5 times after DBD plasma treatment. The MHSDBD also generates atmospheric-pressure plasma in ambient air, however, it is the result of a combination of two methods of obtaining non-isothermal plasma at atmospheric pressure; gas flow and dielectric insertion into the discharge space. One of the advantages of MHSDBD is the possibility of driving it under atmospheric pressure at low power. The MHSDBD plasma has been adapted for the modification of small planar objects and for the need for patterned surface treatment [34,35]. This is the first application of this plasma source in combination with WPCs. However, in a case study from 2016, a similar plasma source, diffuse coplanar surface barrier discharge (DCSBD), was used to improve the adhesive properties of WPC formulated with 50% HDPE. The obtained results showed a 100% increase in the bond shear strength, and a 30% change in the failure mode, from adhesive to cohesive, was observed [29]. MHSDBD is in many ways similar to the industrially available DCSBD plasma source. This similarity is in its construction, i.e., used materials and geometry of electrode system, and physical parameters such as temperature or plasma power density. However, MHSDBD has proven to be much more effective in the treatment of topologically challenging materials thanks to its higher effective thickness plasma layer. Since the main parameters such as electrode geometry, plasma temperature and plasma power density are similar to DCSBD, we have been able to suggest the same treatment times, as in our previous work with DCSBD in 2016. The DCSBD plasma source was not suitable for WPC surface modification since the surfaces of the selected materials have an embossment which imitates the look of real wood, the depth of the embossment varied for each tested WPC and ranged between 0.05–0.5 mm.

The focus of this study is assessing the surface treatment effectiveness. The main aim is to determine whether the selected plasma treatment can be used for different types of WPCs with the same or better results compared to more traditional methods, i.e., mechanical and chemical treatments. The effect is investigated on bonded assemblies of three types of WPCs in combination with a moisture-curing permanently elastic one-component polyurethane adhesive. The adhesive is a part of one of the most common systems intended for façade applications [35,36]. The shear strength of a single lap joint under tensile stress and adhesion of bonded joint under axial tensile stress were measured with a tensile test, to verify the impact of the selected surface treatment methods.

2. Materials and Methods

2.1. Materials Selection

The adhesive system used in this study is produced by Dinol GmbH and is intended only for façade bonding. It consists of the moisture-curing permanently elastic one-component polyurethane pre-polymer adhesive Dinitrol F500LP Polyflex, the primer Dinitrol Multiprimer 550 that acts as a bonding promoter for the surfaces of adherends, and Dinitrol 520 Cleaner that degreases the bonded surfaces. To ensure durable and efficient joints, all components have to be used as recommended by the manufacturer. According to the information provided in the technical data sheet, Dinitrol F500LP Polyflex has a tensile strength of 9.0 MPa with a maximum elongation at break of 600% and a shear strength after 7 days of 5.5 MPa [37].

The material representing the load-bearing substructure was selected with the intent to avoid the premature failure of joints due to their incompatibility with the chosen adhesive system. The EN AW-2011 aluminium alloy with a tensile strength of 295.0 MPa and a yield strength of 195.0 MPa was chosen. The thickness of the material selected for the shear strength test was 5 mm, see Figure 1; the samples used in adhesion testing were 15 mm thick, see Figure 2.

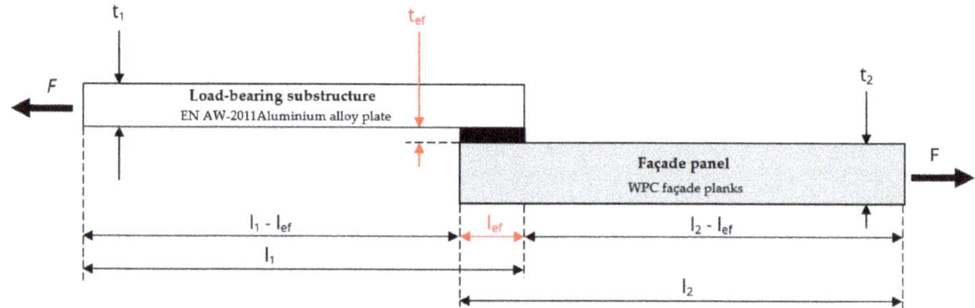

Figure 1. Single lap joint test sample geometry—cross section.

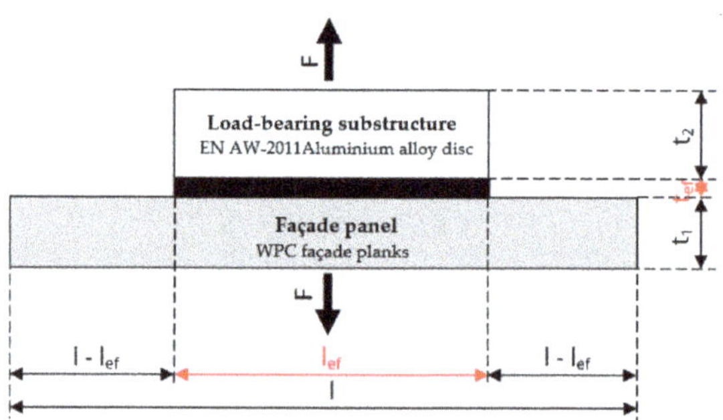

Figure 2. Adhesion test sample assembly—cross section.

The first tested WPC cladding (referred to as WPC_45/45 in the text) was 21 mm thick. The ratio of components in the tested material was 45 wt % beech wood flour, 45 wt % PVC (specifically VYNOVATM S6502 high molecular weight and high porosity vinyl chloride homopolymer) and 10 wt % additives [38]. A material density of 1279.0 kg/m^3 was determined experimentally, the bend strength was 37.9 MPa, and information about tensile and shear strength were not declared by the manufacturer.

The second type of WPC cladding was material described in the text as WPC_60/30, which was 12 mm thick. The tested product was made of 60 wt % wood flour (hardwood, the exact type was not declared by the manufacturer), 30 wt % high-density polyethylene (referred to as HDPE) and 10 wt % additives [39]. Its bend strength was 21.7 MPa and tensile and shear strength were not declared by the manufacturer. The material density was 1210 kg/m^3, however, a density of 1230 kg/m^3 was determined experimentally.

The last tested material, referred to as WPC_50/38, was 9 mm thick and made of 50 wt % poplar wood flour, 38 wt % HDPE and 12 wt % additives (light stabilizer, coupling agent, anti-ageing component, UV retardant and colorant) [40]. The bend strength was 15.0–17.0 MPa, tensile strength around 4.9 MPa and shear strength of 2.2 MPa. A material density of 1250 kg/m^3 was determined experimentally. Some product sheets did not contain information about material density. However, as Klyosov verified [4], the ratio of additives, in particular that of coupling agents, may significantly increase material density, and most importantly, it affects the material's ability to absorb liquid. Specification of additives was provided only by one manufacturer, that is why the volumetric density was determined. The higher the volumetric mass density of WPC, the worse the wettability of the surface may be expected.

2.2. Contact Angle Measurement

Prior to the manufacturing of the test samples it was necessary to determine the wettability of the surface of the selected materials. For all the selected materials a simple wettability test was performed according to ČSN EN 828 (it is the national equivalent to EN 828:2013. The wettability test was also repeated after the application of the chosen surface modification. 10 drops of distilled water were placed on the surface of the material. The technical standard for water recommends a volume of 2–6 μL [41], a volume of 3 μL was used.

2.3. Surface Treatment

2.3.1. Mechanical Treatment

The mechanical modification of bonded surfaces was performed using P40 grid sandpaper. A layer of c. (0.25 ± 0.10) mm thickness was removed from the test samples. The thickness of the layer that was removed was determined using a 150 mm XTline P13430 digital Vernier caliper (XTline s.r.o., Velké Meziříčí, Czech Republic) with a rated accuracy of 0.01 mm and measured at three points. Microscopic examination of surface topography was not performed. All test samples were roughened by the same person to avoid any distortion of the results.

2.3.2. Chemical Treatment

The ratio of components was determined based on the calculations of mass concentration of the solution, 10 g of 98% NaOH and 88 g of distilled water were used to achieve the desired solution concentration. The quantity was measured using digital scales with an accuracy of 0.01 g. The surface of the adherend was covered with the solution for c. 3 min. Subsequently, the surface was dried with blotting paper.

2.3.3. Physical Treatment

The MHSDBD generated atmospheric-pressure plasma on a surface area of 18×18.9 mm. The contactless testing mode was used in this experiment. The samples were held by hand and always treated in the central part of the MHSDBD by gentle movements, see Figure 3. Plasma exposure time for all types of WPCs was 10 s. The power input of the MHSDBD during operation was 30 W in a flow air mode of 8.0 l/min.

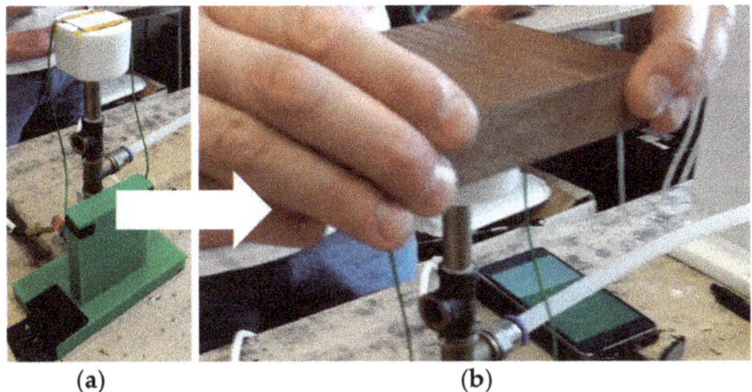

(a)   (b)

Figure 3. Plasma surface treatment: (**a**) Portable source of Multi-Hollow Surface Dielectric Barrier Discharge plasma; (**b**) Surface Treatment of the test sample (here with WPC_45/45).

2.4. Specimen Preparation

All WPCs were supplied in the form of planks, the original dimensions of which had to be adjusted to suit the requirements of the selected testing methods. The two most common destructive test methods were selected. The joints were tested for tensile and shear strength. For testing the adhesion of bonded joints at tensile stress, recommendations given in the ČSN 73 2577 standard were respected (the national standard describes steps similar to pull-off adhesion testing). All samples consisted of two components—the chosen façade cladding and the load-bearing substructure. The cladding was cut into squares of side length $l/b = 100$ mm, and the load-bearing substructure was represented by an aluminium disc with a circular cross-section area of $A_{ef} = 2\ 500$ mm^2 [42]. The dimensions recommended in the ČSN EN 1465 standard [43] were used for test samples to determine the shear strength of a single lap joint under tensile stress. The test specimens were again made of two components, both with the dimensions $b/b_{ef} = 25$ mm and $l_1/l_2 = 100$ mm. One component represented the load-bearing substructure, the other the façade cladding. A standardized design of a single lap joint sample was used with $l_{ef} = 12.5$ mm.

The recommended thickness of the adhesive joint was 3 mm. To avoid influencing the results negatively, this parameter was strictly followed since it is general knowledge that the thickness of the adhesive layer has a significant effect on the resulting mechanical properties of the bonded joint [24,27,44–49].

When a physical or chemical treatment was applied to WPC surfaces, a cleaning agent was used before the surface treatment. Where mechanical treatment was applied, surfaces were cleaned after the roughening to remove all debris. To one half of the test samples (i.e., 5 samples) primer was subsequently applied. On the second half of the samples, the primer was not applied. This procedure would determine the effectiveness of the selected surface modifications as well as the effect of the primer on the joint's efficiency and strength. The test samples were kept in a standard, dry and clean environment, at constant temperature (23 ± 2) °C and humidity (55 ± 10) % for 28 days, and left to cure statically.

2.5. Strength Test

A Heckert FP 10/1 tearing machine (ZwickRoell LP, Kennesaw, GA, USA) was used to record the development of deformations in the tested joints in relation to the applied load and time. The testing range of the machine is from 0 to 10 kN. Axial load was always applied on the samples. For this purpose molds were designed and made for the samples to be attached to the jaws of the tearing machine. The strain rate was set at 5 mm/min. The displacement of the test samples was recorded using a HBM 1-WA/100 MM-T inductive sensor (with a maximum deviation of 0.15%, Hottinger Baldwin Messtechnik GmbH, Darmstadt, Germany), which was placed on the cross member of the tearing machine, see the test setup in Figure 4.

An HBM Spider8 measuring station and catman® (V2.1) software (both Hottinger Baldwin Messtechnik GmbH, Darmstadt, Germany) were used for data recording. The load and joint elongation were recorded at a 5 Hz data storage frequency. The tests were carried out at a temperature of (20 ± 5) °C and relative humidity of (50 ± 20)%.

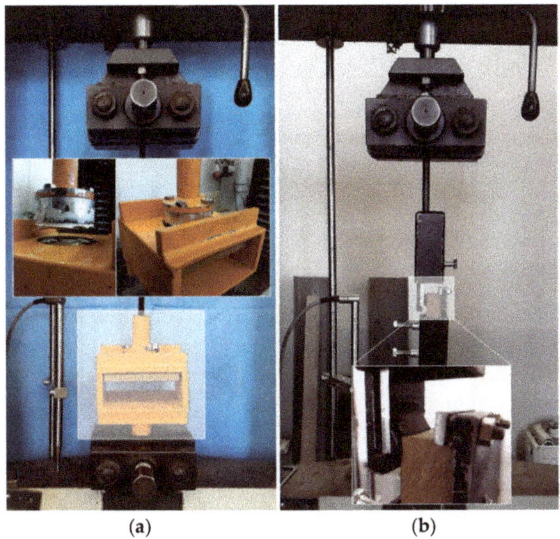

(a)   (b)

Figure 4. Test setup: (**a**) tensile test; (**b**) shear test.

2.6. Data Analysis

Contact angles were determined by capturing images of water drops and evaluating them using ImageJ software and the Contact Angle extension module. The output of this simple testing method was the determination of the arithmetic average of the contact angles α and of the wettability of the selected surfaces.

The failure modes of test samples were also evaluated. A scale based on recommendations presented in the ČSN ISO 10365 standard [50] (the national equivalent to ISO 10365) and in the international technical standard ASTM D 5573 [51] was prepared. For all tested combinations of joints, the predominant failure mode was determined.

Based on the recommendations of the aforementioned standards, the tensile stress (σ_{adh} in MPa) and shear strength of a bonded SLJ under tensile stress (τ in MPa) was determined. For each selected set of measured and calculated values, an arithmetic average was calculated, together with standard deviation. Subsequently, the coefficient of variation was calculated using the given values. The values of the coefficient over 20% indicate an ineffective surface treatment method. The elongation of the bonded joints was continuously recorded during the tests. Using the recorded values, it was possible to calculate the tensibility (δ in %) of a bonded joint. As only the change in length was monitored during the tests, we can further talk about relative elongation. The tensibility is the expression of the relative elongation in percentages.

3. Results and Discussion

3.1. Contact Angle Measurement

The results presented in Table 1 show that it is not always possible to improve wettability using surface modification. Based on these results, it can be assumed that the chosen mechanical surface modification will be the least appropriate option for improving the adhesive properties, however, Oporto et al. [18] monitored similar results. They examined a 20% decrease of surface energy after mechanical surface modification. Nevertheless, they observed a 60% and 80% increase in shear strength. The contact angle measured after physical and chemical modification showed more than a 40% improvement in surface wettability.

Table 1. Average contact angle α of WPC surfaces WITHOUT and WITH surface treatment [1].

Surface Treatment/WPC Type	WPC_45/45 [°]	WPC_60/30 [°]	WPC_50/38 [°]
Without Treatment	90.94 ± 2.96	80.60 ± 4.29	98.89 ± 2.97
Mechanical (P40)	104.95 ± 2.26	103.83 ± 1.47	118.77 ± 1.28
Physical (MHSDBD)	63.51 ± 2.15	50.99 ± 2.61	45.29 ± 1.09
Chemical (10% NaOH)	67.98 ± 3.55	67.43 ± 3.03	67.13 ± 1.94

[1] The average values are a summary of ten measurements conducted for each WPC cladding before and after surface modification.

3.2. Failure Mode

The purpose of this assessment was to determine the predominant type of failure mode for each set of test samples and, if possible, to determine whether the selected method of surface modification has any effect on the failure mode of the bonded joint. To fully understand the adhesive properties of bonded surfaces, a description of their failure modes is necessary, as even under the same testing conditions, the same stress rate and with the application of the same adhesive, entirely different failure modes may occur in various materials.

The results presented in Tables 2 and 3 show that only three types of failure occurred: AF–adhesive failure; SF–substrate (adherend) failure and NF–the sample was not broken.

Table 2. Predominant failure mode of samples WITH and WITHOUT primer after tensile test [1].

Surface Treatment/ WPC Type	WPC_45/45		WPC_60/30		WPC_50/38	
	With	Without	With	Without	With	Without
Without Treatment	AF 100%	AF 100%	AF 100%	AF 100%	AF 100%	AF 100%
Mechanical (P40)	SF 80%	AF 100%	AF 100%	AF 100%	AF 100%	AF 100%
Physical (MHSDBD)	NF 60%	AF 100%	AF 100%	AF 100%	AF 100%	AF 100%
Chemical (10% NaOH)	AF 80%	AF 100%	AF 100%	AF 100%	AF 100%	AF 100%

[1] The predominant failure mode was determined for one set of samples (i.e., 5 samples with or without primer).

Table 3. Predominant failure mode of samples WITH and WITHOUT primer after shear test [1].

Surface Treatment/ WPC Type	WPC_45/45		WPC_60/30		WPC_50/38	
	With	Without	With	Without	With	Without
Without treatment	AF 100%	AF 100%	AF 100%	AF 100%	AF 100%	AF 100%
Mechanical (P40)	AF 80%	AF 100%	AF 100%	AF 100%	AF 100%	AF 100%
Physical (MHSDBD)	SF 100%	AF 100%	AF 80%	AF 100%	AF 100%	AF 100%
Chemical (10% NaOH)	SF 80%	AF 100%	AF 100%	AF 100%	AF 100%	AF 100%

[1] The predominant failure mode was determined for one set of samples (i.e., 5 samples with or without primer).

Adhesive failure occurred in almost all cases, although, in some combinations very high joint strength was observed. The monitored failure modes to some extent also disproved hypothesis that suitable surface treatment might compensate for primer usage. The adhesive failure mode suggests poor bondability of the bonded surface. In practice, this type of failure would pose considerable danger if it occurred on an actual façade. While cohesive failure occurs gradually, adhesive failure and the subsequent fall of the cladding is often very fast.

In combination with the WPC_60/30, only the adhesive failure mode was observed. Even though this failure of the bonded joint was dominant, the deterioration of the bonded surface was detected in all examined cases and wood fibers were visible, see the example in Figure 5 (WPC_60/30). This result is similar to the conclusion presented by Oushabi et al. [32], who stated that 10 wt % NaOH surface treatment resulted in surface damage and fiber degradation. In the presented case, due to this damage, slight penetration of the primer into the modified surface was observed. In Figure 5 (WPC_60/30 a)), the effect of chemical surface treatment is visible. All observed adhesive failures occurred in the interface between the primer and the WPC cladding. In combination with WPC_45/45, other types of failure than adhesive failure of the joint were also observed. In a few cases, the bonded joint was not broken, even though the limit of the tearing machine was reached. In several cases, substrate failure occurred. When multiple failure modes occurred, see example in Figure 5 (WPC_45/45 a)), the failure mode was classified according to the prevalent mode, in the example case it was the adhesion failure mode.

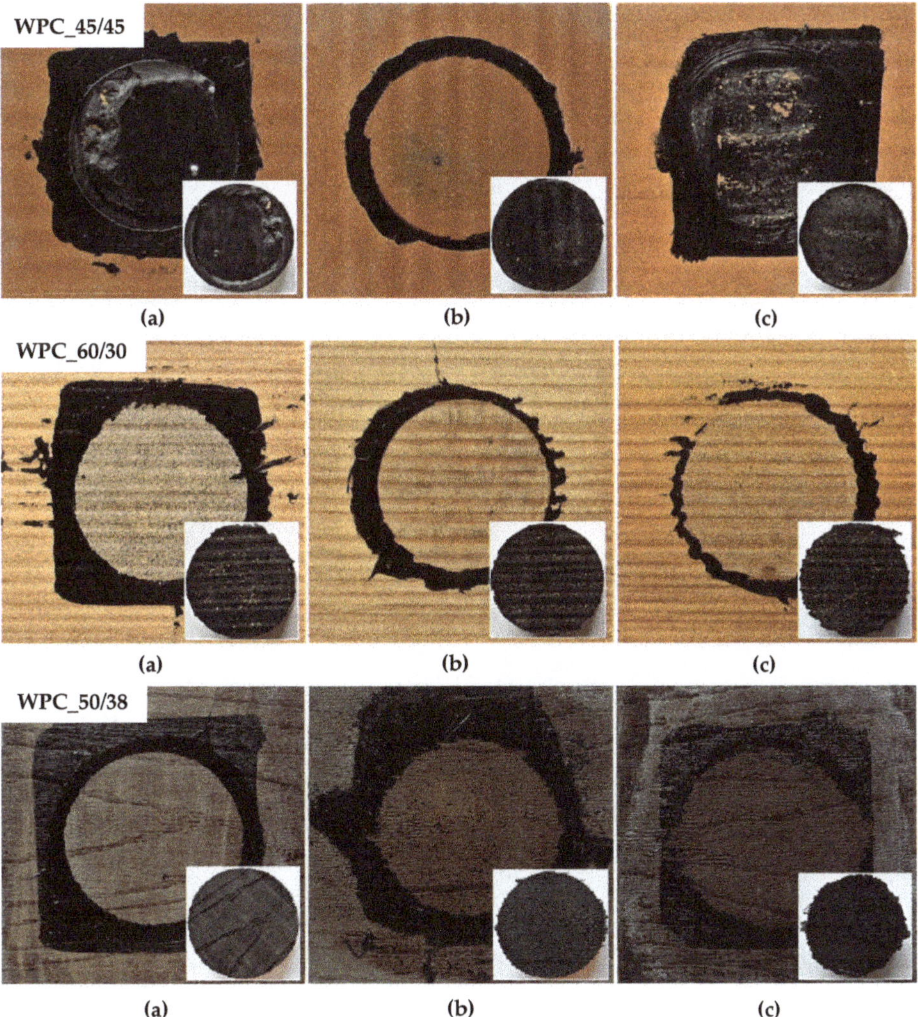

Figure 5. Typical failure mode of test samples after adhesion (tensile) test: WPC_45/45 (**a**) sample with primer, plasma treatment (86% AF; 14% CF); (**b**) sample without primer, chemical treatment (100% AF); (**c**) sample with primer, chemical treatment (100% AF); WPC_60/30 (**a**) sample with primer, chemical treatment (100% AF); (**b**) sample without primer, mechanical treatment (100% AF); (**c**) sample without primer, chemical treatment (100% AF); WPC_50/38 (**a**) sample with primer, mechanical treatment (100% AF); (**b**) sample without primer, plasma treatment (100% AF); (**c**) sample with primer chemical treatment (100% AF).

Similar results were observed after the shear test. However, the more significant effect of chemical surface modification was monitored, as shown in Figure 6. In combination with WPC_45/45, SF was witnessed after the physical and chemical surface modification, see Figure 6 (WPC_45/45 a)). Adhesive failure was observed in all tested combinations with WPC_60/30 and WPC_50/38. Again, the effect of chemical treatment was monitored predominantly in combination with WPC_60/30, however, it did not affect the failure mode. Moreover, slight surface deterioration of WPC_60/30 could be seen in all presented examples after the application of all surface treatments.

Figure 6. Typical failure mode of test samples after shear test: WPC_45/45 (**a**) sample with primer, plasma treatment (100% SF); (**b**) sample with primer, mechanical treatment (74% AF; 26% CF); (**c**) sample without primer, chemical treatment (100% AF); WPC_60/30 (**a**) sample without primer, plasma treatment (100% AF); (**b**) sample with primer, mechanical treatment (100% AF); (**c**) sample with primer, chemical treatment (100% AF); WPC_50/38 (**a**) sample with primer, mechanical treatment (100% AF); (**b**) sample without primer, chemical treatment (100% AF); (**c**) sample with primer plasma treatment (100% AF).

The improvement of adhesive properties was observed in all tested samples with PVC-based WPC, however, the effect was not so prominent in combination with PE-based WPCs. Moreover, although changes were observed in all samples after surface modification, these were not confirmed by failure mode evaluation. The combination of the embossed surface and the HDPE component resulted in both small changes in the surface adhesion and almost 100% occurrence of adhesive failure. The prevalence of the given failure mode in the test samples was ambiguous, since all tested samples had a visibly damaged surface.

The selected surface modifications were completely unsuitable in terms of improving the adhesion properties of WPC_50/38. The poor results can be most likely attributed to the silicon layer (lubricant) that is protecting the surface of the material and polyethylene component. The surface of PVC-based WPC (i.e., WPC_45/45) was modified by all tested treatments. This cladding was the only sample containing this type of thermoplastic and at the same time was the only cladding the treated surface of which was not originally deeply embossed. This allowed the modification of the bonded area to be done more evenly. The deeper surface embossment of PE-based WPCs prevented an even application of surface modifications, and the insufficient surface adhesion in these areas caused premature joint failure.

3.3. Tensile/Shear Stress and Tensibility of Bonded Assemblies

All test samples without surface treatment debonded during the curing period; therefore, only the samples with a treated bonded area were tested and evaluated.

Based on the results of the contact angle measurements and the determination of surface wettability, it can be assumed that all samples with a mechanically modified surface would disintegrate before even being tested. However, this hypothesis was not confirmed by any of the tested combinations. Not even in the case of samples without primer.

In combination with WPC_50/38, the best results were achieved after mechanical surface treatment, even though the bond strength was very small compared to the other tested combinations, see Tables 4 and 5. The results achieved with the WPC_50/38 cladding were the worst of all the tested combinations. It was observed that this cladding had a very thick protective silicone-based layer (lubricant), compared to the other WPCs, which protects the surface against any damage after its extrusion. None of the tested surface modifications removed a sufficient thickness of this layer; therefore, any significant improvement in adhesive properties was not monitored. A similar observation was reported by Oporto et al. [18].

In some cases with WPC_45/45 no failure was recorded as the loading limit of the testing machine was exceeded. This referred to the samples with primer and plasma surface modification. The bonded joint was very durable even if the primer coating was not applied. The recorded strength as well as maximum joint elongation at break were diametrically different compared to other tested WPCs. It is also the only sample of the composite that was composed of PVC, the other two types of cladding contained HDPE. This fact influenced the effectiveness of the selected surface modifications. The comparison of results presented in Figure 7 shows that MHSDBD plasma treatment was also the most effective method in combination with WPC_60/30.

Table 4. Average tensile stress σ_{adh} (in MPa), variation coefficient (in %) and tensibility (in %) of samples WITH and WITHOUT primer [1].

WPC Type/ Surface Treatment		Mechanical (P40)		Physical (MHSDBD)		Chemical (10% NaOH)	
		With	Without	With	Without	With	Without
WPC_45/45	σ_{adh}	2.62	1.21	3.87	2.34	3.43	1.16
	VC [2]	12.99	12.34	10.87	8.34	10.94	4.44
	δ	127.52	51.21	191.67	79.79	160.58	46.27
WPC_60/30	σ_{adh}	1.35	0.91	1.79	1.35	1.21	1.05
	VC	17.79	11.45	7.41	6.81	13.10	13.26
	δ	87.67	59.17	87.54	64.54	87.67	71.63
WPC_50/38	σ_{adh}	1.02	0.73	0.54	0.63	0.29	0.34
	VC	8.08	8.08	6.88	24.06	19.22	19.65
	δ	58.40	58.52	45.81	52.60	38.63	40.98

[1] The average values are a summary of five measurements conducted for each tested combination. [2] VC is an abbreviation of variation coefficient.

Table 5. Average shear stress τ (in MPa), variation coefficient (in %) and tensibility (in %) of samples WITH and WITHOUT primer [1].

WPC Type/ Surface Treatment		Mechanical (P40)		Physical (MHSDBD)		Chemical (10% NaOH)	
		With	Without	With	Without	With	Without
WPC_45/45	τ	2.91	0.93	5.11	1.21	5.63	0.56
	VC [2]	18.58	0.78	11.68	11.27	7.54	8.70
	δ	62.41	24.95	95.38	33.46	79.93	16.76
WPC_60/30	τ	2.71	0.83	3.94	0.74	2.00	0.78
	VC	5.80	14.54	13.25	4.77	11.44	7.60
	δ	47.92	25.04	76.44	23.44	46.24	29.01
WPC_50/38	τ	1.95	0.53	0.95	0.27	0.62	0.29
	VC	12.07	12.73	29.77	38.39	27.14	24.23
	δ	44.29	17.11	36.31	10.70	15.99	8.88

[1] The average values are a summary of five measurements conducted for each tested combination. [2] VC is an abbreviation of variation coefficient.

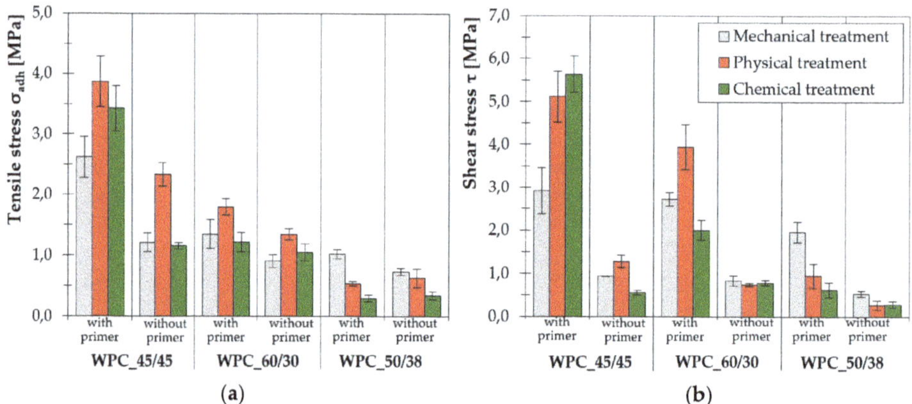

Figure 7. Comparison of average stress values: (**a**) in tension and (**b**) in shear after surface treatment. The results of samples with and without primer are compared. The standard error in the mean is presented in the error bar.

The results of the tests determining the shear strength of SLJs under tensile stress are clearer in showing that using all the components of a facade mounting system is essential. Figure 7b shows that the differences between the samples with and without primer are diametrical. Although the bonded area was sufficiently modified in all examined alternatives, there was a significant increase in strength in the test samples coated with primer.

The effect of the primer and increase in bond strength is clear from the comparison presented in Figure 8. The difference in joint strength was compared. The higher the percentage value the more important the primer coating is to achieve an efficient bond. The negative values presented in Figure 8a in combination with WPC_50/38 indicate that samples without primer achieved better results. The application of primer did not affect the bond tensile strength and efficiency. The negative values also indicated that the surface wettability was not increased by the modification.

Comparable results were achieved by physical modification. The MHSDBD plasma treatment appears to be the most suitable option for WPC_45/45 as well as for WPC_60/30. A different exposure period might have improved the adhesive properties of the surface even more. On the other hand, it seems to be that the selected chemical surface treatment was the least effective.

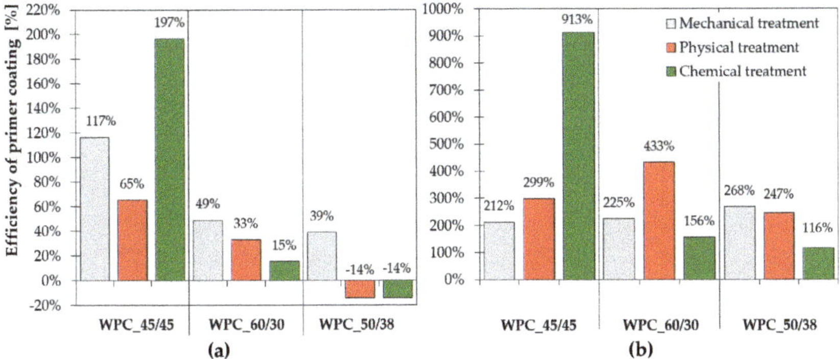

Figure 8. Comparison of surface treatment efficiency: (**a**) tensile test and (**b**) shear test. The higher the percentage is the more important the application of primer coating and the less effective the surface treatment. The negative values indicate that the results of samples without primer were better.

The importance and necessity of using all the components of the mounting system is obvious. The measured values of the strength of the bonded test samples on which primer was not used are too low, and the differences in comparison with the samples containing the primer are quite ambiguous. Nevertheless, some isolated results show that the choice of an appropriate surface modification may increase the effectiveness of a bonded joint even without using the primer coating, and this is certainly a good direction to follow in the future.

While for the samples with WPC_45/45, the selected modifications were effective in both testing methods since high ultimate stress values were achieved; in the case of WPC_60/30 and WPC_50/38, the effect of surface modification was not so clear. Moreover, the 'ideal' deformation of the bonded joint during the tests occurred only in combination with WPC_45/45, when the bonded joint transformed progressively from elastic deformation to plastic deformation and the disintegration of the joint appeared only after the cohesive or substrate failure. In other cases, this phenomenon was rare, as depicted in Figure 9. The average stress-strain curves of all tested samples shown in Figure 9 (samples after tensile test) and in Figure 10 (samples after shear test), present the effectiveness of physical treatment. The MHSDBD plasma source successfully modified bonded surfaces of all tested types of WPCs. The tensile strength (adhesion) of plasma treated samples increased by 100% compared to the untreated samples even when primer was not used.

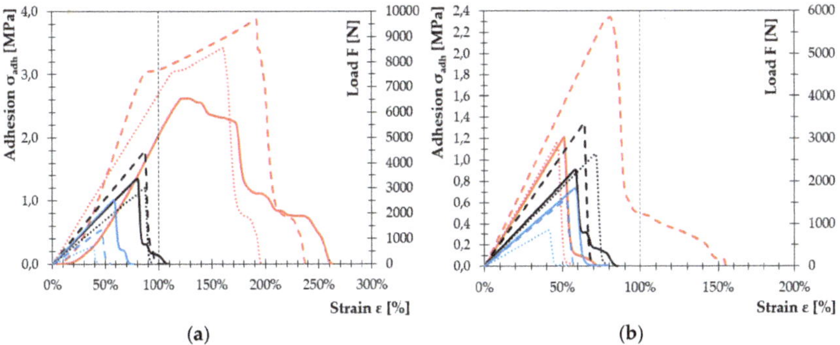

Figure 9. Comparison of average stress–strain curves from tensile test: (**a**) samples with primer and (**b**) samples without primer. Color marking: RED—WPC_45/45; BLACK—WPC_60/30; BLUE—WPC_58/30. Solid lines are samples with mechanical treatment, dashed lines are samples with physical treatment and dotted lines are samples with chemical treatment.

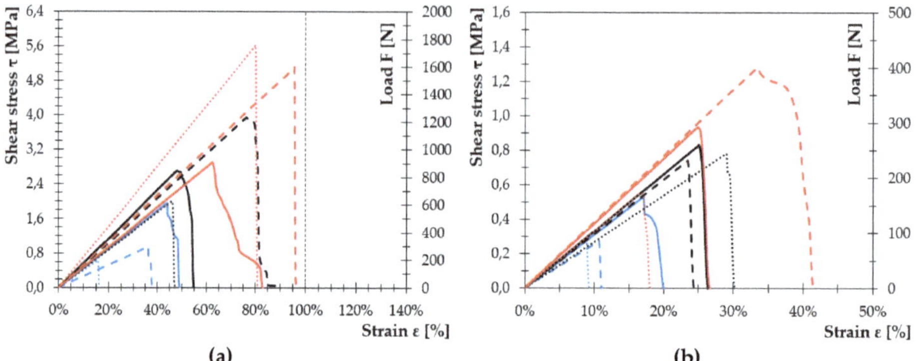

Figure 10. Comparison of average stress–strain curves from shear test: (**a**) samples with primer and (**b**) samples without primer. Color marking: RED—WPC_45/45; BLACK—WPC_60/30; BLUE—WPC_58/30. Solid lines are samples with mechanical treatment, dashed lines are samples with physical treatment and dotted lines are samples with chemical treatment.

4. Conclusions

In this study, a moisture-curing permanently elastic one-component polyurethane pre-polymer adhesive was used to study different surface treatments of bonded areas and its effect on the strength of the adhesive joint. The main purpose was to evaluate the effect of three different surface treatment methods: A mechanical, a physical and a chemical treatment, on the adhesive properties of WPCs façade cladding. Three different formulations of WPC were tested: one PVC-based and two PE-based (HDPE) WPCs. Two destructive standardized test methods were selected to provide sufficiently detailed information for the analysis of the effect of surface treatment on the adhesive properties of the WPC cladding:

- An improvement in the adhesive properties was demonstrated by the increase in the strength of the bonded joint after tensile as well as shear tests. In one combination, the achieved strength was nearing the maximum strength of the adhesive system declared by the manufacturer.
- A positive effect of surface modifications on the failure mode was observed. In two combinations substrate failure was the predominant failure mode after the tensile test.
- The selected surface treatments were more suitable for PVC-based WPCs than for PE-based WPCs. Furthermore, it was confirmed that the effect of the thermoplastic component of WPC on its adhesion properties is dominant.
- The effect of the wood fibers in WPCs was minor in all tested combinations. The assumption that a higher content of wood flour would have a positive effect on the adhesive properties was not confirmed.
- The most consistent results were achieved after physical modification of the bonded surface. However, different exposure period for each tested material seems to be necessary. Physical modification using the MHSDBD plasma source can be considered universal surface treatment for the selected WPCs. It is also the cleanest and the least time-consuming method.
- The presented results confirm the well proven fact that the adhesive properties of the surface of WPC materials are very poor, and without modification of the bonded surfaces, their bonding is almost impossible, or rather the effectiveness of the joint is negligible.

The tests have shown that selected adhesive/mounting system is suitable for the bonding of WPCs cladding, however, prior verification of the compatibility of selected materials is required. This conclusion has proved to be decisive in terms of the statement presented in the introduction, i.e., that adhesive-bonded joints are more suitable alternative than traditional mechanical joints. The necessity to conduct a detailed and often financially demanding experimental assessment of the compatibility of selected materials, frequently with unsatisfying results, is a difficult obstacle to overcome when arguing e.g., with investors and/or clients about why they should choose adhesive bonding to anchor WPCs cladding. For these reasons, the plasma treatment is in our opinion, the best surface modification method for WPCs.

Author Contributions: Conceptualization, B.N. and J.Š.; methodology, B.N., P.L. and J.K.; formal analysis, B.N.; investigation, B.N.; resources, B.N.; data curation, B.N.; writing—original draft preparation, B.N.; writing—review and editing, B.N., P.L. and; J.K.; visualization, B.N.; funding acquisition, B.N.

Funding: This research was funded by Brno University of Technology.

Acknowledgments: The authors would like to thank the Cernak group at Department of Physical Electronics at Masaryk University for supplying the Multi-Hollow Surface Dielectric Barrier Discharge plasma source.

Conflicts of Interest: The authors declare no conflict of interest.

References

1. Park, S.-J.; Seo, M.-K. Chapter 7—Types of Composites. *Interface Sci. Technol.* **2011**, *18*, 501–629. [CrossRef]
2. Gardner, D.J.; Han, Y.; Wang, L. Wood-Plastic Composite Technology. *Curr. For. Rep.* **2015**, *1*, 139–150. [CrossRef]
3. Clemons, C. Wood-plastic composites in the United States: The interfacing of two industries. *For. Prod. J.* **2002**, *52*, 10–18.
4. Klyosov, A.A. *Wood-Plastic Composites*, 1st ed.; John Wiley & Sons, Inc.: London, UK, 2007; pp. 1–49.
5. Spear, M.J.; Eder, A.; Carus, M. 10—Wood polymer composites. In *Wood Composites*, 1st ed.; Ansell, M.P., Ed.; Woodhead Publishing Elsevier Ltd.: Cambridge, UK, 2015; pp. 195–249.
6. Otheguya, M.E.; Gibsona, A.G.; Robinsona, M.; Findonb, E.; Crippsb, B.; Ochoa Mendozac, A.; Aguinaco Castroc, M.T. Recycling of end-of-life thermoplastic composite boats. *Plast. Rubber. Compos.* **2009**, *38*, 406–411. [CrossRef]
7. Kokta, B.V.; Raj, R.G.; Daneault, C. Use of wood flour as filler in polypropylene: Studies on mechanical properties. *Polym.-Plast. Technol. Eng.* **1989**, *28*, 247–259. [CrossRef]
8. Sommerhuber, P.F.; Wang, T.; Krause, A. Wood-plastic composites as potential applications of recycled plastics of electronic waste and recycled particleboard. *J. Clean. Prod.* **2016**, *121*, 176–185. [CrossRef]
9. Gordon, J.E. *The New Science of Strong Materials (or Why You Don't Fall through the Floor)*, 1st ed.; Princeton University Press: Princeton, NJ, 1998; p. 179.
10. Simons, H.R.; Weith, A.J.; Shack, W. *Extrusion of Plastics, Rubber, and Metals*, 1st ed.; Reinhold Publishing: New York, NY, USA, 1952; p. 454.
11. Kraus, E.; Baudrit, B.; Heidemeyer, P.; Bastian, M.; Stoyanov, O.V.; Starostina, I.A. Problems in Adhesion Bonding of WPC. *Polym. Res. J.* **2015**, *9*, 327–335.
12. Friedrich, D.; Luible, A. Measuring the wind suction capacity of plastics-based cladding using foil bag tests: A comparative study. *J. Build. Eng.* **2016**, *8*, 152–161. [CrossRef]
13. Chen, Y.; Stark, N.M.; Tshabalala, M.A.; Gao, J.; Fan, Y. Weathering Characteristics of Wood Plastic Composites Reinforced with Extracted or Delignified Wood Flour. *Materials* **2016**, *9*, 1–12. [CrossRef] [PubMed]
14. Stark, N.M.; Matuana, L.M.; Clemons, C.M. Effect of processing method on surface and weathering characteristics of wood-flour/HDPE composites. *J. Appl. Polym. Sci.* **2004**, *93*, 1021–1030. [CrossRef]
15. Cheng, R.; Zhang, L.; Li, Y. Study on bonding properties of PVC-based WPC bonded with acrylic adhesive. *J. Adhes. Sci. Technol.* **2012**, *26*, 2729–2735. [CrossRef]
16. Soury, E.; Behravesh, A.H.; Rouhani Esfahani, E.; Zolfaghari, A. Design, optimization and manufacturing of wood-plastic composite pallet. *Mater. Des.* **2009**, *30*, 4183–4191. [CrossRef]
17. Mohamadzadeh, A.; Rostampour Haftkhani, A.; Ebrahimi, G.; Yoshihara, Y. Numerical and experimental failure analysis of screwed single shear joints in wood plastic composite. *Mater. Des.* **2012**, *35*, 404–413. [CrossRef]

18. Oporto, G.S.; Gardner, D.J.; Bernhardt, G.; Neivandt, D.J. Characterizing the mechanism of improved adhesion of modified wood plastic composite (WPC) surfaces. *J. Adhes. Sci. Technol.* **2007**, *21*, 1097–1116. [CrossRef]

19. Gramlich, W.M.; Gardner, D.J.; Neivandt, D.J. Surface treatments of wood-plastic composites (WPCs) to improve adhesion. *J. Adhes. Sci. Technol.* **2006**, *20*, 1873–1887. [CrossRef]

20. Moghadamzadeh, A.; Rahimi, H.; Asadollahzadeh, M.; Hemmati, A.R. Surface treatment of wood polymer composites for adhesive bonding. *Int. J. Adhes. Adhes.* **2011**, *31*, 816–821. [CrossRef]

21. Banea, M.D.; da Silva, L.F.M. Adhesively bonded joints in composite materials: An overview. *Proc. Inst. Mech. Eng.* **2009**, *223*, 1–18. [CrossRef]

22. Ebnesajjad, S. *Adhesives Technology Handbook*, 2nd ed.; William Andrew: New York, NY, USA, 2008; p. 387.

23. Petrie, E.M. *Handbook of Adhesives and Sealants*, 2nd ed.; The McGraw-Hill Companies, Inc.: New York, NY, USA, 2007; p. 765.

24. Da Silva, L.F.M.; Öchsner, A.; Adams, R.D. *Handbook of Adhesion Technology*, 1st ed.; Springer: Berlin, Germany, 2011; p. 1568.

25. Nečasová, B.; Liška, P.; Šlanhof, J. Research summary: Analysis of selected adhesive systems intended for facade bonding. In *Advances and Trends in Engineering Sciences and Technologies II. In Proceedings of the 2nd International Conference on Engineering Sciences and Technologies, ESaT 2016, Vysoké Tatry, Slovakia, 29 June–1 July 2016*; Al Ali, M., Platko, P., Eds.; CRC Press/Balkema: London, UK, 2017; pp. 573–578.

26. Wolkenhauer, A.; Avramidis, G.; Hauswald, E.; Militz, H.; Viöl, W. Plasma Treatment of Wood–Plastic Composites to Enhance Their Adhesion Properties. *J. Adhes. Sci. Technol.* **2008**, *22*, 2025–2037. [CrossRef]

27. Kuczmaszewski, J. Fundamentals of Metal-Metal Adhesive Joint Design. Ph.D. Thesis, Lublin University of Technology, Lublin, Poland, 2006; p. 199.

28. Ebnesajjad, S. *Surface Treatment of Materials for Adhesive Bonding*, 2nd ed.; William Andrew, Elsevier: London, UK, 2014; p. 341.

29. Nečasová, B.; Liška, P.; Kelar, J. Study on Surface Treatments of Modified Wood Plastic Composite (WPC) to Improve Adhesion. *Appl. Mech. Mater.* **2016**, *861*, 96–103. [CrossRef]

30. Nečasová, B.; Liška, P.; Šlanhof, J. Determination of Bonding Properties of Wood Plastic Composite Façade Cladding. *J. News Eng.* **2015**, *3*, 5–13.

31. Saleema, N.; Sarkar, D.K.; Paynter, R.W.; Gallant, D.; Eskandarian, M. A simple surface treatment and characterization of AA 6061 aluminum alloy surface for adhesive bonding applications. *Appl. Surf. Sci.* **2012**, *261*, 742–748. [CrossRef]

32. Oushabi, A.; Sair, S.; Oudrhiri Hassani, F.; Abboud, Y.; Tanane, O.; El Bouari, A. The effect of alkali treatment on mechanical, morphological and thermal properties of date palm fibers (DPFs): Study of the interface of DPF–Polyurethane composite. *S. Afr. J. Chem. Eng.* **2017**, *23*, 116–123. [CrossRef]

33. Agarwal, S.; Gupta, R.K. Chapter 5—Improving the mechanical properties of wood plastic composites. In *Applied Researchers in Polysaccharides*; Alsewailem, F.D., Ed.; Research Signpost: Kerala, India, 2015; pp. 103–124.

34. Sihelnik, S. Activation and Cleaning of Glass Using Non-Thermal Plasma at Atmospheric Pressure. Master's Thesis, Masaryk University, Brno, Czech Republic, 2018.

35. Chlupová, S.; Kelar, J.; Slavíček, P. Changing the surface properties of ABS plastic by plasma. *Plasma Phys. Technol.* **2017**, *4*, 32–35. [CrossRef]

36. Homola, T.; Krumpolec, R.; Zemánek, M.; Kelar, J.; Synek, P.; Hoder, T.; Černák, M. An Array of Micro-hollow Surface Dielectric Barrier Discharges for Large-Area Atmospheric-Pressure Surface Treatments. *Plasma Chem. Plasma Process.* **2017**, *37*, 1149–1163. [CrossRef]

37. Auto-Color. Available online: http://www.a-c.cz/index.php?/dinitrol/stavebnictvi-lepeni-fasadnich-panelu/pu-lepidla-pro-lepeni-fasadnich-panelu/f500lp_fp-dinitrol-f500lp-polyflex/flypage.tpl.html (accessed on 11 January 2019).

38. Perwood. Available online: https://www.perwood.cz/en/ (accessed on 11 January 2019).

39. Woodplastic. Available online: https://www.woodplastic.eu/terasy/forest/ (accessed on 11 January 2019).

40. Nextwood. Available online: https://www.nextwood.cz/plotovky-a-nosniky/ (accessed on 11 January 2019).

41. ČSN EN 828. *Adhesives—Wettability—Determination by Measurement of Contact Angle and Surface Free Energy of Solid Surface*; Sorting No. 66 8621; Czech Office for Standards, Metrology and Testing: Prague, Czech Republic, 2013; p. 16.
42. ČSN 73 2577. *Test for Surface Finish Adhesion of Building Structures to the Base*; Sorting No. 32691; Czech Office for Technical Approvals and Measurements: Prague, Czech Republic, 1981; p. 4.
43. ČSN EN 1465. *Adhesives—Determination of Tensile Lap-Shear Strength of Bonded Assemblies*; Sorting No. 66 8510; Czech Office for Standards, Metrology and Testing: Prague, Czech Republic, 2009; p. 12.
44. Dukes, W.A.; Bryant, R.W. The Effect of Adhesive Thickness on Joint Strength. *J. Adhes.* **1969**, *1*, 48–53. [CrossRef]
45. Davies, P.; Sohier, L.; Cognard, J.-Y.; Bourmaud, A.; Choqueuse, D.; Rinnert, E.; Créac'hcadec, R. Influence of adhesive bond line thickness on joint strength. *Int. J. Adhes. Adhes.* **2009**, *29*, 724–736. [CrossRef]
46. Banea, M.D.; da Silva, L.F.M.; Campilho, R.D.S.G. The Effect of Adhesive Thickness on the Mechanical Behavior of a Structural Polyurethane Adhesive. *J. Adhes.* **2015**, *91*, 331–346. [CrossRef]
47. Lijuan, L.; Chenguang, H.; Toshiyuki, S. Effect of adhesive thickness, adhesive type and scarf angle on the mechanical properties of scarf adhesive joints. *Int. J. Solids Struct.* **2013**, *50*, 4333–4340. [CrossRef]
48. Boutar, Y.; Naïmi, S.; Mezlini, S.; da Silva, L.F.M.; Hamdaoui, M.; Ben Sik Ali, M. Effect of adhesive thickness and surface roughness on the shear strength of aluminium one-component polyurethane adhesive single-lap joints for automotive applications. *J. Adhes. Sci. Technol.* **2016**, *30*, 1913–1929. [CrossRef]
49. Akhavan-safar, A.; Ayatollahi, M.R.; da Silva, L.F.M. Strength prediction of adhesively bonded single lap joints with different bondline thickness: A critical longitudinal strain approach. *Int. J. Solids Struct.* **2017**, *109*, 189–198. [CrossRef]
50. ČSN ISO 10365. *Adhesives. Designation of Main Failure Patterns*; Sorting No. 66 8509; Czech Institute for Technical Approvals: Prague, Czech Republic, 1995; p. 5.
51. ASTM D 5573. *Standard Practice for Classifying Failure Modes in Fiber-Reinforced-Plastic (FRP) Joints*; ASTM International: West Conshohocken, PA, USA, 2005; p. 3.

Article

One-Pot Processing of Regenerated Cellulose Nanoparticles/Waterborne Polyurethane Nanocomposite for Eco-friendly Polyurethane Matrix

Soon Mo Choi [1,2], Min Woong Lee [3] and Eun Joo Shin [3,*]

[1] School of Chemical Engineering, Yeungnam University, 280 Daehak-Ro, Gyeongsan, Gyeongbuk 38541, Korea; smchoi@ynu.ac.kr

[2] Regional Research Institute for Fiber & Fashion Materials, Yeungnam University, 280 Daehak-Ro, Gyeongsan, Gyeongbuk 38541, Korea

[3] Department of Organic Materials and Polymer Engineering, Dong-A University, 550-37 Nakdong-daero, Saha-gu, Busan 49315, Korea; tnlswkqjqtk@naver.com

* Correspondence: sejoo6313@dau.ac.kr; Tel.: +82-51-200-7343

Received: 30 January 2019; Accepted: 15 February 2019; Published: 18 February 2019

Abstract: Regenerated cellulose nanoparticles (RCNs) reinforced waterborne polyurethanes (WPU) were developed to improve mechanical properties as well as biodegradability by using a facile, eco-friendly approach, and introducing much stronger chemical bonding than common physical bonding between RCNs and WPU. Firstly, RCNs which have an effect on improving the solubility and stability of a solution, thereby resulting in lower crystallinity, were fabricated by using a NaOH/urea solution. In addition, the stronger chemical bond between RCNs and WPU was here introduced by regarding at which stage in particular added RCNs worked best on strengthening their bond in the process of WPU synthesis. The chemical structure, mechanical, particle size and distribution, viscosity, and thermal properties of the resultant RCNs/WPU nanocomposites were investigated by Fourier transform infrared analysis (FTIR), Zeta-potential analysis, viscometer, thermogravimetric analysis (TGA), Instron, and dynamic mechanical analysis (DMA). The results of all characterizations indicated that the RCNs/WPU-DMF associated with the addition of RCNs in DMF-dispersed step resulted in more effectively crosslinked between WPU and nano-fillers of nanocellulose particles in the dispersion than Acetone and Water-dispersed steps, thereby attributing to novel interactions formed between RCNs and WPU.

Keywords: regenerated cellulose nanoparticles (RCNs); waterborne polyurethanes (WPU); nanocomposites; polyols; isocyanates; one-pot processing.

---

1. Introduction

The reinforcement of nanoparticles within a continuous polymeric phase to improve the thermal and mechanical properties of resultant nanocomposite has been attracting attention in high functional applications such as electronics, tissue engineering, and civil [1–4]. In particular, the mechanical properties depend on the constituents of the nanocomposite as well as the phase morphology. Polyurethanes (PUs) consisting of both isocyanate (hard segments) and polyol (soft segments) are one of the polymers class with the widest applications by engineering polymer structure, thereby having various desirable properties. In spite of the strong advantage related to their versatile applicability, the use of organic solvents has been a controversial one to prepare PU-based composites. Consequently, waterborne polyurethanes (WPUs) have been replaced by conventional PUs associated with the use of organic solvents owing to growing concerns about environmental issues, health, toxicity, and the extended application caused by substitution of it. Because of global limitations on the release of volatile organic compounds (VOCs) into the atmosphere, eco-friendly products have gained popularity in diverse industries over recent years [5–7]. Viewed from this angle, water is the best medium in order to synthesize chemicals due to it being safe, nontoxic, and a very low in price. The synthesis process of WPU is conducted in water which functions not only as an emulsifier, but also solvent reacting with WPUs pre-polymer. Therefore, the WPU has been considered specific materials to a particular global condition. This is a group of multiblock copolymer elastomeric materials with alternating polyether or polyester soft segments and rigid urethane hard segments [8]. Due to the thermodynamic differences between the hard and soft segments, phase separation often occurs during cooling. The hard segment domain acts as a physical intersection point and as a filler with a higher modulus than the soft segment matrix. This network molecular structure exhibits excellent flexibility and elasticity. Nonetheless, the improvement of the low thermal stability and mechanical properties of WPU is still a matter of further discussion [9,10]. We here focused on the introduction of regenerated cellulose nanoparticles (RCNs) produced from microcrystalline cellulose (MCC) isolated from diverse natural resources in order to overcome the mentioned drawbacks. RCNs have received peculiar attention due to their unique features such as biodegradability, biocompatibility, renewability, light weight, high aspect ratio, and, especially, abundance [7,8]. However, many researchers have striven to concentrate on solving the biggest challenges about dispersing cellulose in aqueous solutions with no use of surfactant and/or modification. Accordingly, NaOH-based solutions and urea were introduced for improving both its solubility and the stability of solution. This mechanism can be explained by the strong interaction with NaOH to reduce the aggregation of cellulose molecules through the formation of new hydrogen bonds between cellulose and NaOH [7,11,12]. The improvement of its solubility in NaOH/urea aqueous solvents results in low degree of crystallinity. Also, the dispersion of RCNs within polymeric phase enables to enhance both low thermo-stability and biodegradability of polymeric phase, WPUs [13–17]. This strategy provides a potential for dispersing of RCNs into WPU through incorporating RCNs into the polyols which has a decisive influence on the the final WPU product characteristics [18]. We previously developed RCNs/WPU nanocomposites through this approach, and also investigated the effect of RCNs incorporation on their performance and properties.

Furthermore, in this study, we strengthened the bond of RCNs with WPU matrix within the nanocomposite through the adding of RCNs in each process sequence of WPU synthesis, respectively. In addition, it was found what a particular stage added by RCNs worked best on strengthening the bond among all processes. Therefore, the final purpose is to develop high-stretchable WPU films with good biocompatibility by introducing not most of physical bond, but much stronger chemical bond between RCNs and WPUs.

2. Experimental

2.1. Materials

MCC, Polycaprolactone diol (M_n = 1250), isophorone diisocyanate (IPDI), dimethylol propionic acid (DMPA), and triethylamine (TEA) was purchased from Sigma-Aldrich (St. Louis, MO, USA) Corporation and used as received without further purification. Urea, NaOH, and other chemicals used were all AR grade. Celluase enzyme obtained from Aspergillusniger was purchased from Sigma. *N*,*N*-dimethylformamide (Alfa Aesar, Haverhill, MA, USA) and Acetone (JUNSEI, Tokyo, Japan) were used without further purification.

2.2. Preparation of RCNs

The preparation of regenerated cellulose nanoparticles (RCNs) is described in previous paper [7], which were prepared from MCC through the method described by Adsul et al. [19]. Obtaining RCNs aqueous dispersion with solid contents was about 1 wt %. To compare properties of RCNs with commonly used cellulose nanocrystals (CNCs), the CNCs were obtained via acid hydrolysis of MCC. This process was already published several times by authors [20]. Briefly, the MCC and

H_2SO_4(64 wt %) were mixed at 45 °C for 30 min. Hydrolysis was stopped by pouring cold deionized water, and the suspension was washed by centrifugation. The resulting suspension was dialyzed against deionized water to a pH about 8 to obtain a RNC aqueous dispersion having a solids content of about 0.5 wt %.

2.3. Preparation of WPU and RCNs/WPU Nanocomposites Solutions

The preparation of WPU and RCNs/WPU is also described in previous paper [7] in detail. The acetone process [21] is a process mainly used to control the viscosity of PU polymer during the polymerization. Acetone process is often used because it is inert with the WPU forming reactions, can be mixed with water, and has a low boiling point. The advantage of acetone process includes not only obtaining a homogeneous solution but also wide range of structure and emulsion, high quality and reliable reproducibility of end products.

Subsequently, the 1.5 g amount of RCNs was added with different ways in the WPU synthetic process, which was shown in Scheme 1. In the first method, the RCNs DMF suspension was used, of which suspension was done from water to acetone then from acetone to DMF by several successive centrifugation steps at 12,000 rpm and 10 °C for 20 min. The dispersion was added after the IPDI addition process, stirred for 2 h, and the reaction was allowed to continue for 2 more h. And then DMF removed by vacuum between 60 to 80 °C, which was coded as RCNs/WPU-DMF. In the second method, the RCNs acetone suspension was used, of which suspension was done from water to acetone by several successive centrifugation steps as same as first method of RCNs/WPU-DMF. Finally, a stable suspension in acetone was obtained through 10 min ultrasonic treatment. It was added to the acetone pouring step of the WPU synthesis, which was coded as RCNs/WPU-acetone. In the third method, the RCNs aqueous suspension was added instead of pure water during the dispersion step of the pre-polymer. It coded as RCNs/WPU-water. The WPU prepolymer molecules with isocyanate groups have cross-linked effectively with the hydroxyl groups on the RCNs.

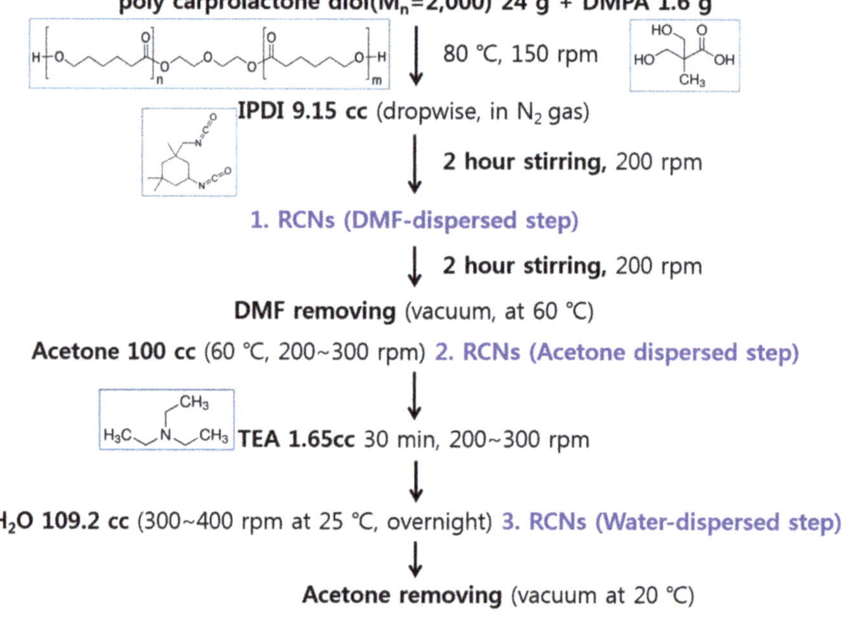

Scheme 1. Regenerated cellulose nanoparticles (RCNs)/waterborne polyurethane (WPU) nanocomposites synthetic process.

2.4. Preparation of RCNs/WPU Nanocomposite Films

RCNs/WPU nanocomposite films were prepared by casting process. RCNs/WPU aqueous dispersion was sonicated for 1 h before casting and poured on glass plate, which was set square molding type by Teflon tape with 0.32 mm thickness. Then drying was conducted in a chamber at 25 °C and 50% relative humidity for 24 h, and it was progressed in oven at 80 °C for 1 h and in vacuum oven at 40 °C for 24 h, respectively. The thickness of the film used for the tensile strength was 0.16 ± 0.02 mm.

3. Analysis of RCNs/WPU Composite Solutions and Films

Transmission electron microscope (TEM): Samples were sonicated for 30 min and completely dispersed in distilled water. The droplets of the sample dispersion were deposited on a polycarbonate film supported on a copper lattice. After drying at room temperature, 1×10^{-9} (Gun), 8.82×10^{-8} (Column) Torr and air, a transmission electron microscope (JEM-2010, Jeol, Tokyo, Japan) was collected at an acceleration voltage of 200 kV.

Brookfiled viscosity: The viscosity of the WPU and RCNs/WPU dispersions was measured by a DV2T viscometer (Brookfiled, Middleborough, MA, USA). Experiments were carreid out at 12, 30, or 60 rpm of the spindle at 25 °C, respectively.

Zata-potential analysis: The zata potential of the dispersion was measured by diluting the dispersion to 0.5 wt % with deionized water before measurement and then measuring the electrical potential at 25 °C with a Nano-ZS 90 zata potential analyzer (Malvern Instrument Co., Ltd, Worcester, UK) Respectively.

Particle size distribution: The average particle size and the its distribution of the WPU were measured using a Nano-ZS 90 zata-potential analyzer (Malvern Instrument Co., Ltd, Worcester, UK). 0.1 mL of the WPU was diluted with 3 mL deionized water. The sample was added to a deionised water tank the pinhole of 200 μm and measured at 25 °C.

Fourier transform infrared analyses (FT-IR): FTIR spectroscopy was observed at room temperature on a Nicolet FTIR spectrometer (Perkin-Elmer, Waltham, MA, USA). Powdered RCNs and RCN/WPU were crushed with KBr and compressed to produce pellets.

UV-vis absorption spectrum: Transmittance of WPU, RCNs/WPU nanocomposite was measured by UV-vis spectrometer (SHIMADZU UV-3600, Tokyo, Japan) with the scan wavelength from 800 nm to 400 nm and 1nm per step. The film was shaped to 10 mm × 50 mm × 1 mm.

Thermogravimetric analysis (TGA): Thermogravimeter (TGA Q500, TA instruments, New Castle, DE, USA) was used to measure the weight loss of the RCNs, WPU, and RCN/WPU nanocomposite under a nitrogen atmosphere. The samples were heated from 30 to 600 °C at a heating rate of 10 °C/min. Generally, 10 mg samples were used for thermogravimetric analysis.

Dynamic mechanical analysis (DMA): The dynamic mechanical thermal properties of the film samples were measured at 1 Hz with a DMA-Q800 (TA Instruments, New Castle, DE, USA) at a heating rate of 4 °C /min from −80 to 80 °C. The size of the films sample were 20 mm × 6 mm × 0.2 mm for the DMA measurements.

Measurements of mechanical properties: The tensile strength was measured by Autograph tester (Instron 4201, Sidmazu, Tokyo, Japan). The width and length of the samples were set to 5 and 20 mm, respectively, with dumbbell shape, respectively, and the test speed was set to 100 mm/min.

4. Results and Discussion

4.1. Properties of RCNs Nanoparticles

Figure 1 shows TEM images of regenerated cellulose nanoparticles (RCNs). They are not a rod-like cellulose nanocrystal from H_2SO_4 but spherical shape which was regenerated from NaOH/urea system. The role of hydroxyl ions of NaOH is described in previous paper [7]

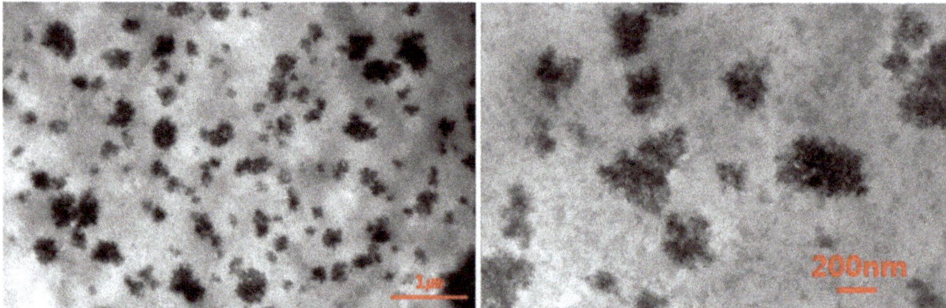

Figure 1. TEM micrographs of spherical RCNs prepared by hydrolyzing microcrystalline cellulose (MCC) with a NaOH-based aqueous solution.

This difference was leads to either reduction in intensity or increase of certain IR bands characteristic of cellulose crystalline domains or considered as an amorphous band. Figure 2 shows FTIR spectroscopy to explain difference in degree of crystallinity between RCNs and cellulose nanocrystals (CNCs) from H_2SO_4 method. The FTIR absorption band at 1430 and 893 cm^{-1} were correspond to CH_2 bending vibration and C–O–C stretching vibration at β-(1,4)glycosidic linkage, they are known as the crystallinity band and amorphous band in cellulose, respectively [19]. A significant reduction in the intensity of the crystallinity band (1430 cm^{-1})and clearly increased in the intensity of the amorphous band(893 cm^{-1}) in the FTIR spectrum of RCNs suggests that the crystallinity of RCNs lower than that of CNCs. The crystallinity of RCNs and CNCs was quantitatively compared by calculating the crystallinity index of RCNs and CNCs using following Equation [22].

$$CrR = A_{1430}/A_{893} \tag{1}$$

where, A1430 and A893 correspond to absorbance at 1430 cm^{-1} (crystallinity band) and 893 cm^{-1} (amorphous band), respectively. The calculated values of crystallinity index for RCNs and CNCs were 0.5 and 0.87, respectively. The significant reduction in the intensity of FTIR bands at 1055, 1107, and 1161 cm^{-1} in RCNs was found, which corresponds to reduction inter- and intra- molecular hydrogen bonding in RCNs. A newly appeared bands at 993 cm^{-1}, observed in amorphous cellulose [22], was found not in CNCs but RCNs.

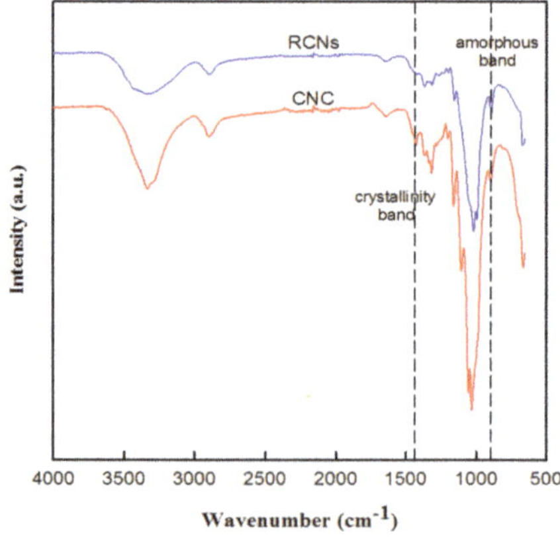

Figure 2. FT-IR spectroscopy bands of RCNs prepared by hydrolyzing MCC with a NaOH-based aqueous solution and CNCs by H_2SO_4 method.

4.2. Characteristics of the RCNs/WPU Nanocomposites Dispersions

The type of polyurethane water dispersion varies depending on the ionic bond that can be generated when mixing the ionic additive in the post-treatment process. An important and practical type of waterborne PU (WPU) is the anionic type. The WPU has a pendant ionized carboxylic acid group and is synthesized by a four step process as follows. First, the macromonomer diisocyanate is prepared by reacting the excess diisocyanate with a long chain polyol. Then carboxylic acid-containing macromonomer diisocyanate is prepared through dimethylolpropionic acid (DMPA) is incorporated into the backbone of macromonomer diisocyanate. The next step, carboxylic acid was neutralized with tertiary amine. Finally, the anionic WPU prepolymer is vigorously stirred in water. The chain extenders transform the residual isocyanate groups in water into urea bonds to produce an anionic WPU that is stably dispersed. In synthesizing WPU, water plays the role of chain extender to react with terminal isocyanate groups. For this reason, WPU synthesized without addition of a chain extender may be used in eco-friendly or biodegradable applications [5].

Based this procedure, the RCNs were added as above shown in Scheme 2 (right) (case of RCNs/WPU-DMF and RCNs/WPU-Acetone). In the liquid suspension, interfacial polar interactions are possible between the RCNs molecules and the liquid polymeric components. There are also strong H-bonding forces between themselves. The low viscosity of the liquid medium allows the RCNs molecules to move due to thermal fluctuations. The molecules can spatially rearrange, and form large structures as shown in Scheme 2 (left). The dispersions states and properties are shown in Table 1.

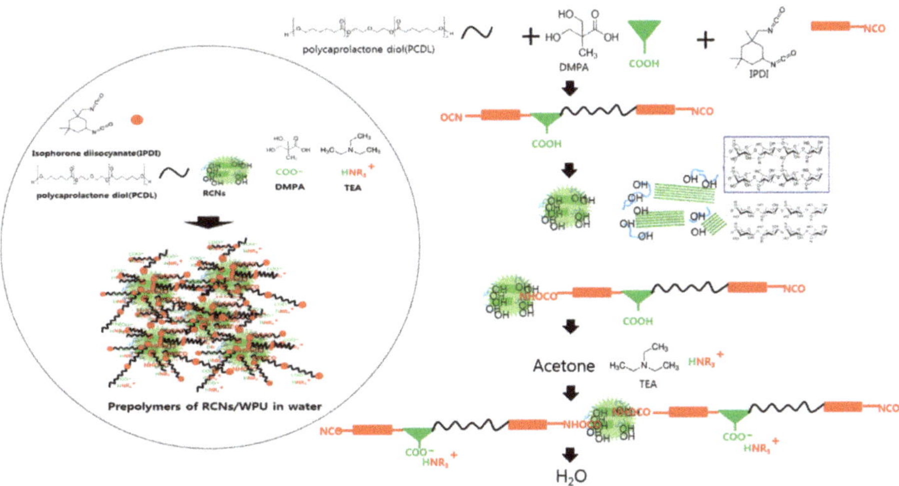

Scheme 2. Final dispersion of prepolymer of WPU/RCNs in water (left) and dispersion of prepolymer of WPU/RCNs in water/acetone (right).

Table 1. Properties of the WPU and RCNs/WPU nanocomposites dispersions obtained with various methods.

Samples	Solid Contents (wt %)	Viscosity (mPa·s)	Mean Particle Size (nm)	Zeta Potential (mV)	Appearance
WPU	25.9	23.8	322	−49.5	
RCNs/WPU-DMF	28	34.8	335	−42.5	
RCNs/WPU-Acetone	29.3	41.4	896	−39.7	
RCNs/WPU-Water	25	62.0	1041	−40.3	

The mean particle size distribution of the WPU dispersions was 322 nm, and that of RCNs/WPU nanocomposites was larger than that of WPU. The particle formation was affected by amount of residual free NCO groups, which can react with added water to result in urea group and larger particles [23]. The orders of average particle size for nanocomposite dispersions were RCNs/WPU-DMF, RCNs/WPU-Acetone and RCNs/WPU-Water. This change could results from the ionic and hydrogen bonding interactions between WPU and nanocellulose particles contains much amorphous regions.

The WPU aqueous dispersion has a negative zeta potential of −49.5 mV. The particles of the WPU dispersion were stabilized in water by COO– ions of neutralized DMPA. Thus, the WPU particles represent negative surface charge and zeta potential. The zeta potentials of the nanocomposites dispersions become less negative due to the addition of nanocellulose particles. The hydroxyl group of nanocellulose would interact with NCO– of WPU pre-polymer chain end and urethane bond was formed. The dispersion of RCNs/WPU-DMF shows the lowest particle size distribution, followed by RCNs/WPU-Acetone, and RCNs/WPU-Water.

The Brookfield viscosity of the RCNs/WPU nanocomposites was higher than that of WPU due to its nanocomposites between WPU pre-polymers and nanocellulose particles. With higher mean particle size, the viscosity of nanocomposites was increased in the order DMF, acetone, and water-dispersed. In case of RCNs/WPU-Water, the nanocelluose particles were inserted in last step of synthesis process of WPU, which acts as chain extender between WPU pre-polymer in synthesis. This produced increasing the mean particle size of RCNs/WPU nanocomposites.

Figure 3 shows IR spectroscopy of WPU, RCNs/WPU-DMF, RCNs/WPU-Acetone, and RCNs/WPU-Water. In the WPU+RCNs spectrums, we could observed the stronger appearance of composited WPU spectrum than neat WPU at around 1731, 1664, 1365, 1307, 1196, and 1062 cm^{-1}, which were associated with the absorption of urethane hydrogen-bonded carbonyl groups (C=O), urea hydrogen-bonded carbonyl groups (C=O), cellulose CH stretching band, cellulose CH$_2$ stretching band, cellulose C–C ring vibration, and cellulose C–O–C ring vibration [24], respectively. The results indicated in Table 2.

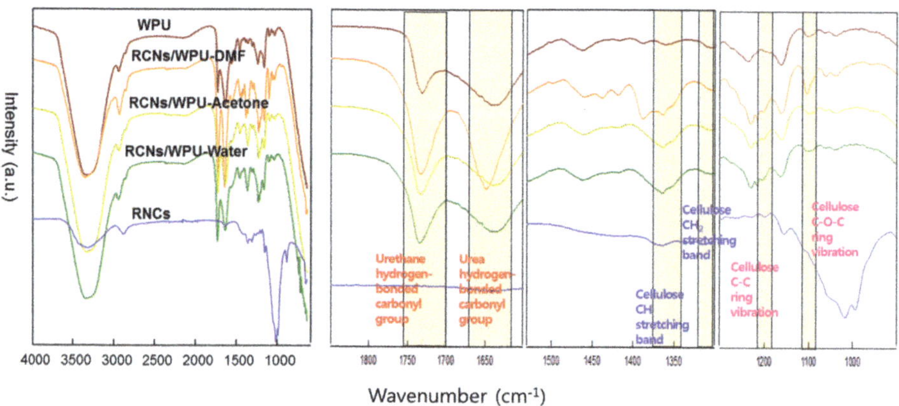

Figure 3. FT-IR spectroscopy bands of WPU, RCNs/WPU nanocomposites, and RCNs.

Table 2. Curve-fitting results of WPU and RCNs/WPU nanocomposites FTIR spectra.

Related Group	C=O Stretching Regions of Urethane Group		C=O Stretching Regions of of Urea Group		Cellulose CH Stretching Band		Cellulose CH$_2$ Stretching Band		Cellulose C-C Ring Band		Cellulose C-O-C Ring Vibration	
Location (cm^{-1})	1731		1664		1365		1307		1196		1062	
Items / Samples	intensity	area	intensity	area	intensity	area	intensity	area	Intensity	area	Intensity	area
WPU	16.2	388	10.1	55	2.1	43.5	0.8	15.3	0.5	5.1	0.7	7.1
WPU/RCNs-DMF	25.7	840	26.6	1170	9.9	475.1	2.4	39.7	1.5	19.4	2.0	223.0
WPU/RCNs-Acetone	21.0	640	10.8	575	8.4	303.3	2.2	15.3	1.2	9.2	1.7	16.3
WPU/RCNs-Water	18.6	565	11.6	649	7.3	260.5	1.3	20.4	1.3	14.8	0.9	8.0

The nanocomposite WPU + RCNS can prove very efficient dispersion of cellulose nanoparticles by IR results. Pei et al. [25] has already demonstrated covalent bond formation between cellulose nanocrystals and polyurethane through FTIR spectra. The absorption of the carbonyl group (C=O) is judged to be due to the reaction of hydroxyl group of cellulose and isocyanate group of WPU. The peak intensity and area about hydrogen bond of WPU was order of RCNs/WPU-DMF > RCNs/WPU-Acetone > RCNs/WPU-Water. The RCNs/WPU-DMF was most effective method for in-situ polymerization of WPU. From the nanocomposite with nanocellulose, however the RCNs/WPU-Acetone appeared to be more effective, which was estimated the intensity of related nanocellulose bands at 1302, 1196, and 1664 cm^{-1}. From the RCNs/WPU manufacturing process point of view, the method RCNs/WPU-Acetone is simpler than the method RCNs/WPU-DMF, of which polymerization method is troublesome to evaporate DMF at an intermediate stage.

4.3. Characteristics of the RCNs/WPU Nanocomposites Films

Transparency: The UV-Vis transmission spectra of the different composite films were tested and the results are shown in Figure 4. The transmittance of WPU film was highest in the visible region with a value of about 88.8% at 600 nm. The RCNs/WPU composite films have poorer visible light transmittance than the WPU film. The order of transmittance were WPU > RCNs/WPU-DMF > RCNs/WPU-Acetone > RCNs/WPU-Water. This result is related to tendency of mean particle size of RCNs/WPU nanocomposites. With increase particle size, the film looks opaque.

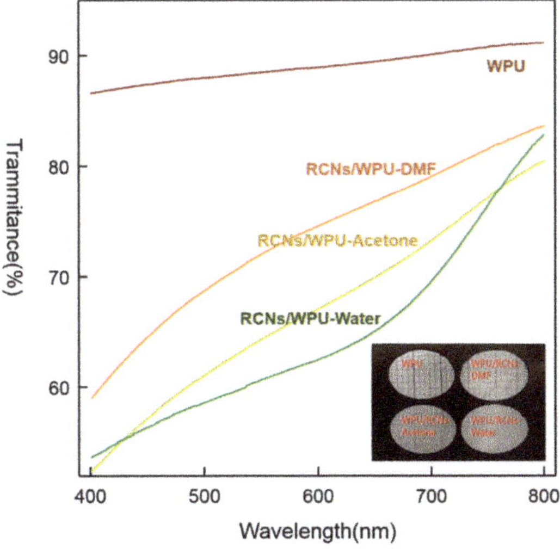

Figure 4. UV-visible spectrums of WPU, RCNs/WPU nanocomposites films.

Tensile, dynamic mechanical properties: Static / cyclic loading and dynamic mechanical analysis (DMA) tensile tests were conducted to study the toughness and softening of RCNs/WPU nanocomposites. The stress–strain (S–S) curves of the WPU and RCNs/WPU nanocomposites are shown in Figure 5A.

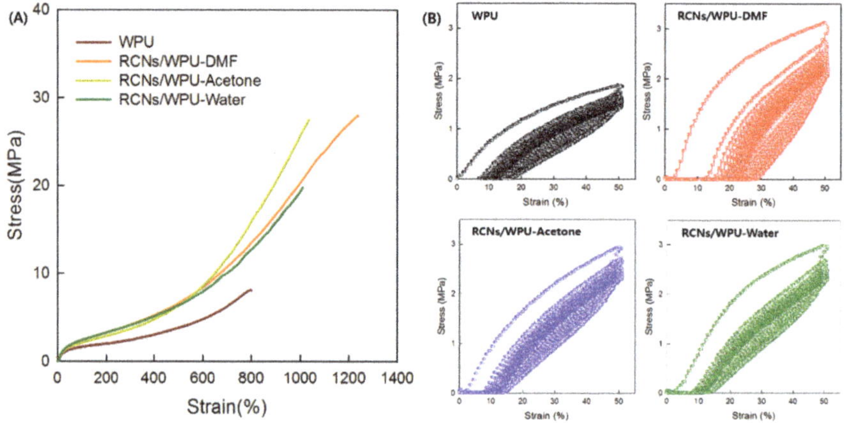

Figure 5. (**A**) Stress–strain, Young's moduli (inset) and (**B**) Tensile recovery properties curves of WPU, RCNs/WPU nanocomposites films.

The resulting values, such as the tensile strength, strain-to-failure, and Young's modulus are summarized in Table 3. There is a remarkable development in the tensile strength associated with increasing strain simultaneously. In the case of the neat WPU, it had an S–S curve with 800% strain and 7.8 MPa tensile strength at break, the RCNs/WPU nanocomposites containing 1 wt % RCNs exhibits stronger stress and higher elongations than WPU film. The tensile strength increases from 7.8 MPa for neat WPU to 27.9 MPa for RCNs/WPU-DMF or RCNs/WPU-Acetone. This is more than 4 times the neat WPU value. The strain value for the RCNs/WPU nanocomposites reaches the highest value of 1235% for RCNs/WPU-DMF with containing 1 wt % RCNs, which value is almost 2 times higher than that of neat WPU. There is a tendency that the strain of the nanocomposite is further increased after manufacture of RCNs nanocomposites. The Young's modulus of the RCNs/WPU nanocomposites increases to 10.7 MPa for RCNs/WPU-Water compare to 1.8 MPa for the neat WPU. This increase is due to the stiffening of the polymers. Polyurethane is block copolymer with alternating soft and hard segments. In this study, the blocks of the WPUs consist of IPDI and PCDL, which are role as hard and soft segments, respectively. The soft phases are derived from the polyols linked by IPDI and roles of elastomeric properties. The hard phase for the IPDI segment with high polar urethane bonds is bounded by the nanocellulose particles. The domains in which these nanoparticles are introduced acts as physical crosslinking agent and acts as high modulus filler. The nanocellulose particles were introduced in the reactor before increasing viscosity (DMF-dispersed type) of the pre-polymer; at this step, there are free NCO groups at the ends of the low molecular polyurethane chains. This result explained that later addition of nanocellulose of DMF-dispersed steps induces an increase in the effective cross-link by homogeneous dispersion of the nanoparticles in the nanocomposites than Acetone and Water-dispersed step. This linkage can be attributed to new interactions formed between WPU and RCNs. For the Acetone dispersion step, the nanocellulose was added to the viscosity control process of the polymer with acetone, which is finished formation of pre-polymer of polyurethane. Nanocellulose particles have many hydroxyl groups on surface, which roles as graft-point of polyurethane chains as Scheme 2. In case of Water-dispersed step, the nanocelluloses were introduced during aqueous dispersion step of the pre-polymer, which roles as chain extender in WPU synthesis process. This part appears as a hard segment and contributes to the high modulus of the WPU composite matrix through hydrogen bonding. The role as chain extender of nanocellulose particles for RCNs/WPU synthesis process is mainly formed by Water-dispersed step. Although the effects of reinforcement WPU by cellulose nanoparticles were accomplished, the roles of nanoparticles were likely different in detail according to when they introduced. Comparing with cellulose nanocrystal or nanofibers for reinforcement, semi-crystalline nanocellulose particles improved the strain property of RCNs/WPU without stress property damage.

Table 3. Characterization Data for WPU and RCNs/WPU nanocomposites.

Samples	Transmittance [a] (%)	Young's Modulus E [b] (MPa)	Tensile Strength σ [b] (MPa)	Strain at Break ε [b] (%)	Residual Strain ε [b] (%)	
					Cycle 1 [c]	Cycle 20 [c]
WPU	88.5	1.8±0.2	7.8±1.3	796.2±13	12–13	21–22
RCNs/WPU-DMF	71.8	27.9±2.5	27.9±2.5	1234.9±18	10–11	22–23
RCNs/WPU-Actone	64.4	27.5±1.6	27.5±1.6	1031.5±22	9–10	18–19
RCNs/WPU-Water	60.8	19.7±2.2	19.7±2.2	1009.2±20	9–10	18–19

[a] Measured by UV–vis spectroscopy from 450 to 660. [b] Mechanical properties fixed on ASTM D1708 microtensile bars. [c] Percent of plastic deformation calculated after 1 cycle or 20 cycles from 0 to 50% strain at 130 mm min^{-1}.

To understand the elastomeric recovery and deformation of neat WPU and the RCNs/WPU nanocomposites, a cyclic tensile test was conducted. The tensile loading and unloading cycles performed at a maximum strain of 50% with a constant speed at 130 mm·m^{-1}. The loops of hysteresis for cycles 20 are shown in Figure 5B. The large hysteresis loop showed unrecoverable deformation in the first cycle, and followed by very small hysteresis with negligible residual strain values in the next 19 cycles. This is desirable phenomenon for an elastomeric material like PU (Table 3). The large hysteresis region of the first loops in the RCNs/WPU nanocomposites represents the stretch-induced softening due to the increased volume fraction of the effective soft phases as a result of conversion from hard domains to soft phases during the initial alignment of the microstructures [26]. However, RCNs/WPU-Water and RCNs/WPU-Acetone displays a decrease in softening comparing to RCNs/WPU-DMF, as shown in Figure 5B. This result can be explained that entanglement and the networks from addition of RCNs in hard segments, which induced enhancing mechanical properties of the RCNs/WPU nanocomposite. These results are in good agreement with the tendency of the Young's modulus (E), as shown in Table 3.

Dynamic mechanical analysis (DMA) of the WPU and RCNs/WPU nanocomposite were demonstrated that the effect of the temperature and the frequency on the dynamic elastic properties. The storage modulus of the samples is shown in Figure 6. The modulus of the RCNs/WPU nanocomposites are slightly higher than that of the neat WPU at room temperature (20 °C), which increases from 0.17 MPa for neat WPU to 0.37 MPa for RCNs/WPU-DMF. The hardening mechanism is attributed to increasing the hard segments in the RCNs/WPU nanocomposites and effective cross-link density of the RCNs in the nanocomposites. The neat WPU showed a sharply decreased modulus during the glass-rubber transition. However, the decreasing tendency of storage modulus for the RCNs/WPU nanocomposite is slowly reduced because of the confined network structure, providing remarkable thermomechanical stability up to 50 °C. Also, the T_g values of the WPU soft segments represents the degree of relaxation of the PCDL phases, which shifted to a lower temperature from −37.3 (neat WPU) to −44.6 °C (RCNs/WPU-DMF). This phenomenon is incompatible with the result of complexing with highly crystalline nanocellulose [25] Generally, when highly crystalline nanoparticles (CNs) are introduced, the hydrogen bonded carbonyl groups among the hard segments formation increased and strong association with CNs of the hard segments [25]. However, since the RCN has a semi-crystalline phase, it maintains a soft chain even in the hard segment of WPU, so that the damping due to the external environment is formed at a lower temperature than neat WPU. This means that the cross-coupling between the WPU molecule and the RCN increases the movement of the WPU chain, thereby increasing the damping capacity.

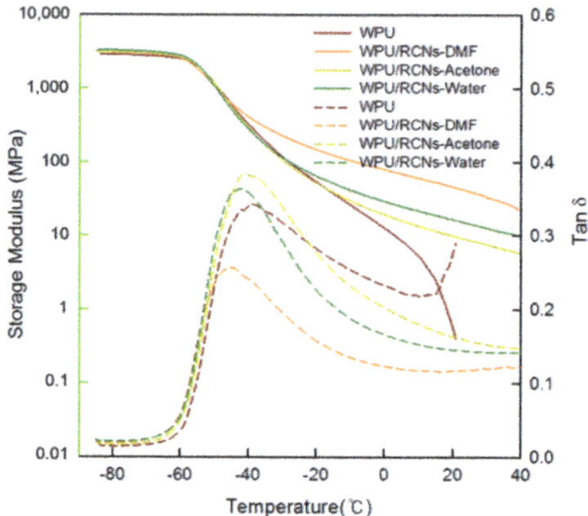

Figure 6. DMA properties curves of WPU, RCNs/WPU nanocomposites films.

Thermal properties: Figure 7 shows the TGA curves for WPU, RCNs and RCNs/WPU nanocomposite with different loadings steps of cellulose nanoparticles. The thermogram of RCNs shows 2 steps at around 150–300 °C and 350–400 °C, which were corresponding to dehydration and decomposition of glycosyl units, and oxidation and breakdown to lower molecular weight gaseous products respectively, followed by the formation of a char. The curve of WPU shows 2 steps, first decomposition stage (from approximately 280–320 °C) results from urethane/urea-bond degradation of hard-segments; second decomposition stage (from approximately 320–380 °C) corresponding of polyol degradation of soft-segments. The RCNs/WPU nanocomposite have 2 steps degradation at 280–400 °C due to decomposition of nanocellulose particles as chain extender in hard segment in RCNs/WPU nanocomposite and a maximum degradation at about 400 °C over due to WPU chains, which is much higher than the degradation temperature of pure WPU (320–380 °C). This significant enhancement of thermal resistance by the presence of RCNs can be attributed to the formation of a confined structure in the RCNs/WPU nanocomposites. Not only the T_{max} of RCNs/WPU nanocomposite is higher than that of WPU, but also the start temperature of decomposition is different with loadings steps of cellulose nanoparticles. The difference between RCNs-Acetone and RCNs-Water was not significant, but RCNs-DMF showed higher thermal stability than other samples. The role of nanocellulose particles as thermos-stability enforcement of nanocomposites is most effect in RCNs/WPU-DMF.

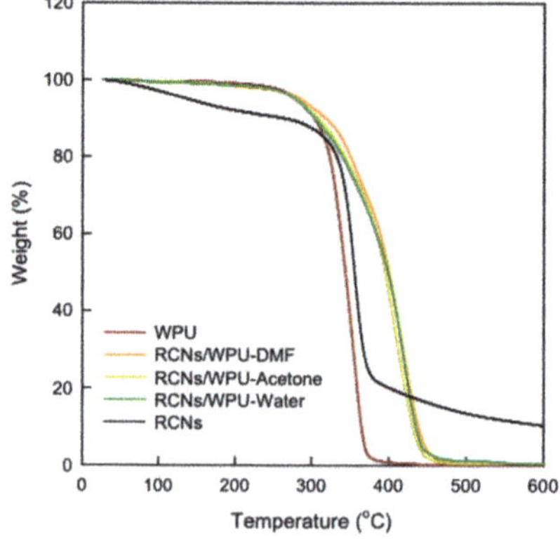

Figure 7. TGA properties curves of RCNs, WPU, RCNs/WPU nanocomposites films.

5. Conclusions

In this study, RCNs with improving solubility and stability of solution, thereby resulting in lower crystallinity were fabricated by using a NaOH-based solution with urea. Additionally, the stronger chemical bond between RCNs and WPU was here founded by investigating which stage in particular added by RCNs worked best on strengthening their bond. Finally, we could develop RCNs reinforced WPU nanocomposites with biodegradability as well as enhancing mechanical properties through a facile, eco-friendly approach fittable for global politics. FTIR indicated that the RCNs/WPU nanocomposites using a DMF dispersed step has the strongest intensity of hydrogen bonds between cellulose and isocyanate groups attributed by the absorption of carbonyl groups (C=O). Tensile test also proved that the DMF-dispersed step induce an increase in the effective crosslink density by homogeneous dispersion of the nanofillers in the nanocomposites, with 27.9 MPa of 4 times more improvement value, 1235% of strain-to-failure, 2 times higher value. The moduli were also higher than that of the neat WPU matrix illustrating that the value of G' of the nanocomposites increases from 0.17 MPa for neat WPU to 0.37 MPa for RCNs/WPU-DMF, and the T_g of the WPU soft segments were shifted to a lower temperature from -37.3 to -44.6 °C for the neat WPU and RCNs/WPU-DMF, because the breaking of hydrogen bonded carbonyl groups in the hard segments being essential for inhibiting of HD formation, accordingly causing the enhancement in the mobility of the poly(ether) chain in RCNs/WPU, comparing with the neat WPU sample. TGA results displayed that the role of RCNs as thermos-stability enforcement of the resultant nanocomposites was the most effective in the DMF-dispersed step.

In conclusion, the above all results signified that the properties involving the mechanical, thermal, and chemical specific features, of RCNs/WPU nanocomposites by using the DMF-dispersed step were remarkably improved, comparing with acetone or water-dispersed steps. Therefore, fabricated nanocomposites have a great potential for further applications such as environmentally friendly artificial leather processes, coatings, and biomedical fields.

Author Contributions: Conceptualization, E.J.S.; Methodology, E.J.S.; Software, M.W.L.; Validation, E.J.S., S.M.C. and M.W.L.; Formal Analysis, S.M.C.; Investigation, S.M.C.; Resources, E.J.S.; Data Curation, E.J.S.; Writing-Original Draft Preparation, E.J.S., S.M.C.; Writing-Review & Editing, E.J.S., S.M.C.; Visualization, S.M.C.; Supervision, E.J.S.; Project Administration, E.J.S., S.M.C.

Funding: This research was supported by the Basic Science Research Program of the National Research Foundation of Korea (NRF) funded by the Ministry of Education (2015R1C1A2A01056027 and 2016R1A6A3A11930280).

Conflicts of Interest: The authors declare no conflict of interest.

References

1. Tian, Y.; Zhang, H.; Zhang, Z. Influence of nanoparticles on the interfacial properties of fiber-reinforced-epoxy composites. *Compos. Part A. Appl. Sci. Manuf.* **2017**, *98*, 1–8. [CrossRef]
2. Zhang, C.; Zhao, J.; Rabczuk, T. The interface strength and delamination of fiber reinforced composites using a continuum modeling approach. *Compos. B. Eng.* **2018**, *137*, 225–234. [CrossRef]
3. Skovsgaard, S.P.H.; Jensen, H.M. Constitutive model for imperfectly bonded fibre-reinforced composites. *Compos. Struct.* **2018**, *192*, 82–92. [CrossRef]
4. Li, H.; Zhang, B. A new viscoelastic model based on generalized method of cells for fiber-reinforced composites. *Int. J. Plast.* **2015**, *65*, 22–32. [CrossRef]
5. Zhou, X.; Li, Y.; Fang, C.; Li, S.; Cheng, Y.; Lei, W.; Meng, X. Recent Advances in Synthesis of Waterborne Polyurethane and Their Application in Water-based Ink: A Review. *J. Mater. Sci. Technol.* **2015**, *31*, 708–722. [CrossRef]
6. Shin, E.J.; Choi, S.M. Advances in Waterborne Polyurethane-Based Biomaterials for Biomedical Applications. *Adv. Exp. Med. Biol.* **2018**, *1077*, 251–283.
7. Shin, E.J.; Choi, S.M.; Lee, J.W. Fabrication of regenerated cellulose nanoparticles/waterborne polyurethane nanocomposites. *J. Appl. Polym. Sci.* **2018**, *135*, 46633–46641. [CrossRef]

8. Shin, E.J.; Choi, S.M. High Performance Fabrics Using Nanocellulose. *Trends Text. Eng. Fash. Technol.* **2018**, *3*, 1–3.

9. Velankar, S.; Cooper, S.L. Microphase Separation and Rheological Properties of Polyurethane Melts 2. Effect of Block Incompatibility on the Microstructure. *Macromolecules* **2000**, *33*, 382–394. [CrossRef]

10. Hsu, S.; Tang, C.; Tseng, H. Biocompatibility of poly(ether)urethane-gold nanocomposites. *J. Biomed. Mater. Res.* **2006**, *79A*, 759–770. [CrossRef]

11. Tao, J.; Yan, Y.; Xu, W. Physical Characteristics and Properties of Waterborne Polyurethane Materials Reinforced with Silk Fibro in Powder. *J. Polym. Sci. Part B: Polym. Phys.* **2010**, *48*, 940–950. [CrossRef]

12. Xiong, B.; Zhao, P.; Hu, K.; Zhang, L. The dissolution of cellulose in NaOH-based aqueous system by two step process. *Cellulose* **2014**, *18*, 237–245.

13. Moon, R.J.; Martini, A.; Nairn, J.; Simonsen, J.; Youngblood, J. Cellulose nanomaterials review: Structure, properties and nanocomposites. *J. Chem. Soc. Rev.* **2011**, *40*, 3941–3994. [CrossRef] [PubMed]

14. Liu, H.; Cui, S.; Shang, S.; Wang, D.; Song, J. Properties of rosin-based waterborne polyurethanes/cellulose nanocrystals composites. *Carbohydr. Polym.* **2013**, *96*, 510–515. [CrossRef] [PubMed]

15. Park, S.H.; Oh, K.W.; Kim, S.H. Reinforcement effect of cellulose nanowhisker on bio-based polyurethane. *Compos. Sci. Technol.* **2013**, *86*, 82–88. [CrossRef]

16. Wang, Y.; Tian, H.; Zhang, L. Role of starch nanocrystals and cellulose whiskers in synergistic reinforcement of waterborne polyurethane. *Carbohydr. Polym.* **2010**, *80*, 665–671. [CrossRef]

17. Rico, M.; Rodríguez-Llamazares, S.; Barral, L.; Bouza, R.; Montero, B. Processing and characterization of polyols plasticized-starch reinforced with microcrystalline cellulose. *Carbohydr. Polym.* **2016**, *149*, 83–93. [CrossRef]

18. Xiaohua, K.; John, W.; Liyan, Z.; Jonathan, M.C. The preparation and characterization of polyurethane reinforced with a low fraction of cellulose nanocrystals. *Prog. Org. Coat* **2018**, *125*, 207–214.

19. Adsul, M.; Soni, S.K.; Bhargava, S.K.; Bansal, V. Facile Approach for the Dispersion of Regenerated Cellulose in Aqueous System in the Form of Nanoparticles. *Biomacromolecules* **2012**, *13*, 2890–2895. [CrossRef]

20. Santamaria-Echart, A.; Ugarte, L.; Garcia-Astrain, C.; Arbelaiz, A.; Corcuera, M.A.; Eceiza, A. Cellulose nanocrystals reinforced environmentally-friendly waterborne polyurethane nanocomposites. *Carbohydr. Polym.* **2016**, *151*, 1203–1323. [CrossRef]

21. Kim, B.K. Aqueous polyurethane dispersions. *Colloid. Polym. Sci.* **1996**, *274*, 599–611. [CrossRef]

22. Ciolacu, D.; Ciolacu, F.; Popa, V. Amprphous cellulose–Structure and Characterization. *Cellul. Chem. Technol.* **2011**, *45*, 13–21.

23. Santamaria-Echart, A.; Ugarte, L.; Arbelaiz, A.; Barreiro, F.; Corcuera, M.A.; Eceiza, A. Modulating the microstructure of waterborne polyurethanes for preparation of environmentally friendly nanocomposites by incorporating cellulose nanocrystals. *Cellulose* **2017**, *24*, 823–834. [CrossRef]

24. Kumara, A.; Negia, Y.S.; Bhardwaja, N.K.; Choudhary, V. Synthesis and characterization of methylcellulose/PVA based porous composite. *Carbohyd. Polym.* **2012**, *88*, 1364–1372. [CrossRef]

25. Pei, A.; Malho, J.M.; Ruokolainen, J.; Zhou, Q.; Berglund, L.A. Strong Nanocomposite Reinforcement Effects in Polyurethane Elastomer with Low Volume Fraction of Cellulose Nanocrystals. *Macromolecules* **2011**, *44*, 4422–4427. [CrossRef]

26. Qi, H.J.; Boyce, M.C. Constitutive model for stretch-induced softening of the stress-stretch behavior of elastomeric materials. *J. Mech. Phys. Solids* **2004**, *52*, 2187–2205. [CrossRef]

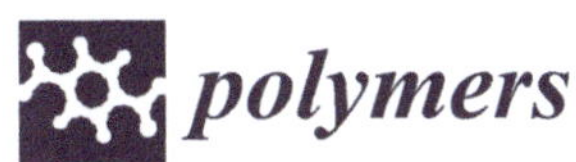 

Article

Synthesis and Morphological Control of Biocompatible Fluorescent/Magnetic Janus Nanoparticles Based on the Self-Assembly of Fluorescent Polyurethane and Fe$_3$O$_4$ Nanoparticles

Botian Li [1,4], Wei Shao [1], Yanzan Wang [1], Da Xiao [1], Yi Xiong [3], Haimu Ye [1], Qiong Zhou [1,*] and Qingjun Jin [2,4,*]

1   Department of Materials Science and Engineering, China University of Petroleum, Beijing 102249, China; botian.li@cup.edu.cn (B.L.); shao222333@sina.com (W.S.); wangyanzan_cup@163.com (Y.W.); 18811372167@163.com (D.X.); yehaimu@cup.edu.cn (H.Y.)
2   Research Institute of Chemical Defense, Academy of Military Sciences, Beijing 102205, China
3   College of Safety and Ocean Engineering, China University of Petroleum, Beijing 102249, China; xiongyi@cup.edu.cn
4   Key Laboratory of Advanced Materials of Ministry of Education of China, Department of Chemical Engineering, Tsinghua University, Beijing 100084, China
*   Correspondence: zhouqiong_cn@163.com (Q.Z.); jinqingjun502@163.com (Q.J.); Tel.: +86-10-8973-3200

Received: 23 January 2019; Accepted: 1 February 2019; Published: 5 February 2019

Abstract: Functionalized Janus nanoparticles have received increasing interest due to their anisotropic shape and the particular utility in biomedicine areas. In this work, a simple and efficient method was developed to prepare fluorescent/magnetic composite Janus nanoparticles constituted of fluorescent polyurethane and hydrophobic nano Fe$_3$O$_4$. Two kinds of fluorescent polyurethane prepolymers were synthesized by the copolymerization of fluorescent dye monomers, and the fluorescent/magnetic nanoparticles were fabricated in one-pot via the process of mini-emulsification and self-assembly. The nanostructures of the resulting composite nanoparticles, including core/shell and Janus structure, could be controlled by the phase separation in assembly process according to the result of transmission electron microscopy, whereas the amount of the nonpolar segments of polyurethane played an important role in the particle morphology. The prominent magnetic and fluorescent properties of the Janus nanoparticles were also confirmed by vibrating magnetometer and confocal laser scanning microscope. Furthermore, the Janus nanoparticles featured excellent dispersity, storage stability, and cytocompatibility, which might benefit their potential application in biomedical areas.

Keywords: Janus particle; morphology; fluorescent polyurethane; magnetic nanoparticle; cytocompatibility

---

1. Introduction

In the past decades, with the development of biomedicine technologies, fluorescent/magnetic nanoparticles (FMNPs) have received increasing attention, as they have shown great superiority in fluorescent labeling and magnetic response [1,2], and could be potentially applied in protein separation [3], drug delivery [4], bio-imaging [5], and so on. Usually, FMNPs consist of fluorescent materials and magnetic nanoparticles such as Fe$_3$O$_4$ nanocrystals [6,7]. Although many kinds of fluorescent quantum dots derived from Cd and Eu have been emphatically studied to constitute the FMNPs [8,9], these inorganic quantum dots are principally involved in disadvantages of the tedious preparation process and the threat of heavy metal ions with potential toxicity, which limit their further applications for in vivo research [10,11]. Therefore, in most cases, organic fluorescent dyes are considered as the alternative owing to their numerous color species and relatively low cost. Generally, the biological toxicity of dyes could be reduced by the covalent bonding with polymer chain, because the migration of the chromophores is greatly inhibited in the so-called covalently colored polymer; meanwhile, the solvent resistance and the light fastness could also be improved [12,13].

Magnetic Janus particles are of particular utility and interest, which has long been motivated by their enormous applications in micromotors [14,15], panel displays [16], biological imaging [17], and magneto separation [18]. So far, many great efforts have been made to prepare fluorescent/magnetic Janus particles with tunable size, shape, and microstructure [19–21], including the microfluidic method [22,23], electrospray method [24], and emulsion method [25]. The two former methods, requiring specialized instruments for preparation, can only produce larger spheres in micro-meter size, while the emulsion method is commonly feasible for the nanoparticles. For example, Gao et. al advocated a microemulsion method to prepare the fluorescent/magnetic Janus nanoparticles (FMJNPs) for magnetolytic therapy of cancer cells using pyrene as a fluorescent probe [26]. However, the addition of small molecule dye introduced the possibility of relatively poor stability and high toxicity. To solve this problem, polymeric fluorescent dyes were introduced in FMJNPs, and the approaches based on the seed emulsion polymerization were developed [27,28], but because of the inhibition of magnetite and fluorophores on polymerization, these methods were confined by low monomer conversion. Furthermore, particle structure was difficult to control. As a consequence, it is still a challenge to develop a facile, versatile, and controllable method to prepare biocompatible FMJNPs.

Polyurethane (PU) is well known as a polymer material with good biocompatibility. Previously, we have reported a series of studies concerning the incorporation of fluorophores into polymer chains to prepare covalently colored PU emulsion [29,30]. In this article, a facile method based on mini-emulsion was developed to fabricate FMJNPs composed of covalently colored fluorescent PU and magnetic nanoparticles. The fluorescent PU prepolymer was synthesized by polymerization of fluorescent monomers derived from 1,8-naphthalimide (NA). After the addition of the hydrophobic Fe_3O_4 nanoparticles, the composite nanoparticles with core-shell and Janus structure were produced by the self-assembly method following mini emulsification in one-pot, and their nanostructure could be regulated by changing the non-polar segment ratio of the PU polymer to generate FMJNPs. The resulting Janus particles featured uniform particle size with narrow distribution, as well as promising cytocompatibility and colloidal stability. Moreover, this method was envisaged to be effective and versatile for FMJNPs derived from other hydrophobic polymers, hence it might be potentially applied in the fabrication of various FMJNPs with biocompatibility.

2. Materials and Methods

2.1. Materials

We purchased 4-bromo-1,8-naphthalic anhydride, 2-amino-1,3-propanediol, sodium methoxide, diethylamine, cupric acetate monohydrate, dibutyltin dilaurate, sodium dodecyl sulfate (SDS), and 1,2-hexadecanediol (HDO) from Shanghai Aladdin Co. Castor oil (CO), 1,4-butanediol (BDO), isophorone diisocyanate (IPDI), and trimethylol propane (TMP) were supplied by Tianjin Chemicals & Reagents. The solvents, such as tetrahydrofuran (THF), dimethyl sulfoxide (DMSO), chloroform, and methanol, were purchased from Beijing Chemicals. All the reagents were analytically pure and used without further treatment.

2.2. Characterizations

The synthesized compounds were characterized using ^{1}H NMR (JEOL JNM-ECA600, Japan) and FT-IR (Bruker TENSOR II, Germany). The spectroscopy of fluorescent polymer was measured on a UV/vis spectrometer (Pgeneral T6, Beijing, China) and a luminescence spectrometer (RF-5301PC, Shimadzu, Japan). The nanoparticles were characterized using transmission electron microscopy

(TEM) 200kV (JEOL JEM 2100, Japan), SEM (FEI Quanta 200F, Netherlands), X-ray diffraction (XRD) (Bruker AXS, D8 Focus, Germany), and dynamic light scattering (DLS) (Anton-Paar, Litesizer 500, Austria). The measurements of magnetic properties were conducted on a BKT-4500 vibrating sample magnetometer (Xinke, Beijing, China). The conversion of fluoresent dye monomer was determined according to the integral proportion of time dependent curves, which was measured by gel permeation chromatography (GPC) equipped with double detectors (refractive index, UV/vis) (Shimadzu, Japan).

Migration fastness was characterized by extracting the PU latex film with dichloromethane in a Soxhlet extractor for 5 h. The extraction rate was calculated as follows: E = ($d2/d1$) × 100%, where $d1$ and $d2$ denote the dye amount used in the recipe and the extracted dye amount measured by UV analysis, respectively.

Cytocompatibility experiments were carried out using the WST-1 assay. The human intestinal epithelial cells were incubated with the samples for a certain amount of time, after which 10 µL of cell proliferation reagent WST-1 (4-[3-(4-iodophenyl)-2-(4-nitrophenyl)-2H-5-tetrazolio]-1,3-benzene disulfonate) was added to each well and incubated at 37 °C under 7% CO_2 in an incubator for 2 h. The absorbance was measured in a microplate reader (Bio-Rad) at 450 nm. The optical microscopic pictures were obtained on a microscope (Olympus IX73).

2.3. Synthesis of 4-bromo-N-(2-hydroxy-1-hydroxymethylethyl)-1,8-naphthalimide (BHHNA)

A total of 4.17 g (0.015 mol) of 4-bromo-1,8-naphthalic anhydride and 2.43 g (0.027 mol) of 2-amino-13-propanediol were dissolved in 120 mL ethanol, after which the solution was stirred and heated to 78 °C. Then, after 4 h reaction, the solvent was evaporated under reduced pressure at 50 °C. Finally, the yellow crude product was washed three times with deionized water, and then dried in a vacuum oven for 12 h to obtain 4.647 g bright yellow product. Yield: 88%.

^{1}H NMR (600 MHz, DMSO-d_6, TMS, δ): 8.57(d, 1H, NA 5-H), 8.53(d, 1H, NA 7-H), 8.33(d, 1H, NA 2-H), 8.22(d, 1H, NA 3-H), 8.01(d × d, 1H, NA 6-H), 5.22(m, 1H, $-CH(CH_2-OH)_2$), 3.82–3.96(m, 4H, $-CH(CH_2-OH)_2$). FT-IR (KBr, v/cm^{-1}): 3431(O–H stretch), 1726(C=O stretch), 1592, 1500 (NA skeleton vibration) (Figure S1).

2.4. Synthesis of 4-methoxyl-N-(2-hydroxy-1-hydroxymethylethyl)-1,8-naphthalimide (MHHNA)

3.5 g (0.01 mol) BHHNA and 1.62 g (0.03 mol) sodium methoxide were added to a flask containing 60 mL methanol, the mixture was stirred and heated to 65 °C, after 4 h reaction the solvent was evaporated under reduced pressure at 40 °C, resulting the white solid. Afterward, the crude product was washed for three times with deionized water, then dried in a vacuum oven for 12 h to give 2.35 g product. Yield: 78%.

^{1}H NMR (600 MHz, DMSO-d_6, TMS, δ): 8.53 (d, 1H, NA 5-H), 8.49 (d, 1H, NA 7-H), 8.45 (d, 1H, NA 2-H), 7.83 (d × d, 1H, NA 6-H), 7.33 (d, 1H, NA 3-H), 5.22 (m, 1H, $-CH(CH_2-OH)_2$), 4.13 (s, 3H, $-O-CH_3$), 3.81–3.94 (m, 4H, $-CH(CH_2-OH)_2$) (Figure S2). FT-IR (KBr, v/cm^{-1}): 3431 (O–H stretch), 2937 (CH$_3$, C–H stretch), 1688 (C=O stretch) (Figure S3).

2.5. Synthesis of 4-diethylamino-N-(2-hydroxy-1-hydroxymethylethyl)-1,8-naphthalimide (DHHNA)

A total of 2.57 g (0.0075 mol) of BHHNA, 5.47 g (0.075 mol) of diethylamine, and 0.4 g (0.002 mol) of cupric acetate monohydrate were dissolved in 20 mL DMSO, and the solution was stirred and heated to 120 °C. Then, after 12 h reaction, it was cooled to room temperature and precipitated in 150 mL deionized water. The solid was filtered and washed three times, and then dried in vacuum. The crude product was purified by silica column chromatography using a mixture of dichloromethane/ethyl acetate (1:3) as eluent to give 1.165 g yellow product. Yield: 45%.

^{1}H NMR (600 MHz, DMSO-d_6, TMS, δ): 8.45 (d, 1H, NA 5-H), 8.43 (d, 1H, NA 7-H), 8.36 (d, 1H, NA 2-H), 7.78 (d × d, 1H, NA 6-H), 7.35 (d, 1H, NA 3-H), 5.22 (m, 1H, $-CH(CH_2-OH)_2$), 3.78–3.96 (m, 4H, $-CH(CH_2-OH)_2$), 3.39 (q, 4H, $-N-(CH_2-CH_3)_2$), 1.10 (t, 6H, $-N-(CH_2-CH_3)_2$) (Figure S4).

FT-IR (KBr, v/cm^{-1}): 3451 (O–H stretch), 2965, 2884 (saturated C–H stretch), 1689 (C=O stretch), 1589, 1480 (NA skeleton vibration) (Figure S5).

2.6. Synthesis of PU Prepolymer

According to the recipe listed in Table 1, IPDI, hydroxyl chain extenders, and THF were charged into a three-necked flask equipped with thermometer, reflux condenser, and electric mechanical stirrer, and one drop of dibutyltin dilaurate was added as catalyst. The flask was maintained at 65 °C in an oil bath for 4 h under stirring, following which the reaction system was cooled to 50 °C, and PU prepolymers were obtained after removal of THF by rotary evaporation under reduced pressure.

Table 1. Recipe for the preparation of the polyurethane (PU) prepolymer. MHHNA—4-methoxyl-N-(2-hydroxy-1-hydroxymethylethyl)-1,8-naphthalimide; TMP—trimethylol propane; CO—castor oil; DHHNA—4-diethylamino-N-(2-hydroxy-1-hydroxymethylethyl)-1,8-naphthalimide; IPDI—isophorone diisocyanate; HDO—1,2-hexadecanediol; BDO—1,4-butanediol; THF—tetrahydrofuran.

Prepolymer	IPDI/g	CO/g	TMP/g	HDO/g	BDO/g	Fluorescent dye/g	THF/g
PU-TMP	2.22	0	0.158	0	0.534	0	10
PU-CO	2.22	0.75	0	0	0.592	0	10
PU-MHHNA	2.22	0.4	0	0.2	0.572	0.016 [a]	10
PU-DHHNA	2.22	0.4	0	0.2	0.572	0.018 [b]	10

[a] MHHNA, 5.3×10^{-2} mmol. [b] DHHNA, 5.3×10^{-2} mmol.

2.7. Preparation of Composite Nanoparticles

The hydrophobic magnetic nanoparticles (HMNPs) were synthesized by the thermal decomposition of iron oleate in the presence of oleic acid, according to the literature [31].

In the principal fabrication procedure of composite nanoparticles, taking PTCP-20 as an example, 20 mg of HMNPs and 80 mg of PU-TMP were dissolved in 2.8 g CHCl$_3$. After the addition of 5 mL SDS aqueous solution (0.6 mg/mL), the mixture was emulsified for 10 min by ultrasonication (200 W) in an ice bath to give a mini-emulsion. Subsequently, the emulsion was slowly evaporated in a rotary evaporator at 55 °C for 15 min until no chloroform vaporized, resulting a dark yellow latex of PTCP-20 (20% HMNPs content). PCCP-*c*, PMCP-*c*, and PDCP-*c* were similarly prepared by utilizing PU-CO, PU-MHHNA, and PU-DHHNA, respectively, wherein *c* denoted HMNPs content (*c* = 0, 10, 20, 30) (Table 2).

Table 2. Fabrication formula of composite nanoparticles. HMNPs—hydrophobic magnetic nanoparticles; PTCP—PU-TMP magnetic composite nanoparticle; PCCP—PU-CO magnetic composite nanoparticle; PMCP—PU-MHHNA magnetic composite nanoparticle; PDCP—PU-DHHNA magnetic composite nanoparticle.

Sample	PU Prepolymer/mg		HMNPs/mg
PTCP-20	PU-TMP	80	20
PCCP-20	PU-CO	80	20
PMCP-0		100	0
PMCP-10	PU-MHHNA	90	10
PMCP-20		80	20
PMCP-30		70	30
PDCP-0		100	0
PDCP-10	PU-DHHNA	90	10
PDCP-20		80	20
PDCP-30		70	30

3. Results and Discussion

3.1. Characterization of Fluorescent Polyurethane

As illustrated in Scheme 1, two fluorescent dyes (MHHNA, DHHNA) were synthesized by a two-step procedure of imidization and nucleophilic substitution. Both of them consisted of 1,8-naphthalimide chromophore, but offered contrasting fluorescent colors due to the different electron-donating substituents. These dyes were employed in the synthesis of polyurethane by reacting with isocyanate group as chain extenders, thus introducing the fluorescent units into macromolecular chain to produce covalently colored fluorescent PU prepolymers containing isocyanate end group (Scheme 2). The polymerization conversion of fluorescent monomers measured by GPC method showed results of 98.5% (MHHNA) and 97.3% (DHHNA), suggesting that most of the fluorescent chromophores were covalently grafted on PU (Figures S6 and S7), and that the covalently colored fluorescent prepolymer was successfully prepared. To characterize the migration fastness of fluorophores, the fluorescent PU films PMCP-0 and PDCP-0 were extracted in dichloromethane for 5 h, both giving extraction rates below 2%. Because dichloromethane was a good solvent for MHHNA and DHHNA, it could be deduced that the covalent bonding with the polymer chain significantly improved the migration fastness of the fluorophore.

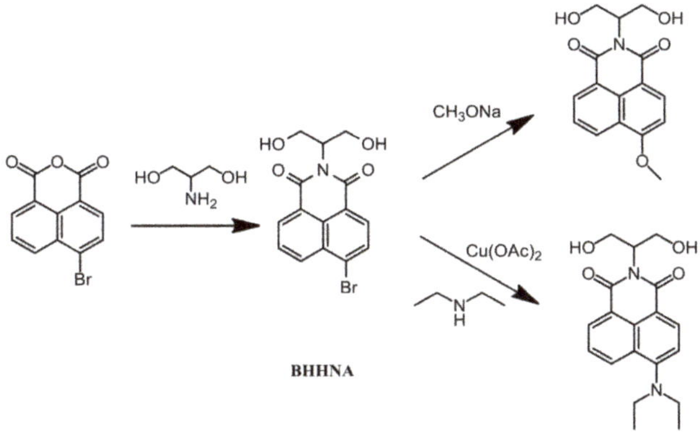

Scheme 1. Synthetic route of fluorescent monomers. BHHNA—4-bromo-N-(2-hydroxy-1-hydroxymethylethyl)-1,8-naphthalimide; MHHNA—4-methoxyl-N-(2-hydroxy-1-hydroxymethylethyl)-1,8-naphthalimide; DHHNA—4-diethylamino-N-(2-hydroxy-1-hydroxymethylethyl)-1,8-naphthalimide.

The UV/vis spectroscopy of the fluorescent monomers and the fluorescent prepolymers were determined at 6.6×10^{-2} mM fluorophore concentration in the solutions of both monomer and prepolymer. MHHNA and PU-MHHNA were colorless with almost no absorption in the visible light range (>400 nm), while DHHNA and PU-DHHNA were yellow under visible light because of their red-shifted $n \rightarrow \pi^*$ absorption band. As shown in Figure 1, the UV/vis absorption spectra of PU-MHHNA were almost consistent with that of MHHNA with the maximum absorption wavelength (λ_{max}) at 362 nm, and DHHNA and PU-DHHNA had the same UV/vis profile with λ_{max} at 419 nm. These results indicated that the fluorescent chromophores did not change their chemical structures during polymerization. In the study of fluorescent properties, it was revealed that the fluorescent spectra of MHHNA and PU-MHHNA were completely identical at 1.66×10^{-3} mM fluorophore concentration, their fluorescent excitation and emission spectra presented a good mirror-image relationship, as well as their maximum fluorescent excitation wavelength (λ_{ex}) and emission wavelength (λ_{em}) exhibited at 362 and 433 nm, respectively, with Stokes shift of 71 nm (Figure 2A,B). Using quinine sulfate as reference, the fluorescence quantum yields of MHHNA and PU-MHNNA were measured to be 0.71. Meanwhile, λ_{ex} (421 nm) and λ_{em} (522 nm) of PU-DHHNA were also the same as that of DHHNA, thus generating a larger Stokes shift (101 nm), because the diethylamino group possessed stronger electron donating ability than the methoxy group, causing a higher intramolecular charge transfer rate and a smaller energy gap between HOMO and LUMO [32,33]. Furthermore, it is worth noting that at the same fluorophore concentration (3×10^{-2} mM), PU-DHHNA displayed enhanced fluorescence intensity in emission and excitation spectra, and its quantum yield (0.15) was considerably increased compared with DHHNA (0.08) (Figure 2C,D). This result might be ascribed to the fact that DHHNA monomers probably underwent aggregation-caused quenching (ACQ) in relatively higher concentration by interacting with each other through intermolecular charge transfer interaction, yet the steric hindrance of PU-DHHNA chain hindered the ACQ effect between the fluorophores, thereby increasing the fluorescence quantum yield. Consequently, the polymerization of fluorescent monomers could not only prevent the migration of fluorophore, but also increase the fluorescence intensity.

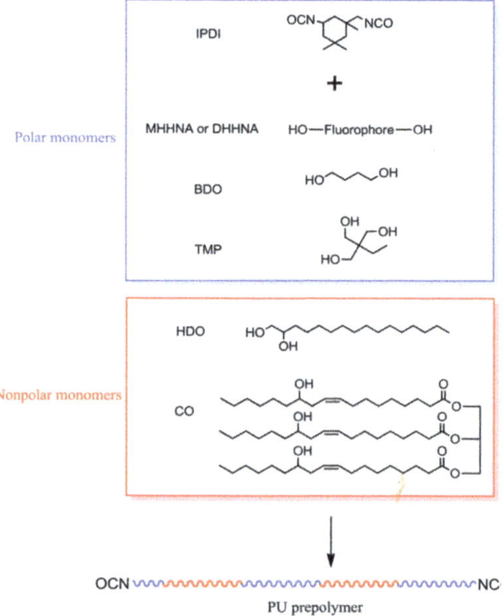

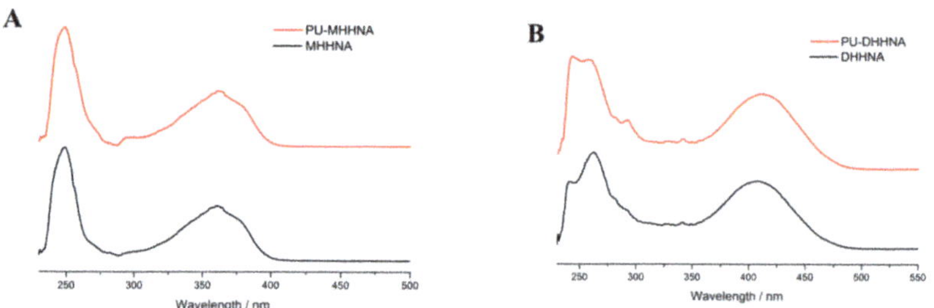

Scheme 2. Illustration of polyurethane (PU) prepolymer synthesis. TMP—trimethylol propane; CO—castor oil; IPDI—isophorone diisocyanate; HDO—1,2-hexadecanediol; BDO—1,4-butanediol.

Figure 1. UV/vis spectra of fluorescent dye (4-methoxyl-N-(2-hydroxy-1-hydroxymethylethyl)-1,8-naphthalimide (MHHNA), 4-diethylamino-N-(2-hydroxy-1-hydroxymethylethyl)-1,8-naphthalimide (DHHNA)) and corresponding fluorescent polyurethane (PU) prepolymer in ethyl acetate solution.

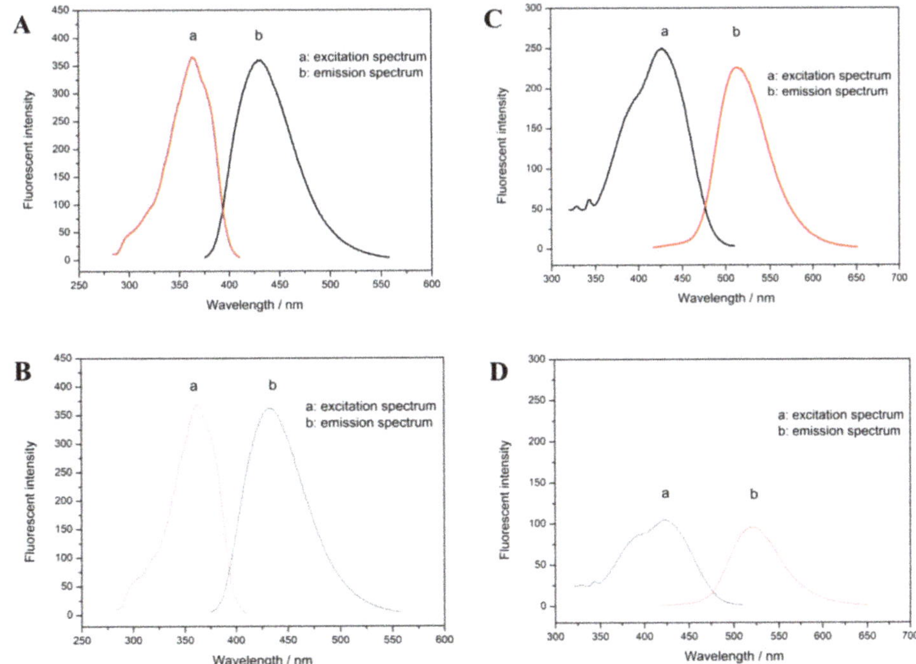

Figure 2. Fluorescence spectra of (**A**) PU-MHHNA, (**B**) MHHNA, (**C**) PU-DHHNA, and (**D**) DHHNA in ethyl acetate solution.

3.2. Fabrication of Composite Nanoparticles

Hydrophobic magnetic nanoparticles prepared by thermal decomposition of ferric oleate had good solubility in nonpolar solvents such as toluene, chloroform, or n-hexane, and could form magnetic liquid in these solvents through surface solvation, while they expectedly precipitated in polar solvents such as ethanol or water because of their poor solubility (Figure S8). It was believed that the surface coating of oleic acid by coordination imparted hydrophobicity to the magnetic nanoparticles. From TEM observation, HMNPs exhibited relatively uniform spherical particles, their particle size was analyzed by nano-measure software to give the frequency of diameter in the TEM image. As indicated in Figure 3, the size of HMNPs was mostly in the range of 11–17 nm with an approximately normal distribution. In the FT-IR spectrum (Figure 4A), the absorption peaks located at 2923 and 2850 cm^{-1} should be assigned to the saturated C–H stretching, the signals at 1707 cm^{-1} stemmed from the oleate C=O vibration, and the intensive peak at 576 cm^{-1} could be attributed to the stretching vibration of the Fe–O bond, suggesting that HMNPs were composed of magnetite and oleic acid. By comparing with the standard XRD pattern, one could ascertain that the crystal structure of HMNPs was the same as that of magnetite (Figure 4B), except that the peaks of HMNPs were significantly broadened because of the grain refinement. According to the Debye–Scherrer formula—D = K·λ/B·cosθ, where Scherrer constant K = 0.89, X-ray wavelength λ = 0.15406 nm, diffraction angle θ = 17.8° (311), and half-peak width B = 0.0091 rad (311)—the average grain size D could be calculated as 15.8 nm, which was slightly larger than the diameter of the most probable distribution (14 nm) in the TEM result, which implied that some particles with a larger size (>22 nm) might not be counted by TEM measurement. Because the critical dimension of paramagnetic Fe$_3$O$_4$ (single domain) was about 30 nm at room temperature [34], that part of the magnetic particles with a larger size could generate the coercivity in HMNPs; accordingly, the remanence occurred when the external magnetic field was removed (Figure 8).

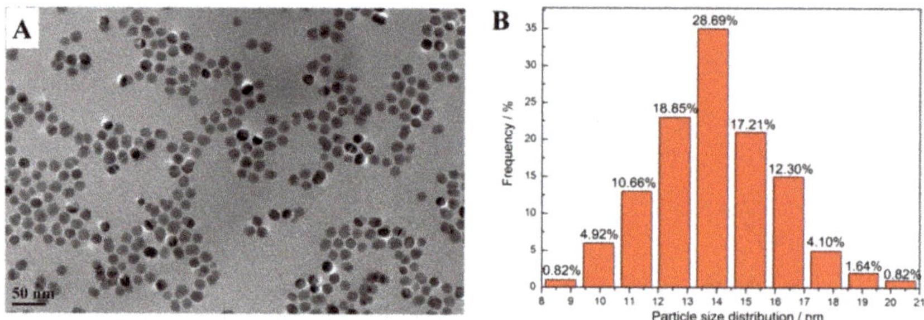

Figure 3. (**A**) Transmission electron microscopy (TEM) image of the hydrophobic magnetic nanoparticles (HMNPs) and (**B**) the frequency of the particle size.

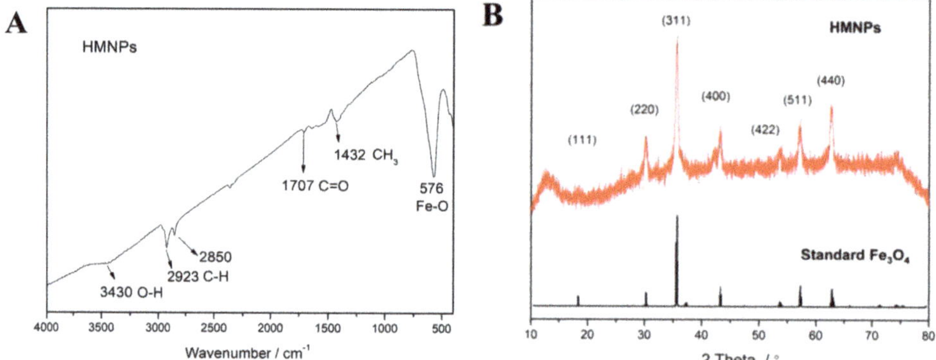

Figure 4. (**A**) FT-IR spectrum of HMNPs and (**B**) X-ray diffraction (XRD) pattern of HMNPs compared with standard XRD of Fe_3O_4.

The composite particles were fabricated via two steps including mini-emulsification and self-assembly in solvent evaporation. First, the PU prepolymer and HMNPs were dissolved in chloroform to form a dilute solution with low viscosity, then an aqueous solution of very few SDS was added and the mixture was homogenized by ultrasonication to obtain mini-emulsion, finally the solvent was evaporated under reduced pressure and the composite particles were formed through the assembly of polyurethane and HMNPs. As chloroform volatilized, PU prepolymers were exposed to the aqueous phase, then their molecular weight gradually increased by the chain extending reaction of isocyanate group and H_2O. Simultaneously, because HMNPs were hydrophobic and extremely arduous to diffuse into the aqueous phase, they would only remain in the nanospheres with PU macromolecules to form inorganic–organic nanocomposites. Notably, to promote the stability of mini-emulsion, it was usually necessary to employ hydrophobic agents to balance the Laplace pressure and prevent the Ostwald ripening [35]; however, in this work, HMNPs in droplet were intrinsically hydrophobic, thus a narrow distribution of particle size could be achieved without the addition of other hydrophobic agents such as hexadecane or hexadecanol.

3.3. Nanostructure of Composite Nanoparticles

To investigate the morphology and the nanostructure, the composite nanoparticles with different ingredients were investigated using the TEM and SEM methods, and it was revealed that the proportion of PU nonpolar and polar segments played an important role in the assembly structure of nanocomposites. PU-TMP without nonpolar segments was first tried in the fabrication of PTCP-20. As shown in Figure 5A, the HMNPs could not be effectively encapsulated in the nanospheres and they mostly distributed on the surface, thus the further aggregation of HMNPs resulted in the precipitation and delamination of latex. In contrast, after the addition of PU-CO containing unsaturated hydrophobic segments, most HMNPs were embedded inside the nanoparticles PCCP-20 (Figure 5B), and the storage stability of the latex was obviously improved. However, it was shown in Figure 5B and Table 3 that the particle size of PCCP-20 prepared from PU-CO was much larger because of the increase of hydrophobicity in composite particles, and the size distribution of PCCP-20 became broad, which might be ascribed to the fact that abundant multifunctional monomer CO increased the crosslinking degree of the prepolymer, which not only broadened the molecular weight distribution, but also increased the viscosity of the organic phase in mini-emulsion to retard homogeneous emulsification. To solve this problem, a bifunctional diol HDO was employed to introduce the hydrophobic segment into PU, replacing a proportion of CO in the recipe of PMCP and PDCP (Table 1). In Figure 5C, it was seen that almost all nanoparticles of PMCP-20 displayed manifest Janus-type structure because of the phase separation, and HMNPs remained inside the nanoparticles, indicating the formation of FMJNPs. Similarly, the Janus nanostructures of PDCP-20 were completely consistent with PDMP-2, as shown in Figure 5D. Moreover, the SEM image of PMCP-20 presented in Figure 6 demonstrated the FMJNPs were well distributed and their smooth surface inferred that HMNPs located in the particles rather than on the surface. The investigation of the colloidal properties measured by DLS further revealed that the Janus nanoparticles of PMCP-20 and PDCP-20 enjoyed a uniform particle size distribution with reduced polydispersity index in comparison with PCCP-20 (Table 3).

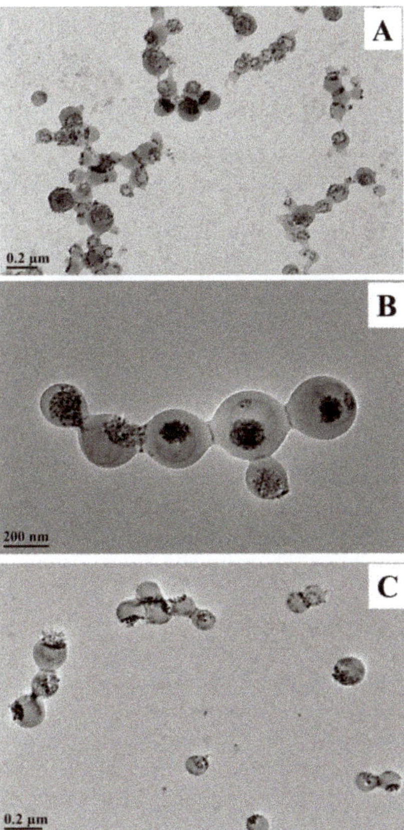

Figure 5. *Cont.*

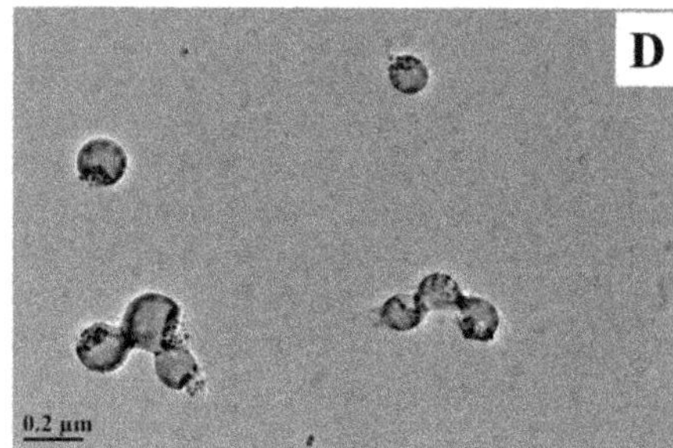

Figure 5. TEM images of (**A**) PU-trimethylol propane (TMP) magnetic composite nanoparticle (PTCP-20), (**B**) PU-castor oil (CO) magnetic composite nanoparticle (PCCP-20), (**C**) PU-MHHNA magnetic composite nanoparticle (PMCP-20), and (**D**) PU-DHHNA magnetic composite nanoparticle (PDCP-20).

Table 3. The colloidal properties of the composite nanoparticles.

Sample	Mean Diameter/nm	Polydis	Zeta Potential/mV
PTCP-20	135.5	0.143	−67.3
PCCP-20	189.0	0.259	−65.1
PMCP-10	125.7	0.052	−69.5
PMCP-20	128.4	0.083	−68.8
PMCP-30	129.0	0.095	−64.1
PDCP-10	130.2	0.064	−59.7
PDCP-20	122.8	0.090	−69.9
PDCP-30	128.8	0.121	−66.8

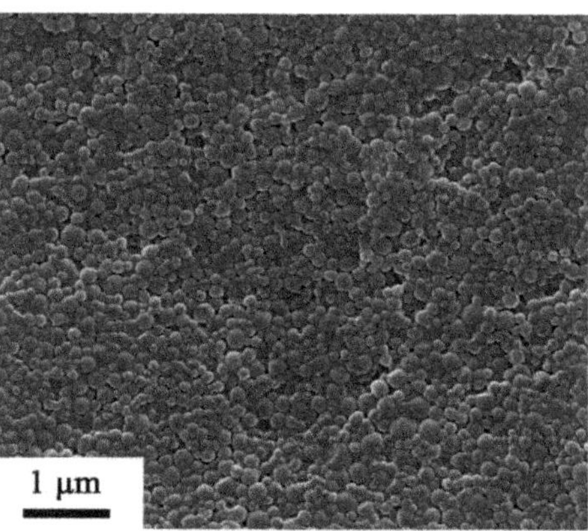

Figure 6. SEM image of PMCP-20.

From these results, it was inferred that the compatibility between HMNPs and PU was of importance in the formation of stable composite nanoparticles by reducing the interface energy, the compatibility arose from the hydrophobic interaction between PU nonpolar segments and oleic acid coated HMNPs. On the basis of the different proportions of polar and nonpolar segments in PU, three structures of composite nanoparticles could be formed in the assembly process (Scheme 3), and are classified as follows: (1) Nanoparticles coated with HMNPs, in which PU constituted of all polar segment had almost no affinity with HMNPs, leading to the severe phase separation, which excluded the HMNPs out of the sphere. With the aid of surfactant, the hydrophobic nano Fe_3O_4 could only disperse at the solid–water interface, and its combination with the nanoparticle was instable (Figure 5A). (2) Nanoparticles with HMNPs/PU as core/shell structure, in which PU consisted of considerable nonpolar segments from CO or HDO, which suggested it promising interfacial compatibility with HMNPs. Because of their hydrophobicity, the nonpolar segments of PU and HMNPs were inclined to be distributed in the core of composite nanoparticles. (3) Janus nanoparticles, where PU contained fewer nonpolar segments. The polymer phase was less compatible with HMNPs, thereby the phase separation of PU polar polymer and HMNPs occurred during the solvent evaporation process, resulting in the Janus-type particles. Accordingly, it was beneficial to utilize a suitable amount of nonpolar chain extender in the synthesis of PU prepolymer for the preparation of FMJNPs.

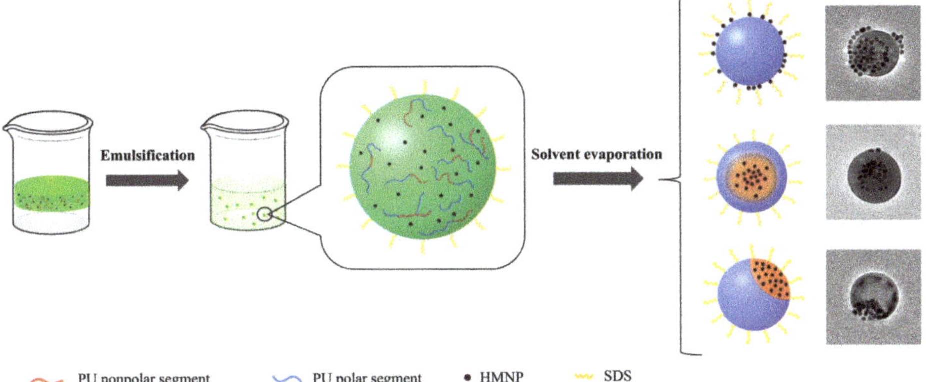

Scheme 3. Illustration of the preparation of the composite nanoparticles. HMNP—hydrophobic magnetic nanoparticle; SDS—sodium dodecyl sulfate.

Furthermore, the particle structures of PMCP with different HMNP contents were examined by TEM. In Figure 7A, HMNPs accounted for only 10% of the total weight in PMCP-10, and most of them were encapsulated in the nanoparticles, despite the fact that they were not evenly dispersed because of the phase separation. The polymer spheres without the composition of HMNPs could also be distinguished as marked by the arrow. With HMNP content increased to 30% (PMCP-30), most of the nanoparticles were entirely filled with HMNPs, triggering a lower ratio of Janus nanoparticles (Figure 7B). The recipe with HMNPs content higher than 30% was still applicable in preparation, but the stability of the product decreased with the increasing HMNPs. Therefore, with the optimum HMNP content of 20% for Janus particles, PMCP-20 and PDCP-20 were selected for further study, and their fluorescent magnetic latexes both displayed excellent storage stability, showing neither the demulsification of the latex nor the precipitation of Fe_3O_4 during five months of storage.

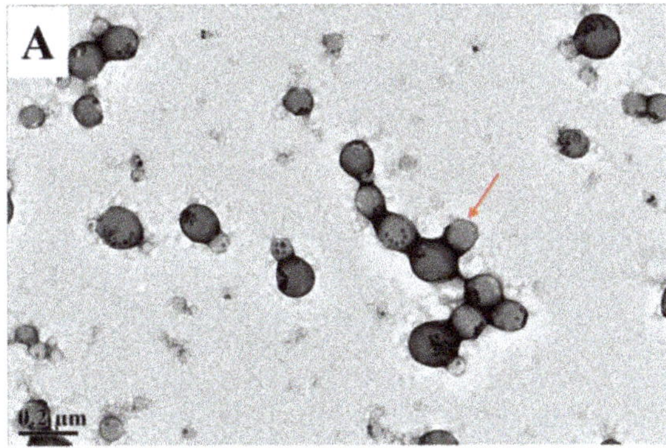

Figure 7. *Cont.*

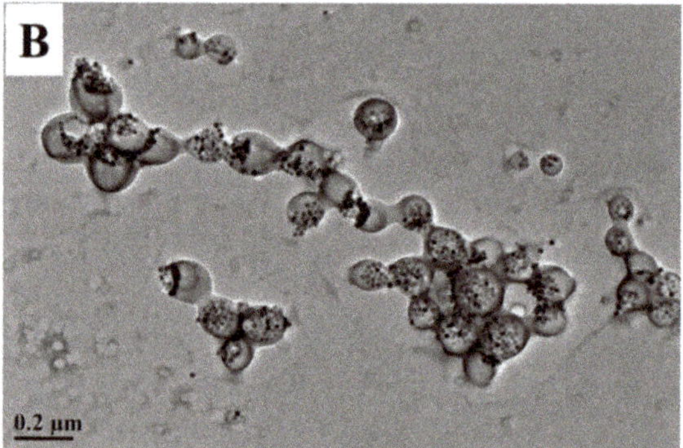

Figure 7. TEM images of (**A**) PMCP-10 and (**B**) PMCP-30.

3.4. Magnetic and Fluorescent Properties of Composite Nanoparticles

The magnetic hysteresis loops of PMCP-10, PMCP-20, and PMCP-30 were determined in order to investigate their magnetic properties. In Figure 8, the saturated magnetization of pure HMNPs was 26.5 emu/g, while that of the PMCP-10, PMCP-20, and PMCP-30 was measured as 2.6 emu/g, 5.4 emu/g, and 7.7 emu/g, respectively, which were linearly related to their HMNP contents of 10%, 20%, and 30%, suggesting that the HMNPs coated in nanoparticles persisted the original properties of magnetic response. In addition, as analyzed above via the TEM and XRD methods, most HMNPs were superparamagnetic with a particle size below 20 nm, resulting in the exceedingly low coercivity of the samples (indicated by the transverse–axis intercept in Figure 8).

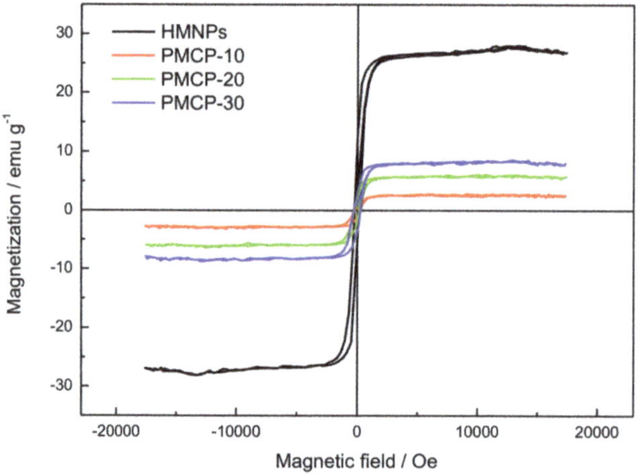

Figure 8. The hysteresis loops of HMNPs and the composite nanoparticles with different HMNP content.

Under visible light, both PMCP-20 and PDCP-20 appeared as yellow homogenous latex, whereas their fluorescent colors under UV light corresponded with that of their dye monomers. As shown in Figure 9, the microscopic observation of PMCP-20 and PDCP-20 was conducted by the laser confocal microscope (405 nm exciting light), where the FMJNPs emitting bright fluorescence behaved in an intensive Brownian motion in dilute latex because of their small particle size. Meanwhile, their magnetic response could be directly examined by an external magnetic field (Figure 10). By magnetic attraction, PMCP-20 and PDCP-20 latex gradually became transparent within approximately 16 min (Figure S9), yet the precipitated FMJNPs were able to be quickly re-dispersed by 20 s ultrasonication. In comparison, sodium oleate was also tried as the emulsifier in mini-emulsification to substitute SDS; however, after magnetic attraction, the resulting nanoparticles with zeta potential of −37 mV were unable to be dispersed by sonication. This suggested that the strong ionization of the emulsifier was of importance to generate an adequate negative zeta potential, which triggered the repulsion of the nanoparticles and benefitted reversible dispersion after the removal of the magnetic field.

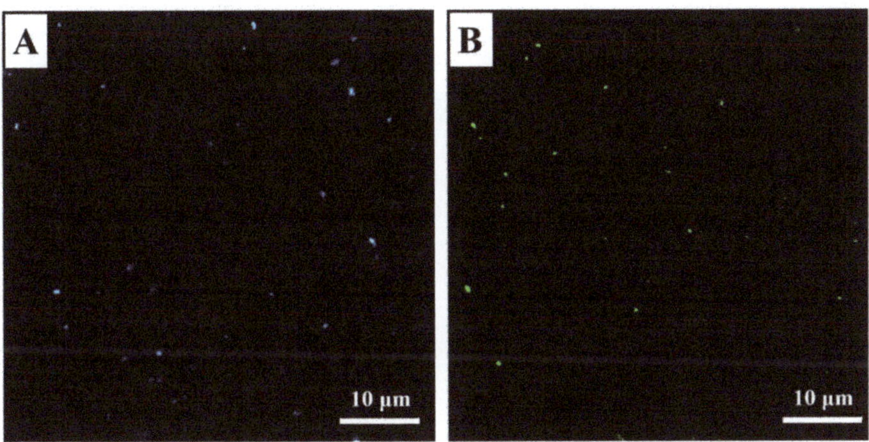

Figure 9. Confocal laser scanning microscopy of (**A**) PMCP-20 and (**B**) PDCP-20.

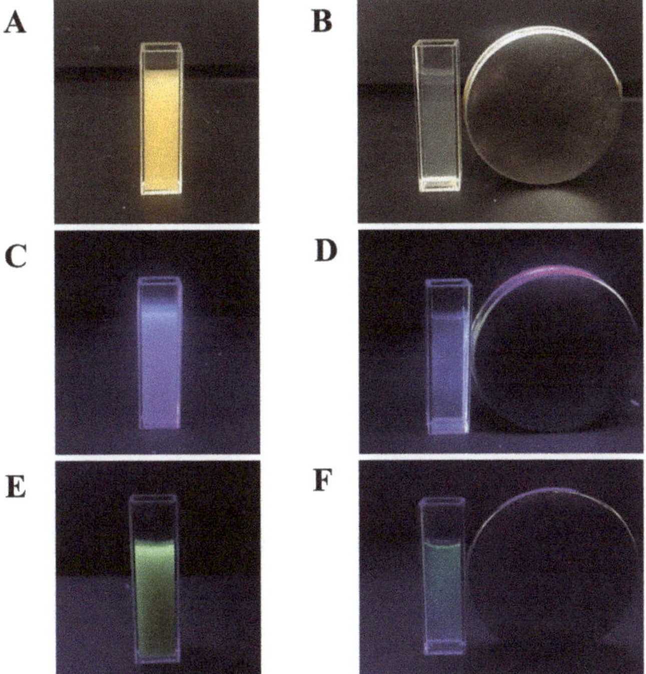

Figure 10. The magnetic response of (**A**,**B**) PMCP-20 latex, (**C**,**D**) PMCP-20 under 365 nm UV light, and (**E**,**F**) PDCP-20 under 365 nm UV light.

3.5. Cytocompatibility

The in vitro cytocompatibility of FMJNPs was determined by the WST-1 method [36]. Herein, the human intestinal epithelial cells were used as model recipients and the cytocompatibility of the nanoparticles was represented by the cell viability and, more exactly, the activity of mitochondrial dehydrogenase in cells. The enzyme could chemically degrade the colorless substrate (WST-1) to a yellow dye. By detecting the absorbance at 450 nm of the yellow product, one can quantitatively measure the activity of mitochondrial dehydrogenase, thus investigating the toxicity or the cytocompatibility of the sample. As illustrated in Figure 11, compared with the blank cell specimen, the sample incubated with PMCP-20 and PDCP-20 (1.5 mg/mL) had relatively similar plots of time–absorbance, indicating that the FMJNPs were nontoxic and had no influence on the cell activity. Despite that the addition of yellow PMCP-20 and PDCP-20 resulted in higher absorbance at 450 nm than the blank sample, the t-test values of PMCP-20 and PDCP-20 against the blank were calculated as 0.432 and 0.143, respectively, being greater than the *p*-value (0.05), so there was statistically considered to be no significant difference (n.s.) between the samples and the blank.

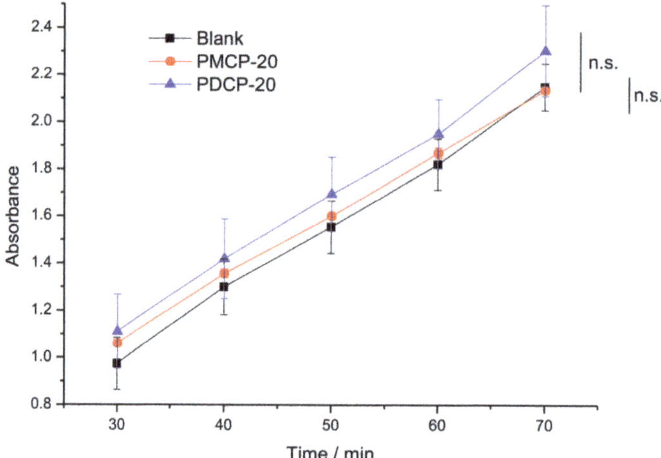

Figure 11. The time–absorbance plot of the blank cell and the cell incubated with PMCP-20 and PDCP-20.

In addition, from the optical microscopy (Figure S10), the cells incubated with PMCP-20 and PDCP-20 maintained normal growth without the collapsed morphology of dead cells, which again supported the outstanding cytocompatibility of the fluorescent magnetic nanoparticles as expected. Accordingly, we envisaged that the FMJNPs prepared by mini-emulsion method might be employed in different areas such as enzyme separation, cell labeling, drug delivery, imaging probe, and so on.

4. Conclusions

In this work, two kinds of fluorescent PU prepolymers were firstly synthesized by copolymerization of fluorescent dye monomers MHHNA and DHHNA. It was found that the migration fastness and the fluorescent intensity of fluorophores were improved by the covalent bonding with the polymer chain. Then, the nanoparticles composed of fluorescent PU and hydrophobic Fe_3O_4 nanoparticles were efficiently fabricated by mini-emulsification and self-assembly process. Noteworthily, the composition proportion of nonpolar segments of PU played an important role in regulating the nanostructure of products including Janus and core/shell structure, while Janus nanoparticles could be controllably produced as a result of the phase separation between PU and Fe_3O_4. By vibrating magnetometer and confocal laser scanning microscope, the prominent magnetic properties and fluorescent properties of the FMJNPs were confirmed. They displayed responses to the magnetic attraction and ultrasonication for reversible aggregation–dispersion behavior. Last but not least, the excellent cytocompatibility of the FMJNPs was proven by the WST-1 method, and the cell incubated with PMCP-20 and PDCP-20 grew normally. Accordingly, these nanoparticles enjoyed potential applications in the biomedical field, and this work might be meaningful for the facile fabrication of various FMJNPs with versatile functions.

Supplementary Materials: The supplementary materials are available online at http://www.mdpi.com/2073-4360/11/2/272/s1.

Author Contributions: B.L. and Q.J. conceived and designed the experiments; W.S., Y.W., and D.X. performed the experiments; Y.X. and H.Y. analyzed the data; B.L. and Q.Z. wrote the paper.

Funding: This work was financially supported by the National Key R&D Program of China (No. 2017YFC0804500, No. 2016YFC0303708) and the Science Foundation of China University of Petroleum-Beijing (No. 2462017YJRC008).

Acknowledgments: The authors gratefully acknowledge Chong Zhang, Nan Su, and Yi Wang for their kind help in cell toxicity experiments.

Conflicts of Interest: The authors declare no conflict of interest.

Abbreviations

FMNP	fluorescent/magnetic nanoparticle
FMJNP	fluorescent/magnetic Janus nanoparticle
HMNP	hydrophobic magnetic nanoparticle
BHHNA	4-bromo-N-(2-hydroxy-1-hydroxymethylethyl)-1,8-naphthalimide
MHHNA	4-methoxyl-N-(2-hydroxy-1-hydroxymethylethyl)-1,8-naphthalimide
DHHNA	4-diethylamino-N-(2-hydroxy-1-hydroxymethylethyl)-1,8-naphthalimide
PU	polyurethane
TMP	trimethylol propane
CO	castor oil
PU-TMP	polyurethane from trimethylol propane
PU-CO	polyurethane from castor oil
PU-MHHNA	polyurethane from 4-methoxyl-N-(2-hydroxy-1-hydroxymethylethyl)-1,8-naphthalimide
PU-DHHNA	polyurethane from 4-diethylamino-N-(2-hydroxy-1-hydroxymethylethyl)-1,8-naphthalimide
PTCP	PU-TMP magnetic composite nanoparticle
PCCP	PU-CO magnetic composite nanoparticle
PMCP	PU-MHHNA magnetic composite nanoparticle
PDCP	PU-DHHNA magnetic composite nanoparticle
HDO	1,2-hexadecanediol
BDO	1,4-butanediol
NA	1,8-naphthalimide
SDS	sodium dodecyl sulfate
IPDI	isophorone diisocyanate
THF	tetrahydrofuran
DMSO	dimethyl sulfoxide
XRD	X-ray diffraction
TEM	transmission electron microscopy
SEM	scanning electron microscopy
DLS	dynamic light scattering
GPC	gel permeation chromatography

References

1. Wen, C.Y.; Xie, H.Y.; Zhang, Z.L.; Wu, L.L.; Hu, J.; Tang, M.; Wu, M.; Pang, D.W. Fluorescent/magnetic micro/nano-spheres based on quantum dots and/or magnetic nanoparticles: Preparation, properties, and their applications in cancer studies. *Nanoscale* **2016**, *8*, 12406–12429. [CrossRef] [PubMed]
2. Bigall, N.C.; Parak, W.J.; Dorfs, D. Fluorescent, magnetic and plasmonic-Hybrid multifunctional colloidal nano objects. *Nano Today* **2012**, *7*, 282–296. [CrossRef]
3. Wang, G.N.; Su, X.G. The synthesis and bio-applications of magnetic and fluorescent bifunctional composite nanoparticles. *Analyst* **2011**, *136*, 1783–1798. [CrossRef] [PubMed]
4. Ganipineni, L.P.; Ucakar, B.; Joudiou, N.; Bianco, J.; Danhier, P.; Zhao, M.N.; Bastiancich, C.; Gallez, B.; Danhier, F.; Préat, V. Magnetic targeting of paclitaxel-loaded poly(lactic-co-glycolic acid)-based nanoparticles for the treatment of glioblastoma. *Int. J. Nanomed.* **2018**, *13*, 4509–4521. [CrossRef] [PubMed]
5. Yan, K.; Li, H.; Li, P.H.; Zhu, H.E.; Shen, J.; Yi, C.F.; Wu, S.L.; Yeung, K.W.; Xu, Z.S.; Xu, H.B.; et al. Self-assembled magnetic fluorescent polymeric micelles for magnetic resonance and optical imaging. *Biomaterials* **2014**, *35*, 344–355. [CrossRef] [PubMed]
6. Xi, P.X.; Cheng, K.; Sun, X.L.; Zeng, Z.Z.; Sun, S.H. Magnetic Fe_3O_4 nanoparticles coupled with a fluorescent Eu complex for dual imaging applications. *Chem. Commun.* **2012**, *48*, 2952–2954.
7. Mahmoudi, M.; Shokrgozar, M.A. Multifunctional stable fluorescent magnetic nanoparticles. *Chem. Commun.* **2012**, *48*, 3957–3959.
8. Wang, M.; Fei, X.F.; Lv, S.W.; Sheng, Y.; Zou, H.F.; Song, Y.H.; Yan, F.; Zhu, Q.L.; Zheng, K.Y. Synthesis and characterization of a flexible fluorescent magnetic Fe_3O_4@SiO_2/CdTe-NH_2 nanoprobe. *J. Inorg. Biochem.* **2018**, *186*, 307–316. [CrossRef]
9. Sun, Y.Q.; Wang, D.D.; Zhao, T.X.; Jiang, Y.N.; Zhao, Y.Q.; Wang, C.X.; Sun, H.C.; Yang, B.; Lin, Q. Fluorescence-magnetism functional EuS nanocrystals with controllable morphologies for dual bioimaging. *ACS Appl. Mater. Interfaces* **2016**, *8*, 33539–33545. [CrossRef]
10. Resch-Genger, U.; Grabolle, M.; Cavaliere-Jaricot, S.; Nitschke, R.; Nann, T. Quantum dots versus organic dyes as fluorescent labels. *Nat. Methods* **2008**, *5*, 763–775. [CrossRef]
11. Jackeray, R.; Abid, C.K.; Singh, G.; Jain, S.; Chattopadhyaya, S.; Sapra, S.; Shrivastav, T.G.; Singh, H. Selective capturing and detection of Salmonella typhi on polycarbonate membrane using bioconjugated quantum dots. *Talanta* **2011**, *84*, 952–962. [CrossRef] [PubMed]
12. Li, B.T.; Shen, J.; Liang, R.B.; Ji, W.J.; Kan, C.Y. Synthesis and characterization of covalently colored polymer latex based on new polymerizable anthraquinone dyes. *Colloid Polym. Sci.* **2012**, *290*, 1893–1900. [CrossRef]
13. Li, B.T.; Shen, J.; Jiang, Y.M.; Wang, J.S.; Kan, C.Y. Preparation and properties of covalently colored polymer latex based on a new anthraquinone monomer. *J. Appl. Polym. Sci.* **2013**, *129*, 1484–1490. [CrossRef]
14. Peyer, K.E.; Zhang, L.; Nelson, B.J. Bio-inspired magnetic swimming microrobots for biomedical applications. *Nanoscale* **2013**, *5*, 1259–1272. [CrossRef] [PubMed]
15. Orozco, J.; Pan, G.Q.; Sattayasamitsathit, S.; Galarnyk, M.; Wang, J. Micromotors to capture and destroy anthrax simulant spores. *Analyst* **2015**, *140*, 1421–1427. [CrossRef] [PubMed]
16. Wang, H.H.; Yang, S.Y.; Yin, S.N.; Chen, L.; Chen, S. Janus suprabead displays derived from the modified photonic crystals toward temperature magnetism and optics multiple responses. *ACS Appl. Mater. Interfaces* **2015**, *7*, 8827–8833. [CrossRef] [PubMed]
17. Schick, I.; Lorenz, S.; Gehrig, D.; Schilmann, A.M.; Bauer, H.; Panthöfer, M.; Fischer, K.; Strand, D.; Laquai, F.; Tremel, W. Multifunctional two-photon active silica-coated Au@MnO Janus particles for selective dual functionalization and imaging. *J. Am. Chem. Soc.* **2014**, *136*, 2473–2483. [CrossRef]
18. Kim, S.H.; Sim, J.Y.; Lim, J.M.; Yang, S.M. Magnetoresponsive microparticles with nanoscopic surface structures for remote-controlled locomotion. *Angew. Chem.* **2010**, *49*, 3786–3790. [CrossRef]
19. Kaewsaneha, C.; Tangboriboonrat, P.; Polpanich, D.; Elaissari, A. Multifunctional fluorescent-magnetic polymeric colloidal particles: Preparations and bioanalytical applications. *ACS Appl. Mater. Interfaces* **2015**, *7*, 23373–23386. [CrossRef]
20. Yi, Y.; Sanchez, L.; Gao, Y.; Yu, Y. Janus particles for biological imaging and sensing. *Analyst* **2016**, *141*, 3526–3539. [CrossRef]
21. Liang, F.X.; Liu, B.; Cao, Z.; Yang, Z.Z. Janus colloids toward interfacial engineering. *Langmuir* **2018**, *34*, 4123–4131. [CrossRef]

22. Yin, S.N.; Wang, C.F.; Yu, Z.Y.; Wang, J.; Liu, S.S.; Chen, S. Versatile bifunctional magnetic-fluorescent responsive Janus supraballs towards the flexible bead display. *Adv. Mater.* **2011**, *23*, 2915–2919. [CrossRef] [PubMed]

23. Lan, J.W.; Chen, J.Y.; Li, N.X.; Ji, X.H.; Yu, M.X.; He, Z.K. Microfluidic generation of magnetic-fluorescent Janus microparticles for biomolecular detection. *Talanta* **2016**, *151*, 126–131. [CrossRef]

24. Li, P.; Li, K.; Niu, X.F.; Fan, Y.B. Electrospraying magnetic-fluorescent bifunctional Janus PLGA microspheres with dual rare earth ions fluorescent-labeling drugs. *RSC Adv.* **2016**, *6*, 99034–99043. [CrossRef]

25. Teo, B.M.; Young, D.J.; Loh, X.J. Magnetic anisotropic particles: Toward remotely actuated applications. *Part. Part. Syst. Charact.* **2016**, *33*, 709–728. [CrossRef]

26. Hu, S.H.; Gao, X.H. Nanocomposites with spatially separated functionalities for combined imaging and magnetolytic therapy. *J. Am. Chem. Soc.* **2010**, *132*, 7234–7237. [CrossRef] [PubMed]

27. Rahman, M.M.; Montagne, F.; Fessi, H.; Elaissari, A. Anisotropic magnetic microparticles from ferrofluid emulsion. *Soft Matter* **2011**, *7*, 1483–1490. [CrossRef]

28. Kaewsaneha, C.; Bitar, A.; Tangboriboonrat, P.; Polpanich, D.; Elaissari, A. Fluorescent-magnetic Janus particles prepared via seed emulsion polymerization. *J. Colloid Interface Sci.* **2014**, *424*, 98–103. [CrossRef]

29. Jin, Q.J.; Hu, Y.; Shen, J.; Li, B.T.; Kan, C.Y. A novel 1,8-naphthalimide green fluorescent dye and its corresponding intrinsically fluorescent polyurethane latexes. *J. Coat. Technol. Res.* **2017**, *14*, 571–582. [CrossRef]

30. Jin, Q.J.; Li, B.T.; Zhang, H.; Li, L.X.; Shen, J.; Kan, C.Y. Investigation of covalently colored polyurethane latexes based on novel anthraquinone polyurethane chain extenders. *J. Macromol. Sci. A* **2017**, *54*, 52–59. [CrossRef]

31. Park, J.; An, K.J.; Hwang, Y.S.; Park, J.G.; Noh, H.J.; Kim, J.Y.; Park, J.H.; Hwang, N.M.; Hyeon, T. Ultra-large-scale syntheses of monodisperse nanocrystals. *Nat. Mater.* **2004**, *3*, 891–895. [CrossRef]

32. Sutradhar, T.; Misra, A. Role of electron-donating and electron-withdrawing Groups in tuning the optoelectronic properties of difluoroboron-napthyridine analogues. *J. Phys. Chem. A* **2018**, *122*, 4111–4120. [CrossRef] [PubMed]

33. Zhang, Y.B.; Xia, S.; Fang, M.X.; Mazi, W.; Zeng, Y.B.; Johnston, T.; Pap, A.; Luck, R.L.; Liu, H.Y. New near-infrared rhodamine dyes with large Stokes shifts for sensitive sensing of intracellular pH changes and fluctuations. *Chem. Commun.* **2018**, *54*, 7625–7628.

34. Heider, F.; Dunlop, D.J.; Sugiura, N. Magnetic properties of hydrothermally recrystallized magnetite crystals. *Science* **1987**, *236*, 1287–1290. [CrossRef] [PubMed]

35. Kaewsaneha, C.; Tangboriboonrat, P.; Polpanich, D.; Eissa, M.; Elaissari, A. Facile method for preparation of anisotropic submicron magnetic Janus particles using miniemulsion. *J. Colloid Interface Sci.* **2013**, *409*, 66–71. [CrossRef] [PubMed]

36. Ngamwongsatit, P.; Banada, P.P.; Panbangred, W.; Bhunia, A.K. WST-1-based cell cytotoxicity assay as a substitute for MTT-based assay for rapid detection of toxigenic Bacillus species using CHO cell line. *J. Microbiol. Methods* **2008**, *73*, 211–215. [CrossRef] [PubMed]

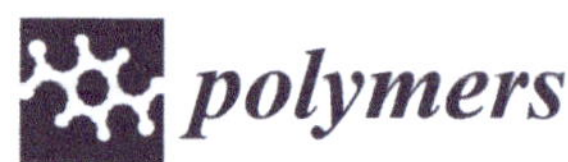 *polymers*

Article

Computational Analysis of Nonuniform Expansion in Polyurethane Foams

D. Niedziela *, I. E. Ireka and K. Steiner

Department Flow and Material Simulation, Fraunhofer Institute for Industrial Mathematics, Fraunhofer-Platz 1, D-67663 Kaiserslautern, Germany; iirex4@gmail.com (I.E.I.); konrad.steiner@itwm.fraunhofer.de (K.S.)
* Correspondence: dariusz.niedziela@itwm.fraunhofer.de

Received: 4 December 2018; Accepted: 21 December 2018; Published: 9 January 2019

Abstract: This paper computationally investigates heterogeneity in the distribution of foam fraction in chemically expanding blown polyurethane foam. The experimentally observed disparity in the volumes of expanded foam when an equal mass of the foaming mixture was injected into tubes of different dimensions motivated this study. To understand this phenomenon, attributed to local variations in the thermal and rheological properties of the expanding system, we explore available data from free-rise foam-expansion experiments in different geometries. Inspired by the mathematical framework for the microstructure modelling of bubble growth in viscous liquids, we study the reacting mixture as a continuum and formulate appropriate mathematical models that account for spatial inhomogeneity in the foam-expansion process. The nonlinear coupled system of partial differential equations governing flow was numerically solved using finite-volume techniques, and the associated results are presented and discussed with graphical illustrations. The proximity of the foaming-mixture core to the external environment and the thickness of a thermal-diffusion layer formed near the bounding geometry was seen to influence the distribution of the foam fraction. Our simulations showed an average spatial variation of about 1.1% in the distribution of solid foam fraction from the walls to the core, as verified with data from μCT scan analysis of the expanded foam. This also reflects the distribution of void fraction in the foam matrix. The models were validated with experimental data, and our results favourably compared with the experiment observations.

Keywords: polyurethane foams; nonuniform expansion; foam fraction distribution; reaction injection molding; chemorheology; finite-volume method

1. Introduction

The commercial relevance of flexible or rigid polyurethane (PU) foams has stimulated extensive research interest in their production and product-optimization processes. In their production phase, PU foams exhibit complex behavior initiated by premixing relevant isocyanate and polyol groups in the presence of suitable catalyst and blowing agents. Depending on the type of blowing agent, among many factors affecting the foaming process [1], the resulting PU foam matrix is classified as rigid or flexible foam. However, certain factors, such as mixture rheology, amount of nucleated bubbles in the mixture, rate of depletion of the created/injected gas, or other external control mechanisms, affect their expansion process.

In flexible foams, the reaction between isocyanate and water in a chemically blown system results in the formation of amine and CO_2 gas. This CO_2 gas diffuses into nucleated bubbles in the mixture due to a pressure difference [2], leading to a continuous increase in mixture volume until the reacting water is totally converted. More so, the combined effect of the evolution of mixture viscosity via chain-linking/polymerization, urea formation, as well as bubble rupture also limit the expansion process. This process results in a foam matrix with open cells [3]. On the contrary, physical blowing agents, such as trichlorofluoromethane, cyclopentane, or liquid CO_2, do not react in the mixture [4]. However, due to their relatively low boiling point, the associated hydrocarbon vaporizes (at its boiling temperature) as a result of the exothermic nature of the foaming system [5]. The vaporized gas diffuses into the nucleated bubbles in the mixture, leading to the expansion of the reacting mixture and the formation of a foam matrix with closed cells and rigid morphology

The PU foam expansion process may be homogeneous or heterogeneous depending on whether the system is controlled or not. This, in turn, influences pore structure and pore-size distribution [6], as well as the thermophysical and physicomechanical properties of the final foam matrix [7–10]. However, some applications of PU foams require homogeneously distributed bubbles in the final product. Situations arise, for example, in a geotechnical engineering process, where the expanding system cannot be controlled, thereby resulting in a heterogeneous expansion of the foaming mixture [11]. Such inhomogeneity in bubble-size distribution, attributed to the spatial variation of flow properties, results in a subdivided domain consisting of zones where the bubbles grow freely, and zones where their growth is restricted [7,12].

Several successful attempts to predict bubble growth and their size distribution in various liquids from a microscopic (cell model) view have been reported in the literature [7–9,13–18]. For instance, the theoretical modelling and analysis of the evolution of a vapor bubble expanding in a shell of power-law (non-Newtonian) fluid by Street et al. [13] illustrated the influence of the shear thinning, melt viscosity, and molecular diffusivity of the blowing agent on the initial growth of the bubble. In addition, the mass and momentum transport process of the fluid was also shown to significantly affect the initial growth rate of the nucleated bubble [13]. In a related study [7], the effect of blowing-agent concentration and gelling (cream) time on the expanding bubble in a PU foam mixture, as well as bubble-size distribution, was presented. To understand the influence of fluid viscoelasticity on expanding bubbles, Feng and Bertelo [8] carried out an extensive study on bubble expansion in a viscoelastic (Oldroyd-B) fluid. Adopting the cell model, they simulated and discussed the effect of gas depletion and the proximity of neighboring bubbles in the physically blown polymer, and predicted the evolution and size distribution of the bubbles in reasonable agreement with available experimental data.

In their work, Amon and Denson [14] presented a mathematical framework for a system of expanding bubbles, with each bubble enclosed in a shell of liquid containing supersaturated gas. Relevant features that allow for the possible extension of the proposed model to the macroscopic investigation of bubble growth in liquids were analyzed. Based on the cell model [13,14] and neglecting surface tension effects, Bruchon and Coupez [15] numerically tracked the evolution of the radius of a single bubble in a Newtonian fluid and further carried out 2D and 3D simulations of a finite number of bubbles expanding in a pseudoplastic fluid. Although the cell model provides relevant qualitative information on bubble dynamics in liquids [14], it reduces the foam-structure formation to the resolution of tracking the evolution of the radius of a limited number of bubbles within the liquid [15]. This becomes more complicated and (numerically) expensive when an increasingly large number of bubbles [15,16], as in the case of PU foam formation, expands heterogeneously under nonisothermal conditions.

Motivated by the theoretical framework in References [13–15] and the challenge in Reference [15], we sought to understand and predict the nonuniform expansion observed in a thermally uncontrolled PU foam-formation process. We propose a macroscale (continuum) model that accounts for local variations in the foam-expansion process. We adopted the modelling approach of Reference [19], which summarizes the specie consumption with the Kamal law [20] for the degree of cure, with an adequately modified expansion source term and accounting for local contributions of the mixture temperature and viscosity to the expanding system. Although this approach does not quantify the bubbles in the foam, it gives a qualitative description of the distribution of void fractions in the domain. Of particular importance in this study was to investigate and understand the observed volume variation of the expanded foam when an equal mass of the reacting foam was injected into different cylinders in free-rise experiments. In this regard, three foam-expansion experiments in different geometries were studied and simulated. With graphical illustrations, we present our results and validate them against available experimental data. Furthermore, we carried-out μCT scan analysis of the expanded foam matrix and compared our observations with results from our simulation. Our results show qualitative agreement with observations from the experiments.

The rest of this paper is structured as follows. In Section 2, we briefly describe the experimental setup and present a mathematical framework for the nonuniform expansion source term adopted in our simulations. The results and corresponding discussions are presented in Section 3, and we conclude the study in Section 4.

2. Experiment and Mathematical Framework

Following the experiment and discussions in Reference [19] on the free-rise PU foam-expansion process, equal masses (77 and 37 g) of a reacting mixture containing isocyanate and precursors for rigid foam with some water are injected into different cylindrical tubes in a series of experiments at room temperature. Each pair had a diameter of 56 and 112 mm, and a height of 812 and 203 mm, respectively. In related reaction injection molding (RIM) experiments, the same mixture was injected into a rectangular mold of $500 \times 50 \times 40$ mm (Figure 1) under a fixed wall temperature of 55 °C. The expanding foam in each tube was monitored, and their heights and volumes were recorded in time with the aim of understanding the effect of the mold or geometry conditions on the expanding PU-foam system, particularly since the free-rise experiments in the cylinders resulted in a foam matrix of different volumes. This suggested a possible influence of the geometry conditions on the foam-expansion process.

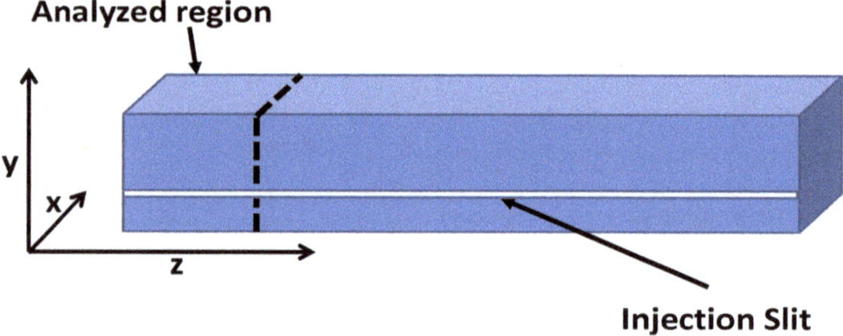

Figure 1. Schematic description of the geometry of the closed rectangular mold in the reaction injection molding (RIM) experiments. The marked region corresponds to the part of the expanded foam analyzed with μCT scans, (see Section 3.2).

Assuming that the foaming mixture is a quasihomogeneous continuum with a constant rate law for the degree of cure, the experiments presented above were modelled mathematically, subject to appropriate boundary conditions. The equations governing the transport of mass, energy, and degree of polymerization follow directly from Reference [19] (see Appendix A). However, the source term in the equation governing the conservation of mass is adequately restructured to account for possible nonuniformity in the expanding system.

The mathematical formulations of Street et al. [13], and Bruchon and Coupez [15] from the conservation of mass and momentum transport describe the dynamics of a bubble of radius R expanding in a pool of viscous liquid under quasistatic motion. Adopting the continuity of stresses at the surface of the bubble and neglecting the surface-tension effect, the evolution of the bubble radius was summarized by:

$$\frac{\dot{R}}{R} = \frac{P_b - P_l}{4\mu_l^{eff}}. \tag{1}$$

where P_b and P_l are the pressure in the bubble and liquid, respectively. μ_l^{eff} is the effective viscosity of the surrounding liquid, and $\dot{R}$ is the rate of change of the radius of the bubble as it expands in time. Equation (1) implies that, for a given bubble, expanding in a liquid with viscosity μ_l^{eff} under isothermal condition

$$P_b - P_l \propto \mu_l^{eff} \frac{\dot{R}}{R}. \tag{2}$$

The equation governing the conservation of mass of the expanding PU foam relates flow velocity **v** to foam volume in time ($V(t)$) (see Reference [19]) by:

$$\nabla \cdot \mathbf{v} = \frac{1}{V(t)} \frac{dV(t)}{dt} = S_p. \tag{3}$$

S_p here is the expansion source term. We adopted a growth model [19,21,22] for the volume of the expanding foam, so that at any time t, foam volume $V(t)$ is described by

$$V(t) = A \exp\left(-\tilde{\pi}\left[(t + t^*)^{-\tilde{\varepsilon}}\right]\right) + \gamma, \tag{4}$$

where t^* is the time when the foam starts to expand after being injected into the mold. Constants A, $\tilde{\varepsilon}$, $\tilde{\pi}$, and γ are estimated from available volume-expansion experiment data.

In a thermally controlled (adiabatic) foam-expansion setup, the source term, denoted here as S_p^{ad}, is ideally obtained from volume $V(t)$ of the expanding PU foam under spatially uniform temperature conditions. Consequently, this results in the homogeneous expansion of the foaming mixture at constant pressure. Hence, suppose there are n nucleated spherical bubbles in such an expanding system, each bubble with an R radius, then one can show that the contribution of S_p^{ad} to individual bubbles relate to the radius of each bubble by

$$\frac{1}{V(t)} \frac{dV(t)}{dt} \equiv \left(\frac{\dot{R}}{R}\right)_{add} = \beta S_p^{ad}, \tag{5}$$

where $\beta = 1/3$ (see Appendix B). Therefore, under adiabatic conditions, Equation (1) can be rewritten as:

$$P_b - P_l = 4\beta\mu_l^{ad} S_p^{ad} \tag{6}$$

Here, μ_l^{ad} is the mixture viscosity under adiabatic temperatures.

Thermal conditions in chemorheological fluids significantly affect the rate of reaction, which has direct consequence on the evolution of the chemoviscosity of the reacting fluid [19,23]. More so, in a thermally uncontrolled (nonadiabatic) foaming system, the temperature of the expanding foam spatially varies. Under this condition, temperature measurements at the core is often approximated to be adiabatic since temperatures in this region are less diffusive. However, beyond a progressive thermal diffusion layer in the region close to the boundaries, temperature conditions are nonadiabatic. Hence, expanding bubbles in such regions would lead to restricted growth due to the combined effect of temperature, reaction rate, and mixture effective viscosity (μ_l^{eff}) [7,12].

Dividing both sides of Equation (6) by $4\mu_l^{eff}$ and assuming that the pressure difference between the bubble and the surrounding liquid is locally the same under adiabatic or nonadiabatic temperature conditions, we have:

$$\left[\frac{P_b - P_l}{4\mu_l^{eff}} = \frac{\dot{R}}{R}\right] \equiv \beta S_p = \frac{4\beta\mu_l^{ad} S_p^{ad}}{4\mu_l^{eff}}. \tag{7}$$

Therefore, considering local thermal influence on the growth of individual bubbles reflected in the local viscosity of the expanding PU foam, we define a linearly averaged local effective viscosity μ_l^{eff} by

$$\mu_l^{eff} = (1-\alpha)\mu_l^{ad} + \alpha\mu_l^{nad}, \quad \text{with } \mu_l^{nad} \geq \mu_l^{ad}. \tag{8}$$

where μ_l^{nad} is the viscosity of the foaming mixture under a nonadiabatic condition, and thermal sensitivity parameter α is to be determined (see Section 3.3.1 for details). Hence, Equation (7) becomes:

$$\beta S_p = \frac{4\beta\mu_l^{ad}S_p^{ad}}{4((1-\alpha)\mu_l^{ad} + \alpha\mu_l^{nad})}, \tag{9}$$

Rearranging Equation (9), we obtain:

$$S_p = \frac{S_p^{ad}}{1 + \alpha\left(\dfrac{\mu_l^{nad}}{\mu_l^{ad}} - 1\right)}. \tag{10}$$

The equation governing mixture viscosity μ_m as in Reference [19] is given by

$$\mu_m = \mu_{oo}\exp\left(\frac{E_\mu}{RT}\right) \cdot H(\zeta) \cdot F(\varphi_g), \tag{11}$$

where μ_{oo} is constant, E_μ is the foam activation energy, T is the temperature, and R is the rate constant. $H(\zeta)$ and $F(\varphi_g)$ are, respectively, the contributions of the degree of cure/polymerization ζ and the absorbed gas fraction φ_g to the viscosity of the mixture. We assume for simplicity that these contributions are locally uniform so that by substituting Equations (11) in Equation (10), and simplifying the resulting expression, we obtain

$$S_p = \frac{S_p^{ad}}{\Gamma_\mu} \tag{12}$$

where local damping factor Γ_μ is given by

$$\Gamma_\mu = 1 + \alpha\left(\exp\left(\frac{E_\mu}{R}\left(\frac{1}{T^{nad}} - \frac{1}{T^{ad}}\right)\right) - 1\right) \tag{13}$$

This implies that nonlocal parameter α controls the influence of temperature, at a constant reaction rate, on the expansion of the foam. Therefore, at any thermal conditions, the right-hand side of Equation (3) is adequately replaced with Equation (12).

In the next section, we present and discuss the results obtained from both the experiments and our simulations. Further comparison against the experimental data are carried out.

3. Results and Discussion

We commence this section by presenting some observations from the experimental data and follow this up with the results from our simulations. Details of the numerical method adopted in resolving the system of nonlinear partial differential equations governing the expansion process are presented in Reference [19].

3.1. Experiment Observations: Free-Rise Foam Expansion in the Cylinders

Results from the free-rise experiments, carried out by our colleagues at the Institute of Lightweight Structures, Chemnitz University of Technology, Germany (Table 1 and Figure 2) reveal a possible influence of geometry constriction on the expanding foams.

Table 1. Experiment observations showing the final average volume estimated from several experiments.

Tube Diameter	Injected Mass	Final Volume
(mm)	(g)	1e^6 (mm^3)
112	77	1.622
56	77	1.401
112	37	0.74547
56	37	0.6452

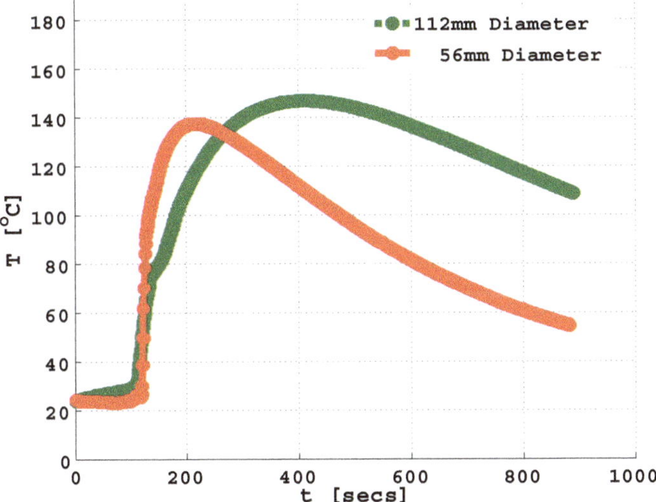

Figure 2. Temperature measurement at a point with equal volumetric reading of 1.25 L from the graduated cylinders. This corresponds to a height of 507.5 mm in the 56 mm diameter tube and 127 mm in the 112 mm diameter tube, respectively.

The final volume of the expanded PU foam in the 56 mm cylinder was observed to be lower in comparison with that of the 112 mm tube though equal initial masses of 77 and 37 g were injected into each pair of cylinders. This unexpected behavior suggests an interplay between the heat transfer in the reacting system and the proximity of the external environment to the core of the expanding PU foam. In the case where the distance from the core to the bounding surface was small, the rapid heat loss in time (Figure 2) enhanced the reduction in the final volume of the expanded foams.

3.2. μCT Scan Analysis

To investigate the distribution of solid foam fraction in the matrix, which conversely depicts the void fraction, μCT scan images of a section of the foam with a 50 × 50 × 40 mm dimension from the rectangular mold (Figure 3a) were studied. The corresponding scanned images were reconstructed and analyzed with GeoDict digital material laboratory software [24]. The reconstructed image (Figure 3b) was divided into 15 bins of equal sizes in each direction and the average solid fraction of the foam in each bin was obtained.

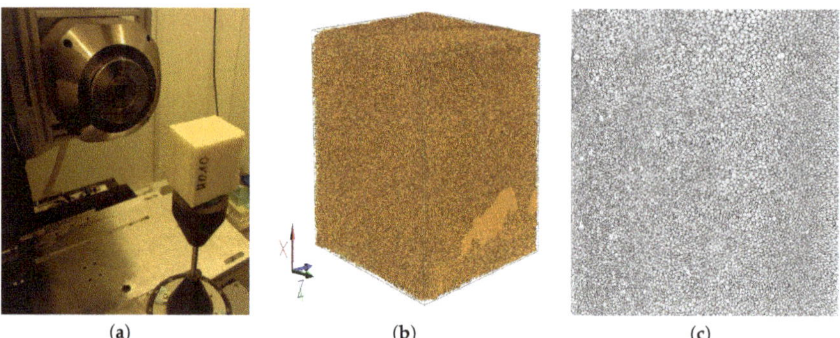

Figure 3. (**a**) Setup for the μCT scan data capture. (**b**) Reconstructed image of the μCT scan raw data using GeoDict. (**c**) Transversal binarised image of CT scan. Slice taken from middle point along the z-axis.

In Figure 4, about 1.2% variation in the solid foam fraction was observed along the x-axis, and about 1.8% along the y-axis. The x and y-axis are taken to be perpendicular to the entrant (y-z plane) and the base (x-z plane), respectively. However, the distribution of the solid fraction remains almost constant, with a variation of about 0.3%, in the z-direction perpendicular to (x-y plane) from the side wall. We noted that the high solid fractions at the base (Bin 1) along the y-axis resulted from the fact that this region got filled first during the injection molding, with some of the material remaining there at lower temperatures throughout the expansion period, thereby resulting in a more dense region in the domain.

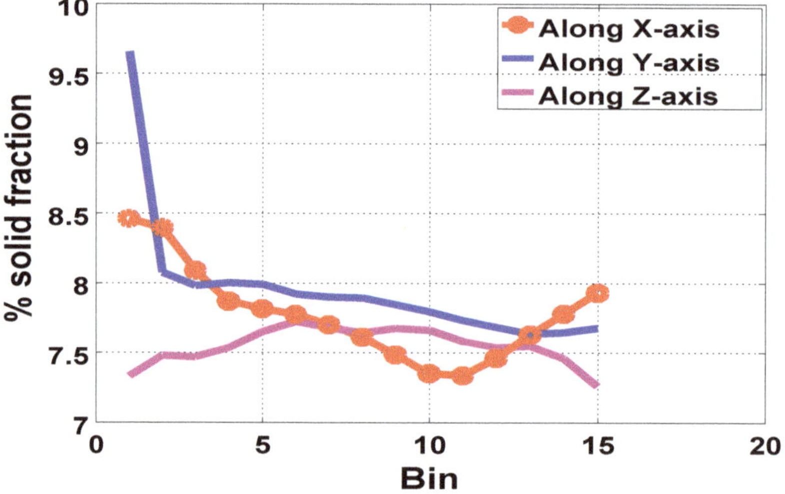

Figure 4. Percentage average volume fraction in each bin along each axis of the reconstructed μCT scan image.

Since the proportion of the void in a given material conversely relates to the solid fractions, the distribution of void fractions in the foam matrix, which reflects pore- or bubble-size distribution, is also deductible from Figure 4. More voids were observed toward the core when compared to regions near and at the walls in the expanded foam. Hence, relatively bigger bubbles concentrated at the core when compared to the region near the walls. We therefore surmised that the higher temperature at the core significantly contributes to the expansion rate of the nucleated bubbles in this region.

This spatiotemporal variations in temperature and material viscosity induce a nonuniform expansion in the foam, resulting in the variation of bubble size and their distribution across the mold.

In the next section, we numerically explore the reported inhomogeneity observed in the expanded PU foam from experiments, and attempt to address the fundamental questions motivating this study:

- Why does equal mass of PU foam mixture injected into tubes of different dimensions in the injection molding experiment (without overflow) result in expanded foams with different volumes?
- How can this observed nonuniformity be accounted for when modelling and simulating PU foam-expansion processes in other geometries, different from the cylinders?

3.3. Numerical Results and Discussion

All numerical implementations in this section were carried out with our inhouse FOAM solver based on the finite-volume method, and resident in the Complex Rheology Simulation (CoRheoS) platform developed at the Fraunhofer ITWM. Unless otherwise stated, all relevant input material values for our simulations were adopted from Reference [19].

3.3.1. Estimating α and Adiabatic Source Term S_p^{ad}

With the volume data from the free-rise expansion experiments in both cylinders, we obtained appropriate parameters for the fit functions (see Equation (4)). Here, fit volumes V_{112} and V_{56} describe the volume of the expanding PU foam in the cylinders with 112 and 56 mm radius, respectively (Figure 5). Following the methodology described in Reference [19], we computed the corresponding S_p^{112} for volume V_{112} and S_p^{56} for V_{56}. Due to the unavailability of the relevant adiabatic measurements for the expanding PU foam, we estimated the value for α and numerically obtained the adiabatic source term S_p^{ad} required for our simulations (Figure 6). We conjectured that any given PU foam of the same mixture under adiabatic conditions, irrespective of geometry, would expand with the same volume in time. So, from Equation (12) in the form

$$S_p^b = \frac{S_p^{ad}}{1 + \alpha \left(\exp \left(\frac{E_\mu}{R} \left(\frac{1}{T^b} - \frac{1}{T^{ad}} \right) \right) - 1 \right)}, \quad b = (112, 56). \tag{14}$$

and rearranging the resulting expressions, we obtain

$$\alpha = \frac{\hat{S}_p - 1}{\hat{S}_p + \exp \left(\frac{E_\mu}{R} \left(\frac{1}{T^{56}} - \frac{1}{T^{ad}} \right) \right) - \hat{S}_p \exp \left(\frac{E_\mu}{R} \left(\frac{1}{T^{112}} - \frac{1}{T^{ad}} \right) \right) - 1}, \tag{15}$$

where $\hat{S}_p = \frac{S_p^{112}}{S_p^{56}}$.

Using the simulated average temperatures $< T >$ in each tube over the duration of the free-rise experiment, we obtained α from Equation (15), which varies in time, and computed $\hat{S}_p^{ad}$ as the average between the S_p^{ad} calculated from S_p^{112}, and that calculated from S_p^{56} using the same α, so that

$$S_p^{ad} = \left\langle S_p^{112} \Gamma_\mu^{112}, S_p^{56} \Gamma_\mu^{56} \right\rangle, \tag{16}$$

serve as input to our simulation. With one set of values for the duo, we achieved very good correlation between the volumes from our simulations and those obtained in the free-rise experiments for all considered geometries. However, we noted that approximating S_p^{ad} became unnecessary if at least one set of the experimental data was obtained from an adiabatic setup. In such a case, we would only need to obtain the appropriate value for α.

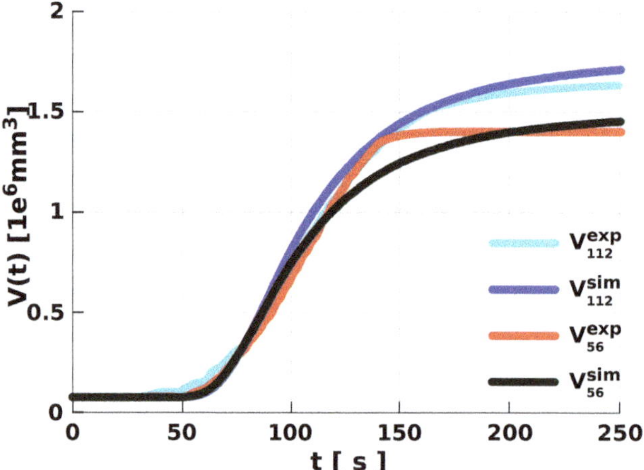

Figure 5. Fit functions for the volume data from the polyurethane (PU) foam-expansion experiments in the 112 mm with fit parameters values $A_{112} = 1.7041$, $\bar{\varepsilon}_{112} = 3{,}249{,}603$, $\bar{\pi}_{112} = 3.293393$ and $\gamma_{112} = 0.076$, and the 56 mm tubes with parameter values $A_{56} = 1.4288$, $\bar{\varepsilon}_{56} = 3{,}049{,}603$, $\bar{\pi}_{56} = 3.3$ and $\gamma_{56} = 0.076$.

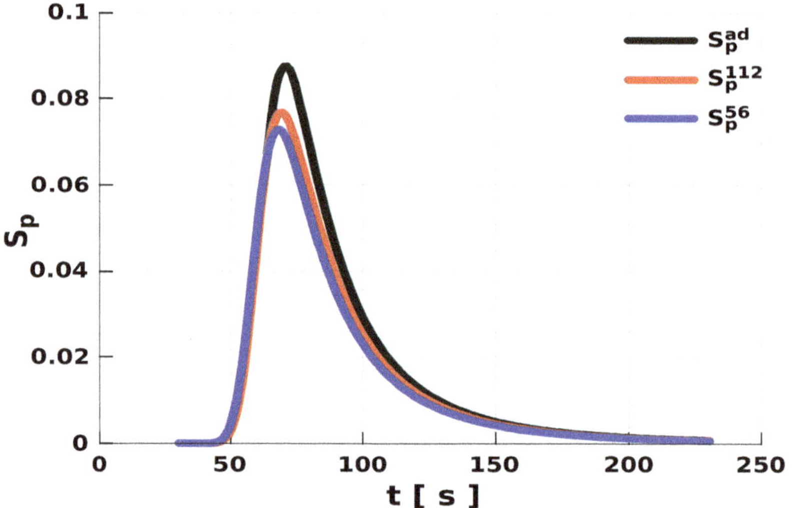

Figure 6. Graphical representation of the expansion source term for each experiment and the estimated S_p^{ad} for the adiabatic expansion.

3.3.2. Influence of Thermal Conditions on the Foam-Expansion Process

The significant difference between the adiabatic and nonadiabatic temperatures in the 56 mm tube (Figure 7a) is attributed to the propinquity of the external environment to the core of the expanding foam.

This proximity results in rapid heat loss through the walls of the cylinder due to diffusion. In the 112 mm tube, on the other hand, temperatures at the core were close to adiabatic conditions up to a thermal diffusive layer (near the walls), where heat loss gradually becomes significant (Figure 7b).

The time variation of α in Figure 8a induces a local damping effect (Figure 8b), which is a consequence of the spatial difference in temperatures. Regions with temperature values close to the adiabatic temperature experience minimal or no damping. In addition, depending on the closeness of the bounding surface to the core and the local temperatures in the expanding mixture, the damping effect increases with an increase in α (see Figure 8a,b). Therefore, at higher α values, the mixture in the tube with the 56 mm diameter becomes less expansive when compared with the 112 mm tube, thereby resulting in a significant decrease in the final volume of the expanded foam in the smaller tube.

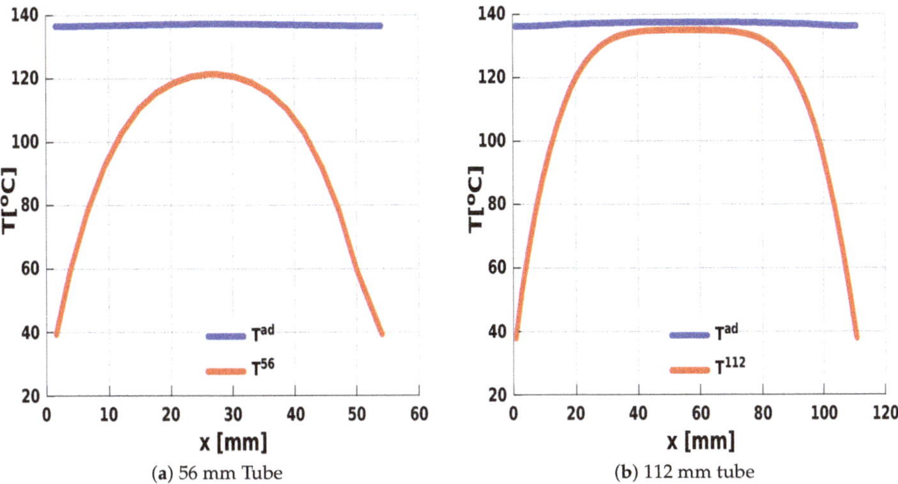

(a) 56 mm Tube

(b) 112 mm tube

Figure 7. Adiabatic and nonadiabatic temperature profiles along an axis across the tubes (with maximum temperatures in the core) at a fixed time during the expansion process (**a**) in the 56 mm tube and (**b**) in the 112 mm tube.

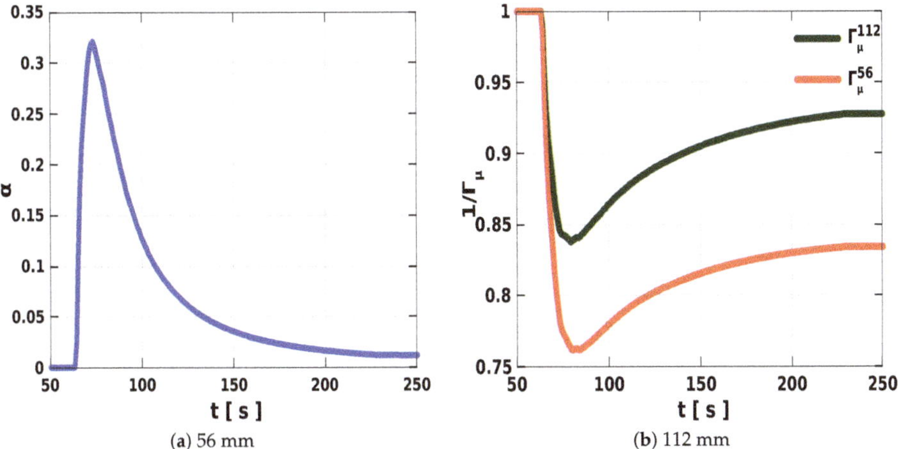

(a) 56 mm

(b) 112 mm

Figure 8. Plots showing the (**a**) variation of the computed α in time and (**b**) the associated damping factor $1/\Gamma_\mu$ in each tube as a consequence of thermal disparieties in the cylinders.

To further consolidate our proposed mathematical framework for expansion source term S_p in Equation (12), we compared the data for the final volumes of the expanded foams from the free-rise experiments with results from our simulations (Table 2). In addition, we simulated a supplementary experimental setup for a free-rise foam-expansion experiment conducted with 77 g of the same PU foam material in a cylindrical tube with 84 mm diameter using the same α and S_p^{ad} obtained from our previous calculations (Equations (15) and (16)). The results showed very good correlation with data from experiments (see Table 2).

Table 2. Comparison for the final foam volumes from the experiment and our simulations for uniform and nonuniform expansion models using the computed S_p^{ad}.

		Simulation				Experiment
		$\alpha = 0$		$\alpha > 0$		
d	Mass	Vol Function	Vol	S_p	Vol	Vol
(mm)	(g)	V(t)	$1e^6$ (mm^3)	S_p^{ad}	$1e^6$ (mm^3)	$1e^6$ (mm^3)
112	77	V_{112}	1.63667	S_p^{ad}	1.61	1.622
112	37	V_{112}	0.813583	S_p^{ad}	0.7250	0.74547
56	77	V_{56}	1.41645	S_p^{ad}	1.44822	1.401
56	37	V_{56}	0.696319	S_p^{ad}	0.664682	0.6452
		Result from simulation of supplementary experiment				Supplementary Experiment
84	77			S_p^{ad}	1.55188	1.5657

3.3.3. Spatial Inhomogeneity in the Expanded PU Foams: Simulation Results

In our previous study [19], the expanding foam was assumed to undergo nonlocal expansion, which resulted in spatially homogeneous foam fraction within the expanded foam. However, accounting for spatial inhomogeneity in the expansion source term (S_p) (i.e., $\alpha > 0$), the local variations which arises as a consequence of the spatial changes in temperature and the mixture viscosity influences the distribution of foam fraction in the PU foam system, Figure 9a,b.

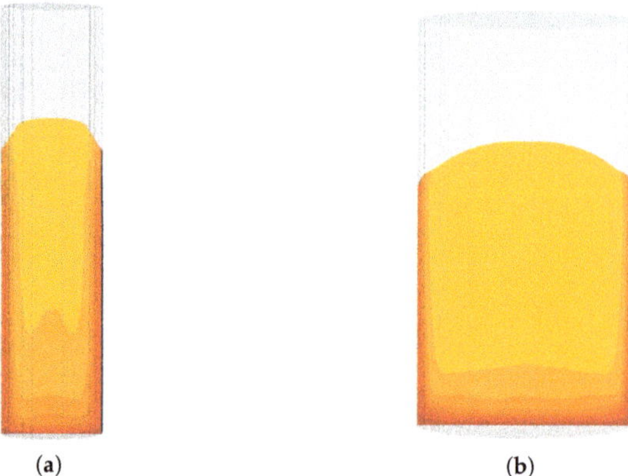

(a)   (b)

Figure 9. Spatial variation of solid foam fraction in the expanded foam (**a**) in the 56 mm diameter tube and (**b**) in the 112 mm diameter cylinder. The deep red region indicates higher solid fractions.

This lower temperature at the wall causes an increase in local viscosity. Therefore, the interplay between temperature, viscosity, and local expansion results in restricted growth of the foaming mixture near the walls. This further leads to the formation of densely packed foam fraction around those regions. However, since thermal conditions are higher in the core, the material in this region freely expands.

This corresponds to the lighter colored regions in Figure 9, where foam fractions are lower compared to the deep-red parts with higher solid fractions.

The symmetry in flow direction during free-rise expansion in the tubes induces a symmetric distribution of the foam fractions across the tube (Figure 10a,b). More so, the solid fraction of the foamed material was observed to decrease with height around the bounding walls and in the core.

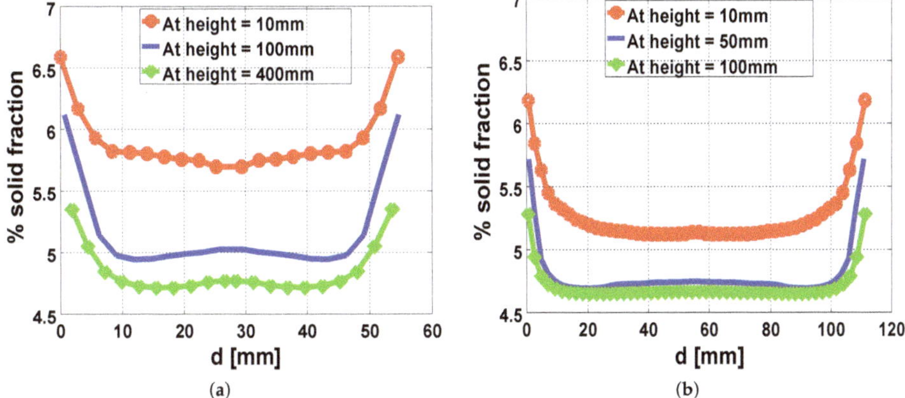

Figure 10. Simulation results showing the symmetric distribution of the solid foam fraction across the cylinder at different heights (**a**) in the 56 mm tube and (**b**) in the 112 mm tube.

The observed disparity in the fractions of the expanded foam (Figure 11a,b) directly follows from the temperature conditions in the corresponding setup. Provided the expanding system is non-adiabatic the thickness of the heat diffusion layer near the walls, attributed to the proximity of the core of the expanding foam to the external environment, results in significant increase in the local viscosities around those regions. This diffusion effect is more prominent in the 56 mm cylinder when compared to the 112 mm cylinder. Consequently, the foaming material in the smaller tube experiences more growth restriction around the walls in both cases of the injected masses (37 gm and 77 gm). Therefore, the higher values of the solid foam fractions in the constricted geometry result in denser foam with a reduction in the final volumes of the expanded foams, as seen in the 56 mm cylinder.

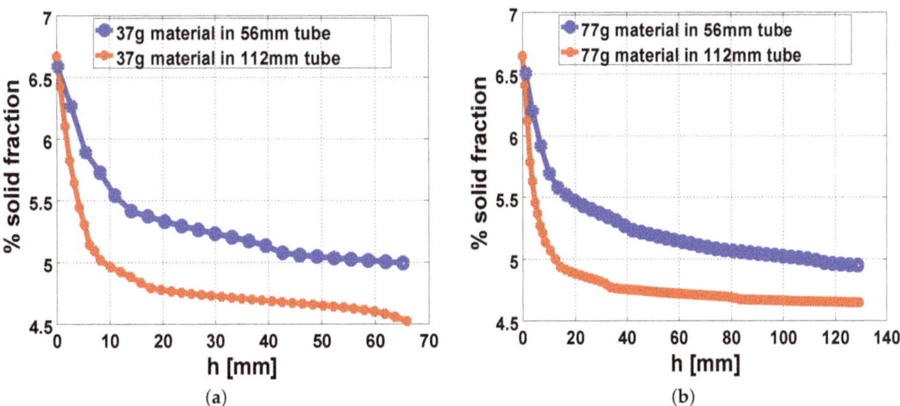

Figure 11. Disparity in foam-fraction distribution from the center of the base to a point in the core for both cylinders, with injected material mass of (**a**) 37 g and (**b**) 77 g.

Furthermore, we assessed our modelling technique and confirmed its efficacy by validating our results with those from the μCT scan analysis of the expanded foam in the rectangular geometry. Using the same values for α and S_p^{ad} obtained from the tube simulations, we simulated the injection molding process of PU foam in a rectangular mold as described in Section 2. We particularly investigated the distribution of the solid fraction and compared our simulation results with the data from μCT scan analysis (Figure 12a–c). In Figure 12a along the x-axis, we observed a similar variation (in foam-fraction distribution) of about 1.2% in both cases and, along the y-axis, a variation of about 1.3%–1.8% was observed (Figure 12b). A variation of about 0.2%–0.3% was also observed along the z-axis, as shown in Figure 12c. The asymmetry observed in the distribution of the solid fractions in Figure 12 is due to the asymmetry in the flow of the reacting mixture along the flow direction during the foam-expansion process. This observation qualitatively agrees with those from the experiments.

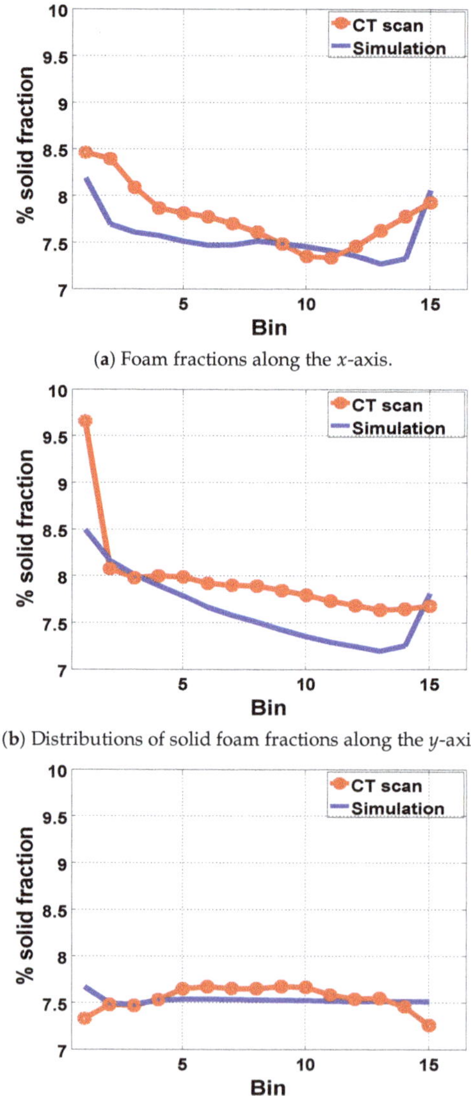

(a) Foam fractions along the x-axis.

(b) Distributions of solid foam fractions along the y-axis.

(c) Solid foam fractions along the z-axis.

Figure 12. Comparison of the percentage volume fraction from the experiment and those from our simulations along each axis.

Finally, we adopted the obtained models and simulated the reaction injection molding of the PU foams in a wavelike geometry. The comparison between the final results from the simulations and those obtained from the experiment showed very good agreement (Figure 13).

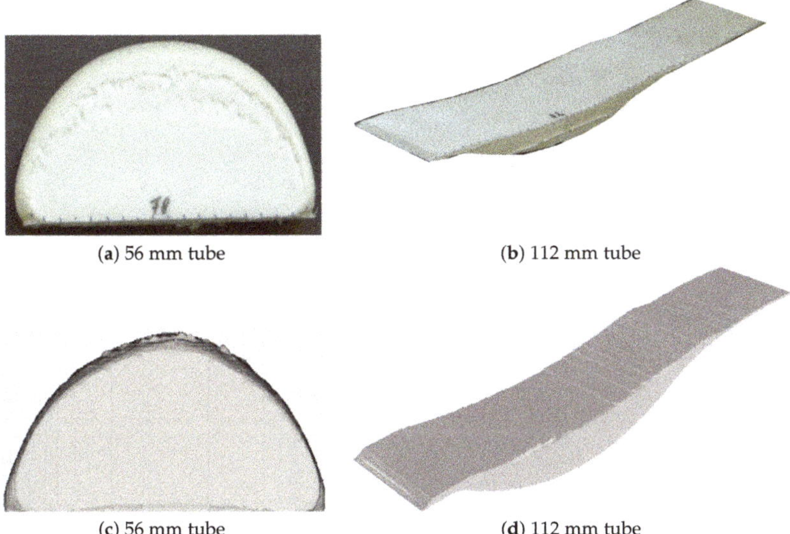

(**a**) 56 mm tube   (**b**) 112 mm tube

(**c**) 56 mm tube   (**d**) 112 mm tube

Figure 13. Comparison between simulation and experiment for reaction injection molding in a wavy mold.

4. Conclusions

Inspired by the cell model for a unit bubble expanding in a viscous liquid, we proposed a mathematical framework that accounts for spatial inhomogeneity in a thermally uncontrolled (nonadiabatic) system of expanding polyurethane foam. We surmised that, under adiabatic conditions, a given foam material would expand in the same manner, irrespective of mold geometry. Hence, with the understanding that temperature distribution in a nonadiabatic system plays significant role in the local expansion of the foaming material, the source term driving foam expansion is structured to depend on the variation between adiabatic and nonadiabatic temperatures. With one setup for the expansion source term, calibrated for adiabatic expansions, we studied all flow conditions considered in this work. To understand the degree of nonuniformity in the expanding foam, we modelled and simulated free-rise PU foam-expansion experiments in cylindrical tubes of different diameters, and the injection molding of the same foam in a rectangular mold. A fundamental issue in this study was to investigate the observed difference in the volumes of the expanded PU foam when equal amounts of mass were injected in different geometries.

The interplay between temperature, material viscosity, and the gap between the core and the external environment were observed to affect the distribution of solid foam fractions in the expanded PU foam. In a nonadiabatic system, the thickness of the thermal diffusion layer near the walls resulted in the appearance of a viscous layer that restricts the expansion process around such locations. Then, a densely packed foam fraction formed in the regions near the wall. This phenomenon was more pronounced in constricted geometries, where the core of the expanding foam was relatively closer to the external environment. Therefore, the observed disparity in the volumes of the expanded foam in the free-rise experiments in the cylinders resulted from the combined effects of temperature, viscosity, and the proximity of the foam core to the bounding surface. To validate our models, we compared the results from our simulations with the volume data of the expanded foams, obtained from the free-rise experiments in the cylinders. Our results showed good correlation with those from the experiments.

Images from μCT scans of the expanded foam from the rectangular mold were reconstructed and studied using GeoDict digital material laboratory software. The reconstructed images were analyzed for the distribution of foam fractions in the mold. Analysis showed an average spatial variation of about 1.1% in the foam fraction from the walls to the core of the foam matrix. This observation favorably compared with the results from our injection molding simulation.

This study serves as a platform for our ongoing studies on foam-expansion processes in porous media, with applications in reinforced structures with complex geometry.

Author Contributions: Conceptualization, D.N., I.E.I. and K.S.; methodology, D.N., I.E.I.; software, D.N.; validation, D.N., I.E.I. and K.S.; formal analysis, D.N. and I.E.I.; investigation, D.N., I.E.I. and K.S.; writing–original draft preparation, I.E.I.; writing–review and editing, D.N. and I.E.I.; visualization, D.N. and I.E.I.; project administration, D.N. and K.S.

Funding: This research received no external funding.

Acknowledgments: We thank our colleagues at the Institute of Lightweight Structures, Chemnitz University of Technology, Germany, for their fruitful co-operation and for the relevant experimental data they provided us for the study. This study was partly supported by the Fraunhofer High Performance Center for Simulation- and Software-Based Innovation.

Conflicts of Interest: The authors declare no conflict of interest.

Appendix A. Governing Equations

Conservation of Mass

$$\nabla \cdot \mathbf{v} = \frac{1}{V(t)}\frac{dV}{t} = S_p, \tag{A1}$$

Conservation of Momentum

$$\frac{\partial \rho \mathbf{v}}{\partial t} + \nabla \cdot (\rho \mathbf{v}\mathbf{v}) = -\nabla P + \nabla \cdot (\eta_m \mathbf{D}) + \rho \mathbf{g}, \tag{A2}$$

Conservation of Energy

$$\rho C_p \left(\frac{\partial T}{\partial t} + \mathbf{v}\cdot\nabla T\right) = \nabla \cdot (k\nabla T) + \frac{1}{2}(\eta_m \mathbf{D} : \mathbf{D}) + \rho H_R \frac{d\zeta}{dt}. \tag{A3}$$

Degree of Polymerization

$$\frac{d\zeta}{dt} = (k_1 + k_2\zeta^m)(1-\zeta)^n, \tag{A4}$$

k_1, k_2 m, and n are constants. ζ is the degree of cure/polymerization. ρ is density, P is pressure, $\mathbf{D}$ is deformation tensor, $\mathbf{v}$ and $\mathbf{g}$ are, respectively, flow velocity and the gravitational force. C_p, H_R, and k are the specific heat capacity, heat of reaction, and thermal conductivity, respectively.

Appendix B. Relationship between Radius R and Expansion Source Term S_p

Suppose there are n bubbles in the expanding PU foam system, and each bubble of radius $R(t)$ expands uniformly in time with volume V(t). We can write the total volume of the expanding bubbles as

$$\sum_{i=1}^{n} \frac{1}{V_i}\frac{dV_i}{dt} = \sum_{i=1}^{n} \frac{3}{4\pi R_i^3}\frac{d}{dt}\left(\frac{4}{3}\pi R_i^3\right), \tag{A5}$$

since the expansion is uniform, so that $R_i's$ are equal in time; then, with some algebraic manipulations, it suffices to write

$$\frac{1}{V}\frac{dV}{dt} = \frac{3}{R}\frac{dR}{dt}. \tag{A6}$$

Polymers **2019**, *11*, 100

Therefore,

$$\left[\frac{1}{V}\frac{dV}{dt} = S_p\right] \equiv \frac{3}{R}\frac{dR}{dt} = S_p. \tag{A7}$$

So that

$$\frac{1}{R}\frac{dR}{dt} = \frac{1}{3}S_p \tag{A8}$$

References

1. Spina, R. Technological characterization of PE/EVA blends for foam injection molding. *Mater. Des.* **2015**, *84*, 64–71. [CrossRef]
2. Bikard, J.; Bruchon, J.; Coupez, T.; Silva, L. Numerical Simulation of 3D Polyurethane expansion during manufacturing process. *Colloids Surf. A Physicochem. Eng. Asp.* **2007**, *309*, 49–63. [CrossRef]
3. Lee, S.T.; Ramesh, N.S. *Polymeric Foams Mechanisms and Materials*; CRC Press LLC: Boca Raton, FL, USA, 2004.
4. Ashida, K. *Polyurethane and Related Foams: Chemistry and Technology*; CRC Press Taylor & Francis Group: Boca Raton, FL, USA, 2007.
5. Seo, D.; Youn, J.R. Numerical Analysis on Reaction Injection molding of Polyurethane Foam by Using Finite Volume Method. *Polymer* **2005**, *46*, 6482–6493. [CrossRef]
6. Wang, X.; Pan, Y.; Shen, C.; Liu, C.; Liu, X. Facile Thermally Impacted Water-Induced Phase Separation Approach for the Fabrication of Skin-Free Thermoplastic Polyurethane Foam and Its Recyclable Counterpart for Oil–Water Separation. *Macromol. Rapid Commun.* **2018**, *39*, 1800635. [CrossRef]
7. Niyogi, D.; Kumar, R.; Gandhi, K.S. Modeling of Bubble-Size Distribution in Free Rise Polyurethane Foams. *AIChe J.* **1992**, *38*, 1170–1184. [CrossRef]
8. Feng, J.J.; Bertelo, C.A. Prediction of bubble growth and size distribution in polymer foaming based on a new heterogeneous nucleation model. *J. Rheol.* **2004**, *48*, 439–462. [CrossRef]
9. Goel, S.K.; Beckman, E.J. Nucleation and Growth in Microcellular Materials: Supercritical CO_2 as Foaming Agent. Materials, Interfaces and Electrochemical Phenomena. *AIChe J.* **1995**, *41*, 357–367. [CrossRef]
10. Kim, S.; Shin, H.; Rhim, S.; Rhee, K.Y. Calibration of hyperelastic and hyperfoam constitutive models for an indentation event of rigid polyurethane foam. *Compos. Part B* **2019**, *163*, 297–302. [CrossRef]
11. Buzzi, O.; Fityus, S.; Sasaki, Y.; Sloan, S. Structure and properties of expanding polyurethane foam in the context of foundation remediation in expansive soil. *Mech. Mater.* **2008**, *40*, 1012–1021. [CrossRef]
12. Villamizar, C.A.; Han, C.D. Studies on structural foam processing II. Bubble dynamics in foam injection molding. *Polym. Eng. Sci.* **1978**, *18*, 699–710. [CrossRef]
13. Street, J.R.; Fricke, A.L.; Reiss, L.P. Dynamics of Phase Growth in viscous, Non-Newtonian Liquids. *Ind. Eng. Chem. Fundam.* **1971**, *10*, 54–64. [CrossRef]
14. Amon, M.; Denson, C.D. A study of the Dynamics of Foam Growth: Analysis of the Growth of Closely spaced spherical Bubbles. *Polym. Eng. Sci.* **1984**, *24*, 1026–1034. [CrossRef]
15. Bruchon, J.; Coupez, T. A numerical strategy for the direct 3D simulation of the expansion of bubbles into a molten polymer during a foaming process. *Int. J. Numer. Meth. Fluids* **2008**, *57*, 977–1003. [CrossRef]
16. Geier, S.; Piesche, M. Macro and Micro-Scale Modeling of Polyurethane Foaming Processes. *AIP Conf. Proc.* **2014**, *1593*, 560. [CrossRef]
17. Bikard, J.; Bruchon, J.; Coupez, T.; Vergnes, B. Numerical prediction of the foam structure of polymeric materials by direct 3D simulation of their expansion by chemical reaction based on a multidomain method. *J. Mater. Sci.* **2005**, *40*, 5875–5881. [CrossRef]
18. Youn, J.R.; Park, H. Bubble growth in reaction injection molded parts foamed by ultrasonic excitation. *Polym. Sci. Eng.* **1999**, *39*, 457–468. [CrossRef]
19. Ireka, I.E.; Niedziela, D.; Schäfer, K.; Tröltzsch, J.; Steiner, K.; Helbig, F.; Chinyoka, T.; Kroll, L. Computational Modelling of the Complex Dynamics of Chemically Blown Polyurethane Foam. *Phys. Fluids* **2015**, *27*, 113102. [CrossRef]
20. Kamal, M.R. Thermoset Characterization for Modality Analysis. *Polym. Eng. Sci.* **1974**, *14*, 231–239. [CrossRef]
21. Savageau, M.A. Growth Equations: A General Equation and a Survey of Special Cases. *Math. Biosci.* **1980**, *48*, 267–278. [CrossRef]

22. McCormick, N.G. A Proposal on Nature's Time-Scale. *Nature* **1965**, *208*, 334–336 . [CrossRef]
23. Garschke, C.; Parlevliet, P.P.; Weimer, C.; Fox, B.L. Cure kinetics and viscosity modelling of a high-performance epoxy resin film. *Polym. Test.* **2013**, *32*, 150–157. [CrossRef]
24. Andrä, H.; Combaret, N.; Dvorkin, J.; Glatt, E.; Han, J.; Kabel, M.; Keehm, Y.; Krzikalla, F.; Lee, M.; Madonna, C.; et al. Digital rock physics benchmarks—Part I: Imaging and segmentation. *Comput. Geosci.* **2013**, *50*, 25–32.

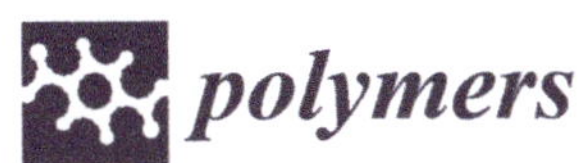 

Article

Effects of Tung Oil-Based Polyols on the Thermal Stability, Flame Retardancy, and Mechanical Properties of Rigid Polyurethane Foam

Wei Zhou [1,2], Caiying Bo [1,3], Puyou Jia [1,3], Yonghong Zhou [1,3,4] and Meng Zhang [1,3,4,*]

[1] Institute of Chemical Industry of Forestry Products, CAF, 16 Suojin North Road, Nanjing 210042, China; vvzhou90@163.com (W.Z.); newstar2002@163.com (C.B.); jiapuyou@163.com (P.J.); yhzhou777@163.com (Y.Z.)

[2] Key Lab of Forest Chemical Engineering, SFA, 16 Suojin North Road, Nanjing 210042, China

[3] Key Lab of Biomass Energy and Material, Jiangsu Province, 16 Suojin North Road, Nanjing 210042, China

[4] Co-Innovation Center of Efficient Processing and Utilization of Forest Resources, Nanjing Forestry University, 159 Longpan Road, Nanjing 210037, China

* Correspondence: zhangmeng@icifp.cn; Tel.: +86-025-8548-2520

Received: 20 November 2018; Accepted: 24 December 2018; Published: 30 December 2018

Abstract: A phosphorus-containing tung oil-based polyol (PTOP) and a silicon-containing tung oil-based polyol (PTOSi) were each efficiently prepared by attaching 9,10-dihydro-9-oxa-10-phosphaphenanthrene (DOPO) and dihydroxydiphenylsilane (DPSD) directly, respectively, to the epoxidized monoglyceride of tung oil (EGTO) through a ring-opening reaction. The two new polyols were used in the formation of rigid polyurethane foam (RPUF), which displayed great thermal stability and excellent flame retardancy performance. The limiting oxygen index (LOI) value of RPUF containing 80 wt % PTOP and 80 wt % PTOSi was 24.0% and 23.4%, respectively. Fourier transfer infrared (FTIR), Nuclear Magnetic Resonance (NMR) and thermogravimetric (TG) analysis revealed that DOPO and DPSD are linked to EGTO by a covalent bond. Interestingly, PTOP and PTOSi had opposite effects on T_g and the compressive strength of RPUF, where, with the appropriate loading, the compressive strengths were 0.82 MPa and 0.25 MPa, respectively. At a higher loading of PTOP and PTOSi, the thermal conductivity of RPUF increased while the RPUF density decreased. The scanning electron microscope (SEM) micrographs showed that the size and closed areas of the RPUF cells were regular. SEM micrographs of the char after combustion showed that the char layer was compact and dense. The enhanced flame retardancy of RPUF resulted from the barrier effect of the char layer, which was covered with incombustible substance.

Keywords: tung oil; DOPO; dihydroxydiphenylsiane; flame retardant; rigid polyurethane foam

1. Introduction

The depletion of petroleum reserves is driving the development of polymers and polymer additives from sustainable bio-materials. Polymer materials made from renewable sources have attracted attention around the world [1]. Vegetable oils are particularly promising renewable chemicals for the preparation of polyols and rigid polyurethane foam (RPUF) due to their renewability, universal availability, and environmental compatibility. Vegetable oils have double bonds, ester groups and other reactive sites that can be modified to form new polyol structures [2]. Vegetable oils such as castor oil, palm kernel oil [3,4], rapeseed oil [5], and tung oil [6] have been used to produce RPUF. RPUF has been widely used in materials for building insulation due to its excellent mechanical performance. However, a limitation to the use of RPUF is flammability, which is related to its cellular structure and low density [7,8].

In recent years, much effort has been devoted to increasing the flame-retardant properties of RPUF. Conventional methods of flame-retardant treatment for RPUF include the incorporation of flame-retardant additives based on phosphorus, silicone, nitrogen, and halogen compounds. While these flame-retardant additives impart excellent flame retardancy to RPUF, the introduction of larger amounts of flame retardants may have an adverse effect on the mechanical and physical characteristics of RPUF [9]. Furthermore, flame retardant additives may easily mobilize out of the RPUF matrix into the exterior [10]. When burning, flame retardants that contain halogens release a lot of harmful gases and bring about environmental pollution. More importantly, organohalogen compounds fragmenting from the polymers are stable and able to persist and bioaccumulate in the environment, and thus may pose a risk to human health [11]. Consequently, an environmentally friendly reactive flame retardant is needed. The main benefit of reactive flame retardants compared to other flame-retardant additives is that the flame-retardant molecule is linked to the RPUF structure with a covalent bond [12].

The compound 9,10-dihydro-9-oxa-10-phosphaphenanthrene (DOPO) [13] is an environmentally friendly reactive flame retardant that has attracted considerable attention because of its outstanding thermal stability and ability to resist oxidation and hydrolysis [14]. Due to its aromatic and phenanthrene ring structure, DOPO is more thermally and chemically stable than standard organic phosphate, and little toxic gas is released during its combustion [15]. The flame-retardant ability of DOPO has been attributed to the formation of a reactive substance such as PO^-, which is released to the gas phase where it engages with H· and OH· groups and disrupts the combustion process [16]. DOPO exhibits prominent gas-phase flame retardant properties during RPUF combustion [17,18]. Modification of RPUF through the incorporation of silicon-containing flame retardants is also recognized as an environmentally friendly method that reduces flammability [19]. Silicone materials show good flame-retardant properties through the formation of char barriers (Si–C, Si–O) at the surface of the polymer at high temperatures [20]. There are two ways to enhance the flame retardancy of RPUF, one of which is the addition of the inorganic silicon such as nano-silica and whisker silicon oxide [21], and the other of which is the introduction of silicon into the structure of polyols using 3-aminopropylmethyldiethoxysilane (APTES), *N*-(*β*-aminoethyl)-*γ*-aminopropylmethyl dimethoxysilane (KH-602), and SiO_2 [22–24]. This has resulted in the development of novel reactive flame retardants for RPUF. Tung oil-based flame-retardant polyols have not been reported previously.

The goal of this work was to develop an innovative RPUF based on the incorporation of a phosphorus-containing tung oil-based polyol (PTOP) and a silicon-containing tung oil-based polyol (PTOSi). The molecular structure of the compounds and the thermal properties of the two new tung oil-based flame-retardant polyols were characterized using FTIR, ^{1}H NMR, ^{31}P NMR, and TG. RPUFs were characterized for their thermal and flame-retardant behaviors using several analytical techniques.

2. Materials and Methods

2.1. Materials

Tung oil (TO) was provided by Jiangsu Qianglin Bioenergy Material Co., Ltd (Qianglin, China). Glycerol, sodium methoxide and ethyl acetate were obtained from Nanjing Chemical Reagent Co., Ltd (Nanjing, China). Acetic acid was purchased from Nanjing Chemical Reagent Co., Ltd (Nanjing, China). Hydrogen peroxide (30 wt %) was purchased from Shanghai Linfeng Chemical Reagent Co., Ltd (Shanghai, China). Other materials such as P-toluenesulfonic acid monohydrate (TsOH·H₂O), 9,10-dihydro-9-oxa-10-phosphaphenanthrene (DOPO), dihydroxydiphenylsilane (DPSD) and dibutyltin dilaurate (DBTDL) were provided by Aladdin Industrial Corporation (Shanghai, China). Polyether polyol (PPG4110) was purchased from Jiangsu Qianglin Bioenergy Material Co., Ltd (Jiangsu, China); and its hydroxyl value was 403 mg KOH/g and its viscosity was 3.6 Pa·s at 25 °C. Polyaryl polyisocyanate (PAPI) was provided by Yantai Wanhua Polyurethane Co., Ltd (Yantai, China) with 30.3 wt % NCO. Foam stabilizer (AK8804) was purchased from Jiangsu Maysta Chemical Co., Ltd (Jiangsu, China). Foaming agent (HFC-365mfc) was obtained from Dongguan Changze Chemical Co., Ltd (Dongguan, China). Deionized water was made at the laboratory.

2.2. Synthesis of GTO and EGTO

The synthesis of the tung oil monoglyceride (GTO) was carried out as described in a previous study [25]. The epoxidized monoglyceride of tung oil (EGTO) was prepared as follows: 100 g of GTO (0.46 mol carbon-carbon double bond), CH_3COOH (0.92 mol) and $TsOH \cdot H_2O$ (0.046 mol) were added to a round-bottom flask. Hydrogen peroxide solution (0.69 mol H_2O_2) was added dropwise over a period of 1 h and the reaction was allowed to proceed for 4 to 5 h at a temperature of 55 °C. The EGTO product was neutralized with 1 mol/L sodium hydrocarbonate solution and then extracted with ethyl acetate. The ethyl acetate solvent was removed by reduced pressure distillation. EGTO was obtained and its epoxy value was 3%. The synthesis of GTO and EGTO is illustrated in Scheme 1.

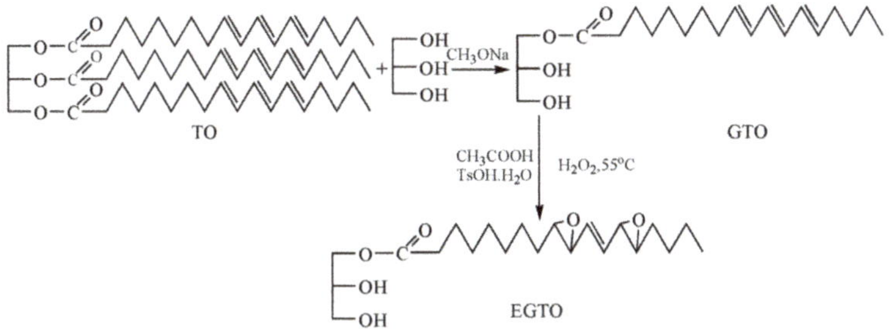

Scheme 1. Synthesis route of tung oil monoglyceride (GTO) and epoxidized monoglyceride of tung oil (EGTO).

2.3. Synthesis of PTOP and PTOSi

The reaction equations for PTOP and PTOSi are shown in Schemes 2 and 3 and the parameters for PTOP and PTOSi are summarized in Table 1. The reaction temperature and time of PTOP were 130 °C and 120 min, respectively. After the reaction, the product was allowed to cool to room temperature and to stand for 48 h, after which a small amount of unreacted DOPO precipitated as a white solid. The brown clear transparent liquid at the top was washed to neutral with distilled water, which was distilled by vacuum distillation. The conversion rate of DOPO was obtained by comparing the change in the epoxy value of PTOP with that of EGTO. EGTO and DPSD were blended and heated to 140 °C for 2 h. After the product was cooled down to room temperature and allowed to stand for 48 h, a small amount of the white solid of the unreacted DPSD subsided, the top brown clear transparent liquid was washed to neutral with distilled water and water was removed by reduced pressure distillation. The conversion rate of PTOSi was obtained by comparing the change in the epoxy value of PTOSi with that of EGTO.

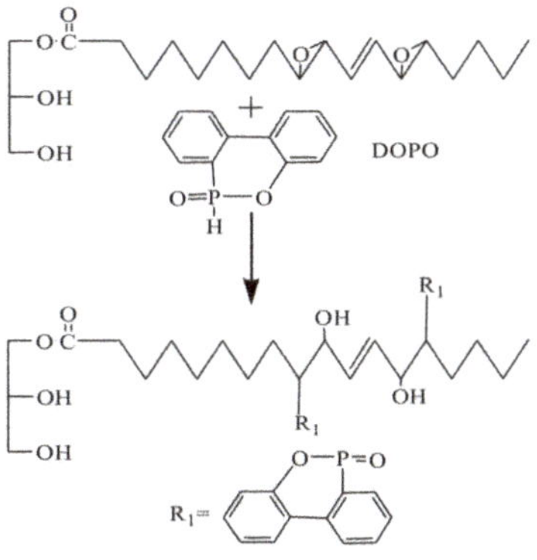

Scheme 2. Synthesis route of phosphorus-containing tung oil-based polyol (PTOP).

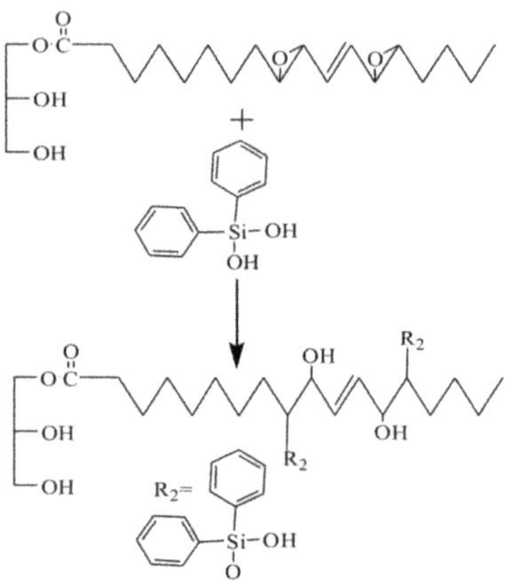

Scheme 3. Synthesis route of silicon-containing tung oil-based polyol (PTOSi).

Table 1. The parameters of PTOP and PTOSi.

Sample	DOPO (mol)	EGTO/Epoxy Group (g/mol)	DPSD (mol)	Conversion Rate (%)	Viscosity (Pa·s)	P (%)	Si (%)	Hydroxyl Value (mg KOH/g)
PTOP	0.072	100/0.1875	0	95.85	2.19	1.85	0	339.96
PTOSi	0	100/0.1875	0.084	95.48	2.42	0	1.95	317.93

2.4. Preparation of RPUF

RPUF was prepared using a one-step method. The raw materials and formulations used in the preparation of RPUF are listed in Table 2. Firstly, Part A was made up of AK8804, DBTDL, blowing agent, water, PPG4110 and flame-retardant polyols. Then, the mixture of PAPI (Part B) and Part A was stirred for 2 min at approximately 2000–2500 rpm to ensure homogeneous mixing. At the end, the mixture was poured into a mold and allowed to cure at 80 °C for 24 h. The P content of blended polyols of RPUF/P60 and RPUF/P80 were 1.11% and 1.48%, respectively, and the Si content of blended polyols of RPUF/Si60 and RPUF/Si80 were 1.14% and 1.52%, respectively.

Table 2. Formulation of rigid polyurethane foam (RPUF).

Sample	Content Weight (g)							
	PPG4110	PTOP	PTOSi	AK8804	DBTDL	Water	HFC-365mfc	PAPI
Neat RPUF	100	0	0	2.4	1.2	1.2	20	120
RPUF/P60	40	60	0	2.4	1.2	1.2	20	120
RPUF/P80	20	80	0	2.4	1.2	1.2	20	120
RPUF/Si60	40	0	60	4.0	0.28	0.17	20	100
RPUF/Si80	20	0	80	4.0	0.11	0.11	20	100

2.5. Characterization

2.5.1. Physical Parameter Analysis

The hydroxyl value and acid value of the sample were determined according to ASTM S957-86 and ASTM D4662-03, respectively. The epoxy value of the sample was tested according to GB/T 1677-2008. The viscosity of the sample was determined based on ASTM D2983 using a rotary Viscometer DVS+ (Brookfield, Middleboro, MA, USA).

2.5.2. Gel Permeation Chromatography (GPC)

The reaction progress of PTOP was tracked by GPC (Waters, Milford, MA, USA). GPC spectra were obtained at a temperature of 30 °C and flow rate of 1 mL/min. The parameters of the column were as follows: mixed PL gel 300 × 718 mm, 25 μm. The solvent was THF.

2.5.3. Fourier Transform Infrared (FTIR)

FTIR spectra of each specimen was measured using a Nicolet iS10 FTIR spectrometer (Madison, WI, USA). The wavenumber range was from 500 to 4000 cm^{-1} at a resolution of 4 cm^{-1}.

2.5.4. Nuclear Magnetic Resonance (NMR)

Nuclear magnetic resonance (^{1}H NMR) spectra and ^{31}PNMR spectra were measured using a BRUKER AV-300 (300MHz) (Bruker BioSpin AG Facilities, Fallanden, Switzerland). Deuterated chloroform was used as the solvent and tetramethylsilane (TMS) and H_3PO_4 were used as internal standards.

2.5.5. Scanning Electron Microscope (SEM)

The microstructure of the RPUF and RPUF residue was determined by SEM (3400N, Hitachi, Tokyo, Japan).

2.5.6. Thermogravimetric Analysis (TGA)

The thermal stability of the specimen was tested using a NETZSCH TG 209F3 (Gebruder-Netzsch-Straße, Germany). Each specimen was heated from 40 to 800 °C under an N2 atmosphere with a flow rate of 20 mL/min and a heating rate of 10 °C/min in alumina crucible.

2.5.7. Differential Scanning Calorimeter (DSC)

The thermal property of RPUF was analyzed by DSC on a NETZSCH DSC 200F3 (Gebruder-Netzsch-Straße, Germany). Samples of about 10 mg were weighed and sealed in the aluminum DSC pans and placed in the DSC cell. They were first cooled for 0.5 min at −60 °C, and then heated from −60 to 200 °C at the rate of 20 °C/min under nitrogen atmosphere. They were kept at 200 °C for 0.5 min to eliminate the previous heat history and subsequently cooled to −60 °C at 20 °C/min. Lastly, they were heated again to 200 °C at 20 °C/min.

2.5.8. Limit Oxygen Index (LOI) Test

The limiting oxygen index test was performed according to ASTM D4986 and using an LOI Meter (PX-01-005, Nanjing Analytical Instrument Factory Co., Ltd, Nanjing, China). The dimensions of the specimen were 100 × 10 × 10 mm^3. Each sample was run twice and the average result was calculated.

2.5.9. Cone Calorimeter Test (CTT)

CTT was carried out by employing a cone calorimeter (FTT2000, Fire Testing Technology, West Sussex, UK) according to ISO5660-1. The dimensions of the sample were 20 × 100 × 100 mm^3, which was enveloped in aluminum foil. FTT2000 provided the external heat flux at 35 kW/m^2 horizontally. Each specimen was detected two times and the average result was obtained.

2.5.10. Compressive Strength Test

The compressive strength of RPUF was measured according to ASTM D1621. It was measured using a CMT4000 universal testing machine (Shenzhen Suns Technology Stock Co., Ltd, Shenzhen, China) where a speed of 2 mm/s of compressive force was adopted and each sample was tested 5 times and the average result was obtained.

2.5.11. Thermal Conductivity Test

The thermal conductivity of RPUF was examined using a thermal conductivity analyzer (JB-DZDR-P, Shanghai Jiubin Instrument Co., Ltd, Shanghai, China) according to ASTM C518. The precision of the sample was $50 \times 50 \times 20$ mm^3.

2.5.12. Density Test

The density of RPUF was measured according to ASTM D1622. The measurement of the sample was $50 \times 50 \times 50$ mm^3, and the density value was obtained from the average of five repeated tests.

3. Results

3.1. Synthesis of PTOP

The reaction progress of PTOP was monitored by GPC at different times. In Figure 1a, the peak at 17.25 min was attributed to DOPO, and the broad peak from 10 min to 15.40 min was assigned to PTOP, while the narrow peak from 15.40 min to 16.50 min was assigned to EGTO. Due to the excessive amount of EGTO, the peak representing EGTO was evident during the PTOP reaction process. The DOPO peak almost disappeared at 2 h, which implies that the reaction between DOPO and EGTO was completed. It can be seen from the Figure 1e,d that no DOPO remained after 2 h at 130 °C and 140 °C. High temperatures may cause other side reactions to occur. Thus, the synthesis of PTOP was conducted for 2 h at a temperature below 130 °C.

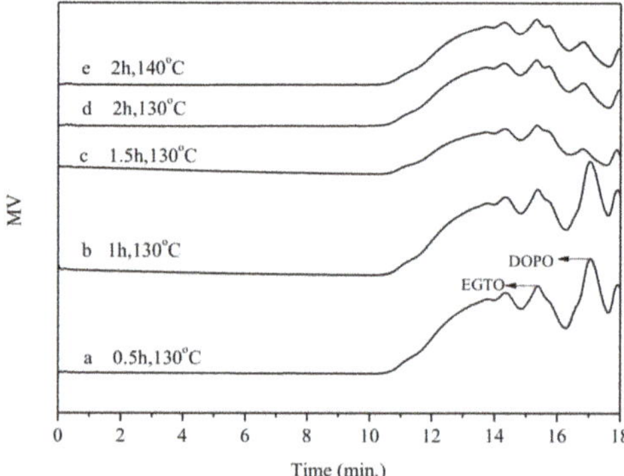

Figure 1. The Gel Permeation Chromatography (GPC) curves of time.

3.2. FTIR of EGTO, PTOP, and PTOSi

Figure 2 presents the FTIR spectra of EGTO, PTOP, and PTOSi. Hydroxyl absorption in EGTO appeared at 3440 cm^{-1}, while the –CH$_2$– stretching vibration peaks appeared at 2930 and 2850 cm^{-1}. The stretching vibration at 900 cm^{-1} belonged to the epoxy group in EGTO [26]. The characteristic peak for P(O)–H stretching of DOPO in PTOP at 2428 cm^{-1} disappeared, while a broad peak appeared at 3340 cm^{-1}, indicating that the epoxy ring in EGTO opened and a hydroxyl group was formed. Simultaneously, the characteristic –OH peak in PTOP occurred around 3440 cm^{-1}, while the absorption peak at 1590 cm^{-1} was attributed to P–Ph, and the absorbing peaks of P=O appeared at 1232 and 1198 cm^{-1}. The absorption peaks around 1043 and 918 cm^{-1} corresponded to P–O–C and P–O–Ph vibrations, which are characteristics of EGTO and DOPO, respectively [27]. The absorption peaks on the FTIR spectrum of PTOSi were consistent with observations from the literature [28]. The presence of a pronounced peak at 1450 cm^{-1} characteristic of benzene was significant. Peaks at 1070–1130 cm^{-1} corresponded to C–O and Si–O bonding [22], and the stretching vibration of two peaks at 880 and

730 cm^{-1} were assigned to the Si–C and C–H bonds of benzene. All the above changes confirmed that PTOP and PTOSi were formed via the ring opening reaction.

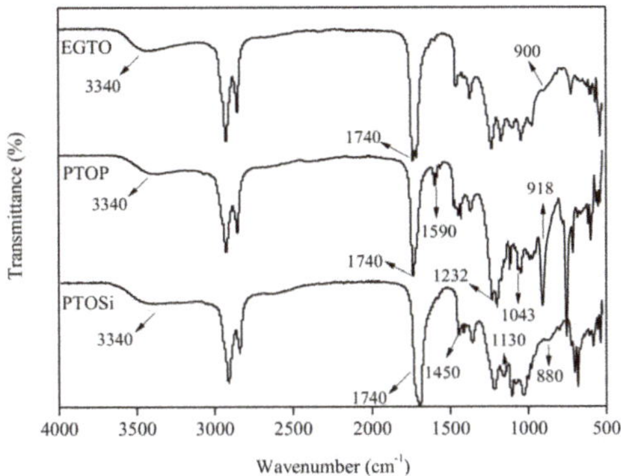

Figure 2. Fourier transfer infrared (FTIR) spectra of EGTO, PTOP, and PTOSi.

3.3. ^{1}H NMR and ^{31}P NMR

Figure 3a presents peaks at 0.9 ppm and 2.5 ppm, which were attributed to methyl protons (peak 1) and fatty acid protons [(–CH$_2$) –CO–] (peak 5), respectively. While protons in –CH$_2$– (peak 6) from glycerol appeared from 4.0 to 4.1 ppm, the peak at 5.3 to 5.4 ppm was attributed to the carbon–carbon double bond. The peak at 3.3 ppm in the EGTO spectra indicates that the epoxy group (peak 3) was formed after epoxidation. The peak at 1.9 ppm corresponded to the hydrogen (peak 8) attached to the carbon adjacent to –OH and the peak that appeared at 2.0 ppm represented the hydrogen atoms adjacent to the ester bond [–O–CO–] (peak 7). In Figure 3b, the chemical shifts at 7.2–8.2 ppm (peak 10) indicated the Ar–H of PTOP, while the absorption peaks at 8.65 ppm (peak 11) and 3.60 ppm (peak 12) represented the P–C–H and –OH of PTOP, respectively. All of these characteristic ^{1}H NMR bands matched the PTOP structure. In Figure 3c, the peaks at 7.2–8.2 ppm (peak 14) were associated with the Ar–H of PTOSi, while the peak at 3.4 ppm (peak 17) was characteristic of the –OH of DPSD that formed from the epoxy group via a ring opening reaction. The peak at 2.8 ppm (peak 16) was assigned to Si–O–CH–, while the –Si–OH peak appeared at 2.66 ppm (peak 15). Furthermore, PTOP exhibits a single peak at 35.50 ppm while there is no peak at 14.90 ppm in the ^{31}PNMR spectrum (Figure 4). The chemical shift at 14.90 ppm was attributed to DOPO, which reveals that the reaction between DOPO and EGTO was completed.

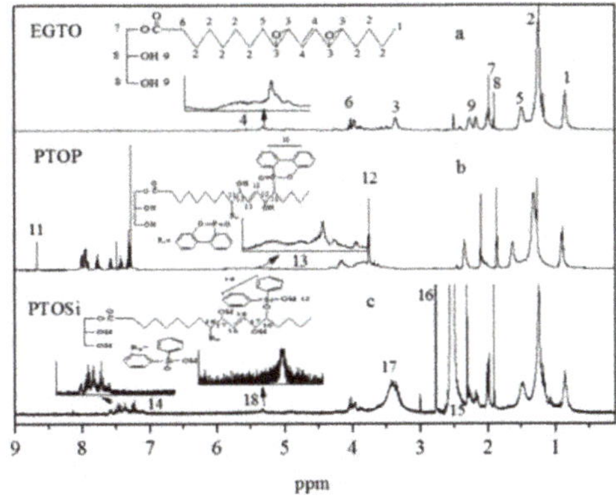

Figure 3. The ^{1}HNMR spectra of EGTO, PTOP, and PTOSi.

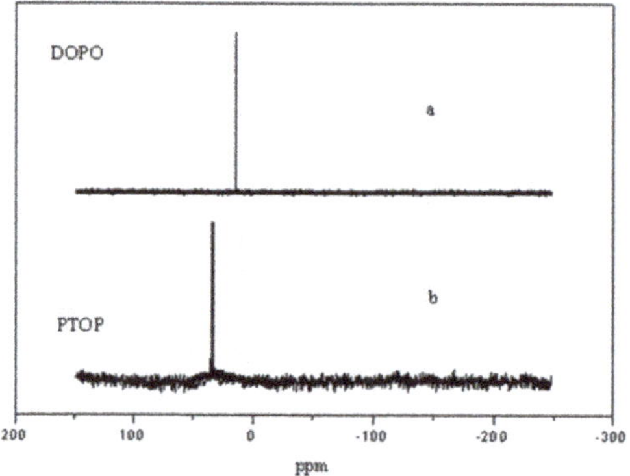

Figure 4. The ^{31}P NMR of DOPO and PTOP.

3.4. SEM Micrographs of RPUF

The SEM micrographs of RPUF are presented in Figure 5, where the main structural areas noted are the cells, cell walls, and wall joints [29]. The size range of the cells increased with an increase in PTOP content. This phenomenon was also observed for RPUF/Si60 and RPUF/Si80 with an increase in PTOSi content. It was noted that a number of foam cell orifices and the size of regular cells in RPUF/P60 were consistent with those in RPUF/Si60. However, for RPUF/P80 and RPUF/Si80, the number of cell windows is decreased. The insufficient cross-linking action of PTOP and PTOSi with low functionality leads to a decrease in the number of closed foam cells, and the collapse of the newly formed foam structure due to the decrease in the number of closed foam cells.

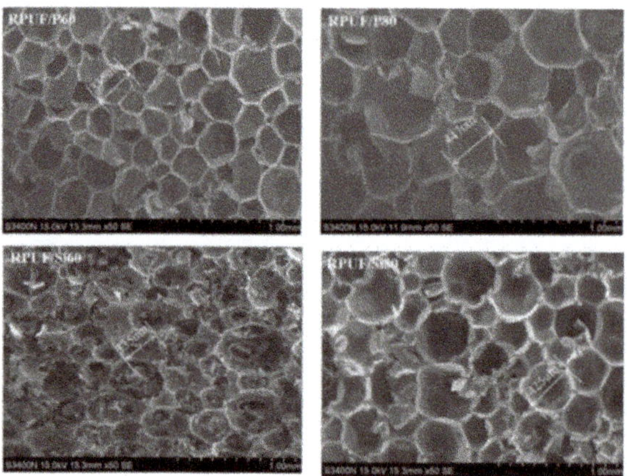

Figure 5. SEM of rigid polyurethane foam (RPUF).

3.5. Thermal Property of PTOP, PTOSi, and RPUF

3.5.1. Thermal stability of PTOP, PTOSi, and RPUF

Figures 6–8 show the TG and DTG curves of PTOP, PTOSi, and RPUF, respectively. T_{onset} represents the temperature at 10% weight loss, T_{max} represents the temperature when the weight loss rate was maximum, and the residue rate was obtained at 600 °C. These parameters are summarized in Tables 3 and 4.

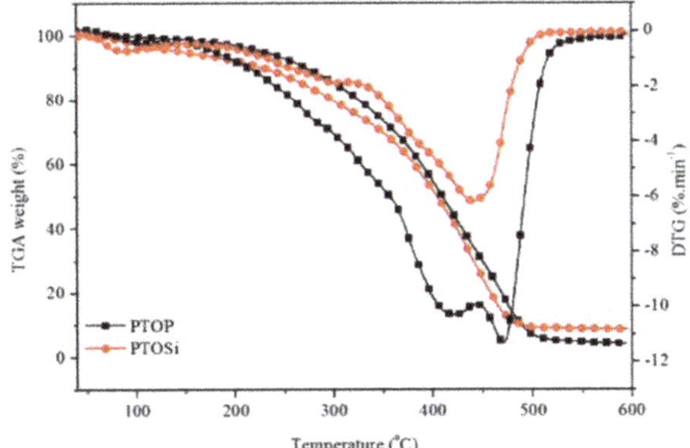

Figure 6. Thermogravimetric (TG) and DTG curves for PTOP and PTOSi.

Table 3. The parameters of TG and DTG of PTOP and PTOSi.

Sample	N$_2$ Atmosphere		The Maximum Decomposition Rate (%/min)	Residue Rate (wt %)
	T_{onset} (°C)	T_{max} (°C)		
PTOP	272.40	468.10	11.31	4.13
PTOSi	228.20	440.10	6.18	8.58

As can be seen from Figure 6, PTOP exhibited a higher T_{onset} than PTOSi, which was attributed to the low temperature of the initial DPSD degradation. Moreover, its T_{max} was higher than that of PTOSi, while the residue yield of PTOP was less than that of PTOSi. Silicon retained in the chemical bond could be acted as a barrier that prevents the transfer of heat and mass in the condensed phase [30]. These results indicate that the thermal property of PTOP is higher than that of PTOSi.

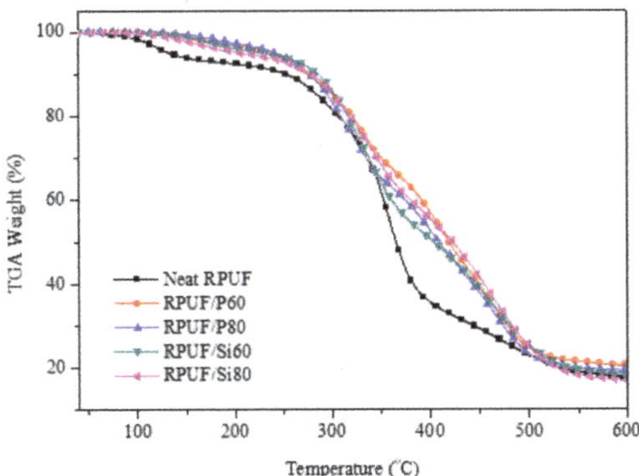

Figure 7. TG curves for RPUF.

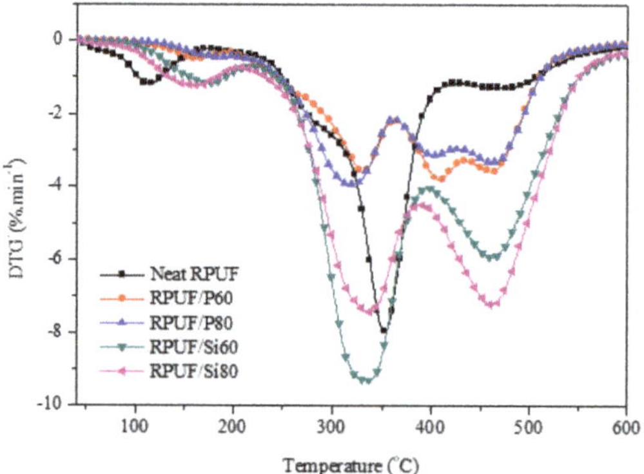

Figure 8. DTG curves for RPUF.

Table 4. The parameters of TG and DTG of RPUF.

Sample	T_{onset} (°C)	T_{max} (°C)			Residue Rate (wt %)
		Step I	**Step II**	**Step III**	
Neat RPUF	313.10	354.30	478.10	/	17.54
RPUF/P60	282.60	330.90	408.70	461.40	20.48
RPUF/P80	261.80	319.60	402.30	460.60	19.20
RPUF/Si60	291.40	335.10	463.70	/	18.03
RPUF/Si80	288.50	338.60	464.50	/	17.00

Thermal degradation of RPUF/P60 and RPUF/P80 showed four stages at 100 to 200 °C, 200 to 350 °C, 350 to 450 °C, and 450 to 600 °C, respectively. The first stage was attributed to the entrapment of volatile compounds in closed cells, while the other stages were attributed to the hard and soft segments of polyurethane. With an increase in the PTOP content of RPUF, the residue rate at 600 °C after complete decomposition becomes smaller, indicating that the degradation products of DOPO have mainly decomposed to form volatile products. The initial decomposition temperature of RPUF/P80 was lower than that of RPUF/P60, and the decreased T_{onset} could be attributed to the prior chemical decomposition of DOPO compared to PPG4110. Step I of T_{max} was associated with the urethane unit, which was considered as the hard segment in the polyurethane network. Step II of T_{max} was attributed to the soft segments, which were mainly made up of long chains of PPG4110 and PTOP. While step III of T_{max} was affected by the aromatic components in RPUF, the aromatic components were released within the temperature range of 400 to 600°C [31]. The condensation, chain scission, and cross-linking reactions of PTOP and the decomposition mechanism of PTOP in RPUF were summarized in that acid source, and polyphosphate compound could be produced via degradation in the step III.

Two stages showed in the degradation process of neat RPUF, RPUF/Si60, and RPUF/Si80. The initial thermal temperature and T_{max} of neat RPUF were highest among the five samples, due to the hydroxyl functionality of PPG4110 was higher than those for PTOP and PTOSi, the relative molecular weight and the cross-linking yield of neat RPUF were highest. While the maximum decomposition rate was higher than these for other samples, results reveal that the neat RPUF was degraded quickly when the temperature was higher than the initial thermal temperature.

The initial thermal decomposition and residue rates of RPUF/Si60 and RPUF/Si80 were similar to those for RPUF/P60 and RPUF/P80, while the change in following characteristic parameters such as T_{max} were contrary. The degradation of RPUF/Si60 and RPUF/Si80 was divided into two steps. The original degradation was ascribed to the urethane unit in the cross-linking network. Substances containing benzene structures and Si-containing groups in the RPUF were degraded when the temperature was raised.

3.5.2. DSC of RPUF

Thermal property of RPUF was also studied by DSC, the corresponding heating thermograms are shown in Figure 9. The molecular chain of polyurethane foam was generally composed of the soft segment and hard segment. At the room temperature, soft segment was in an elastomeric state, which has an effect on low temperature performance, flexibility, organic solvent resistance and the weather-ability on RPUF. While the hard segment was in a glassy or crystalline state, which affects the melting point, thermal stability, hardness and modulus of RPUF. The temperature range of glassy transition of RPUF was from 29 to 102 °C, the T_g of RPUF/P60, RPUF/P80, RPUF/Si60, and RPUF/Si80 were 70.72, 65.08, 73.17, and 77.95 °C, respectively. The soft segment was made up of the tung oil-based flame retardant polyols and PPG4110, the hard segment was made up of isocyanate. The reason why the T_g of RPUF was less when increasing the amount of PTOP was that PTOP could be acted as the soft segment. On the contrary, with increasing the amount of PTOSi, the T_g of RPUF was higher due to PTOSi could be acted as soft segment.

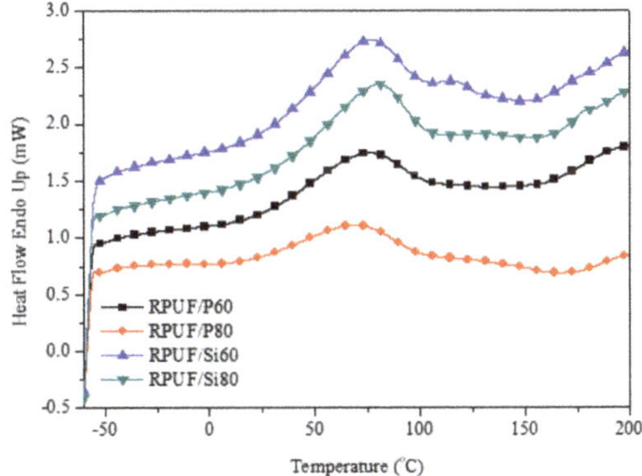

Figure 9. Differential Scanning Calorimetry (DSC) of RPUF.

3.6. Flame Retardancy

3.6.1. Limiting Oxygen Index (LOI) Test of RPUF

Air usually contains almost 21% O_2, so materials with an LOI value of 21% or less will burn quickly in air [6]. With the accession of PTOP and PTOSi, the LOI value of RPUF increased, indicating that both of these are flame retardant polyols. The LOI values for RPUF are listed in Table 5, where it can be seen that they are higher than that of neat RPUF, which is 19.0%. With increasing amounts of PTOP and PTOSi, the LOI value increased from 21.3% for RPUF/P60 to 24.0% for RPUF/P80, while a lower increase in the LOI value was observed for RPUF/Si60 and RPUF/Si80 from 21.0% to 23.4%. This result was attributed to the large number of PO^-, PO_2^-, and $P_2O_4^-$. Free radicals that were generated and eliminated the combustible gas during the gaseous phase. Meanwhile, phosphate and its derivatives were formed from PO^-, PO_2^-, and $P_2O_4^-$, and served as a barrier to control the transfer of heat and combustible gas during the burning process [24]. During the combustion of RPUF/Si60 and RPUF/Si80, silicones can effectively shield via silica residue left when under oxygen at elevated temperatures [32]. Silica plays the role of an "insulating blanket" as a barrier to delay volatile mass transport during the decomposition. In Table 5, the LOI value of RPUF. Based on the LOI value of RPUF, PTOP showed better flame-retardant performance than that of PTOSi.

Table 5. RPUF Limiting Oxygen Index (LOI) values.

Sample	Neat RPUF	RPUF/P60	RPUF/P80	RPUF/Si60	RPUF/Si80
LOI value (%)	19.0	21.3	24.0	21.0	23.4

3.6.2. Fire Behaviors of RPUF

The heat release rate (HRR), peak of heat release rate (PHRR), time to ignition (TII), and total heat released (THR) measured for RPUF during the cone calorimetry test are presented in Table 6.

Table 6. Cone calorimeter testing data for RPUF.

Sample	TII (s)	PHRR (kW·m^{-2})	Time to PHRR (s)	THR (MJ·m^{-2})
RPUF/P60	2	266.03	35, 75	22.78
RPUF/P80	3	259.38	40, 60	18.40
RPUF/Si60	2	285.36	30, 60	22.80
RPUF/Si80	3	255.20	40, 65	21.99

Very similar HRR curves are obtained for RPUF in Figure 10 when the PTOP and PTOSi content is increased, where the curves are seen to decrease but then quickly increase to attain the second HRR peak. RPUF ignited within the first few seconds (about 2–3 s) of exposure to heat burned steadily for 175 to 230 s. The first sharp peaks for RPUF were turned up at 35, 40, 30, and 40 s, respectively. Then, narrow peaks for each RPUF appeared at 75, 60, 60, and 65 s. The protective char layer formed from the first peak in the HRR curves, then the protective char layer was degraded, which is represented by the second peak was [33]. The HRR peaks for RPUF/P80 and RPUF/Si80 were slightly smaller and thinner, while the peak times were lengthened with an increase in the amount of PTOP and PTOSi. This may be attributed to the introduction of PTOP and PTOSi into the soft segments, contributing to the formation of the protective and continuous char layer.

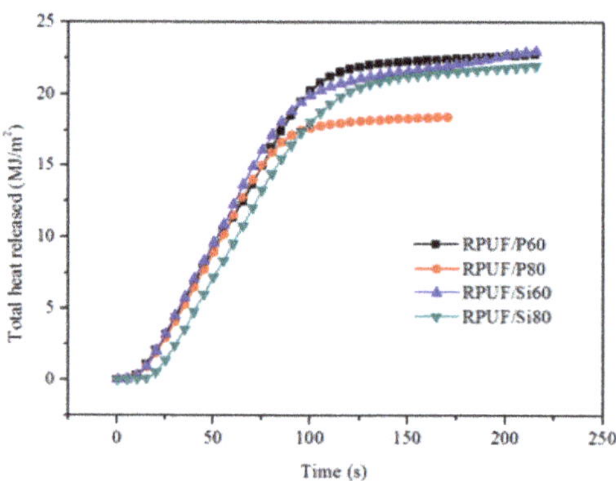

Figure 10. Heat release rate (HRR) curves of RPUF.

THR curves in Figure 11 illustrate that the value of THR decreases in the order of RPUF/P60>RPUF/Si60>RPUF/Si80>RPUF/P80. For example, the THR of RPUF/P80 is 18.40 MJ·m^{-2}, a 19.30% reduction compared to that of RPUF/P60. There are two reasons for this result. On the one hand, heat is taken away by the condensed-phase and gas-phase actions of the volatile aromatic phosphates in RPUF/P60 and RPUF/P80. On the other hand, the barrier layer produced by PTOP and PTOSi can prevent heat, oxygen and flammable substance from transferring. Moreover, non-flammable gas in the gas phase system can also effectively reduce the heat released from RPUF/P80 [34]. In comparing to RPUF/Si60, RPUF/Si80 has the preferable flame-retardant property indicated by both the LOI value and CTT. The excellent flame-retardant behavior of PTOSi was illustrated by its better dispersion within the RPUF and movement towards the RPUF surface during the burning process. A highly flame-retardant char generated from the combustion of PTOSi can reduce the total heat released.

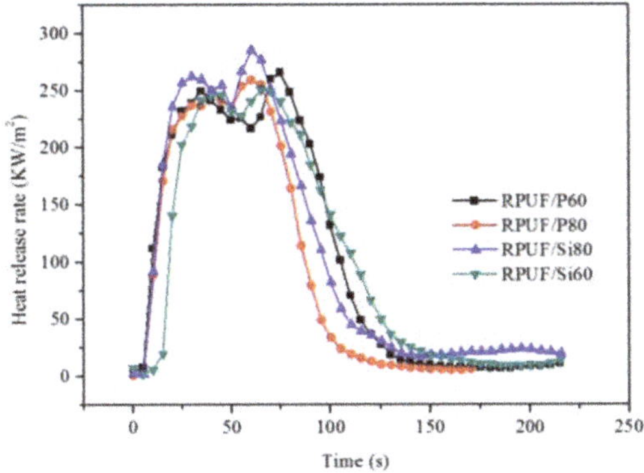

Figure 11. Total heat released (THR) curves of RPUF.

Based on related parameters such as PHRR and THR presented in Table 6, the flame-retardant efficiency of PTOP is higher than that of PTOSi. This result accounts for the fact that PTOP can migrate to the surface of the RPUF during combustion and allow for the formation of the condensed-phase and gas-phase of volatile aromatic phosphates. This indicates that the addition of PTOP and PTOSi can boost the fire-inhibition activity during combustion.

3.6.3. SEM and EDAX of RPUF Residue

To confirm that P-containing compounds and Si-containing compounds were retained in the residue and acted as the flame retardant barrier. The microstructure of the residue after combustion was examined by SEM, RPUF/P60, RPUF/P80, RPUF/Si60, and RPUF/Si80 had the compacted and smooth char. EDAX (energy dispersive spectroscopy) associated with SEM was used to get the elemental composition of residues of RPUF. Table 7 shows the surface composition of the residues obtained by EDAX. As can be observed from Table 7, C and O were the main remaining elements. The char layer of RPUF/P60 and RPUF/P80 can compress well to form a protective phosphatic barrier to resist the transfer of heat and mass during fire. As shown in Figure 12, a number of flaws were found on the surface of RPUF/Si60 and RPUF/Si80, which may be the result of SiO_2 forming and moving onto the surface during burning. PTOP and PTOSi have the effect of promoting the char layers, as compacted char layers may prevent the internal products of pyrolysis from transferring to the flame zone.

Table 7. EDAX data of the residue of RPUF.

Sample	C (wt %)	O (wt %)	P (wt %)	Si (wt %)	Others (wt %)
RPUF/P60	79.67	17.72	1.83	0	0.78
RPUF/P80	83.38	14.35	0.83	0	0.84
RPUF/Si60	77.37	20.61	0	1.13	0.89
RPUF/Si80	79.61	17.08	0	2.31	1.01

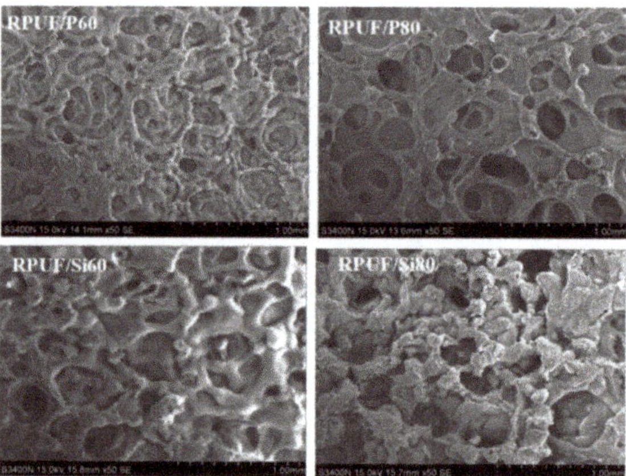

Figure 12. SEM of RPUF residue.

3.7. Mechanical Property of RPUF

Mechanical property is an essential issue for the practical application of RPUF, which was further investigated, such as density, compressive strength and thermal conductivity. In order to eliminate the effect of RPUF density on the compressive strength, the specific compressive strength (compressive strength/density) was used to for comparison.

Table 8 shows that the trend in RPUF density is opposite to the trend noted for compressive strength. With large amounts of PTOP and PTOSi added into the blended polyols component, the PTOP and PTOSi may disperse poorly, which can interfere with the development of the cell and make the cell size larger and the number of closed cells lower (Figure 5). PTOP can reduce the compressive strength of RPUF, while the higher content of PTOSi incorporated into RPUF can increase the compressive strength. A similar result was obtained for σ_{sp}. When the PTOP content was below 60%, PTOP dispersed well in the blended polyols component, and almost no agglomeration appeared. PTOSi with its self-crosslinking function can make the RPUF matrix easily collapsible due to the wall fragility of the foam cell, resulting in the worse value of compressive strength.

Table 8. Physical and mechanical properties of RPUF.

Sample	ρ (kg·m^{-3})	σ (MPa)	σ_{sp} (N.m.kg^{-1})	K (W·mk^{-1})
RPUF/P60	97.08 ± 0.44	0.82 ± 0.02	8446	0.039
RPUF/P80	89.48 ± 0.80	0.75 ± 0.03	8381	0.038
RPUF/Si60	57.09 ± 0.38	0.18 ± 0.02	3152	0.051
RPUF/Si80	52.94 ± 0.32	0.25 ± 0.02	4722	0.044

ρ: density; σ: compressive strength; σ_{sp}: specific compressive strength; K: thermal conductivity.

RPUF is also used as the thermal insulation material in building insulation. The best method for improving the sustainability of building insulation is to diminish the thermal conductivity of RPUF. With an increase in the PTOP and PTOSi content, the thermal conductivity of RPUF decreased as the thermal conductivity is mainly affected by the average cell size. The thermal conductivity of RPUF/Si60 and RPUF/Si80 were higher than that of RPUF/P60 and RPUF/P80 due to the lower closed cell content, which is influenced by the degree of cross-linking, the equivalent weight of blended polyols and the content of various agents. Therefore, the molecular weight of polyurethane in RPUF/P60 and RPUF/P80 was lower than that for RPUF/Si60 and RPUF/Si80.

4. Conclusions

A novel method was developed to synthesize PTOP and PTOSi, which were proved to be effective flame-retardant polyols used in the formation of RPUF with excellent flame retardancy. Different PTOP and PTOSi contents affect the thermal stability, flame retardancy, and mechanical properties of RPUF. Results from FTIR, TG, DSC, and LOI revealed that DOPO and DPSD are linked to EGTO by a covalent bond, and the thermal stability of PTOP was higher than that for PTOSi. The thermal stability and flame retardancy of RPUF increased, with the LOI value of RPUFs with 80% PTOP and PTOSi reaching 24.0% and 23.4%, respectively. PTOP and PTOSi could be acted as the soft segment and hard segment in RPUF, respectively. The compressive strength of RPUF was reduced to 0.75 MPa with 80 wt % PTOP, while the compressive strength of RPUF increased to 0.25 MPa with 80 wt % PTOSi. A comparison of the flame-retardant mechanism of RPUF prepared by PTOP and PTOSi showed that RPUF with PTOP released an acid source and polyphosphate compound while RPUF with PTOSi burned out silicon dioxide on the surface. Based on the LOI value of RPUF, PTOP showed better flame-retardant performance than that of PTOSi.

Author Contributions: W.Z. and M.Z. conceived and designed the experiments; C.B., P.J., and Y.Z. performed the experiments and analyzed the data; W.Z. wrote the paper.

Funding: This research was funded by the National Natural Science Foundation of China grant number [31670577 and 31670578].

Conflicts of Interest: The authors declare no conflict of interest.

References

1. Sylwia, C.; Massimo, F.B.; Jan, K.; Anna, S.; Marcin, M.; Krzysztof, S. Linseed oil as a natural modifier of rigid polyurethane foams. *Ind. Crops Prod.* **2018**, *115*, 40–51.
2. Lligadas, G.; Ronda, J.C.; Galia, M.; Cadiz, V. Renewable polymeric materials from vegetable oils: A perspective. *Mater. Today* **2013**, *16*, 337–343. [CrossRef]
3. Athanasia, A.S.; David, A.C.E.; Celine, C.; Darren, J.M.; Pratheep, K.A. A systematic study substituting polyether polyol with palm kernel oil based polyester polyol in rigid polyurethane foam. *Ind. Crops Prod.* **2015**, *66*, 16–26.
4. Zhou, X.; Sain, M.M.; Oksman, K. Semi-rigid biopolyurethane foams based on palm-oil polyol and reinforced with cellulose nanocrystals. *Compos. Part A Appl. Sci. Manuf.* **2015**, *83*, 56–62. [CrossRef]
5. Zieleniewska, M.; Leszczynski, M.K.; Kuranska, M.; Prociak, A.; Szczepkowski, L.; Krzyzowska, M.; Ryszkowska, J. Preparation and characterisation of rigid polyurethane foams using a rapeseed oil-based polyol. *Ind. Crops Prod.* **2015**, *74*, 887–897. [CrossRef]
6. Shirke, A.G.; Dholakiya, B.Z.; Kuperkar, K. Modification of tung oil-based polyurethane foam by anhydrides and inorganic content through esterification process. *J. Appl. Polym. Sci.* **2017**, *135*, 45786. [CrossRef]
7. Wang, C.; Wu, Y.C.; Li, Y.C.; Shao, Q.; Yan, X.G.; Han, C.; Wang, Z.; Liu, Z.; Guo, Z.H. Flame-retardant rigid polyurethane foam with a phosphorus-nitrogen single intumescent flame retardant. *Polym. Adv. Technol.* **2018**, *29*, 668–676. [CrossRef]
8. Shi, X.X.; Jiang, S.H.; Zhu, J.Y.; Li, G.H.; Peng, X.F. Establishment of a highly efficient flame-retardant system for rigid polyurethane foams based on biphase flame-retardant actions. *RSC Adv.* **2018**, *8*, 9985–9995. [CrossRef]
9. Shi, S.Y.; Neisius, M.; Mispreuve, H.; Naescher, R.; Gaan, S. Flame retardancy and thermal decomposition of flexible polyurethane foams: Structural influence of organophosphorus compounds. *Polym. Degrad. Stab.* **2012**, *97*, 2428–2440.
10. Thirumal, M.; Khastgir, D.; Nando, G.B.; Naik, Y.P.; Singha, N.K. Halogen-free flame retardant PUF: Effect of melamine compounds on mechanical, thermal and flame retardant properties. *Polym. Degrad. Stab.* **2010**, *95*, 1138–1145. [CrossRef]
11. Chattopadhyay, D.K.; Webster, D.C. Thermal stability and flame retardancy of polyurethanes. *Prog. Polym. Sci.* **2009**, *34*, 1068–1133. [CrossRef]

12. Laoutid, F.; Bonnaud, L.; Alexandre, M.; Lopez-Cuesta, J.M.; Dubosi, P. New prospects in flame retardant polymer materials: From fundamentals to nanocomposites. *Mater. Sci. Eng. R Rep.* **2009**, *63*, 100–125. [CrossRef]

13. Perret, B.; Schartel, B.; Stob, K.; Ciesielski, M.; Diederichs, J.; Doring, M.; Kramer, J.; Altstadt, V. Novel DOPO-based flame retardants in high-performance carbon fibre epoxy composites for aviation. *Eur. Polym.* **2011**, *47*, 1081–1089. [CrossRef]

14. Li, J.L.; Li, Z.W.; Wang, H.G.; Wu, Z.J.; Wang, Z.; Li, S.C. Liquid oxygen compatibility and cryogenic mechanical properties of a novel phosphorous/silicon containing epoxy-based hybrid. *RSC Adv.* **2016**, *6*, 91012–91023. [CrossRef]

15. Gaan, S.; Liang, S.Y.; Mispreuve, H.; Perler, H.; Naescher, R. Flame retardant flexible polyurethane foams from novel DOPO-phosphonamidate additives. *Polym. Degrad. Stab.* **2015**, *113*, 180–188. [CrossRef]

16. Granzow, A. Flame retardation by phosphorus compounds. *Acc. Chem. Res.* **1978**, *11*, 177–183. [CrossRef]

17. Zhang, M.; Luo, Z.Y.; Zhang, J.W.; Chen, S.G.; Zhou, Y.H. Effects of a novel phosphorus-nitrogen flame retardant on rosin-based rigid polyurethane foams. *Polym. Degrad. Stab.* **2015**, *120*, 427–434. [CrossRef]

18. Liu, Y.L.; He, J.Y.; Yang, R.J. The thermal properties and flame retardancy of 9, 10-dihydro-9-oxa-10-phosphaphenanthrene-10-oxide (10)-Mg/Polyisocyanate -polyurethane foam composites. *Bull. Chem. Soc. Jpn.* **2016**, *89*, 779–785. [CrossRef]

19. Shi, Y.C.; Wang, G.J. The novel silicon-containing epoxy/PEPA phosphate flame retardant for transparent intumescent fire resistant coating. *Appl. Surf. Sci.* **2016**, *385*, 453–463. [CrossRef]

20. Wang, J.H.; Ji, C.T.; Yan, Y.T.; Zhao, D.; Shi, L.Y. Mechanical and ceramifiable properties of silicone rubber filled with different inorganic fillers. *Polym. Degrad. Stab.* **2015**, *121*, 149–156. [CrossRef]

21. Bian, X.C.; Tang, J.H.; Li, Z.M. Flame retardancy of whisker silicon oxide/rigid polyurethane foam composites with expandable graphite. *J. Appl. Polym. Sci.* **2010**, *110*, 3871–3879. [CrossRef]

22. Zhou, H.F.; Wang, H.; Tian, X.Y.; Zheng, K.; Cheng, Q.T. Effect of 3-Aminopropyltriethoxysilane on polycarbonate based waterborne polyurethane transparent coatings. *Prog. Org. Coat.* **2014**, *77*, 1073–1078. [CrossRef]

23. Zhang, L.; Wang, Y.C.; Liu, Q.; Cai, X.F. Synergistic effects between silicon-containing flame retardant and DOPO on flame retardancy of epoxy resins. *J. Therm. Anal. Calorim.* **2016**, *123*, 1343–1350. [CrossRef]

24. Li, M.L.; Zhong, Y.; Wang, Z.; Fischer, A.; Ranft, F.; Drummer, D.; Wu, W. Flame retarding mechanism of polyamide 6 with phosphorus-nitrogen flame retardant and DOPO derivatives. *J. Appl. Polym. Sci.* **2016**, *133*. [CrossRef]

25. Zhou, W.; Jia, P.Y.; Zhou, Y.H.; Zhang, M. Preparation and characterization of tung oil-based flame retardant polyols. *Chin. J. Chem. Eng.* **2018**. [CrossRef]

26. Zhang, Y.C.; Xu, G.L.; Liang, Y.; Yang, J.; Hu, J. Preparation of flame retardant epoxy resins containing DOPO group. *Thermochim. Acta* **2016**, *643*, 33–40. [CrossRef]

27. Xiong, Y.Q.; Zhang, X.Y.; Liu, J.; Li, M.M.; Guo, F.; Xia, X.N.; Xu, W.J. Synthesis of novel phosphorus-containing epoxy hardeners and thermal stability and flame-retardant properties of cured products. *J. Appl. Polym. Sci.* **2012**, *125*, 1219–1225. [CrossRef]

28. Han, M.S.; Kim, Y.H.; Han, S.J.; Choi, S.J.; Kim, S.B.; Kim, W.N. Effects of a silane coupling agent on the exfoliation of organoclay layers in polyurethane/organoclay nanocomposite foams. *J. Appl. Polym. Sci.* **2008**, *110*, 376–386. [CrossRef]

29. Chen, M.J.; Chen, C.R.; Tan, Y.; Huang, J.Q.; Wang, X.L.; Chen, L. Inherently flame-retardant flexible polyurethane foam with low content of phosphorus-containing cross-linking agent. *Ind. Eng. Chem. Res.* **2014**, *53*, 1160–1171. [CrossRef]

30. Wu, K.; Song, L.; Hu, Y.; Lu, H.D.; Kandola, B.K.; Kandare, E. Synthesis and characterization of a functional polyhedral oligomeric silsesquioxane and its flame retardancy in epoxy resin. *Prog. Org. Coat.* **2009**, *65*, 490–497. [CrossRef]

31. Zhang, W.C.; Li, X.M.; Yang, R.J. Pyrolysis and fire behavior of epoxy resin composites based on a phosphorus-containing polyhedral oligomeric silsesquioxane (DOPO-POSS). *Polym. Degrad. Stab.* **2011**, *96*, 1821–1832. [CrossRef]

32. Hamdani, S.; Longuet, C.; Perrin, D.; Lopez-cuesta, J.M.; Ganachaud, F. Flame retardancy of silicone-based materials. *Polym. Degrad. Stab.* **2009**, *94*, 465–495. [CrossRef]

33. Wu, X.; Wang, L.; Wu, C.; Yu, J.; Xie, L.; Wang, G.; Jiang, P. Influence of char residues on flammability of EVA/EG, EVA/NG and EVA/GO composites. *Polym. Degrad. Stab.* **2012**, *97*, 54–63. [CrossRef]
34. Zhao, B.; Chen, L.; Long, J.W.; Chen, H.B.; Wang, Y.Z. Aluminum hypophosphite versus alkyl-substituted phosphinate in polyamide 6: Flame retardance, thermal degradation, and pyrolysis behavior. *Ind. Eng. Chem. Res.* **2013**, *52*, 2875–2886. [CrossRef]

Article

Effect of Evening Primrose Oil-Based Polyol on the Properties of Rigid Polyurethane–Polyisocyanurate Foams for Thermal Insulation

Joanna Paciorek-Sadowska *, Marcin Borowicz *, Bogusław Czupryński and Marek Isbrandt

Department of Chemistry and Technology of Polyurethanes, Technical Institute, Faculty of Mathematics, Physics and Technical Science, Kazimierz Wielki University, J. K. Chodkiewicza Street 30, 85-064 Bydgoszcz, Poland; czupr@ukw.edu.pl (B.C.); m.isbrandt@ukw.edu.pl (M.I.)
* Correspondence: sadowska@ukw.edu.pl (J.P.-S.); m.borowicz@ukw.edu.pl (M.B.); Tel.: +48-34-19-289 (J.P.-S.); +48-34-19-286 (M.B.)

Received: 5 November 2018; Accepted: 30 November 2018; Published: 3 December 2018

Abstract: The article presents the results of research on the synthesis of a new biopolyol based on evening primrose oil, and its use in the production of rigid polyurethane–polyisocyanurate foams intended for thermal insulation. The obtained biopolyol was subjected to analytical, physicochemical, and spectroscopic tests (Fourier transform infrared (FTIR), ^{1}H NMR, ^{13}C NMR) to confirm its suitability for the synthesis of polyurethane materials. Then, it was used for the partial replacement of the petrochemical polyol in the polyurethane formulation. Obtained rigid polyurethane–polyisocyanurate foams are characterized by a lower apparent density, brittleness, water absorption, and thermal conductivity coefficient λ. In addition, foams modified by biopolyols had a higher content of closed cells and higher aging resistance. The results of the conducted research showed that the use of the biopolyol based on evening primrose oil may be an alternative to petrochemical polyols. The research presented herein is perfectly consistent with the trends of sustainable development and the philosophy of green chemistry.

Keywords: evening primrose oil; biopolyol; synthesis; polyurethane–polyisocyanurate foam; foam properties; thermal insulation

---

1. Introduction

Nowadays, the interest of producers in the chemical industry focuses on technologies using environmental-friendly raw materials [1,2]. In the case of polymer materials, particular emphasis has been placed on the impact of raw materials and products on man and the surrounding environment since the beginning of the 1980s [3]. Current trends in the chemical industry are directed at the use of sustainable products, the addition of raw materials from renewable sources, and the solution of the waste problem at the production planning stage. The application of renewable raw materials in the polyurethane (PU) industry is mainly based on synthesis of new polyol compounds. A wide and huge group of renewable raw materials for the synthesis of polyol compounds (biopolyols) for the production of PU materials are vegetable oils. The most commonly used vegetable oils include: soybean, palm, coconut, linseed, castor, rapeseed, and sunflower oils. This group of raw materials includes, above all, the esters of higher unsaturated fatty acids (oleic, linoleic, linolenic) and glycerol. They also contain a few percent of saturated acids (stearic and palmitic) [4]. However, the molecules of oil must be chemically modified to unlock or incorporate reactive hydroxyl groups, which are capable of reacting with isocyanate groups. The methods of chemical modification of double bonds include reactions of: epoxidation, hydrogenation, oxidation, and the halogenation of unsaturated bonds [5–9]. Only such modification ensures obtaining a biopolyol with properties that enable its application in industry [10]. These methods are developed and used in the polyurethane industry, because they are quite cheap. Raw materials (vegetable oils), which are needed in these technologies, are easily available. Some oils, e.g., linseed or mustard oil, have a higher content of unsaturated bonds; therefore the biopolyols that are based on them have a higher hydroxyl number, and the obtained PU foams are more rigid. Others oils, e.g., olive oil, have a lower content of reactive groups, and obtained polyurethane foams can be more flexible [11]. The properties of the obtained polyurethane material can be controlled by selecting appropriate raw materials [12–19].

A large group of polyurethanes is rigid polyurethane foams. These PU materials are a modern and safe method of thermal and acoustic insulation, due to the ease of application. They enable the exact filling of the space, which will be thermally insulated. They are most often used in light construction as insulation materials. They are also used for the production of cores in sandwich panels and for insulating external installations [20]. The functional requirements of PU foams are connected with their most important application (as good and safe thermal insulation materials), while maintaining good mechanical properties and dimensional stability. The quality of the obtained polyurethane material depends on a few factors, for example: the composition of raw materials, their molar ratio, the conditions of synthesis, the additive compounds, and the application method [21]. The huge possibilities of producing polyurethane materials with designed properties are the result of a wide raw material base, especially in the field of polyol raw materials. Ecological aspects also play an important role for the producers of polyurethane foams. The addition of biopolyols obtained from vegetable oils to the base of renewable raw materials is consistent with the direction of the idea of sustainable development for polymeric materials. This is an interesting alternative to petrochemical polyols [22–26].

Earlier research concerned the synthesis of biopolyols based on vegetable oils, in which the ring-opening agent was 2,2′-mercaptodiethanol [24,27,28]. The aim of the research that is presented in this article was obtaining a new biopolyol based on evening primrose oil and diethylene glycol, and using it for the production of rigid polyurethane–polyisocyanurate foams for thermal insulation application.

2. Materials and Methods

2.1. Materials

Fresh and unrefined evening primrose oil was supplied by Olejarnia Kołodziejewo (Kołodziejewo, Poland) and used for synthesis of the new biopolyol (EPB). The iodine value was 0.658 mol I_2/100 g of oil, the acid value was 8.720 mg KOH/g, and the content of unsaturated fatty acids was 92.7% of all fatty acids. EPB was obtained in accordance with Polish Patent Application P.422888 [28], wherein the first step used: 99.5% acetic acid (Chempur, Piekary Śląskie, Poland), 30% hydrogen peroxide (Chempur, Poland), and 96% sulfuric acid (Chempur, Poland) as an oxidizing system of double bonds; a second step applied: 99% diethylene glycol p.a. (Chempur, Poland) and 96% sulfuric acid (as above) to open the epoxy rings. Anhydrous magnesium sulfate (Chempur, Poland) was used for deactivation of the catalyst and drying the oil-based polyol.

Polyols: biopolyol based on evening primrose oil, Rokopol RF-551 (sorbitol oxyalkylation product, PCC Rokita S.A., Brzeg Dolny, Poland) and technical polyisocyanate Purocyn B (polymeric 4,4′-diphenylmethane diisocyanate supplied by Purinova, Bydgoszcz, Poland) were used as the main raw materials for the production of rigid polyurethane–polyisocyanurate (RPU/PIR) foams. As an additive, agents in foam premixes were used, which included: 33% solution of 1,4-diazabicyclo[2,2,2]octane (DABCO, Alfa Aesar, Haverhill, MA, USA) in diethylene glycol (Chempur, Poland) as a catalyst of urethane bond formation; 33% solution of anhydrous potassium acetate (Chempur, Poland) in diethylene glycol (as above) as a catalyst of trimerization of NCO groups; Tegostab 8460 (Evonik, Essen, Germany) as a silicone surfactant; Solkane HFC 365/227 (Solvay, Brussels, Belgium) as a blowing agent, and Antiblaze TCMP (Albemarle, Charlotte, NC, USA) as a flame retardant.

2.2. Synthesis of Evening Primrose Oil-Based Polyol

The synthesis of a biopolyol based on evening primrose oil (EPB) was carried out in two steps in a glass reactor with a heating jacket, equipped with a reflux condenser, temperature sensor, dropping funnel, and mechanical stirrer. In the first step of synthesis, the evening primrose oil (EPO), acetic acid (AA), and sulfuric acid (SA) were loaded into reactor and heated to 40 °C. Hydrogen peroxide (HP) was gradually added after reaching this temperature. After the addition of HP, the whole mixture was heated to 60 °C. The reaction lasted for three hours. The molar ratio of the reactants calculated with respect to the iodine value of EPO was 1:1:1:0.02 for EPO:AA:HP:SA (mass of the reactants were shown in Table 1). Oil and water phases were separated after reaction. The oil phase was washed with distilled water and dried by solid anhydrous magnesium sulfate. In the first step of the synthesis, epoxidized evening primrose oil (EPEO) was obtained, which was subjected to analytical tests. The reaction scheme was shown in Figure 1.

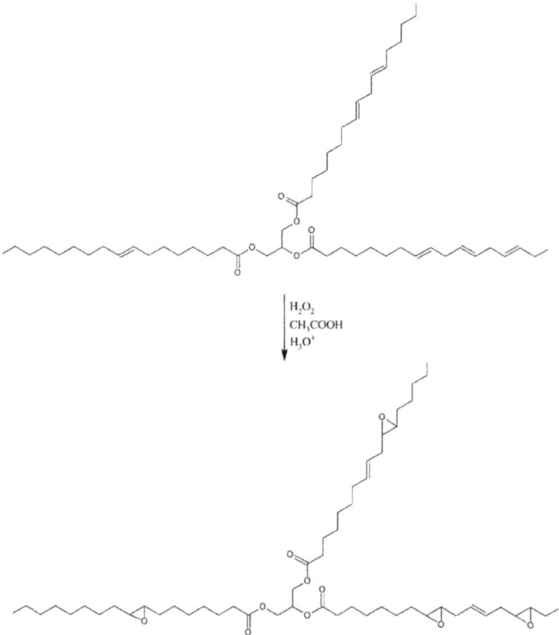

Figure 1. Schematic epoxidation reaction.

The efficiency of the epoxidation reaction of evening primrose oil (first step) is defined by Equation (1):

$$E_1 = \frac{IV_{EPO} - IV_{EPEO}}{IV_{EPO}} \cdot 100\% \tag{1}$$

where: E_1—efficiency of epoxidation reaction, IV_{EPO}—iodine value of evening primrose oil, and IV_{EPEO}—iodine value of epoxidized oil.

In the next step, epoxidized evening primrose oil (EPEO), diethylene glycol (GDE) and the reaction catalyst (SA) were loaded into a glass reactor with a heating jacket, which was equipped with a reflux condenser, temperature sensor, and mechanical stirrer. The molar ratio of reactants calculated with respect to the epoxy value of EPEO was 1:1:0.01 for EPEO:GDE:SA (Table 1). Afterwards, the whole mixture was heated to 100 °C. Reaction was carried out for four hours until the opening of all of the epoxide rings by diethylene glycol. After the synthesis, the obtained biopolyol (EPB) was neutralized by solid anhydrous magnesium sulfate and distilled under vacuum to reduce the water content. The second reaction scheme is shown in Figure 2.

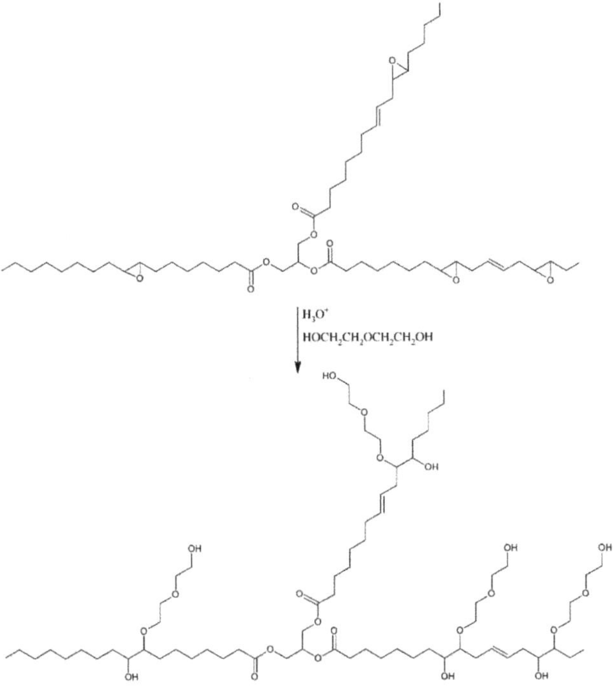

Figure 2. Schematic ring-opening reaction.

The efficiency of the opening reaction of epoxide rings (second step) was defined by Equation (2):

$$E_2 = \frac{EV_{EPEO} - EV_{EPB}}{EV_{EPEO}} \cdot 100\% \tag{2}$$

where: E_2—efficiency of epoxide rings' opening reaction, EV_{EPEO}—epoxy value of epoxidized evening primrose oil, EV_{EPB}—epoxy value of biopolyol.

Table 1. Amounts of the reactants in the two steps of synthesis of the biopolyol based on evening primrose oil (EPB) (g). EPO: evening primrose oil, EPEO: epoxidized evening primrose oil, AA: acetic acid, HP: hydrogen peroxide, SA: sulfuric acid, GDE: diethylene glycol.

	EPO	EPEO	AA	HP	SA	GDE	Efficiency (%)
1st step	1000.00	-	397.11	745.73	13.43	-	41.34
2nd step	-	1000.00	-	-	2.58	271.20	100.00

2.3. Synthesis of RPU/PIR Foams

The formulation of RPU/PIR foam premixes with evening primrose oil-based polyol (EPB) required experimental investigations to determine the optimal composition of additive agents (catalysts, surfactant, flame retardant, and blowing agent). The hydroxyl number was the basis for determining the amount of polyol raw materials in the formulation. These values enabled calculating the mass equivalents (R) of the hydroxyl group in polyols. The sum of mass equivalents of petrochemical polyol and biopolyol was always one. The addition of isocyanate raw material was selected in consideration of the mass equivalent of NCO groups. The ratio of NCO to OH groups in the reaction mixture for RPU/PIR foams was 3:1. An excess of isocyanate raw material was necessary for a reaction between NCO and OH groups (to produce a urethane bond) and the trimerization of three NCO groups (to produce an isocyanurate ring).

The content of additive agents was calculated in relation to the sum of masses of polyols and polyisocyanate (in weight percentages): silicone surfactant (1.7 wt.%), urethane bond catalyst (1 wt.%), isocyanate trimerization catalyst (2.5 wt.%), flame retardant (17 wt.%), and physical blowing agent (12 wt.%). The formulation of RPU/PIR foams was shown in Table 2.

Table 2. Formulation of rigid polyurethane–polyisocyanurate (RPU/PIR) foams. DABCO: 1,4-diazabicyclo[2,2,2]octane.

Foam Symbol	Rokopol RF-551, (R) (g)	EPB (R) (g)	Tegostab 8460 (g)	33% DABCO (g)	33% Potassium Acetate (g)	Antiblaze TCMP (g)	Solkane HFC 365/227 (g)	Purocyn B (R) (g)
EPB2.0	1.0 66.8	0 0	4.59	2.70	6.75	45.90	32.40	3.0 203.2
EPB2.1	0.9 60.12	0.1 15.38	4.74	2.79	6.97	47.48	33.44	3.0 203.2
EPB2.2	0.8 53.44	0.2 30.76	4.89	2.87	7.19	48.86	34.49	3.0 203.2
EPB2.3	0.7 46.76	0.3 46.14	5.03	2.96	7.40	50.34	35.53	3.0 203.2
EPB2.4	0.6 40.08	0.4 61.52	5.18	3.05	7.62	51.82	36.58	3.0 203.2

RPU/PIR foams were obtained at a laboratory scale by using the one-step method, from the two-component system. Component A was obtained as a result of mixing appropriate amounts of the polyol, biopolyol, silicone surfactant, two catalysts, flame retardant, and blowing agent in a polypropylene cup. Component B was a technical polyisocyanate raw material. Components A and B were mixed together and stirred for 10 s with a mechanical stirrer (1800 rpm) in a suitable mass ratio. After that, the mixture was poured into a cuboidal mould with a movable bottom with internal dimensions of 25 cm × 25 cm × 30 cm, where the growth of foam proceeded freely. Five types of foams were obtained in this research: EPB2.0—foam without biopolyol, and EPB2.1–2.4—foams with increasing biopolyol content based on evening primrose oil by the partial replacement of petrochemical polyol. The synthesis of RPU/PIR foams was thrice repeated. Obtained polyurethane materials were thermostated for six hours at 120 °C in a laboratory dryer with forced circulation, after removal from the mold.

2.4. Methods

2.4.1. Analysis of Synthesis Process and Product

Analytical and physicochemical tests were performed on the evening primrose oil, epoxidized oil, and biopolyol. It was aimed to determine its suitability for the synthesis of RPU/PIR foams.

The hydroxyl number (HN) was determined in accordance with the industrial standard of Purinova Ltd. No. WT/06/07/PURINOVA, by an acylation method with acetic anhydride in N,N'-dimethylformamide as a medium. An excess of acetic anhydride after hydrolysis and obtained acetic acid were titrated by using a standard potassium hydroxide solution and phenolphthalein as an indicator.

The acid value (AV) was determined in accordance with PN-EN ISO 660:2010. The analysis was performed by titration of the sample dissolved in a mixture of ethyl ether-ethyl alcohol (1:1) by using the standard solution of potassium hydroxide in ethyl alcohol and phenolphthalein as an indicator.

The iodine value (IV) was determined in accordance with PN-EN ISO 3961:2018-09 by the reaction of unsaturated bonds with iodine monochloride (Wijs solution) and the titration of iodine excess.

The epoxy value (EV) was determined in accordance with PN-EN ISO 3001:2002 by the reaction of epoxy rings with tetraethylammonium bromide and the titration of bromide excess.

The viscosity of the evening primrose oil, epoxidized oil, and biopolyol was determined by using a Fungilab digital rheometer at 20 °C (293 K). The measurements were carried out using a standard spindle (DIN-87) working with the bushing (ULA-DIN-87). A constant temperature was maintained through the thermostat connected to the water jacket of the sleeve.

The densities of the evening primrose oil, epoxidized oil, and biopolyol were measured at 25 °C (298 K) in an adiabatic pycnometer in accordance with ISO 758:1976.

The water content was determined by the Karl Fischer method using a non-pyridine reagent under the trade name Titraqual in accordance with PN-81/C-04959.

The pH value was measured using a Hanna Instruments microprocessor laboratory pH-meter (ORP/ISO/°C) with an RS 22 C connector.

The average molecular weight (M_w) of the biopolyol based on evening primrose oil was determined by gel permeation chromatography (GPC) by using a Knauer chromatograph. The apparatus was equipped with thermostated columns and a refractometer detector. The measurements were made on the basis of calibration, by using polystyrene standards in the range of M_w from 162 g/mol to 25,500 g/mol. Also on the basis of HN and average M_w, the functionality (f) of biopolyol was calculated by Equation (3):

$$f = \frac{M_W \cdot HN}{56100} \tag{3}$$

For confirmation, the course of the synthesis and chemical structure of the obtained oil-based polyol, the evening primrose oil, epoxidized oil, and biopolyol were tested in Fourier transform infrared (FTIR) spectroscopy using Brücker Vector spectrophotometer by KBr technique in the 400 to 4000 cm^{-1} range and in nuclear magnetic resonance spectroscopy ^{1}H NMR and ^{13}C NMR using a Brücker NMR Ascend III spectrometer with a frequency of 400 MHz, in deuterated chloroform, as a solvent.

2.4.2. Foaming Process

The foaming process was analyzed in accordance with ASTM D7487 13e—Standard Practice for Polyurethane Raw Materials: Polyurethane Foam Cup Test [29]. During obtaining RPU/PIR foams, cream, free rise, string gel, and tack free times were measured by an electronic stopwatch.

2.4.3. Properties of RPU/PIR Foams

The obtained rigid polyurethane–polyisocyanurate foams were tested for application as thermal insulation materials.

The apparent density of foams (the ratio of foam weight to its geometrical volume) was determined for cube-shaped samples with a side length of 50 mm in accordance with ISO 845:2006.

Compressive strength was determined by using the universal testing machine Instron 5544 in accordance with ISO 844:2014. The maximum force inducing a 10% relative strain was determined (decreasing of the foam height in relation to the initial height, according to the direction of foam growth).

The brittleness of the foams was determined in accordance with ASTM C-421-61, as a percentage mass loss of 12 cubic foam samples with a side length of 25 mm. Tests were conducted in a standard cuboidal box made of oak wood with dimensions of 190 mm × 197 mm × 197 mm, rotating around the axis at a speed of 60 rpm. The filling of the box during the measurement were 24 normalized oak cubes with dimensions of 20 mm × 20 mm × 20 mm. The brittleness (B) of obtained foams was calculated from Equation (4):

$$B = \frac{m_1 - m_2}{m_1} \cdot 100\% \tag{4}$$

where: m_1—mass of the sample before test (g), and m_2—mass of the sample after test (g).

The flammability of RPU/PIR foams was determined by using three flammability tests: Bütler's combustion test (vertical test) in accordance with ASTM D3014-04; the horizontal combustion test in accordance with PN-EN ISO 3582:2002/A1:2008, and a limited oxygen index Bütler's combustion test, which consisted of burning a foam sample with the dimensions of 150 mm × 20 mm × 20 mm in a vertical column (chimney) with dimensions of 300 mm × 57 mm × 54 mm. Combustion residue (CR) was calculated from Equation (5):

$$CR = \frac{m_b}{m_a} \cdot 100\% \tag{5}$$

where: m_a—mass of the sample before the burning test (g), and m_b—mass of the sample after the burning test (g). The horizontal burning test consisted of determining the susceptibility of foam to flame. The result of this test was an evaluation of material flammability (non-flammable, flammable, or self-extinguishing). A limited oxygen index (LOI) was measured by using Concept Equipment apparatus in accordance with ISO 4589. The percentage limited concentration of oxygen was determined in the mixture consisting of oxygen and nitrogen, which was sufficient to sustain the burning of the sample. *LOI* was calculated according to the Equation (6).

$$LOI = \frac{[O_2]}{[O_2] + [N_2]} \cdot 100\% \tag{6}$$

Absorbability (*A*) and water absorption (*WA*) were determined in accordance with ISO 2896:2001, which was measured after immersion in distilled water for 24 h. Values of these parameters were calculated from Equations (7) and (8):

$$A = \frac{m_A - m_D}{m_D} \cdot 100\% \tag{7}$$

where: m_A—mass of the sample after immersion in distilled water (g), and m_D—mass of the dry sample (g).

$$WA = \frac{m_{WA} - m_D}{m_D} \cdot 100\% \tag{8}$$

where: m_{WA}—mass of the sample after surface drying (g).

Aging resistance of the foams was carried out in thermostating process of cubic samples with a side length of 50 mm in 48 h at a temperature of 120 °C. The result of this test included a change of linear dimensions (Δl), change of geometrical volume (ΔV), and mass loss (Δm). The values of these parameters were calculated in accordance with ISO 1923: 1981 and PN-EN ISO 4590: 2016-11. The formulas for the calculations of Δl, ΔV, Δm are shown in Equations (9)–(11).

$$\Delta l = \frac{l - l_0}{l_0} \cdot 100\% \tag{9}$$

where: l_0—length of the sample before thermostating (according to the direction of foam rise) (mm), and l—length of the sample after thermostating (according to the direction of foam rise) (mm).

$$\Delta V = \frac{V - V_0}{V_0} \cdot 100\% \tag{10}$$

where: V_0—geometrical volume of the sample before thermostating (mm^3), and V—geometrical volume of the sample after thermostating (mm^3).

$$\Delta m = \frac{m_0 - m}{m_0} \cdot 100\% \tag{11}$$

where: m_0—mass of the sample before thermostating (g), and m—mass of the sample after thermostating (g).

The content of closed cells was determined in accordance with PN-EN ISO 4590:2016-11 by using the helium pycnometer AccuPyc 1340 with the FoamPyc option from Micrometrics. This software calculated the content of closed cells based on the measurement of pressure changes in the test chamber.

Thermal conductivity of the foams was determined based on the determination of the thermal conductivity coefficient λ in accordance with ISO 8301. Tests were carried out with the FOX 200 apparatus from LaserComp, in the measurement range of λ equal to 20–100 mW/(m·K). Measurements were performed in the series at intervals of 0.5 s and at an average measuring temperature of 10 °C (temperature of hot plate—20 °C, temperature of cold plate—0 °C).

The foam structure was analyzed by scanning electron microscope (SEM) HITACHI SU8010 (Hitachi High-Technologies Co., Tokyo, Japan). The studies were performed at the accelerating voltage of 30 kV, with the working distance of 10 mm and magnification of 150×. The statistical analysis of cell sizes, wall thickness, and content of cell per area unit was carried out on the basis of obtained micrographs by using ImageJ software (LOCI, Madison, WI, USA).

3. Results and Discussion

3.1. Synthesis of Biopolyol

A new biopolyol based on evening primrose oil was obtained as the result of a two-step synthesis, involving the epoxidation of double bonds and the opening of obtained epoxide rings with diethylene glycol. The use of this oil was due to the high content of unsaturated fatty acids (double bonds). Their high content promotes the production of biopolyols with high hydroxyl numbers, which is desirable in the synthesis of RPU/PIR foams [29]. The changes of raw materials during the synthesis were presented at Figure 3.

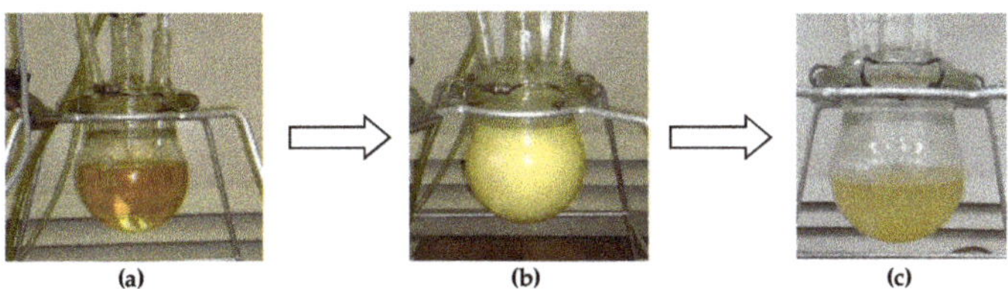

Figure 3. Visual presentation of appearance changes during synthesis (**a**) EPO, (**b**) EPEO, and (**c**) EPB.

The course of the synthesis was controlled by measuring the appropriate analytical parameters, i.e., measuring iodine value and epoxide value in the first step; and measuring the epoxide value and hydroxyl number in the second step. Evening primrose oil, epoxidized oil, and biopolyol were also subjected to basic physicochemical tests. The results of these tests are shown in Table 3.

Table 3. Properties comparison of evening primrose oil, epoxidized oil, and biopolyol.

Parameter	EPO	EPEO	EPB
Smell	earthy	earthy	earthy
Iodine value (mol I2/100 g of fat)	0.658	0.386	0.386
Epoxy value (mol/100 g of fat)	0.000	0.253	0.000
Acid value (mg KOH/g)	8.720	-	0.840
Hydroxyl number (mg KOH/g)	8.720	-	182.410
Viscosity (mPa·s)	170	190	2400
Water content (wt.%)	0	0.6	<0.2
pH (-)	6.5	7.0	7.0
Molecular weight (g/mol)	820	917	1623

A decrease of iodine value and increase of epoxy value was observed as a result of the epoxidation reaction. It was caused respectively by a decrease in the amount of double bonds in the fatty acid residues and an increase in the epoxide rings amount formed in their place. Also, a clear increase in the hydroxyl value was noted. This is due to the attachment of the diethylene glycol molecules to the chain of fatty acids, as a result of the rings' opening reaction. However, the obtained hydroxyl number was lower than theoretical calculations (tHN = 267.479 mg KOH/g). This is due to the side reaction of biopolyol molecules' oligomerization and the esterification of free fatty acids in unrefined oil (which led to a significant decrease in the acid value) [30]. It is noteworthy that the obtained values of biopolyol technological parameters are in the range of industrial standards.

Analysis of evening primrose oil, epoxidized oil, and biopolyol was also carried out in spectroscopic tests (FT-IR, ^{1}H NMR, and ^{13}C NMR).

Infrared spectroscopy analysis of EPO, EPEO, and EPB (Figure 4) showed that these compounds contained characteristic bonds for the structure of fatty acid glycerides. Bands at: 1740 cm^{-1} (stretching) belonged to the C=O groups; 1241, 1163, and 1099 cm^{-1} (stretching) belonged to the C–O bonds of the ester group; 3010 cm^{-1} (stretching) belonged to the C–H bonds in the olefin group; 2950 cm^{-1} (stretching) and 1465 cm^{-1} (deformational) belonged to the C–H bonds in the –CH$_2$- groups; 2925 cm^{-1} (stretching) and 1380 cm^{-1} (deformational) belonged to the C–H bond of the –CH$_3$ groups. Furthermore, the stretching vibration of the C=C bond from the olefin group and the pendulum vibrations of the CH$_2$ groups were observed respectively at 1650 and 725 cm^{-1}. In the EPEO spectrum, the presence of doublet bands coming from epoxy groups at 900 and 850 cm^{-1} was noted. These bands did not occur in the spectra of EPO and EPB. A low-intensity band at 3450 cm^{-1} was also noted. This band belonged to the stretching vibrations of the hydroxyl groups. However, its low intensity suggests that it was a residue of water after purification. The FTIR spectrum of the biopolyol showed a high intensity of band at 3450 cm^{-1}. This suggests the presence of a large number of O–H bonds. In addition, the band of the C–H bond from olefin groups (3010 cm^{-1}) was significantly reduced [31].

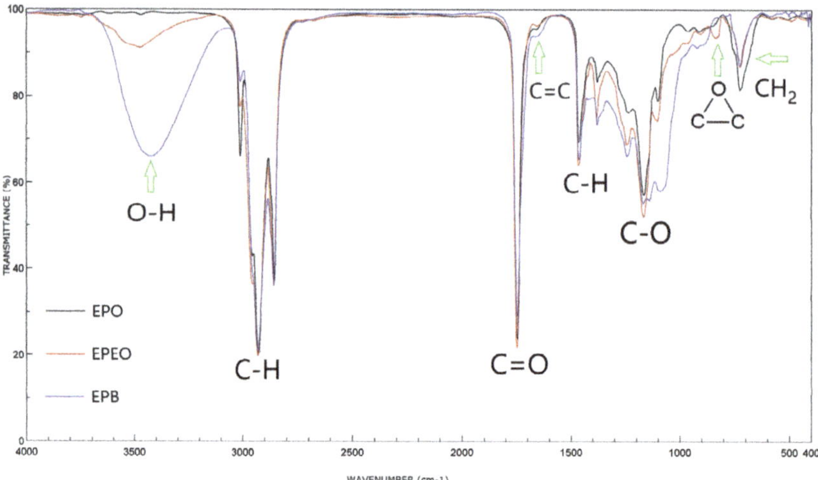

Figure 4. Fourier transform infrared (FTIR) spectra of EPO, EPEO, and EPB.

^{1}H NMR spectra analysis of EPO, EPEO, and EPB (Figure 5a–c) showed characteristic chemical shifts for: 5.35 ppm protons of the olefin groups of fatty acids –**CH=CH**–; 5.25 ppm methane protons of glyceryl –**CH₂**–**CH**–**CH₂**–; 4.12–4.28 ppm methylene protons of glyceryl –**CH₂**–**CH**–**CH₂**–; 2.70–2.80 ppm protons of bis-allyl methylene groups –**CH=CH**–**CH₂**–**CH=CH**–; 2.29–2.34 ppm protons of the α-**CH₂** group to the carbonyl group –**CH₂**-**CO**–; 1.95–2.15 ppm protons of α-**CH₂** groups to the olefin group –**CH₂**–**CH₂**–**CH=CH**–; 1.50–1.60 ppm protons of the β-**CH₂** group to the carbonyl group –**CH₂**–**CH₂**–**CO**–; 1.25–1.40 ppm protons of **CH₂** groups in the fatty acid chain; and 0.86–0.88 ppm protons of ending –**CH₃** groups. A characteristic chemical shift for the epoxy groups was only observed on the EPEO spectrum: 3.1–3.2 ppm protons of epoxy group –**CH**–(O)–**CH**–; 2.85–2.95 ppm protons of α-**CH₂** groups to the epoxy group —**CH**–(O)–**CH**–**CH₂**– and 1.55 ppm protons of α-**CH₂** groups to the epoxy group –**CH**–(O)–**CH**–**CH₂**–**CH₂**–. Additional chemical shifts were only noticed in the EPB spectrum: 3.70–3.80 ppm protons of hydroxyl groups at the end of the chain –**OH**; and 3.42–3.50 ppm protons of α-**CH₂** groups to the hydroxyl group –**CH₂**–**OH** [32].

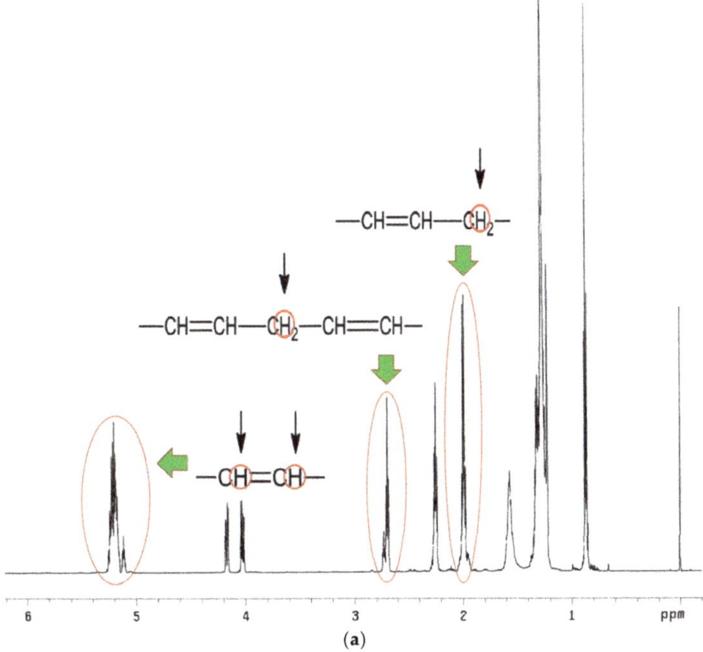

Figure 5. *Cont.*

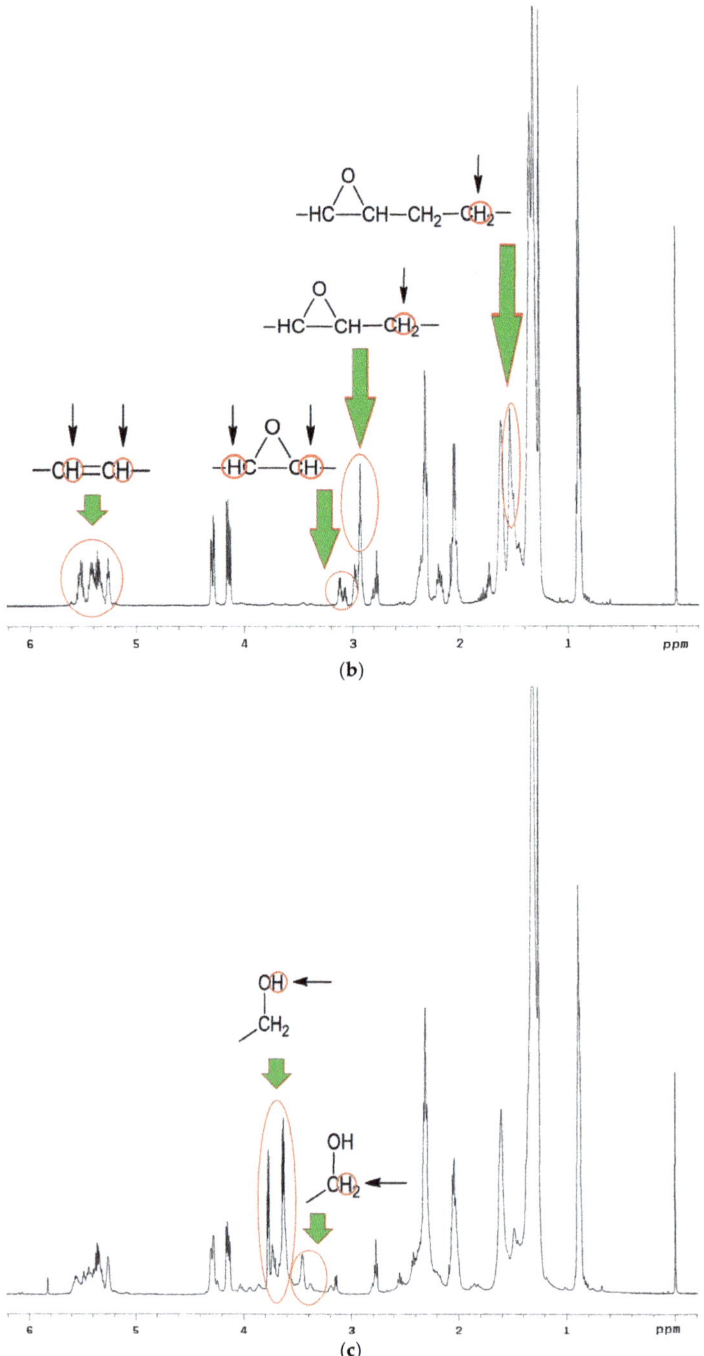

Figure 5. ^{1}H NMR spectra of (**a**) EPO, (**b**) EPEO, and (**c**) EPB.

^{13}C NMR spectra analysis of EPO, EPEO, and EPB (Figure 6a–c) showed characteristic chemical shifts for: 172.80–173.25 ppm carbons of carbonyl groups >C=O; 125.10–132.30 ppm carbons of olefin group of fatty acids –CH=CH–; 68.90 ppm methane carbons of glyceryl –CH$_2$–CH–CH$_2$–; 62.10 ppm methylene carbons of glyceryl –CH$_2$–CH–CH$_2$–; 33.00 ppm carbons of α-CH$_2$ groups to olefin group –CH=CH–CH$_2$–; 31.90 ppm carbons of α-CH$_2$ groups to the carbonyl group –CH$_2$–OOC–CH$_2$–; 27.20–29.80 ppm carbons of CH$_2$ groups in the fatty acid chain; 22.70 ppm carbons of penultimate groups –CH$_2$–CH$_3$; 14.30 ppm carbons of ending groups –CH$_3$. Characteristic chemical shifts for epoxy structures were noted in the EPEO spectrum: 55.21–57.35 ppm carbons of epoxy group –CH–(O)–CH–; 34.00 ppm carbons of α-CH$_2$ groups to the epoxy group –CH$_2$–CH–(O)–CH–. Furthermore, the chemical shift of carbons of α-CH$_2$ groups to a hydroxyl group –CH$_2$–OH in 73.86 ppm was only noticed in Figure 6c [33].

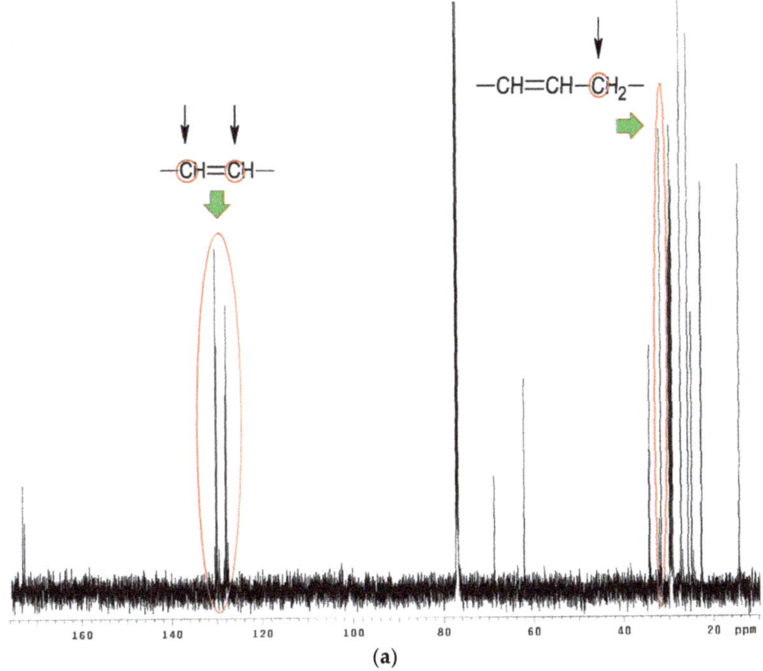

Figure 6. *Cont.*

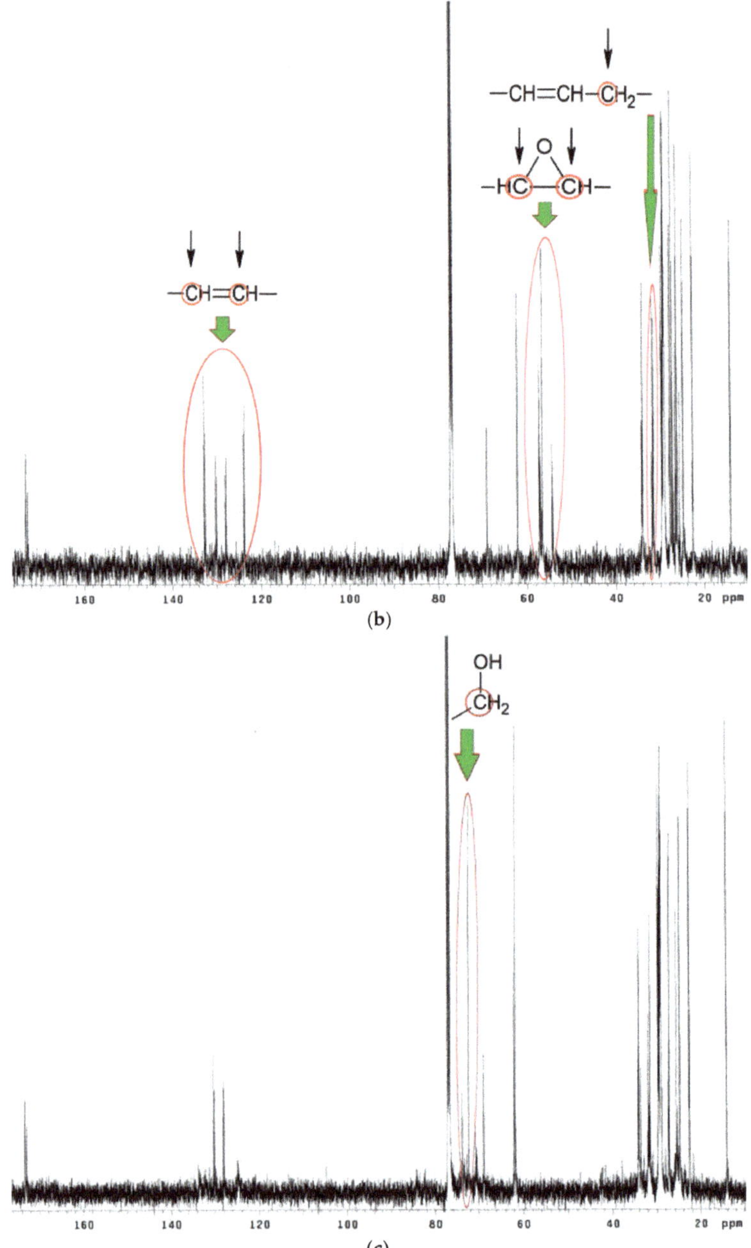

Figure 6. ^{13}C NMR spectra of (**a**) EPO, (**b**) EPEO, and (**c**) EPB.

The spectroscopic tests confirmed the assumed structure of epoxidized oil and a biopolyol based on evening primrose oil presented in Figures 1 and 2.

3.2. Foaming Process

The synthesis of RPU/PIR foams was monitored by measuring the appropriate technological times with an electronic stopwatch. The result of this measurement are shown in Table 4.

Table 4. Processing times of RPU/PIR foams with biopolyol.

Foam Symbol	Cream Time (s)	String Gel Time (s)	Tack Free Time (s)	Free Rise Time (s)
EPB2.0	8	21	23	34
EPB2.1	8	22	24	35
EPB2.2	9	23	25	35
EPB2.3	9	23	25	36
EPB2.4	9	23	25	37

The research showed that the addition of biopolyol based on evening primrose oil did slightly change the technological times. It also meant that an increase in the amount of EPB biopolyol with a higher functionality ($f = 5.3$) than the petrochemical Rokopol RF-551 ($f = 4.5$) does not significant affect the synthesis of RPU/PIR foams.

3.3. Properties of RPU/PIR Foams

The obtained rigid foams with a different content of polyol based on evening primrose oil were tested for application as thermal insulation materials. The foam obtained without biopolyol was used as a reference.

3.3.1. Physicomechanical Properties of RPU/PIR Foams

One of the most important parameters of RPU/PIR foams for thermal insulation application is apparent density. This parameter indirectly influences the other properties of these materials, such as for example, compressive strength in a parallel direction to foam growth and brittleness. In this case, the increase in the content of a plant-based polyol resulted in a decrease of the apparent density from 45.12 kg/m^3 for the reference foam to 40.03 kg/m^3 for the foam with the highest biopolyol content. The main reason for this decreasing was the addition of a component containing long, linear chains (fatty acid residues), which caused a decrease in the packing degree of polyurethane macromolecules (low or non-cross-linking potential).

It also had an influence on the compressive strength of RPU/PIR foams. With the increase in the amount of flexible segments, this parameter decreased from 377 kPa for foam without biopolyol to 335 kPa for foam with the highest content of biopolyol. Dependence between the biopolyol content and the apparent density and compressive strength in the parallel direction to foam growth is shown in the Figure 7.

Despite the decrease in the compressive strength of RPU/PIR foams, the value of this parameter is at a satisfactory level, which for this type of polyurethane materials is above 280 kPa.

Increased flexibility, which was resulting from the addition of linear polyol chains, influenced the decrease in brittleness of obtained RPU/PIR foams. When the evening primrose oil-based polyol was used, a significant decrease of this parameter was noted (from 40.17% for the foam with biopolyol to 22.41% for foam with 0.4 R of biopolyol).

Another important parameter of foams, which are used in thermal insulation applications, are absorbability and water absorption. The former relates to the percentage amount of water in the material after removal from immersion. The second one is the percentage amount of water that stayed inside the foams. In both cases, a decrease in these values was noted. Interdependence between the content of biopolyol based on evening primrose oil, absorbability, and water absorption is shown in Figure 8.

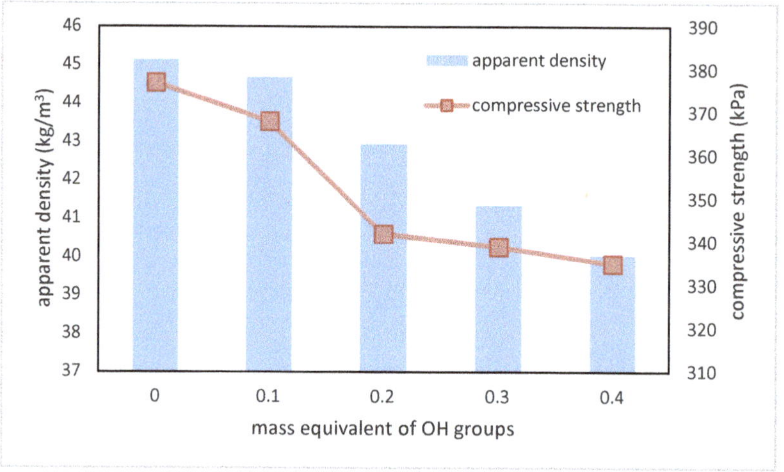

Figure 7. Dependence between EPB content, apparent density, and compressive strength in parallel direction to foam growth.

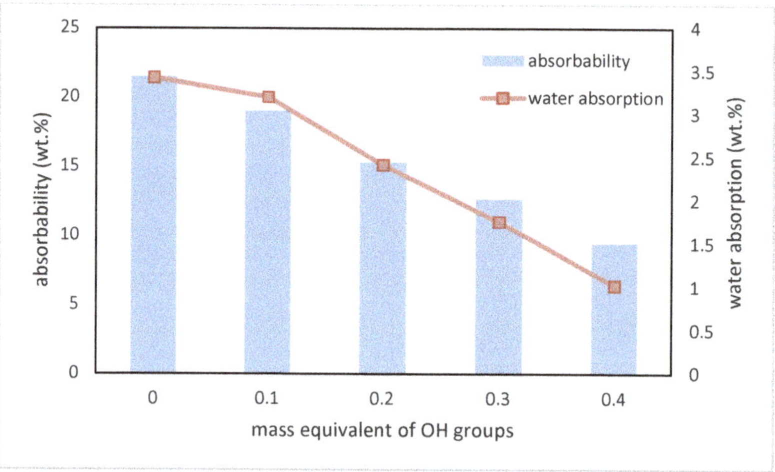

Figure 8. Dependence between EPB content, absorbability, and water absorption.

These decreases were mainly caused by the addition of hydrophobic groups derived from plant oil into the macromolecule of polyurethane. Reducing absorbability and water absorption is very beneficial when these materials are using as thermal insulation. A lack of ability for water to accumulate in foams prevents a multiplication of mold and other microorganisms in the rooms where they are used [34].

3.3.2. Flammability of RPU/PIR Foams

High flammability is a problem for most polymeric materials. In the case of materials used as thermal insulation in civil engineering, this is even more important, because the health and life of people during a fire depend on the flammability of these materials. Therefore, it is indispensable to use the flame-retardant compounds. RPU/PIR foams have an isocyanurate ring in their structure, which affects a partially reduction of flammability in comparison with classic PU foams. Flammability tests of RPU/PIR foams, including Bütler's combustion test, the horizontal combustion test, and the limited oxygen index showed that the new biopolyol did not affect the flammability of these materials (Table 5) [35].

Table 5. Flammability tests of RPU/PIR foams with biopolyol.

Foam Symbol	Combustion Residue (wt.%)	LOI vol % of O^2)	Classification Based on PN-EN ISO 3582:2002
EPB2.0	92.77 ± 0.29	24.1 ± 0.1	
EPB2.1	92.59 ± 0.34	24.1 ± 0.1	
EPB2.2	92.25 ± 0.32	24.0 ± 0.1	self-extinguishing
EPB2.3	92.04 ± 0.27	24.0 ± 0.1	
EPB2.4	91.95 ± 0.36	23.9 ± 0.1	

The presented results of the flammability tests showed a slight decrease in the combustion residue and limited oxygen index values. However, these changes were in the range of measurement error. So, it could be assumed that the use of a biopolyol based on evening primrose oil did not affect the flammability of RPU/PIR foams. This is important information from the application point of view, because the use of an oleochemical raw material did not increase the flammability of the foams. However, the evening primrose oil itself is flammable.

3.3.3. Structure of RPU/PIR Foams

Foam structure was analyzed with regard to EPB2.0 foam without biopolyol (Figure 9a) and foam with the highest content of biopolyol (Figure 9b).

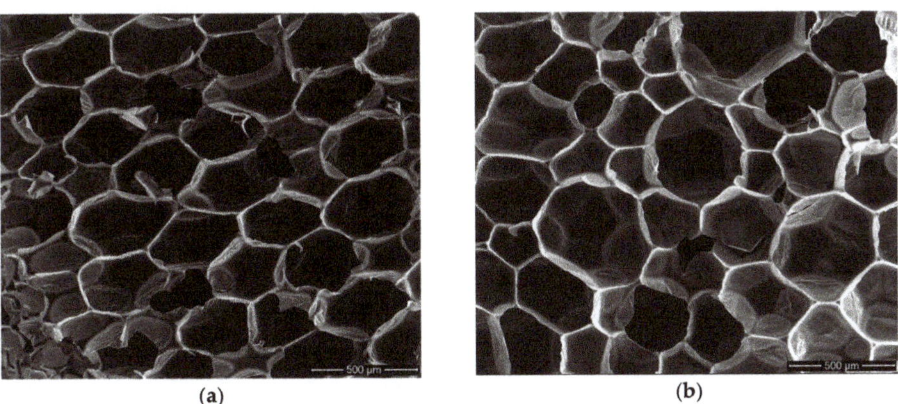

(a)   (b)

Figure 9. SEM micrographs of (**a**) EPB2.0—reference foam; (**b**) EPB2.4—foam with the highest content of biopolyol.

It was ascertained based on analysis of SEM micrographs that the increase in biopolyol content caused a slight increase in the cell size and its wall thickness. The results of this analysis and comparison of EPB2.0 and EPB2.4 micrographs i.e., average cell size, average cell wall thickness, shape of cells, and average amount of cell per area unit are shown in Table 6.

Table 6. Results of SEM micrograph analysis.

Foam Symbol	Cell Size (μm)	Thickness of Cell Wall (μm)	Content of Cell per Area Unit (cell/mm^2)
EPB2.0	258 ± 14	17 ± 2	12 ± 1
EPB2.4	269 ± 17	18 ± 2	11 ± 2

Slight changes in the structure of foams were caused by the presence of long linear molecules from fatty acid residues in biopolyol. The higher elasticity of the polyol segments and the lower cross-linking of the biopolyol allowed the greater migration of an easily volatile blowing agent. Consequently, the RPU/PIR foams with biopolyol based on evening primrose oil had slightly larger cell diameters and thicker cell walls. It affected inter alia the decrease of apparent density, compressive strength, and brittleness.

3.3.4. Thermal Insulation Properties

The most important parameter of the polyurethane materials that are intended for application as thermal insulation is the thermal conductivity coefficient λ. If its value is lower, then the material is a better heat insulator [36]. The content of closed cells in foams is as important as the λ parameter. An increase in the value of this parameter causes a decrease of the λ value. Dependence between the λ value, the content of closed cells, and the content of a biopolyol based on evening primrose oil is presented in Figure 10.

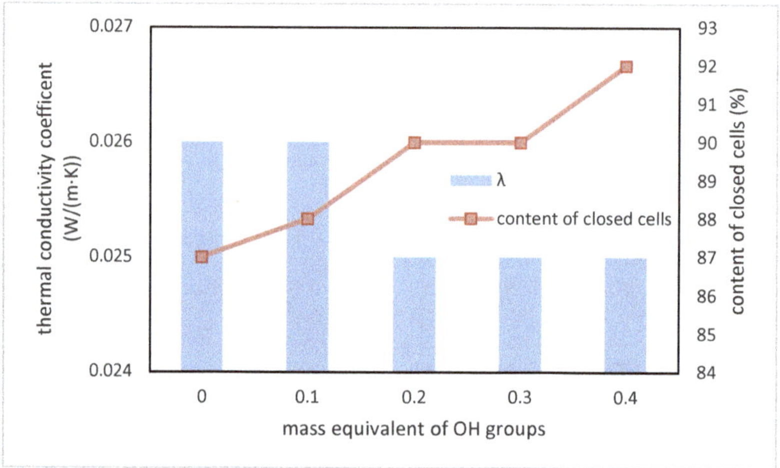

Figure 10. Dependence between EPB content, thermal conductivity coefficient, and content of closed cells.

The presented test results showed that the addition of biopolyol improved the thermal insulation properties of RPU/PIR foams. It is true that the change is slight. However, it confirmed that the biopolyol based on evening primrose oil may be an interesting alternative for petrochemical polyols, because the materials based on it have properties that are not worse than commercial foams.

An important parameter of RPU/PIR foams from the thermal insulation point of view is also aging resistance. The results of accelerated aging tests are shown in Table 7.

Table 7. Results of accelerated aging tests.

Foam Symbol	Change of Linear Dimension (%)	Change of Geometrical Volume (%)	Mass Loss (%)
EPB2.0	+0.96 ± 0.04	+2.75 ± 0.07	5.90 ± 0.12
EPB2.1	+0.92 ± 0.03	+2.61 ± 0.05	5.71 ± 0.06
EPB2.2	+0.91 ± 0.03	+2.53 ± 0.09	5.13 ± 0.10
EPB2.3	+0.87 ± 0.05	+2.46 ± 0.07	4.61 ± 0.09
EPB2.4	+0.85 ± 0.03	+2.39 ± 0.06	4.48 ± 0.15

The addition of vegetable oil-based segments has improved aging resistance. This is due to the fact that the fatty acid glyceride molecule has a higher resistance to external factors, including high temperature, than the polyether polyol. Ether groups contained in Rokopol RF-551 are more susceptible to degradation than ester and ether groups contained in a biopolyol based on evening primrose oil. The reason for this is a greater content of ether bonds in petrochemical polyol (average, one ether bond for every two carbons) than in biopolyol.

4. Conclusions

This paper presented a synthesis of a new oil-based biopolyol for the production of RPU/PIR foams for thermal insulation applications. Evening primrose oil was used as a plant-based raw material, which had a high content of unsaturated fatty acids (unsaturated double bonds). Synthesis was carried out in a two-step method involving the epoxidation of double bonds and opening epoxy rings with diethylene glycol. The obtained biopolyol was characterized by a hydroxyl number of 182.41 mg KOH/g, an acid value of 0.84 mg KOH/g, and a water content of 0.2 wt.%. The new oil-based polyol was used as raw material for the synthesis of rigid polyurethane–polyisocyanurate foams from 0 to 0.4 mass equivalents of OH groups in a mixture with a commercial polyether polyol Rokopol RF-551. RPU/PIR foams were tested for use as thermal insulation materials. The conducted research has shown that the foams based on biopolyol had a slight lower apparent density, compressive strength, and brittleness. A slight improvement in the thermal insulation properties, closed cell contents, and aging resistance was also observed. The price of the finished product is also important. The production costs of this biopolyol are lower than the production costs of petrochemical polyol, which in the economic balance sheet gives a reduction in the price of final RPU/PIR foams. The use of evening primrose oil as an alternative raw material for the synthesis of green polyols offers great opportunities. The appropriate choice of composition for a reaction mixture enables obtaining various products, which may be used in various industries, such as for example in civil engineering, textile, furniture, footwear, or automotive industries. Besides, using plant-based polyols for the production of polyurethane materials is compatible with sustainable development principles and green chemistry philosophy. It enables the partial or complete replacement of petrochemical polyols.

5. Patents

The synthesis of biopolyol based on evening primrose oil was carried out on the basis of the patented method: Polish Patent Application number P.422888.

Author Contributions: Conceptualization, J.P.-S. and M.B.; methodology, M.B.; software, M.I.; validation, J.P.-S.; formal analysis, B.C.; investigation, M.B. and J.P.-S.; data curation, M.I.; writing—original draft preparation, M.B. and J.P.-S.; visualization, M.B. and M.I.; supervision, B.C.

Funding: This research received no external funding.

Conflicts of Interest: The authors declare no conflict of interest.

References

1. Hojabri, L.; Kong, X.; Narine, S.S. Fatty acid-derived diisocyanate and biobased polyurethane produced from vegetable oil: Synthesis, polymerization, and characterization. *Biomacromolecules* **2009**, *10*, 884–891. [CrossRef] [PubMed]
2. Petrović, Z.S.; Guo, A.; Javni, I.; Cvetković, I.; Hong, D.P. Polyurethane networks from polyols obtained by hydroformylation of soybean oil. *Polym. Int.* **2008**, *57*, 275–281. [CrossRef]
3. Baumann, H.; Bühler, M.; Fochem, H. Natural Fats and Oils—Renewable Raw Materials for the Chemical Industry. *Angew. Chem. Int.* **1988**, *27*, 41–62. [CrossRef]
4. De Espinosa, L.M.; Ronda, J.C.; Galia, M.; Cadiz, V. Plant oils: The perfect renewable resource for polymer science? *J. Polym. Sci. Part A Polym. Chem.* **2009**, *47*, 1159–1167. [CrossRef]

5. Datta, J.; Glowinska, E. Chemical modifications of natural oils and examples of their usage for polyurethane synthesis. *J. Elastom. Plast.* **2014**, *46*, 33–42. [CrossRef]

6. Mizera, K.; Ryszkowska, J. Thermal properties of polyurethane elastomers from soybean oil-based polyol with a different isocyanate index. *J. Elastom. Plast.* **2018**, *4*. [CrossRef]

7. Datta, J.; Głowińska, E. Effect of hydroxylated soybean oil and bio-based propanediol on the structure and thermal properties of synthesized bio-polyurethanes. *Ind. Crops Prod.* **2014**, *61*, 84–91. [CrossRef]

8. Miao, S.; Zhang, S.; Su, Z.; Wang, P. Vegetable-oil-based polymers as future polymeric biomaterials. *J. Polym. Sci. Part A Polym. Chem.* **2010**, *48*, 243–250. [CrossRef]

9. Behr, A.; Gomes, J.P. The refinement of renewable resources: New important derivatives of fatty acids and glycerol. *J. Lipid Sci. Technol.* **2010**, *112*, 31–50. [CrossRef]

10. Palaskar, D.V.; Boyer, A.; Cloutet, E.; Le Meins, J.-F.; Gadenne, B.; Alfos, C.; Farcet, C.; Cramail, H. Original diols from sunflower and ricin oils: Synthesis, characterization, and use as polyurethane building blocks. *J. Polym. Sci. Part A Polym. Chem.* **2018**, *50*, 1766–1782. [CrossRef]

11. Akram, D.; Hakami, O.; Sharmin, E.; Ahmad, S. Castor and Linseed oil polyurethane/TEOS hybrids as protective coatings: A synergistic approach utilising plant oil polyols, a sustainable resource. *Prog. Organ. Coat.* **2017**, *108*, 1–14. [CrossRef]

12. Zhoua, X.; Sainab, M.M.; Oksman, K. Semi-rigid biopolyurethane foams based on palm-oil polyol and reinforced with cellulose nanocrystals. *Compos. Part A Appl. Sci. Manuf.* **2016**, *83*, 56–62. [CrossRef]

13. Kadam, H.; Bandyopadhyay-Ghosh, S.; Malik, N.; Ghosh, S.B. Bio-based engineered nanocomposite foam with enhanced mechanical and thermal barrier properties. *J. Appl. Polym. Sci.* **2018**, 47063. [CrossRef]

14. Yu, Z.L.; Jiang, C.; Li, W.; Ma, F.M.; Wei, S.C.; Ruan, M.; Kong, X.; Han, D.Y. Synthesis and Characterization of Polyurethanes from Oleic, Erucic and 10-Undecenoic Acids. *Polym. Renew. Resour.* **2018**. [CrossRef]

15. Wang, C.; Yang, L.; Ni, B.; Wang, L. Thermal and mechanical properties of cast polyurethane resin based on soybean oil. *J. Appl. Polym. Sci.* **2009**, *112*, 1122–1127. [CrossRef]

16. Mekewi, M.A.; Ramadan, A.M.; El Darse, F.M.; Abdel Rehim, M.H.; Mosa, N.A.; Ibrahim, M.A. Preparation and characterization of polyurethane plasticizer for flexible packaging applications: Natural oils affirmed access. *Egypt. J. Pet.* **2017**, *26*, 9–15. [CrossRef]

17. Lubczak, R.; Szczęch, D. Polyurethane foams with starch. *J. Chem. Technol. Biotechnol.* **2018**. [CrossRef]

18. Lubczak, R. Multifunctional oligoetherols and polyurethane foams with carbazole ring. *Pol. J. Chem. Technol.* **2016**, *18*, 120–126. [CrossRef]

19. Chmiel, E.; Lubczak, J. Synthesis of oligoetherols from mixtures of melamine and boric acid and polyurethane foams formed from these oligoetherols. *Polym. Bull.* **2018**, 1–23. [CrossRef]

20. Kurańska, M.; Prociak, A. Environmentally friendly polyurethane-polyisocyanurate foams for applications in the construction industry. *Czasopismo Techniczne Budownictwo* **2014**, *5*, 149–152.

21. Prociak, A.; Rokicki, G.; Ryszkowska, J. *Polyurethane Materials*; Wydawnictwo Naukowe PWN: Warszawa, Poland, 2014. (In Polish)

22. Kirpluksa, M.; Kalnbundea, D.; Benesb, H.; Cabulis, U. Natural oil based highly functional polyols as feedstock for rigid polyurethane foam thermal insulation. *Ind. Crops Prod.* **2018**, *122*, 627–636. [CrossRef]

23. Chen, M.-J.; Wang, X.; Tao, M.-C.; Liu, X.-Y.; Liu, Z.-G.; Zhang, Y.; Zhao, C.-S.; Wang, J.-S. Full substitution of petroleum-based polyols by phosphorus containing soy-based polyols for fabricating highly flame-retardant polyisocyanurate foams. *Polym. Degrad. Stab.* **2018**, *154*, 312–322. [CrossRef]

24. Paciorek-Sadowska, J.; Borowicz, M.; Czupryński, B.; Tomaszewska, E.; Liszkowska, J. Oenothera biennis seed oil as an alternative raw material for production of bio-polyol for rigid polyurethane-polyisocyanurate foams. *Ind. Crops Prod.* **2018**, *126*, 208–217. [CrossRef]

25. Abdel Hakim, A.A.; Nassar, M.; Emam, A.; Sultan, M. Preparation and characterization of rigid polyurethane foam prepared from sugar-cane bagasse polyol. *Mater. Chem. Phys.* **2011**, *129*, 301–307. [CrossRef]

26. Ionescu, M. *Chemistry and Technology of Polyols for Polyurethanes*, 2nd ed.; Smithers Rapra Publishing: Shawbury, UK, 2016.

27. Paciorek-Sadowska, J.; Borowicz, M.; Czupryński, B.; Tomaszewska, E.; Liszkowska, J. New bio-polyol based on white mustard seed oil for rigid PUR-PIR foams. *Pol. J. Chem. Technol.* **2018**, *20*, 24–31. [CrossRef]

28. Paciorek-Sadowska, J.; Borowicz, M.; Czupryński, B.; Liszkowska, J. The Method of Obtaining a Polyol Raw Material for the Synthesis of Rigid Polyurethane-Polyisocyanurate Foams. Polish Patent Application P.422888, 19 September 2017.

29. ASTM International. *Standard Practice for Polyurethane Raw Materials: Polyurethane Foam Cup Test*; ASTM Standard D7487—13e1, 2008; ASTM International: West Conshohocken, PA, USA, 2016. [CrossRef]
30. Dworakowska, S.; Bogdał, D.; Prociak, A. Microwave-Assisted Synthesis of Polyols from Rapeseed Oil and Properties of Flexible Polyurethane Foams. *Polymers* **2012**, *4*, 1462–1477. [CrossRef]
31. Septevani, A.A.; Evans, D.A.C.; Chaleat, C.; Martin, D.J.; Annamalai, P.K. A systematic study substituting polyether polyol with palm kernel oil based polyester polyol in rigid polyurethane foam. *Ind. Crops Prod.* **2015**, *66*, 16–26. [CrossRef]
32. Zhan, G.; Zhao, L.; Hu, S.; Gan, W.; Yu, Y.; Tang, X. A novel biobased resin epoxidized Soybean oil modified cyanate ester. *Polym. Eng. Sci.* **2008**, *48*, 1322–1328. [CrossRef]
33. Zhang, L.; Huang, M.; Yu, R.; Huang, J.; Dong, X.; Zhang, R.; Zhu, J. Bio-based shape memory polyurethanes (Bio-SMPUs) with short side chains in the soft segment. *J. Mater. Chem. A* **2014**, *2*, 11490–11498. [CrossRef]
34. Sripathy, M.; Sharma, K.V. Flammability and Moisture absorption test of rigid polyurethane foam. *Int. J. Sci. Eng. Res.* **2013**, *2*, 1–8.
35. Paciorek-Sadowska, J.; Borowicz, M.; Czupryński, B.; Liszkowska, J. Use of volcanic tuff for production of rigid polyurethane-polyisocyanurate foams. *Przem. Chem.* **2016**, *95*, 42–47. [CrossRef]
36. Zhang, H.; Fang, W.; Li, Y.; Tao, W. Experimental study of the thermal conductivity of polyurethane foams. *J. Appl. Therm. Eng.* **2016**, *115*, 528–538. [CrossRef]

 *polymers*

Article

Synthesis and Properties of Novel Polyurethanes Containing Long-Segment Fluorinated Chain Extenders

Jia-Wun Li [1], Hsun-Tsing Lee [2], Hui-An Tsai [3], Maw-Cherng Suen [4,*] and Chih-Wei Chiu [1,*

[1] Department of Materials Science and Engineering, National Taiwan University of Science and Technology, Taipei 10607, Taiwan; a12352335@gmail.com
[2] Department of Materials Science and Engineering, Vanung University, Jongli, Taoyuan 32061, Taiwan; htlee@mail.vnu.edu.tw
[3] R&D Center for Membrane Technology, Department of Chemical Engineering, Chung Yuan University, Chungli District, Taoyuan 32023, Taiwan; huian@cycu.edu.tw
[4] Department of Fashion Business Administration, LEE-MING Institute of Technology, No. 22, Sec. 3, Tailin. Rd., New Taipei 24305, Taiwan
* Correspondence: sun0414@mail.lit.edu.tw (M.-C.S.); cwchiu@mail.ntust.edu.tw (C.-W.C.); Tel.: +886-2-2909-7811 (ext. 1101) (M.-C.S.); +886-2-2737-6521 (C.-W.C.); Fax: +886-2-2737-6544 (C.-W.C.)

Received: 30 October 2018; Accepted: 19 November 2018; Published: 21 November 2018

Abstract: In this study, novel biodegradable long-segment fluorine-containing polyurethane (PU) was synthesized using 4,4′-diphenylmethane diisocyanate (MDI) and 1H,1H,10H,10H-perfluor-1,10-decanediol (PFD) as hard segment, and polycaprolactone diol (PCL) as a biodegradable soft segment. Nuclear magnetic resonance (NMR) was used to perform ^{1}H NMR, ^{19}F NMR, ^{19}F–^{19}F COSY, ^{1}H–^{19}F COSY, and HMBC analyses on the PFD/PU structures. The results, together with those from Fourier transform infrared spectroscopy (FTIR), verified that the PFD/PUs had been successfully synthesized. Additionally, the soft segment and PFD were changed, after which FTIR and XPS peak-differentiation-imitating analyses were employed to examine the relationship of the hydrogen bonding reaction between the PFD chain extender and PU. Subsequently, atomic force microscopy was used to investigate the changes in the microphase structure between the PFD chain extender and PU, after which the effects of the thermal properties between them were investigated through thermogravimetric analysis, differential scanning calorimetry, and dynamic mechanical analysis. Finally, the effects of the PFD chain extender on the mechanical properties of the PU were investigated through a tensile strength test.

Keywords: fluorinated polyurethanes; chain extender; curve fitting technique; nuclear magnetic resonance; hydrogen bond

---

1. Introduction

Thermoplastic polyurethane (TPU) is a type of block copolymer that is usually synthesized with a soft segment diol, a hard segment diisocyanate, and a chain extender. The incompatibility between soft and hard segments results in microphase separation, the structure of which dominates the mechanical properties and phase morphology of TPU, which has high intensity, toughness, and wear resistance, as well as the properties of both plastics and elastomers [1–4]. Through different proportions of soft and hard segments, polyurethane (PU) can be used to synthesize materials with different properties, which can be applied in various fields [5,6]. Therefore, TPU is of great industrial importance [7]. However, compared with other thermoplastic elastomer materials, the poor thermal stability of TPU [8,9] results in limitations in back-end applications.

The thermal properties of TPU can be improved by using blending [10–12], copolymers [13], and cross-linked structures [14]. Another approach is the functionalized introduction of fluorine-containing chemicals to form fluoroacrylate polyurethane (FPU), which is expected to have properties similar to those of other fluorinated polymers [15], including biocompatibility, excellent environmental stability, hydrolysis resistance, thermal stability [16,17], chemical resistance, low interface free energy, and water and oil resistance [18]. The interaction of organic fluorine is known to involve π–πF, C–F$\cdots$H, F$\cdots$F, C–F$\cdots\pi$F, C–F$\cdots\pi$, C–F$\cdots$M+, C–F$\cdots$C=O, and anion–πF [19]. Therefore, after fluorine is introduced into TPU to form fluoroacrylate thermoplastic polyurethane (FTPU) [20], the stronger interaction of the C–F$\cdots$H hydrogen bond (HB) compared with that of the C=O$\cdots$H HB, together with the interactions of other organic fluorine enable FTPU to possess stronger molecular interactions, resulting in higher rigidity. Moreover, the –CF_2– group of organic fluorine coats the C–C bond with fluorine atoms, which effectively enhances the chemical resistance, barrier properties, and thermal stability of FTPU, leading to wider applications. Liu et al. [21] mixed fluorinated polyether diol with polybutylene adipate in different proportions to produce soft segments, which were then combined with MDI to prepare FTPU. Studies have shown that fluorinated chain extenders with low reactivity affect the final molecular weight of FTPU and can improve thermal stability. Yang et al. [22] successfully introduced 4,4′-[2,2,2-trifluoro-1-(trifluoromethyl)ethylidene] bisphenol into TPU to form FTPU, and their results showed that introducing fluorine-containing chain extenders improved the thermal stability and rigidity of TPU. Furthermore, our laboratory [23,24] successfully completed synthesis and introduced short-segment fluorine-containing chain extenders into the main and side chains of TPU, which enhanced thermal properties, tensile strength, and hydrophobicity. The above studies indicated that fluorine-containing chain extenders can effectively improve the thermal stability and mechanical properties of PU. Moreover, it was found that if a short-segment fluorine-containing chain extender was introduced as a side chain, it was able to effectively increase the tensile strength of FTPU, while the tensile strain and heat stability were lower. If it was introduced as a main chain, it was able to effectively increase the thermal stability and maintain the original breaking strain of PCL and slightly increase the tensile strength. Therefore, in this study, a long-segment fluorine-containing chain extender was introduced and used together with the urethane group to increase molecular interactions, thereby effectively increasing the thermal stability and mechanical properties of FTPU. So, it is hoped to effectively increase the thermal stability and mechanical property at the same time.

In the present study, a novel FTPU was successfully synthesized through the reaction of 4,4′-diphenylmethane diisocyanate (MDI) and polycaprolactone diol (PCL), which formed a PU prepolymer. Subsequently, 1H,1H,10H,10H-perfluor-1,10-decanediol (PFD) was added to the prepolymer to obtain PFD/PU polymers, and their structures were confirmed through Fourier transform infrared spectroscopy (FTIR) and nuclear magnetic resonance (NMR). The resulting PFD/PUs were then subjected to molecular weight determination by using varying amounts of PCL and PFD chain extenders, and the physical and chemical properties of the PFD/PUs were investigated.

2. Experimental

2.1. Materials

1H,1H,10H,10H-Perfluor-1,10-decanediol (PFD) was purchased from Matrix Scientific (Columbia, SC, USA). 4,4′-Diphenylmethane diisocyanate (MDI), polycaprolactone diol (PCL, Mw = 530), and dibutyltin dilaurate (DBTDL) were purchased from Aldrich (St. Louis, MO, USA). *N,N*-Dimethylacetamide (DMAc) was obtained from Mallinckrodt Chemicals (Dublin, Ireland).

2.2. Synthesis of PFD/PUs

A 2-step process was used to polymerize the PFD/PUs. First, MDI, PCL, and DMAc were added to a 500 mL 4-neck reaction flask and heated to 80 °C using a heating mantle. After 2 to 3 drops of dibutyltin dilaurate were added, the solution was mixed using a mechanical stirrer at 200 rpm, and PU prepolymers were formed after 2 h of reaction. In the second step, the PFD chain extenders were dissolved in DMAc and slowly dripped into the reaction flask, and the reaction was continued for 2 h (Scheme 1). In the polymerization, the di-n-butylamine method [25] was used to calculate the NCO content in all steps to monitor the reaction process. The obtained PFD/PU solution was subjected to vacuum defoaming for 2 h, after which it was poured into a serum bottle and stored in a refrigerator for 1 day. Finally, the PFD/PU solution was poured into a Teflon plate and dried in a temperature-programmable circulating oven for 8 h. The recipe, symbols, and theoretical contents of the hard and soft segments for the PFD/PU films are shown in Table 1. This study was only concerned with the relative contents of hard (or soft) segments of PFD/PUs with different PFD contents, so the theoretical hard (or soft) segment content is sufficient. Therefore, theoretical hard and soft segment contents were calculated by using Equations (1) and (2), respectively, as used in the general literature [26].

$$\text{Theoretical hard segment content (wt\%)} = \frac{W_{MDI} + W_{PFD}}{W_{MDI} + W_{PCL} + W_{PFD}} \times 100\% \tag{1}$$

$$\text{Theoretical soft segment content (wt\%)} = 100\% - \text{Theoretical hard segment content (wt\%)} \tag{2}$$

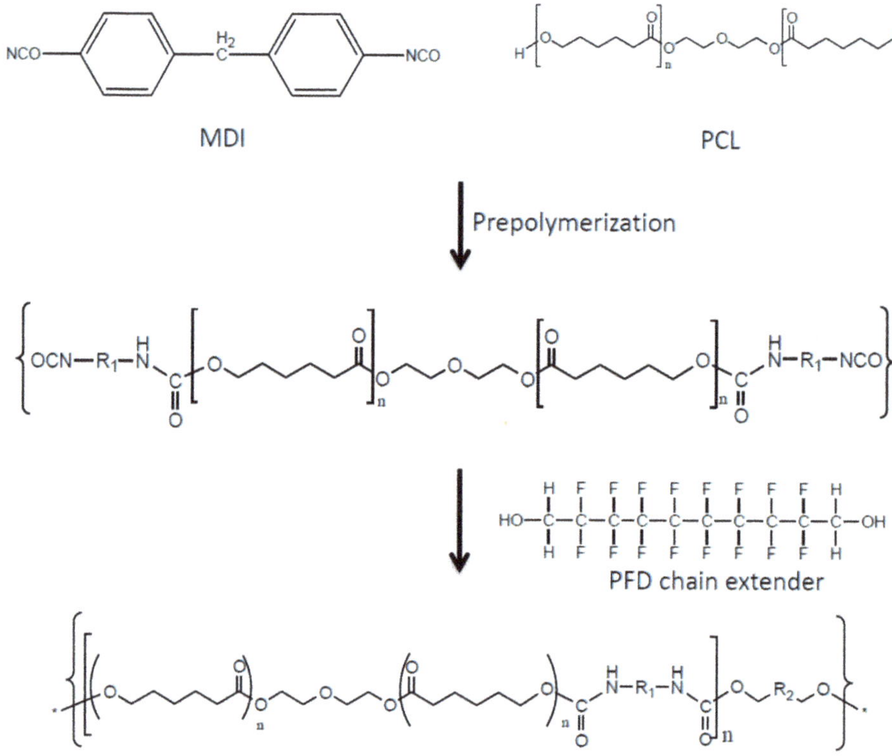

MDI

PCL

Prepolymerization

PFD chain extender

Scheme 1. Formula for 1H,1H,10H,10H-perfluor-1,10-decanediol (PFD)/polyurethanes (PUs).

Table 1. Recipe and theoretical phase compositions of the PFD/PUs. MDI, 4,4′-diphenylmethane diisocyanate; PCL, polycaprolactone diol.

Designation	MDI (moles)	PCL (moles)	PFD (moles)	Theoretical Hard Segment (wt%)	Theoretical Soft Segment (wt%)
PFD/PU-01	4	3.50	0.50	39.91	60.09
PFD/PU-02	4	3.25	0.75	43.89	56.11
PFD/PU-03	4	3.00	1.00	47.92	52.08

W_{MDI}, weight of MDI; W_{PCL}, weight of PCL; W_{PFD}, weight of PFD.

2.3. Advanced Polymer Chromatography System (APC)

The molecular weights of PFD/PUs were characterized using an Acquity APC core system (Waters Corp., Milford, MA, USA) with Tetrahydrofuran (THF) as eluent at a flow rate of 0.8 mL/min. The measurement was carried out at 45 °C.

2.4. Fourier Transform Infrared Spectroscopy (FT-IR)

Fourier transform infrared spectroscopy measurements were performed on a PerkinElmer Spectrum One spectrometer (Waltham, MA, USA). The spectra of the samples were obtained by averaging 16 scans in a range of 4000 to 650 cm^{-1} with a resolution of 2 cm^{-1}.

2.5. ^1H NMR Spectrometer

^{1}H NMR (in DMSO-d$_6$) spectra of the specimens were measured using a Bruker Avance 300 spectrometer (300 MHz; Bruker, Billerica, MA, USA).

2.6. ^{19}F NMR Spectrometer

^{19}F nuclear magnetic resonance (NMR) spectra of the polymers were recorded on a Bruker AVIII HD400 Hz spectrometer (Bruker, Billerica, MA, USA) using DMSO-d$_6$ as a solvent and tetramethylsilane as an internal standard.

2.7. X-Ray Photoelectron Spectroscopy (XPS)

X-ray photoelectron spectroscopy (XPS) measurements were carried out using a Thermo Fisher Scientific (VGS) spectrometer (Waltham, MA, USA). An Al Kα anode was used as the x-ray source (1486.6 eV), and a binding energy range of 0 to 1400 eV was selected for the analysis. The binding energies were calibrated to the C1s internal standard with a peak at 284.8 eV. The high-resolution C1s spectra were decomposed by fitting a Gaussian function to an experimental curve using a nonlinear regression.

2.8. Surface Roughness Analysis

Scanning was performed using a CSPM5500 atomic force microscope from Being Nano-Instruments (Beijing, China), which is generally operated in 2 imaging modes: tapping and contact. The tapping mode was used in this study, and the tip of the oscillation probe cantilever made only intermittent contact with the sample. Regarding the phase of the sine wave that drives the cantilever, the phase of the tip oscillation is extremely sensitive to various sample surface characteristics; therefore, the topography and phase images of a sample's surface can be detected.

2.9. Thermogravimetric Analysis (TGA)

Thermogravimetric analysis was performed on a PerkinElmer Pyris 1 TGA (Perkin Elmer, Waltham, MA, USA). The samples (5–8 mg) were heated from room temperature to 700 °C under nitrogen at a rate of 10 °C/min.

2.10. Differential Scanning Calorimetry (DSC)

Differential scanning calorimetry was performed on a PerkinElmer Jade differential scanning calorimeter (Perkin Elmer, Waltham, MA, USA). The samples were sealed in aluminum pans with a perforated lid. The scans (-50 to 50 °C) were performed at a heating rate of 10 °C/min under nitrogen purging. The glass transition temperatures (T_g) were located as the midpoints of the sharp descent regions in the recorded curves. The melting points were recorded as the peak maximum of the endothermic transition in the second scan. Approximately 5–8 mg of samples were used in all of the tests.

2.11. Dynamic Mechanical Analysis (DMA)

Dynamic mechanical analysis was performed on a Seiko dynamic mechanical spectrometer (model DMS6100) at 1 Hz with a 5 μm amplitude over a temperature range of -50 to 50 °C at a heating rate of 3 °C/min. DMA was conducted in tension mode with specimen dimensions of 20 mm $\times$ 5 mm $\times$ 0.2 mm (L $\times$ W $\times$ H). The T_g was taken as the peak temperature of the glass transition region in the tan δ curve.

2.12. Stress–Strain Testing

Tensile strength and elongation at break were measured using a universal testing machine (MTS QTEST5, model QC505B1, MTS Sys. Corp., Cary, NC, USA). Testing was conducted with ASTM D638. The dimensions of the film specimen were 45 mm $\times$ 8 mm $\times$ 0.2 mm. Every spectrum was tested 3 times and the average value was obtained.

3. Results and Discussion

3.1. Gel Permeation Chromatography Analysis

Figure 1 shows the gel permeation chromatography curves for PFD/PUs synthesized under different PFD proportions. As shown, the weight distributions of the PFD/PUs were unimodal, revealing that the synthesis was complete and without material residues. The molecular weight data are presented in Table 2. The results show that a high PFD content increased the effluent time, and the value of the molecular weight distribution (Mw/Mn; dispersity index) calculated using the PFD/PUs fell within 1.6–1.8. A decline in the viscosity of the PFD/PU polymer solution following the increase in PFD content also occurred during the synthesis. These results reveal that increasing the PFD chain extender content reduced the molecular weight of the FTPU. This was because the carbon chain number and molecular weight of PFD are higher than those of other chain extenders, such as 1,4-butanediol and ethylene glycol used in general PU. Therefore, the reduced molecular weight of the PFD/PUs following an increase in PFD was probably due to the effects of activity. The trend was the same as that in the study by Yang et al. [22], in which an increase in fluorine content reduced the molecular weight.

Table 2. Gel permeation chromatography (GPC) results of PFD/PUs.

Sample	Retention Time of the Peak (min)	$\overline{Mn}$ ($\times 10^4$)	$\overline{Mw}$ ($\times 10^4$)	$\overline{Mw}/\overline{Mn}$
PFD/PU-01	2.97	3.90	6.68	1.7
PFD/PU-02	3.05	3.08	5.46	1.8
PFD/PU-03	3.22	2.31	3.79	1.6

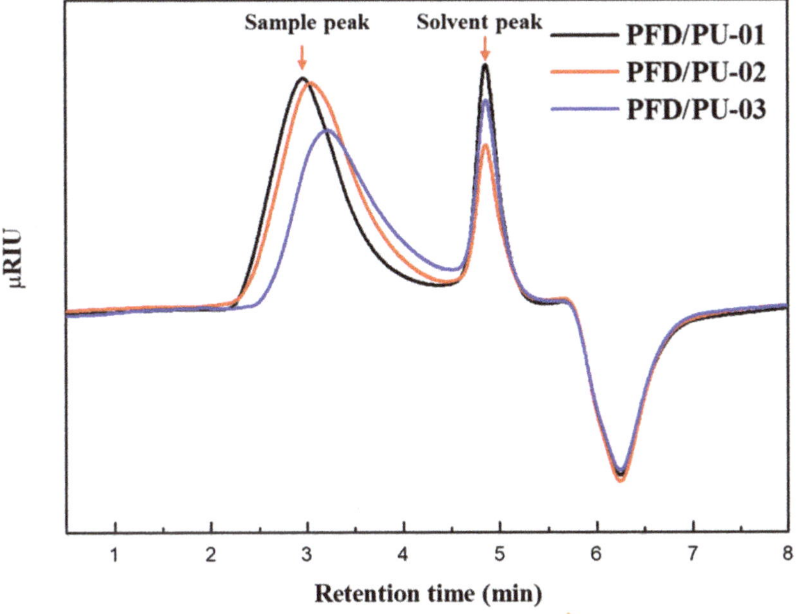

Figure 1. GPC curves of PFD/PUs.

3.2. FTIR

Figure 2a shows the FTIR spectrum of the PFD/PUs at a wavenumber range of 4000–650 cm^{-1}. The spectrum revealed that the polymers had five major common peaks: –NH stretching vibration peak (3333 cm^{-1}), CH$_2$ stretching vibration peak (2925 and 2862 cm^{-1}), C=O (amide I band, near 1727 cm^{-1}), –NH (amide II band, 1534.68 cm^{-1}), stretching vibration peak of the C–F group (1221–1205 cm^{-1}), and C–O stretching vibration peak (near 1098–1071 cm^{-1}). Moreover, no free NCO group was observed at 2240–2275 cm^{-1}. Therefore, MDI had fully reacted with the PCL or PFD chain extender during the synthesis processes and the yields of PFD/PUs were all 100%.

Figure 2b shows the absorption peak within the wavenumber range of 1900–1000 cm^{-1}. FTIR analysis conducted by Yang et al. [27] showed that C=O functional groups obtained from the PU system using the curve-fitting technique included C=O$_{free}$, C=O$_{HB\ disordered}$, and C=O$_{HB\ ordered}$, with C=O$_{HB\ ordered}$ appearing at approximately 1724, 1701, and 1660 cm^{-1}. Wang et al. [28] subjected FPU to an FTIR test, and the data indicated that when the C=O and –NH functional groups appeared at three peaks (free, disordered, and ordered) in the curve fitting, C–O and C–F functional groups produced two peak values (free and HB). The HB percentage of FPU was calculated at 1530 cm^{-1} because the stretching vibration peak of the benzene ring at this wavelength did not overlap with the other peak values. Therefore, the existence of HBs in the PU was proved using the following formula:

$$A\% = \frac{I_H/I_{ref}}{I_H/I_{ref} + I_{free}/I_{ref}} \tag{3}$$

where I_H is the HB strength, I_{free} is the free radical bonding strength, and I_{ref} represents the absorption intensity at 1534 cm^{-1}. Additionally, according to the experimental data in this study, the characteristic peaks of the N–H, C=O, C–F, and C–O groups affected by the HBs also appeared in the PFD/PUs, with H–bonded C=O, H–bonded C–F, and H–bonded C–O located at 1646, 1205, and 1098 cm^{-1}, respectively. This confirmed that a weak HB existed between N–H and C–F.

Figure 3 shows the absorption peaks within the wavenumber range 1240–1190 cm^{-1}. As shown, the three peak values that appeared at 1240–1190 cm^{-1} were amide III, C–F$_{free}$, and C–F$_{HB}$. The HB percentage of N–H···F–C was calculated using Equation (3), with PFD/PU-01, PFD/PU-02, and PFD/PU-03 having an HB percentage of 23.63%, 27.41%, and 31.18%, respectively. Accordingly, an increase in PFD enhanced the HB interaction in the FTPU film.

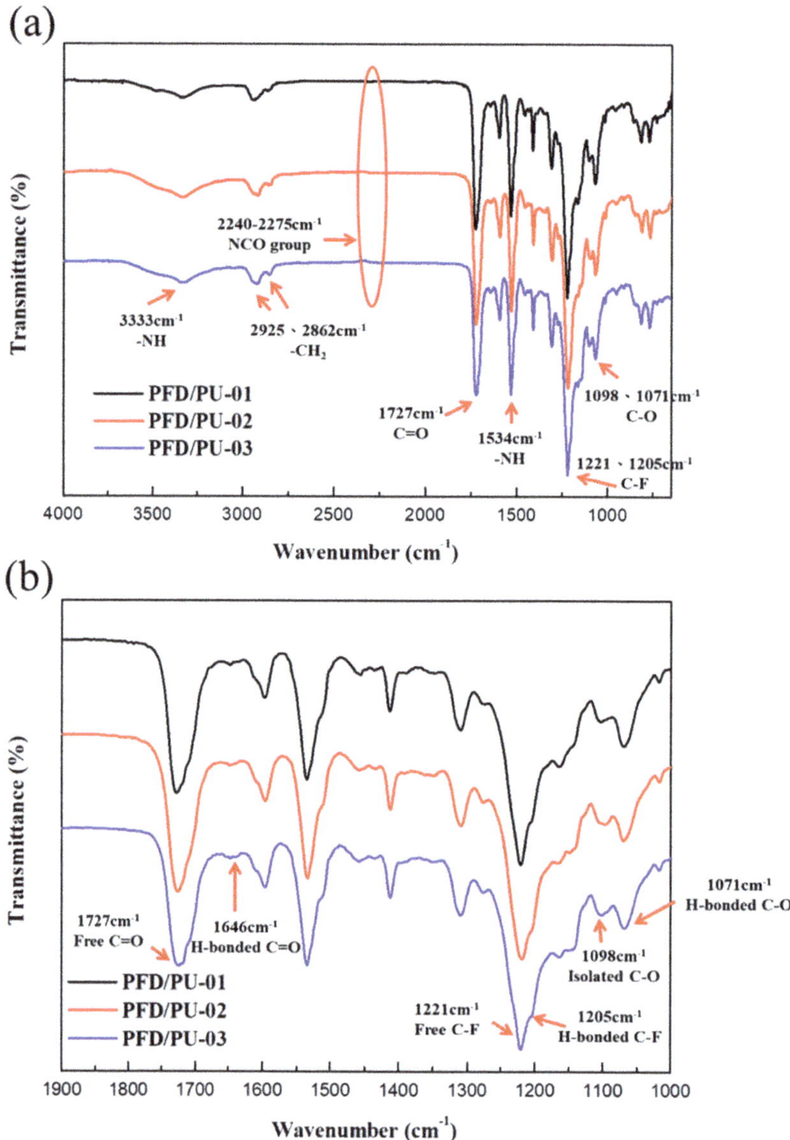

Figure 2. FT-IR spectra of PFD/PUs at wavenumber range of (**a**) 4000–650 cm^{-1} and (**b**) 1900–1000 cm^{-1}.

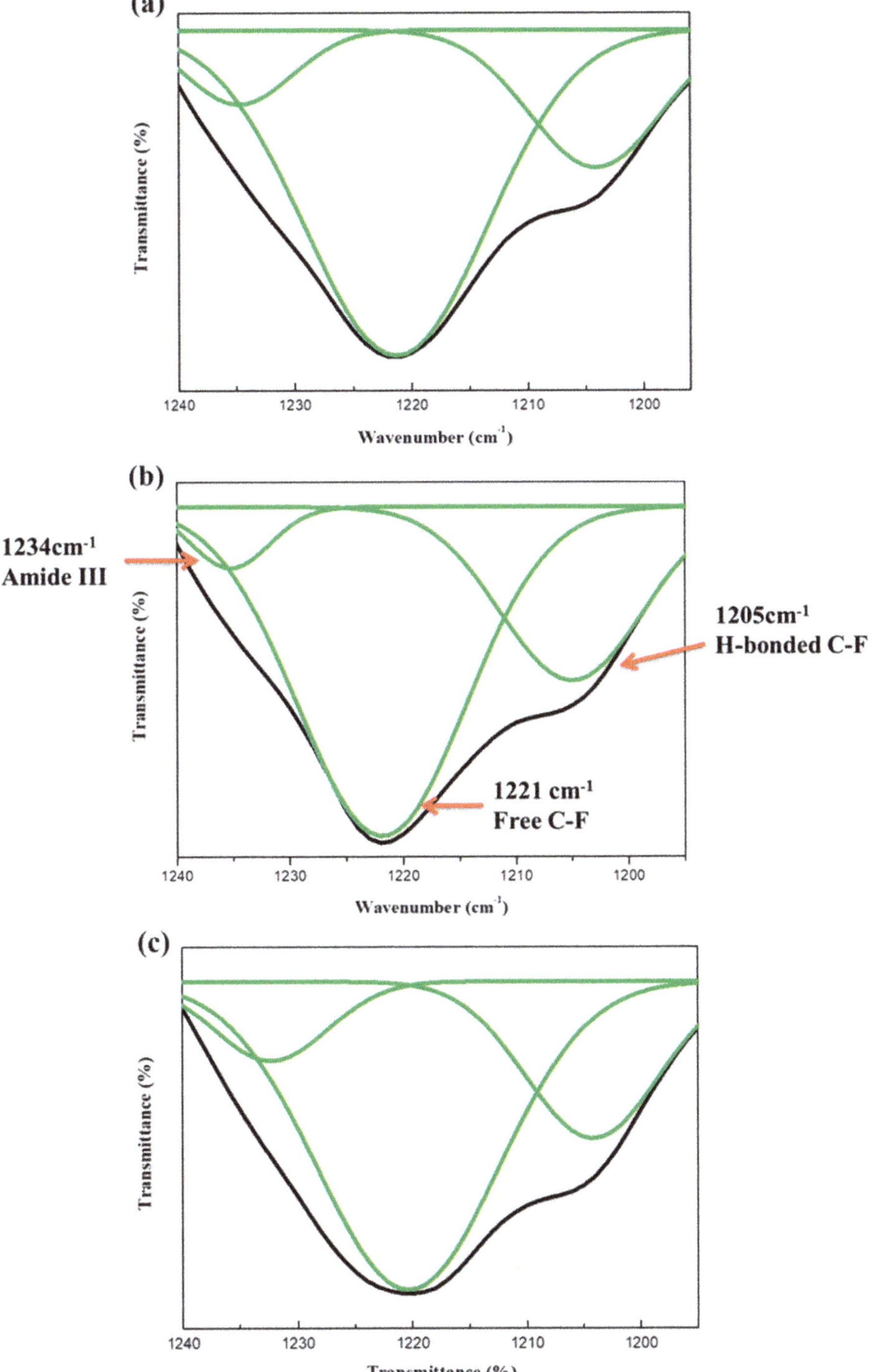

Figure 3. FT-IR spectra of (**a**) PFD/PU-01, (**b**) PFD/PU-02, and (**c**) PFD/PU-03 at the wavenumber range of 1240–1190 cm^{-1}.

3.3. Fluorine-19 NMR

Figure 4 shows the molecular structure of the fluorine parts of the PFD/PUs and the ^{19}F NMR analytical chart for PFD/PU-01. The figure shows three absorption peaks, labeled 1–3. ^{19}F–^{19}F COSY of PFD/PU-01 was performed to accurately analyze F1–F3. Figure 5 shows two signals of F2 (labeled F2 and F2′) and three strong correlations (F2–F4, F2–F2′, and F1–F2′). Fluorine spectrum studies have found that $^4J(F,F)$ is stronger than $^3J(F,F)$ [29], indicating that a strong coupling exists between the next signals (i.e., $^4J(F,F)$). Figure 5 also shows that F–A and F–A′ were the most affected by the other elements and were thus labeled F1. In a similar fashion, the peaks at −119.78 ppm (F1), −121.59 ppm (F2 and F2′), and −123.83 ppm (F3) corresponded to F–A and F–A′; F–C, F–C′, F–D, and F–D′; and F–B and F–B′, respectively. The corresponding positions of fluorine were confirmed, as shown in Figure 6. Such coupling was insufficient to verify that the fluorinated chain extender was attached to the PU, and additional identification through 2D NMR spectroscopy (^{1}H–^{19}F COSY, ^{1}H–^{13}C HMBC) was required. Figure 7 shows the ^{1}H–^{19}F COSY diagram for the PFD/PUs. The spectrum revealed that the H atoms of the CH_2 group in the fluorinated chain extender were located at 4.89 ppm and had a relevant coupling ($^3J(H,F)$) with F1. Additionally, Figure 7 illustrates that the H atoms at 4.89 ppm shared a weak coupling ($^4J(F,F)$) with F3, which again verified that the analysis was correct. Figure 8 shows the ^{1}H–^{13}C heteronuclear multiple bond correlation (HMBC) spectrum of PFD/PU-01. According to the literature, the PU ester O–C=O is located at approximately 153 ppm [30], which revealed that the C=O location corresponded to the location of the H atoms of CH_2 (4.89 ppm). The aforementioned analysis showed that the PFD chain extender had successfully reacted with MDI to form urethane groups.

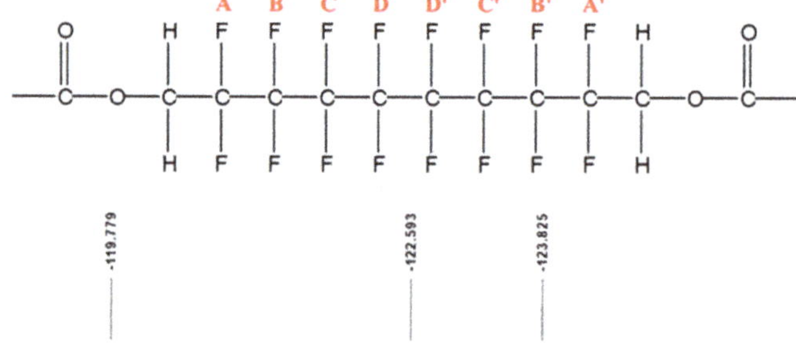

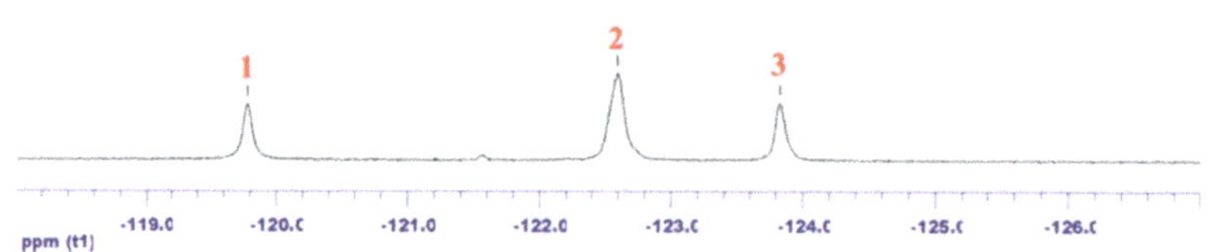

Figure 4. ^{19}F NMR spectrum and molecular structure of PFD/PU-03.

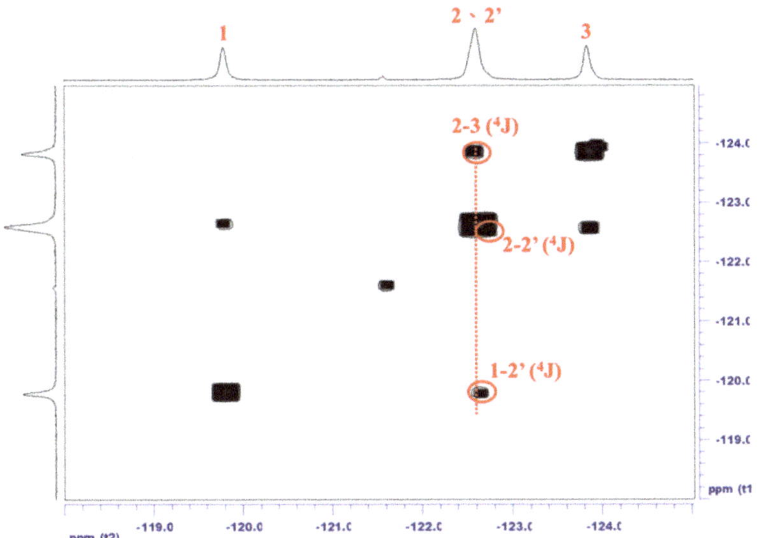

Figure 5. $^{19}\mathrm{F}$–$^{19}\mathrm{F}$ COSY NMR spectrum of PFD/PU-03.

Figure 6. Structure of the biodegradable fluorine-containing polyurethanes.

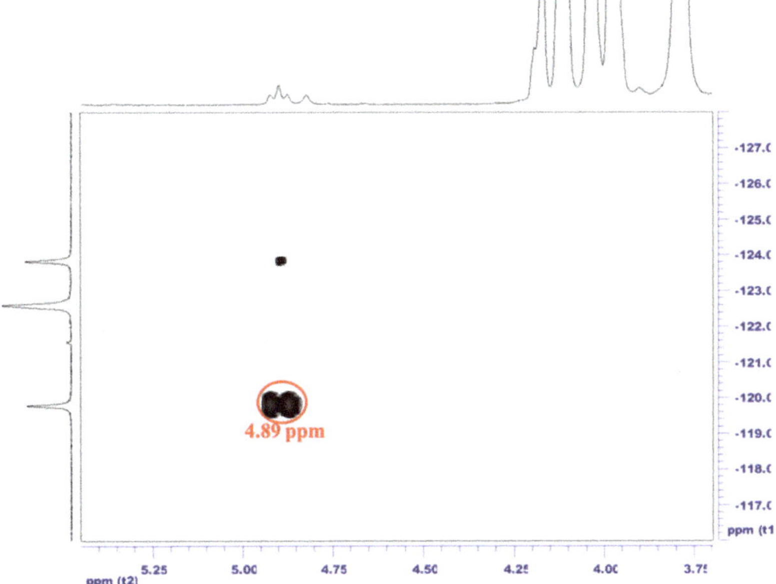

Figure 7. $^{1}\mathrm{H}$–$^{19}\mathrm{F}$ COSY NMR of PFD/PU-03.

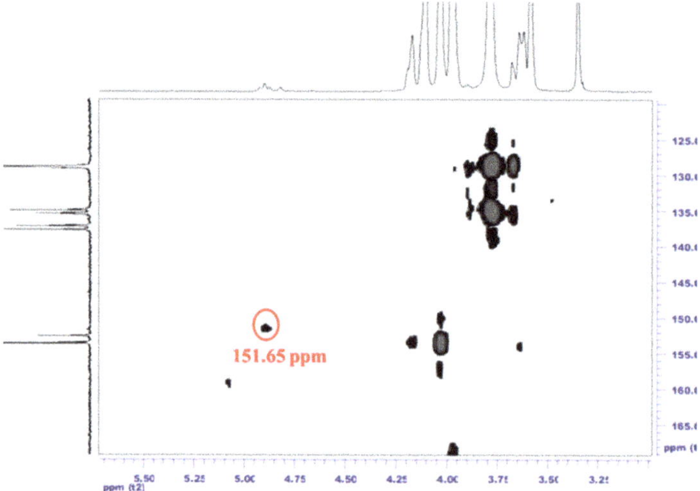

Figure 8. ^{1}H–^{13}C HMBC NMR of PFD/PU-03.

3.4. X-Ray Photoelectron Spectroscopy

Figure 9 shows XPS spectra for PFD/PUs, with each spectrum containing four main peaks: C1s, O1s, N1s, and F1s. The element composition and peak-related properties are listed in Table 3. These findings revealed that the binding energies of C1s, O1s, N1s, and F1s decreased following an increase in PFD content, and the F content increased from 2.31% to 9.47%. Compared with PFD/PU-02 and PFD/PU-03, the F1s binding energies of the C–F bond in PFD/PU-01 exhibited a clear offset from 690 to 688 eV. Accordingly, the molecular interaction in the PFD/PU film changed when the PFD content increased. The O1 elements in PFD/PU-01 and PFD/PU-03 were subjected to an XPS peak-differentiation-imitating analysis (Figure 10). As shown, the O–C=O* binding energies of PFD/PU-01 and PFD/PU-03 were 532.08 and 531.98 eV, respectively. Berger et al. [19] concluded that the interaction of organic fluorine reveals that a dipole–dipole interaction exists between C–F⋯C=O. These results reveal that the lower transfer values of the C1s, O1s, N1s, and F1s binding energies confirmed the interaction between the –C=O group and C–F in the PFD/PUs [31].

Table 3. XPS peak characteristics of PFD/PUs.

Content		Binding Energy (eV)	Atomic Ratio (%)
PFD/PU-01	C1s	289.18	92.19
	O1s	536.83	2.56
	N1s	403.56	2.94
	F1s	690.86	2.31
PFD/PU-02	C1s	286.08	89.36
	O1s	533.85	3.00
	N1s	400.42	3.23
	F1s	688.04	4.41
PFD/PU-03	C1s	284.54	71.45
	O1s	532.23	14.9
	N1s	399.76	4.18
	F1s	688.22	9.47

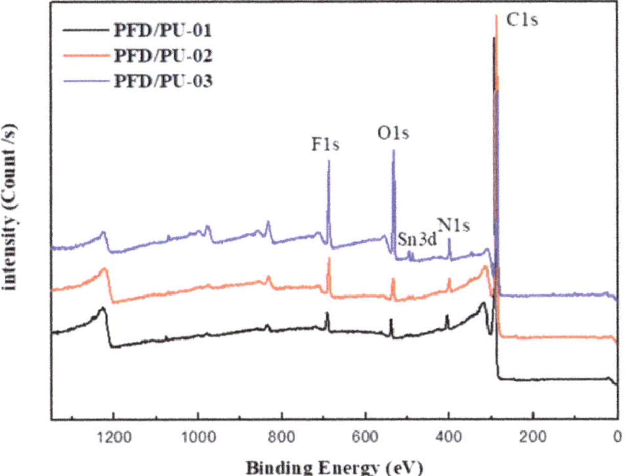

Figure 9. XPS survey spectra of PFD/PUs.

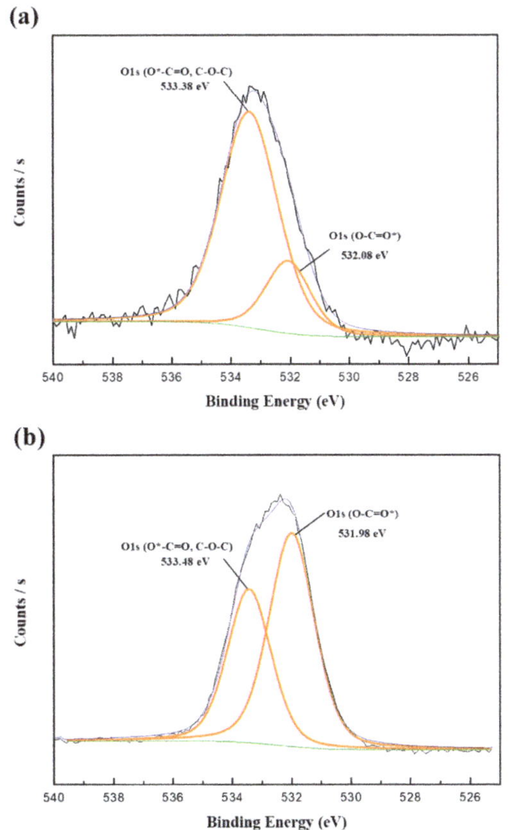

Figure 10. O1s fit theory of (**a**) PFD/PU-01, and (**b**) PFD/PU-03.

Figure 11 shows the XPS peak-differentiation-imitating analysis of C1s plotted for the PFD/PUs with different proportions of the PFD chain extender. The C–C binding energy distributed by the C1s curve in the PFD/PUs was approximately 285.0 eV, and approximately 286 eV for C–O, 287 eV for C–O–C, and 292 eV for C–F_2. The corresponding peak produced by O–C=O was approximately 288 eV [32], and it was distributed to the carbonyl group in the urethane group. Table 3 illustrates that the nitrogen content increased following an increase in PFD and that the amount of nitrogen represented the amount of hard segments. Substantial HB interactions between C–F and N–H may also have existed within the PFD/PUs; thus, fluorine chains were believed to facilitate the pulling of the hard segments to the surface of the PU [33]. According to the figure, the increased PFD content led to a shift in the C–F position from 292 to 293.0 eV, a change in the peak intensity, and an increase in C–N binding energy from 285.88 to 285.95 eV. The reason was that increasing the PFD increased the number of C–F⋯H–N HBs, which in turn increased the C–F binding energy, a result that was consistent with the FTIR analysis. Moreover, the binding energies of C–O, C–O–C, and O–C=O decreased following an increase in PFD content. This may have been caused by introducing long-chain fluothane segments into the PU, which disrupted the original HB reaction of the PU due to a steric hindrance, thereby reducing the binding energy. However, C–F⋯H–N had a greater HB interaction than did C=O⋯H–N because of the high electronegativity of fluorine, and a stronger HB interaction was produced in the PU film.

3.5. Surface Roughness Analysis

The left and right images in Figure 12a–c show the topography and phase data images for PFD/PU-01, PFD/PU-02, and PFD/PU-03, respectively. The PFD/PUs exhibited some continuous protrusions in the topography. The average surface roughness of PFD/PU-01, PFD/PU-02, and PFD/PU-03 was 2.17, 2.72, and 4.45 nm, respectively. The results revealed that the surface roughness increased when the PFD content increased, which caused a rougher FTPU. This phenomenon was attributed to the increase in the hard segments of the FTPU and the interaction between CF_2 and C=O in the hard segments following the increase in PFD. In other words, increasing the PFD chain extender increased the HB interaction in the FTPU film, which in turn caused aggregations or protrusions on the film's surface [24]. Additionally, numerous continuous irregular granular and stripe phases were observed in the phase diagram of the PFD/PUs, which increased as the PFD content increased. These irregular phases revealed that the hard segments were rich in PFD chain extender [34,35], a phenomenon that was consistent with the findings in the XPS spectrum.

3.6. Thermal Properties

Figure 13 illustrates the thermogravimetric analysis (TGA) curve of the PFD/PUs synthesized with different amounts of PFD chain extender. The initial decomposition temperature of the PFD/PUs was defined as T_{onset}, which related to pyrolysis of the FTPU. The data revealed that the T_{onset} of PFD/PU-01, PFD/PU-02, and PFD/PU-03 was 299.2, 305.1, and 308.6 °C, respectively; the thermogravimetric data of PFD/PUs are presented in Table 4. The results show that T_{onset} increased when the PFD chain extender content increased in the PFD/PUs. This could be attributed to the strong bonding energy of –CF_2 (540 kJ/mol), which required relatively high energy to break the bond. Furthermore, the covalent radius of the fluorine atom was equivalent to half the C–C bond length; thus, fluorine atoms shielded the main C–C chain and ensured its stability. Moreover, the interaction between the C=O and –CF_2 groups was verified through FTIR, and the polar bonding of –CF_2 contributed to the formation of the phase separation of hard segments of the PU film in the soft segment [36]. The additional PFD increased the thermal stability of the PU film. The residual weight at 700 °C exhibited an increase when the PFD increased, which consequently reduced the amount of PCL that was required. This resulted in more hard segments, which facilitated carbon formation.

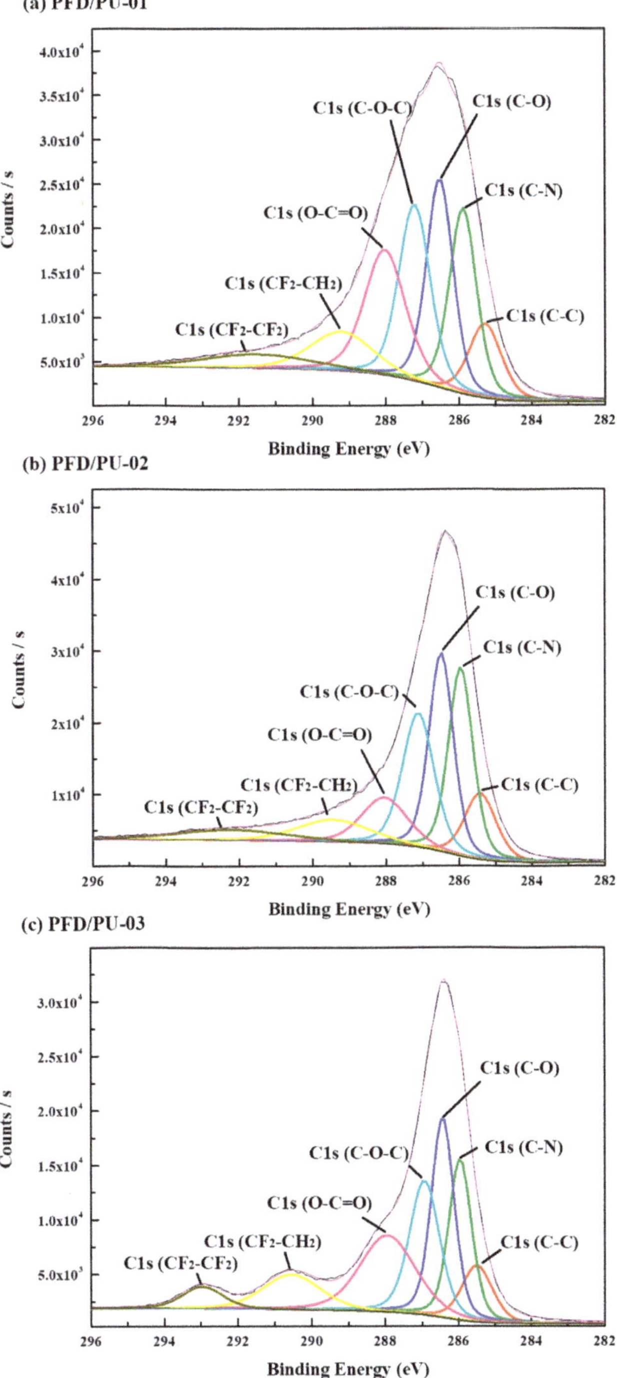

Figure 11. C1s fit theory of PFD/PUs.

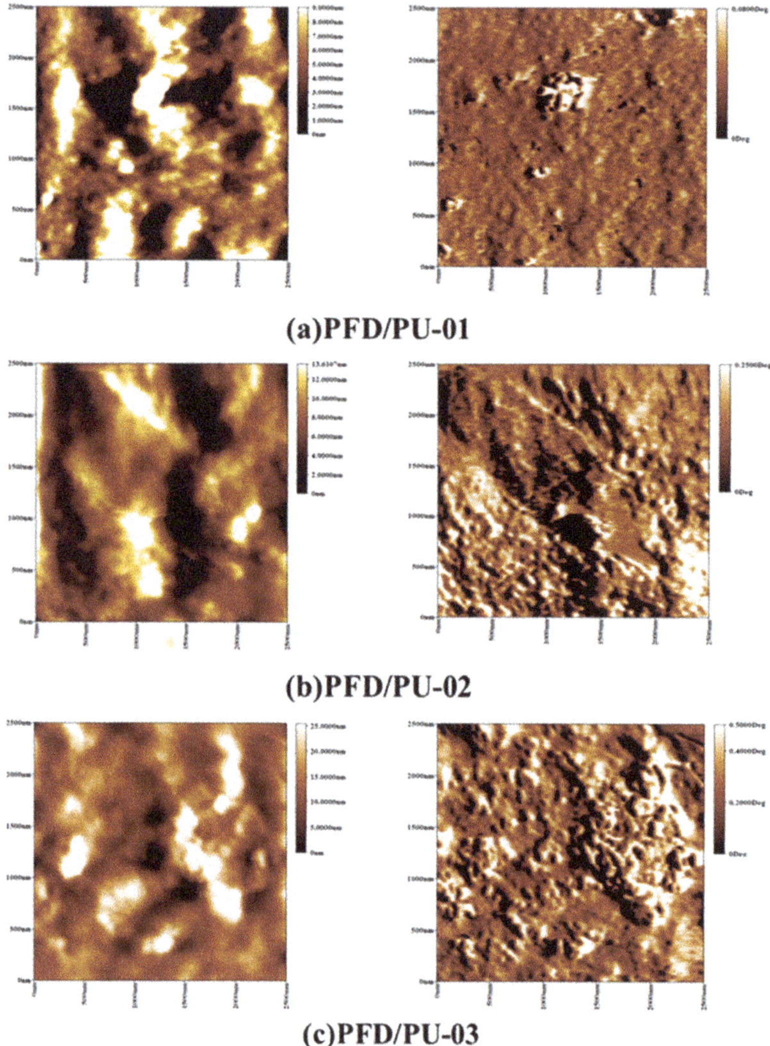

(a)PFD/PU-01

(b)PFD/PU-02

(c)PFD/PU-03

Figure 12. Atomic force microscopy (AFM) topographic and phase images of PFD/PUs: (**a**) PFD/PU-01, (**b**) PFD/PU-02, and (**c**) PFD/PU-03.

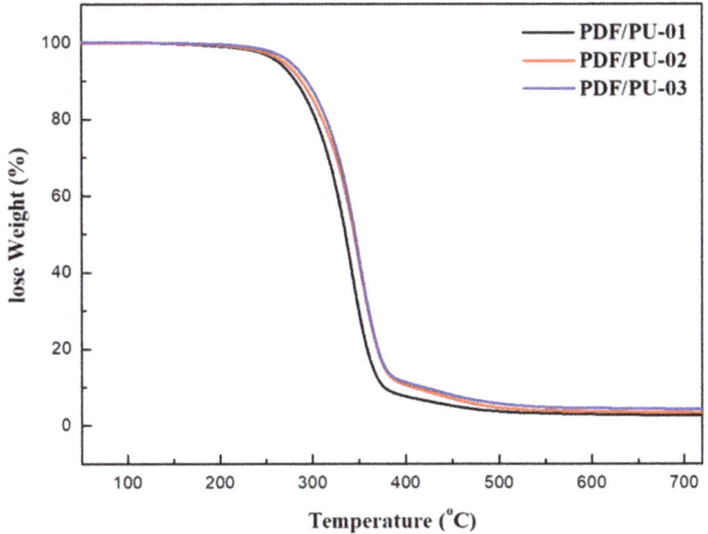

Figure 13. TGA and DTG curves of PFD/PUs.

Table 4. Thermal properties of PFD/PUs. TGA, thermogravimetric analysis; DSC, differential scanning calorimetry.

Designation	TGA		DSC
	T_{onset} (°C)	Residue at 700 °C (%)	T_g (°C)
PFD/PU-01	299.2	2.7	3.7
PFD/PU-02	305.9	4.1	5.6
PFD/PU-03	308.6	4.2	10.3

Figure 14 shows the differential scanning calorimetry thermograms of the PFD/PUs with different PFD content, and the relevant data are displayed in Table 4. The results reveal that the glass transition temperature (T_g) points of PFD/PU-01, PFD/PU-02, and PFD/PU-03 were 3.7, 5.6, and 10.3 °C, respectively. T_g was related to the soft segment that consisted of repeat linkages of reacted alternative MDI and PCL units. Previous FTIR spectra indicated the presence of a strong interaction between the C=O groups in soft segments and the $-CF_2$ groups in hard segments of the PFD/PUs. When PFD content was higher, more $-CF_2$ groups therein would cause a higher interaction that inhibited the segmental chain motion in the PFD/PUs, consequently increasing the T_g of PFD/PUs. In other words, the PFD/PUs with more hard segments or chain extenders would have higher T_g as previously reported [37].

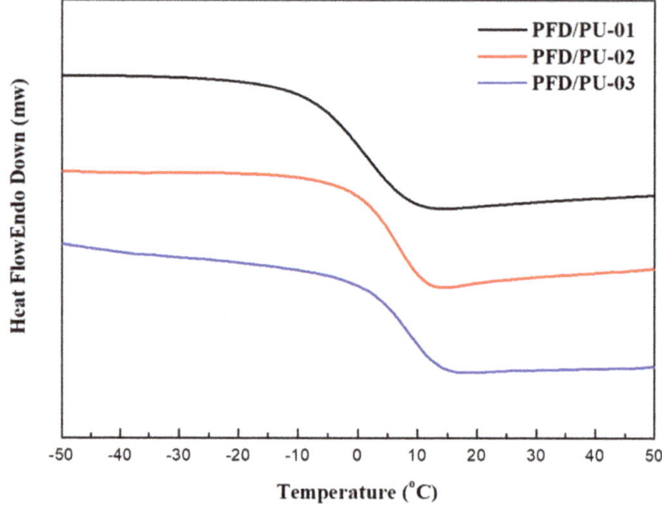

Figure 14. DSC thermograms of PFD/PUs.

3.7. Dynamic Mechanical Analysis

Figure 15 shows the tan δ and loss modulus (E″) of the PFD/PUs with different amounts of PFD chain extenders. The dynamic T_g was defined as T_{gd}. As shown, the T_{gd} of PFD/PU-01, PFD/PU-02, and PFD/PU-0 from the tan δ curve was 7.9, 10.8, and 13.3 °C, respectively, whereas the T_{gd} from the E″ curves was 0.1, 3.6, and 5.4 °C, respectively. The T_g values obtained using different testing methods are listed in Table 5. The results show that the T_{gd} of the PFD/PUs increased as the PFD chain extender content increased. This may have been caused by the inhibition of segmental motion of the PFD/PUs following the increase in hard segments and C–F···H–N HB interactions that increased the T_{gd} of the PFD/PUs, a finding that was similar to the results from the thermal property analysis described previously. The tan δ curves of the PFD/PUs indicate a decrease in tan $δ_{max}$ with an increase in PFD content. This was because of the increased hard segments and influence of the HB interactions following the increased PFD, resulting in more elastic PFD/PUs, because the value of tan δ was obtained by dividing the value of E″ by E′. Therefore, PFD/PU-03, which had the highest fluorine content, showed the lowest peak value. In other words, the hard segments containing PFD units were harder than the soft segments containing PCL units. Additionally, the dipole–dipole interaction between C–F···C=O contributed to the blocking of the segmental activity of the PFD/PUs. In summary, the PFD/PUs with relatively high PFD content were elastic, which suggests that increasing the PFD content improves the rigidity of PFD/PUs.

Table 5. DMA results of PFD/PUs.

Sample	T_{gd} from Tan δ (°C)	Tan $δ_{max}$	T_{gd} from E″ (°C)
PFD/PU-01	7.9	0.756	0.1
PFD/PU-02	10.8	0.668	3.6
PFD/PU-03	13.3	0.638	5.4

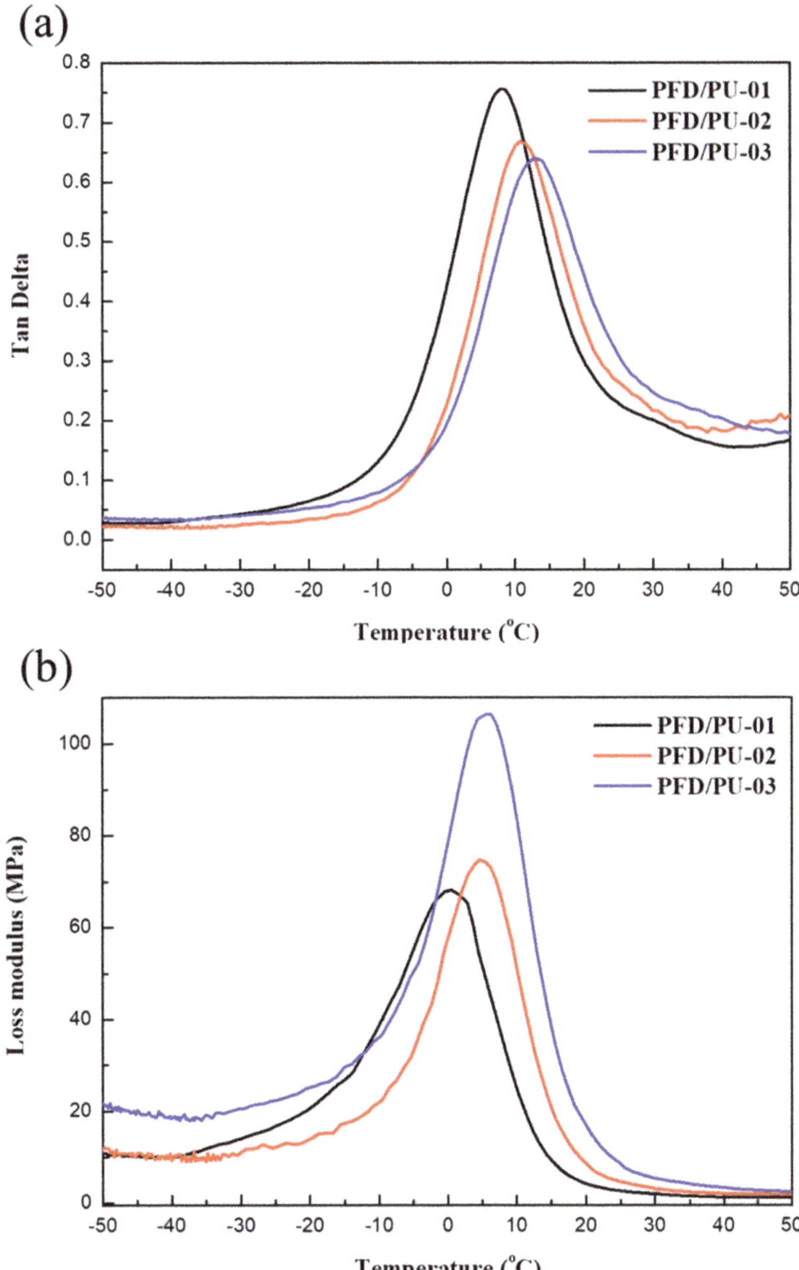

Figure 15. DMA (**a**) tan δ and (**b**) loss modulus (E″) curves of PFD/PUs.

3.8. Tensile Properties

Figure 16 shows the stress–strain curves of the PFD/PUs synthesized with different amounts of PFD chain extender, and the data on their mechanical properties are listed in Table 6. According to the results, PFD/PU-01, PFD/PU-02, and PFD/PU-03 had, respectively, a maximum tensile strength of

8.23, 16.34, and 21.55 MPa; an extension at break of 1630%, 1434%, and 1156%; and a Young's modulus of 0.5, 1.4, and 2.1 MPa. These results reveal that increasing the PFD content increased the tensile strength and Young's modulus. This is due to the following: first, the FTPU film was more rigid when FTPU contained more PFD or hard segments; and second, the HBs produced between –NH and CF_2 in the PFD/PUs inhibited the segmental motion of the PFD/PUs. The result was an increase in tensile strength and Young's modulus in the FTPU film. The results of the mechanical property curve are consistent with those of the dynamic mechanical analysis.

Table 6. Tensile properties of PFD/PUs.

Sample	Tensile Strength (MPa)	Young's Modulus (MPa)	Elongation at Break (%)
PFD/PU-01	8.0 ± 0.2	0.5	1614 ± 34
PFD/PU-02	16.1 ± 0.7	1.4	1427 ± 15
PFD/PU-03	21.0 ± 1.3	2.1	1148 ± 57

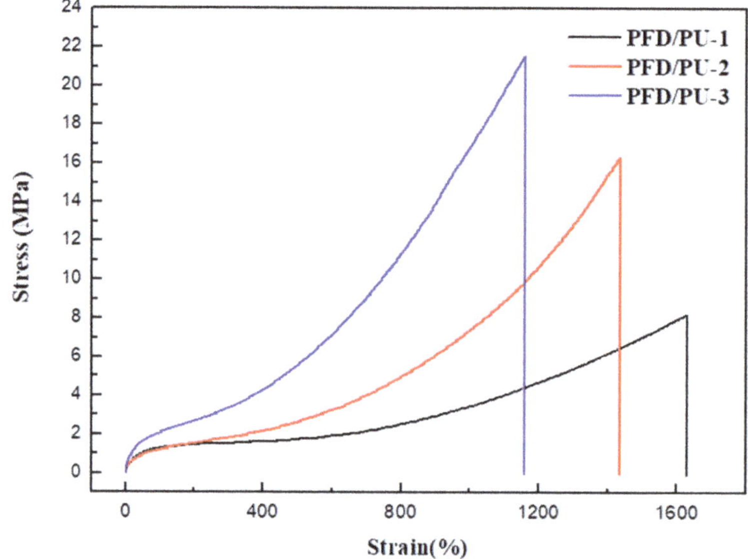

Figure 16. Tensile properties of PFD/PUs.

4. Conclusions

In this study, PFD was introduced into PU to produce FTPU, and ^{1}H NMR, ^{19}F NMR, ^{19}F–^{19}F COSY, ^{1}H–^{19}F COSY, and HMBC confirmed the successful synthesis of PFD/PUs. The results from FTIR and XPS indicate that introducing the CF_2 group into the PFD produced an HB interaction with the –NH group; a high PFD content resulted in more HB interactions in the PFD/PUs. AFM showed that PFD contributed to the microphase separation of the PFD/PUs because of the HB interactions. Thermal property analysis showed that because of the strong binding energy of –CF_2 in the PFD (540 kJ/mol) and because the covalent radius of fluorine atoms is equivalent to half the C–C bond length, increasing the PFD content shielded the main C–C chain and enhanced the thermal stability of the PFD/PUs. Dynamic mechanical analysis and tensile strength tests also confirmed that the PFD extender enhanced the rigidity of the PU film.

Author Contributions: Data curation, H.-A.T.; Formal analysis, J.-W.L.; Methodology, H.-T.L.; Project administration, M.-C.S. and C.-W.C.

Funding: This research received no external funding.

Conflicts of Interest: The authors declare no conflict of interest.

References

1. Alagi, P.; Choi, Y.J.; Hong, S.C. Preparation of vegetable oil-based polyols with controlled hydroxyl functionalities for thermoplastic polyurethane. *Eur. Polym. J.* **2016**, *78*, 46–60. [CrossRef]
2. Schneider, N.S.; Sung, C.S.P.; Matton, R.W.; Illinger, J.L. Thermal Transition Behavior of Polyurethanes Based on Toluene Diisocyanate. *Macromolecules* **1975**, *8*, 62–67. [CrossRef]
3. Paik Sung, C.S.; Schneider, N.S. Structure-property relationships of polyurethanes based on toluene di-isocyanate. *J. Mater. Sci.* **1978**, *13*, 1689–1699. [CrossRef]
4. Guo, Y.; Zhang, R.; Xiao, Q.; Guo, H.; Wang, Z.; Li, X.; Chen, J.; Zhu, J. Asynchronous fracture of hierarchical microstructures in hard domain of thermoplastic polyurethane elastomer: Effect of chain extender. *Polymer* **2018**, *138*, 242–254. [CrossRef]
5. Oprea, S. Effect of Composition and Hard-segment Content on Thermo-mechanical Properties of Cross-linked Polyurethane Copolymers. *High Perform. Polym.* **2009**, *21*, 353–370. [CrossRef]
6. Li, G.; Li, E.; Wang, C.; Niu, Y. Effect of the blocked ratio on properties of natural fiber–waterborne blocked polyurethane composites. *J. Compos. Mater.* **2015**, *49*, 1929–1936. [CrossRef]
7. Bernardini, J.; Cinelli, P.; Anguillesi, I.; Coltelli, M.B.A. Lazzeri, Flexible polyurethane foams green production employing lignin or oxypropylated lignin. *Eur. Polym. J.* **2015**, *64*, 147–156. [CrossRef]
8. Javni, I.; Petrović, Z.S.; Guo, A.; Fuller, R. Thermal stability of polyurethanes based on vegetable oils. *J. Appl. Polym. Sci.* **2000**, *77*, 1723–1734. [CrossRef]
9. Chuang, F.S.; Tsen, W.C.; Shu, Y.C. The effect of different siloxane chain-extenders on the thermal degradation and stability of segmented polyurethanes. *Polym. Degrad. Stab.* **2004**, *84*, 69–77. [CrossRef]
10. Liu, C.H.; Lee, H.T.; Tsou, C.H.; Gu, J.H.; Suen, M.C.; Chen, J.K. In Situ Polymerization and Characteristics of Biodegradable Waterborne Thermally-Treated Attapulgite Nanorods and Polyurethane Composites. *J. Inorg. Organomet. Polym. Mater.* **2017**, *27*, 244–256. [CrossRef]
11. Tayfun, U.; Kanbur, Y.; Abacı, U.; Güney, H.Y.; Bayramlı, E. Mechanical, electrical, and melt flow properties of polyurethane elastomer/surface-modified carbon nanotube composites. *J. Compos. Mater.* **2017**, *51*, 1987–1996. [CrossRef]
12. Verma, G.; Kaushik, A.; Ghosh, A.K. Preparation, characterization and properties of organoclay reinforced polyurethane nanocomposite coatings. *J. Plast. Film Sheet.* **2013**, *29*, 56–77. [CrossRef]
13. Jiang, X.; Gu, J.; Shen, Y.; Wang, S.; Tian, X. New fluorinated siloxane-imide block copolymer membranes for application in organophilic pervaporation. *Desalination* **2011**, *265*, 74–80. [CrossRef]
14. Chen, J.H.; Hu, D.D.; Li, Y.D.; Meng, F.; Zhu, J.; Zeng, J.B. Castor oil derived poly (urethane urea) networks with reprocessibility and enhanced mechanical properties. *Polymer* **2018**, *143*, 79–86. [CrossRef]
15. Baradie, B.; Shoichet, M.S. Novel fluoro-terpolymers for coatings applications. *Macromolecules* **2005**, *38*, 5560–5568. [CrossRef]
16. Lee, S.R.; Kim, M.R.; Jo, E.H.; Yoon, K.B. Synthesis of very low birefringence polymers using fluorinated macromers for polymeric waveguides. *High Perform. Polym.* **2016**, *28*, 131–139. [CrossRef]
17. Saritha, B.; Thiruvasagam, P. Synthesis of diol monomers and organosoluble fluorinated polyimides with low dielectric: Study of structure–property relationship and applications. *High Perform. Polym.* **2015**, *27*, 842–851. [CrossRef]
18. Améduri, B.; Boutevin, B.; Kostov, G. Fluoroelastomers: Synthesis, properties and applications. *Prog. Polym. Sci.* **2001**, *26*, 105–187. [CrossRef]
19. Berger, R.; Resnati, G.; Metrangolo, P.; Weber, E.; Hulliger, J. Organic fluorine compounds: A great opportunity for enhanced materials properties. *Chem. Soc. Rev.* **2011**, *40*, 3496–3508. [CrossRef] [PubMed]
20. Jia, R.P.; Zong, A.X.; He, X.Y.; Xu, J.Y.; Huang, M.S. Synthesis of newly fluorinated thermoplastic polyurethane elastomers and their blood compatibility. *Fiber. Polym.* **2015**, *16*, 231–238. [CrossRef]
21. Liu, T.; Ye, L. Synthesis and properties of fluorinated thermoplastic polyurethane elastomer. *J. Fluorine Chem.* **2010**, *131*, 36–41. [CrossRef]
22. Yang, L.; Wang, Y.; Peng, X. Synthesis and characterization of novel fluorinated thermoplastic polyurethane with high transmittance and superior physical properties. *Pure Appl. Chem.* **2017**, *54*, 516–523. [CrossRef]

23. Wu, C.L.; Chiu, S.H.; Lee, H.T.; Suen, M.C. Synthesis and properties of biodegradable polycaprolactone/polyurethanes using fluoro chain extenders. *Polym. Adv. Technol.* **2016**, *27*, 665–676. [CrossRef]

24. Su, S.K.; Gu, J.H.; Lee, H.T.; Wu, C.L.; Hwang, J.J.; Suen, M.C. Synthesis and properties of novel biodegradable polyurethanes containing fluorinated aliphatic side chains. *J. Polym. Res.* **2017**, *24*, 142. [CrossRef]

25. Dieterich, D. Aqueous emulsions, dispersions and solutions of polyurethanes; synthesis and properties. *Prog. Org. Coat.* **1981**, *9*, 281–340. [CrossRef]

26. Mo, F.; Zhou, F.; Chen, S.; Yang, H.; Ge, Z.; Chen, S. Development of shape memory polyurethane based on polyethylene glycol and liquefied 4, 4′-diphenylmethane diisocyanate using a bulk method for biomedical applications. *Polym. Int.* **2015**, *64*, 477–485. [CrossRef]

27. Yang, W.; Cheng, X.; Wang, H.; Liu, Y.; Du, Z. Surface and mechanical properties of waterborne polyurethane films reinforced by hydroxyl-terminated poly (fluoroalkyl methacrylates). *Polymer* **2017**, *133*, 68–77. [CrossRef]

28. Wang, X.; Xu, J.; Li, L.; Liu, Y.; Li, Y.; Dong, Q. Influences of fluorine on microphase separation in fluorinated polyurethanes. *Polymer* **2016**, *98*, 311–319. [CrossRef]

29. Kysilka, O.; Rybáčková, M.; Skalický, M.; Kvíčalová, M.; Cvačka, J.; Kvíčala, J. HFPO Trimer-Based Alkyl Triflate, a Novel Building Block for Fluorous Chemistry. Preparation, Reactions and 19F gCOSY Analysis. *Collect. Czech. Chem. Commun.* **2008**, *73*, 1799–1813. [CrossRef]

30. Tang, Q.; Gao, K. Structure analysis of polyether-based thermoplastic polyurethane elastomers by FTIR, 1H NMR and 13C NMR. *Int. J. Polym. Anal. Charact.* **2017**, *22*, 569–574. [CrossRef]

31. Kerr, S.; Naumkin, F.Y. Noncovalently bound complexes of polar molecules: Dipole-inside-of-dipole vs. dipole–dipole systems. *New J. Chem.* **2017**, *41*, 13576–13584. [CrossRef]

32. Laoharojanaphand, P.; Lin, T.J.; Stoffer, J.O. Glow discharge polymerization of reactive functional silanes on poly(methyl methacrylate). *J. Appl. Polym. Sci.* **1990**, *40*, 369–384. [CrossRef]

33. Li, N.; Zeng, F.; Wang, Y.; Qu, D.; Hu, W.; Luan, Y.; Dong, S.; Zhang, J.; Bai, Y. Fluorinated polyurethane based on liquid fluorine elastomer (LFH) synthesis via two-step method: The critical value of thermal resistance and mechanical properties. *RSC Adv.* **2017**, *7*, 30970–30978. [CrossRef]

34. Sauer, B.B.; Mclean, R.S.; Thomas, R.R. Tapping mode AFM studies of nano-phases on fluorine-containing polyester coatings and octadecyltrichlorosilane monolayers. *Langmuir* **1998**, *14*, 3045–3051. [CrossRef]

35. Xu, W.; Zhao, W.; Hao, L.; Wang, S.; Pei, M.; Wang, X. Synthesis of novel cationic fluoroalkyl-terminated hyperbranched polyurethane latex and morphology, physical properties of its latex film. *Prog. Org. Coat.* **2018**, *121*, 209–217. [CrossRef]

36. Xu, W.; Zhang, Y.; Hu, Y.; Lu, B.; Cai, T. Synthesis and Characterization of Novel Fluorinated Polyurethane Elastomer Based on 1,4-Bis (4-amino-2-trifluoroformethyloxyphenyl) benzene. *Asian J. Chem.* **2012**, *24*, 3976–3980.

37. Tsou, C.H.; Lee, H.T.; Tsai, H.A.; Cheng, H.J.; Suen, M.C. Synthesis and properties of biodegradable polycaprolactone/polyurethanes by using 2, 6-pyridinedimethanol as a chain extender. *Polym. Degrad. Stabil.* **2013**, *98*, 643–650. [CrossRef]

Article

Investigation on the Short-Term Aging-Resistance of Thermoplastic Polyurethane-Modified Asphalt Binders

Ruien Yu [1,2,*], Xijing Zhu [1,2], Maorong Zhang [1,2] and Changqing Fang [3]

[1] School of Mechanical Engineering, North University of China, Taiyuan 030051, China; zxj161501@nuc.edu.cn (X.Z.); zhangmaorong@nuc.edu.cn (M.Z.)

[2] Shanxi Key Laboratory of Advanced Manufacturing Technology, North University of China, Taiyuan 030051, China

[3] School of Mechanical and Precision Instrument Engineering, Xi'an University of Technology, Xi'an 710048, China; fangcq@xaut.edu.cn

* Correspondence: yuruien@nuc.edu.cn; Tel.: +86-178-3561-0106

Received: 17 September 2018; Accepted: 23 October 2018; Published: 25 October 2018

Abstract: In this reported work, thermoplastic polyurethane (TPU) was used as a reactive polymer modifying agent to prepare a modified-asphalt, using a high-speed shearing method. Physical performance tests of the TPU-modified asphalt were conducted before and after short-term aging, and the aging resistance was examined by the change in materials properties. In addition, low-temperature rheological properties, thermal properties, the high-temperature storage stability, and the aging mechanism of TPU-modified asphalt were also investigated. The results showed that the addition of TPU improved the aging resistance of base asphalt, which was evidenced by the increased penetration ratio and decreased softening point of the asphalt, after aging. Similarly, Fourier Transform infrared (FTIR) spectroscopy results verified that TPU improved the asphalt aging resistance. It was found that the TPU functional groups played a role in improving thermal properties, high-temperature storage stability, and in the dispersion of modified asphalt.

Keywords: modified-asphalt; thermoplastic polyurethane; rheological properties; thermal properties; microstructure

---

1. Introduction

Asphalt is an engineering material used in many industrial sectors of the national economy. Its wide range of uses makes asphalt an irreplaceable product, particularly in road construction and in waterproofing of buildings [1]. However, increased road traffic and the heavier vehicle loads demand that the requirements for road construction is greatly improved, which translates to improving the properties of asphalt materials [2,3]. The elastomer rubber was the earliest asphalt modifier and rubber-modified asphalt offer advantages in terms of low-temperature cracking resistance, elastic properties, and toughness, while reducing traffic noise to improve driving comfort [4,5]. Combining the properties of both rubber and polymer resins, styrene butadiene styrene (SBS) can comprehensively improve the properties of base asphalt and, currently, become the most used, and studied, asphalt modifier [6–9]. While additives help to improve the performance of asphalt, they also create some problems, such as the compatibility of the modifier and the base asphalt [10], stability of the modified-asphalt [11], and the balance between high- and low-temperature properties [12]. To solve these problems, many researchers added nanomaterials into base asphalt or polymer-modified asphalt (PMA) to make up for the deficiencies of PMA in performance. Polacco et al. [13], Zhang et al. [14,15] and other scholars believe that the exfoliated and intercalated structure is formed in a nano-layered material/polymer/asphalt system. They also believe that these structures, especially the exfoliated structure, can separate oxygen and prevent the volatilization of the light-asphalt components, thereby, increasing the aging resistance of asphalt and improving the service life of a modified asphalt binder.

Nanomaterials-modified asphalt technology has made some progress, such as aging resistance, rheological properties, and high-temperature performance. However, due to the complex composition, and viscosity features of asphalt, combined with a huge surface area and high surface energy of nanoparticles, the agglomeration phenomenon of nanomaterials is ubiquitous in a nano-modified asphalt. In addition, the cost of nanomaterials can also limit its engineering application. The thermoplastic polyurethane (TPU) contains many carbamate groups (–NHCOO–), in the main chain, as a typical multi-block copolymer [16]. While improving the performance of asphalt, its compatibility can also be improved by the reaction between the functional groups, in the TPU and the components of the base asphalt [17]. Partal et al. [18–20] have extensively investigated polyurethane prepolymer and isocyanate-based polymer-modified asphalt. TPU is a kind of polymer, with both rigid and flexible properties, as its molecule contains hard and soft segments. As an asphalt modifier, the TPU can not only improve the strength but also the flexibility of the modified asphalt system [21]. Actually, the present literature is lacking in the reporting of TPU-modified asphalt binders. The present study focused on the physical performance, as well as low-temperature rheological properties of TPU-modified asphalt. In addition, the thermal properties, the high-temperature storage stability of TPU-modified asphalt and its aging mechanism were also studied.

2. The Experimental Process

2.1. Materials

Base asphalt 90A was obtained from Xi'an Petroleum & Chemical Corporation (Xi'an, China). Commercial TPU (Desmopan 9380A) was obtained from Bayer AG (Werk Leverkusen, Germany), with a density of 1.110 g/cm^3, and a number-average molecular weight of about 100,000. TPU easily absorbs moisture so the resin was thoroughly dried, before use.

2.2. Sample Preparations

A FLUKO FM300 high shear emulsifier (Shanghai, China) was used to mix the TPU and the base asphalt. The mixing process began with heating 500 g of the base asphalt, until it became pourable. Subsequently, specific quantities of TPU were individually added into the melted asphalt and the mixture was sheared at 150 °C, for 1 h, with a speed of 2500 rpm. Finally, the binders were stirred with a mechanical agitator for another 1 h, in 150 °C. We prepared a series of compositions with 1 wt %, 2 wt %, 3 wt %, and 4 wt % of TPU contents, based on the weight of the base asphalt.

2.3. Aging Procedure

Short-term aging of the base asphalt and the modified-asphalts was conducted, using a rolling thin-film oven test (RTFOT), according to American Society for Testing Materials (ASTM) D2872.

2.4. Tests Procedures

2.4.1. Physical Properties Test

The performance indexes of asphalt samples, such as penetration, softening point, and ductility (5 °C) were obtained from tests conducted in accordance with ASTM D5, ASTM D36, and ASTM D113, respectively. The penetration ratio, softening point increment, and ductility ratio of the asphalt samples before and after aging were used to evaluate the aging resistance of the materials.

2.4.2. Low-Temperature Creep Test

Low-temperature creep tests were performed using a bending beam rheometer (BBR, Cannon Instrument Company, State College, PA, USA), in accordance with ASTM D6648. The creep stiffness (S) and creep rate (m) of the asphalt samples were tested at –24 °C and –18 °C.

2.4.3. Thermal Gravity Test

Thermal gravity test was carried out using a TGA/DSC 1 thermogravimetric analyzer (Mettler Toledo, Zurich, Switzerland), under an air atmosphere, at a scan rate of 15 °C/min, from 100 to 700 °C.

2.4.4. Storage Stability Test

Modified asphalt sample was transferred into a glass tube and stored vertically in an oven, at 163 °C. After 48 h, the tube was removed, cooled to room temperature and then the sample was cut into three equal sections, horizontally. The top and bottom sections were used to evaluate the storage stability of the modified-asphalt, by measuring the softening points. The test was conducted in accordance with ASTM D5976.

2.4.5. Morphology Observation

The sample's morphology was observed using a Nikon 80i fluorescent microscope (Tokyo, Japan). A small amount of the heated asphalt sample was placed on a glass microscope slide and then pressed into a thin layer with a cover glass. After cooling, the sample was inspected at a magnification of 400 times, under a green incident light.

2.4.6. FTIR Test

Infrared spectra of the samples were recorded using a Shimadzu FTIR-8400S spectrometer (Kyoto, Japan), with scans conducted from 400 to 4000 cm^{-1}, with a 4 cm^{-1} resolution. Asphalt samples were individually dissolved in tetrahydrofuran and the solution was painted onto a potassium bromide (KBr) thin plate, for testing.

3. Results and Discussion

3.1. Physical Properties

The penetration, softening point, and ductility of the asphalt samples and their ratio or increment, before and after aging, are shown in Figures 1–3 respectively. Compared with the base asphalt in Figure 1, the penetration of the TPU-modified asphalt sample appeared to be reduced, indicating that the modified-asphalt was toughened and its shear deformation resistance was also enhanced. However, as the TPU content increased, the modified-asphalt became softer. After RTFOT aging, the ratio of residual penetration and the penetration before aging increased and the aging resistance of the asphalt improved. In Figure 2, the TPU-modified asphalts displayed a softening point of 4–6 °C higher than the base asphalt. There was no significant difference between the two, with little improvement in high-temperature performance. With the increase of TPU content, the softening point increment decreased after the RTFOT aging, demonstrating the improved aging resistance of the asphalt. Base asphalt has an excellent ductility, and its low temperature value was 111.7 cm, as shown in Figure 3. However, the light molecular weight (MW) asphalt components, such as saturated and aromatic hydrocarbons were oxidized and volatilized in the following the RTFOT aging, so that the residual ductility was only 4.8% of the original value. This would indicate that, in the field, the brittleness of the asphalt pavement would increase, thereby, decreasing its cracking resistance at a low temperature. In the case of the TPU-modified asphalts, the low temperature ductility increased with an increasing TPU content and the ductility ratio was higher than the base value, but with only a 4 cm of residual ductility, it was still at a very low level.

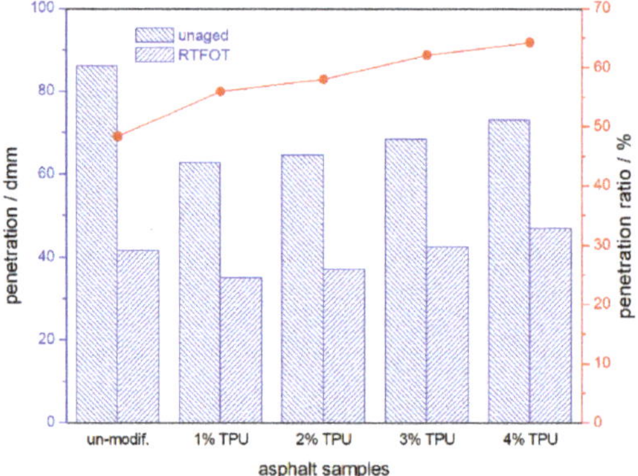

Figure 1. Penetration of the asphalt samples and their residual penetration ratio, after the RTFOT aging.

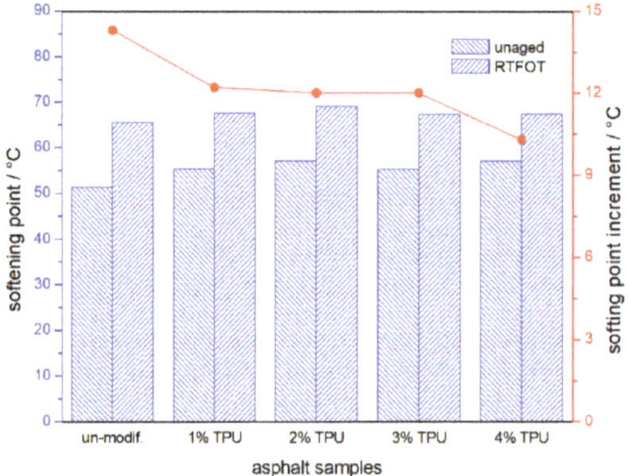

Figure 2. Softening point of the asphalt samples and their increment, after the RTFOT aging.

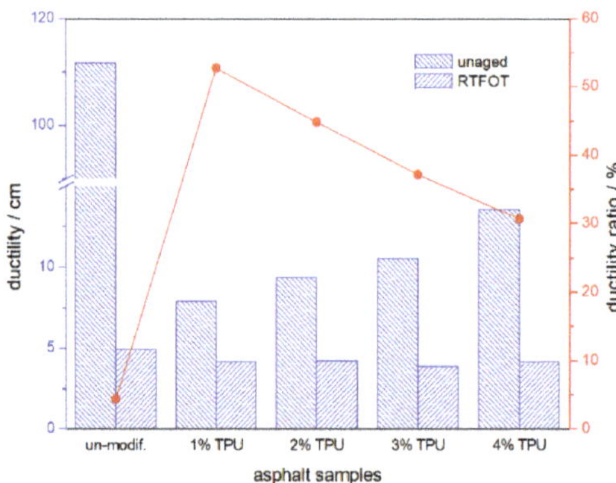

Figure 3. Ductility of the asphalt samples and their residual ductility ratio, after the RTFOT aging.

3.2. Low-Temperature Creep Properties

Figures 4 and 5 represent the BBR experimental results for the asphalt samples stored at $-18\,^{\circ}$C and $-24\,^{\circ}$C, respectively. The addition of TPU appeared to reduce the creep stiffness (S) and improve the creep rate (m), indicating that the inner stress of the asphalt was decreased in the same temperature and loading condition and could also be lost, timely, through the deformation [22]. The improvement in these parameters indicated that the modified-asphalt performed better at a low-temperature and was not prone to cracking. However, when the asphalt samples aged, a series of volatilization, oxidation, polymerization, and even changes in the internal structure of the asphalt would have occurred and the asphalt would have become harder. The low-temperature flexibility was decreased, thus the creep stiffness was increased and creep rate was reduced. Consequently, the comprehensive analysis of S and the m-value, before and after the RTFOT aging, indicated that when the TPU content was 3%, the modified-asphalt had better short-term aging resistance than the other samples.

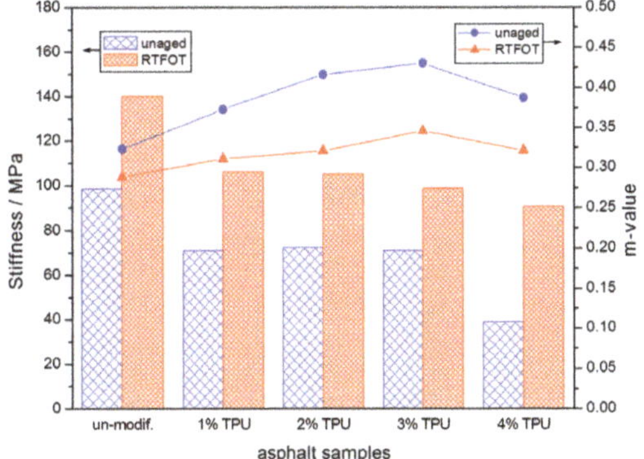

Figure 4. Creep stiffness (S) and creep rate (m-value) of the asphalt samples, before and after the RTFOT aging, at −18 °C.

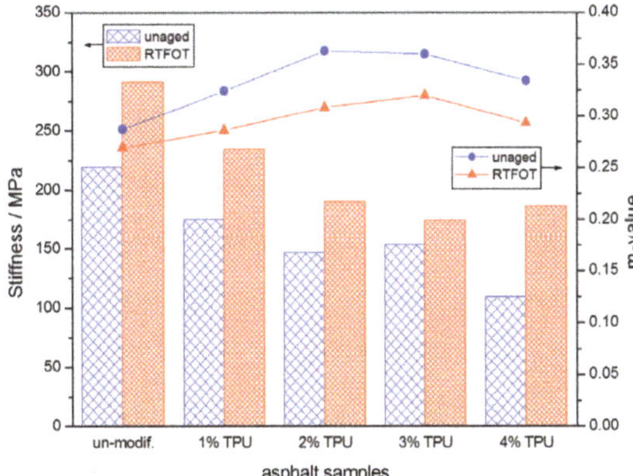

Figure 5. Creep stiffness (S) and creep rate (m-value) of the asphalt samples, before and after the RTFOT aging, at −24 °C.

3.3. Thermal Properties

The Thermal gravity (TG) and differential scanning calorimetry (DSC) curves of the asphalt combustion process are shown in Figures 6 and 7. It is well-known that the components of the asphalt vary in chemical composition and molecular weight [1,23]. This leads to varying physical and chemical properties, in the asphalt, so that the temperature for mass loss of each component is different, consequently, there are several stages in asphalt degradation [24]. Every degradation stage of the asphalt combustion is listed in Table 1, based on Figure 7. The first degradation stage could mainly be attributed to the volatilization of the light-MW components of asphalt, such as the saturated and aromatic hydrocarbons. At this stage, the TPU in the modified-asphalt also began to decompose and absorbed heat, which delayed the time of the asphalt decomposition. With the increasing TPU content, the maximum degradation temperature of the modified-asphalt degradation increased and the thermal stability of the asphalt was improved. The most complex and intense chemical reaction occurred in the second-stage, where the asphalt mass loss was mainly due to the decomposition of resins. The thermal oxidative degradation of resins produced low molecular hydrocarbons that subsequently burned off. Dehydrogenation and polymerization occurred in the residual substance and increased the quantity of the carbonized products, while decreasing the stable free-radical content. The third stage was the degradation and the high-temperature charring of the asphaltene. In base asphalt, the release of volatiles and the combustion of fixed-carbon, rearranged the asphalt molecular structure and formed a dense layer of carbon. As a result, the oxygen and heat could not penetrate the unburned asphalt, which prevented the release of flammable volatiles that remained as a 25–26% of the residue. However, in the TPU-modified asphalt, the decomposition of the TPU caused the asphaltene to regain free radicals and the reactions continued. With an increasing temperature, the hydrogen and methyl of the asphaltene were continually removed and ultimately produced a stable char structure, the residual amount was 1–4%, and the mass-loss stopped. In addition, the derivative thermogravimetric (DTG) analysis curves of TPU-modified asphalt were flatter than the base asphalt's curves, which suggested that the combustion activity of the TPU-modified asphalt had increased and its flame resistance had decreased.

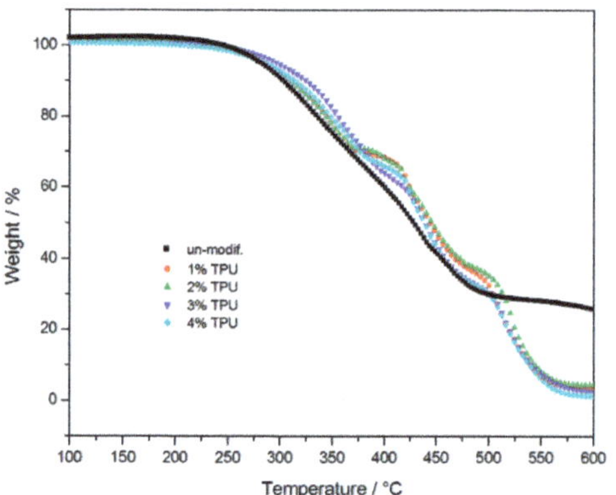

Figure 6. TG curves of base asphalt and modified-asphalts.

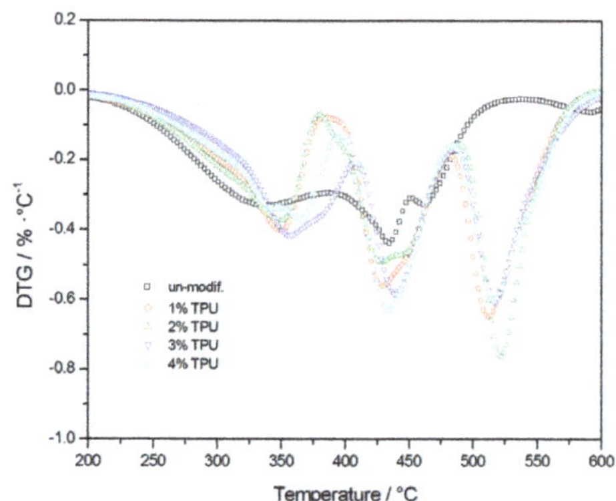

Figure 7. DTG curves of base asphalt and modified-asphalts.

Table 1. Degradation stages in the asphalt combustion process.

Asphalt Samples	Combustion Process		
	Stage 1 (T_{peak})/°C	Stage 2 (T_{peak})/°C	Stage 3 (T_{peak})/°C
un-modif.	282–387 (337)	387–452(434)	452–536 (461)
1% TPU	283–383 (349)	383–482 (428)	482–575 (512)
2% TPU	286–379 (349)	379–488 (428)	488–577 (521)
3% TPU	297–407 (356)	407–485 (439)	485–578 (517)
4% TPU	284–396 (364)	396–483 (433)	483–580 (516)

3.4. Storage Stability

The segregation experiment results of the asphalt samples are shown in Table 2. In contrast to the conventional polymer-modified asphalt, the coalescence of modifying agents, during a high-temperature storage could not occur in the TPU-modified asphalt, which minimized their migration and the subsequent segregation. This was mainly due to the reaction between the –NCO groups in the TPU and the active hydrogen atoms (mainly –OH) in the asphaltene micelles [25], as shown below:

$$R^1\text{–NCO} + R^2\text{–OH} \rightarrow R^1\text{–NHCOO–}R^2$$

Combined with the analysis of FTIR, in Section 3.6, the absorption peak at 2275 cm^{-1} was a characteristic peak of NCO. It was concluded that urethane bonds had formed between the polymer and the asphalt component (mainly asphaltenes and resins). Therefore, asphalt-rich phases had interacted together and formed a crosslinking. The modified-asphalt was uniform in its composition and highly dispersed, and a phase separation was avoided, during the high-temperature storage [26].

Table 2. Softening point in the top and bottom segments of the asphalt samples.

Asphalt Samples	1% TPU	2% TPU	3% TPU	4% TPU
Softening point in top/°C	55.0	55.5	56.6	57.3
Softening point in bottom/°C	55.6	56.3	56.1	56.7
ΔSP [a]	−0.6	−0.8	0.5	0.6

[a] ΔSP means the difference in the softening point, in the top and the bottom segments, of the asphalt samples.

3.5. Morphology

The morphology and dispersion of the TPU polymer in asphalt, before and after the short-term aging, is shown in Figure 8. In Figure 8a, the polymer appears as particles and is dispersed in a continuous phase in the asphalt. After the RTFOT aging, the quantity of the polymer particles had increased, their size had been reduced, and there was a lesser distinction between the polymer and the asphalt phases, as shown in Figure 8b. In Figure 8c, it appeared that the polymer content had increased, as had the polymer particles. The morphology of the 4% TPU-modified asphalt, after aging, is shown in Figure 8d, and it was similar to that of the 2% TPU-modified asphalt. As described above, the TPU could react with some of the asphalt components, making it easier for the TPU to achieve a more uniform dispersion in the asphalt system, than the polyethylene (PE), the SBS, and other non-reactive polymer-modifying agents [27,28]. In addition, the polymer chains are prone to fracture and decomposition in the aging process, thus, the compatibility of the polymer and the asphalt would be further improved and finally a more uniform modified asphalt system would be formed.

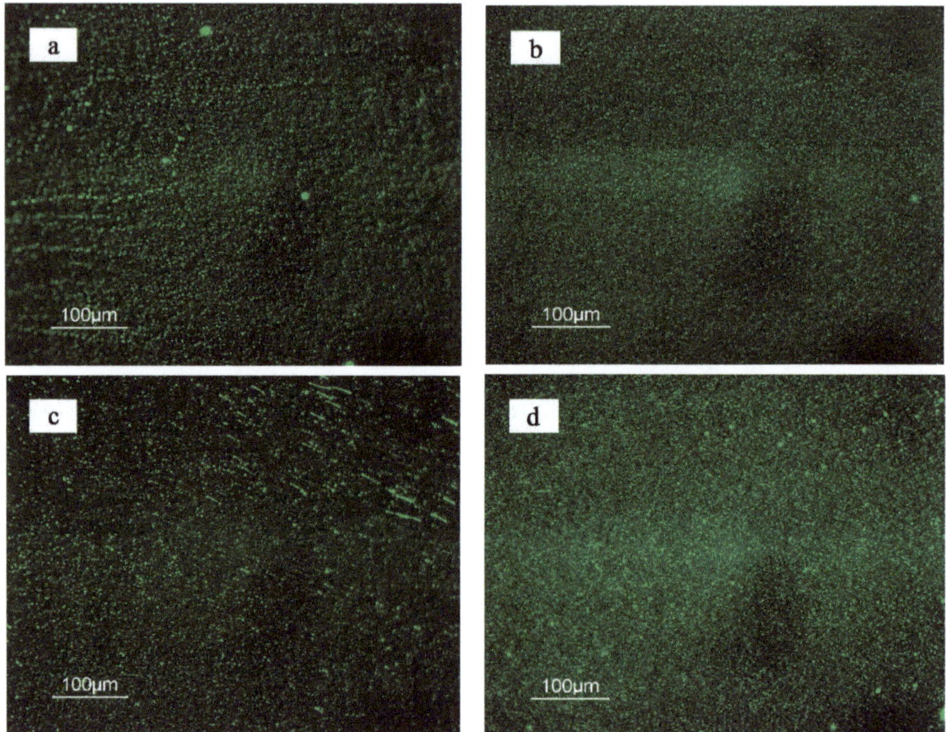

Figure 8. Morphology variation of asphalt samples, before and after aging (**a**: 2% TPU-modified asphalt; **b**: 2% TPU-modified asphalt, after the RTFOT aging; **c**: 4% TPU-modified asphalt; **d**: 4% TPU-modified asphalt, after the RTFOT aging.).

3.6. FTIR Analysis

Figure 9 shows the infrared spectra of the base asphalt, before and after the short-term aging. The peaks at 2932 cm^{-1} and 2854 cm^{-1} were the typical stretching-vibrations of the aliphatic CH$_2$. Aldehyde CH stretching-vibration was observed at 2723 cm^{-1} and the benzene ring vibration, C=C, was seen at 1604 cm^{-1}. The peaks at 1458 cm^{-1} and 1373 cm^{-1} were assigned to the antisymmetric and symmetric deforming-vibrations of the CH$_3$, respectively. Sulfoxide, S=O, stretching-vibration was observed at 1033 cm^{-1}. The peaks at 872 cm^{-1}, 810 cm^{-1}, and 748 cm^{-1} were the stretching-vibrations (out-of-plane) of the C–H in phenyl [29,30]. After the RTFOT aging, a new peak at 1695 cm^{-1} appeared and the intensity of peak at 1033 cm^{-1} was enhanced. We attributed the presence of the carbonyl C=O to the thermal oxidation of the asphalt and the sulfoxides S=O, which resulted from the oxidation of sulfides in the asphalt. These were the characteristic peaks of asphalt-aging. In addition, the enhanced intensity of the peaks at 872 cm^{-1}, 810 cm^{-1}, and 748 cm^{-1} reflected the volatilization of the light alkane components and the relative increment of the fused-ring structured asphaltene, in the asphalt thermo-oxidative aging process. The infrared spectra of the modified-asphalt, before and after the RTFOT aging, could be summarized by the 2% TPU-modified asphalt shown in Figure 10. As described in Section 3.4, the characteristic peak of the NCO, at 2275 cm^{-1}, had appeared. The remaining peaks, after the aging, were similar to that of the base asphalt, except that the characteristic peak at 1695 cm^{-1}, was weaker than that of the base asphalt, indicating the role that was played by the reactive polymer TPU in the aging-resistance of the asphalt.

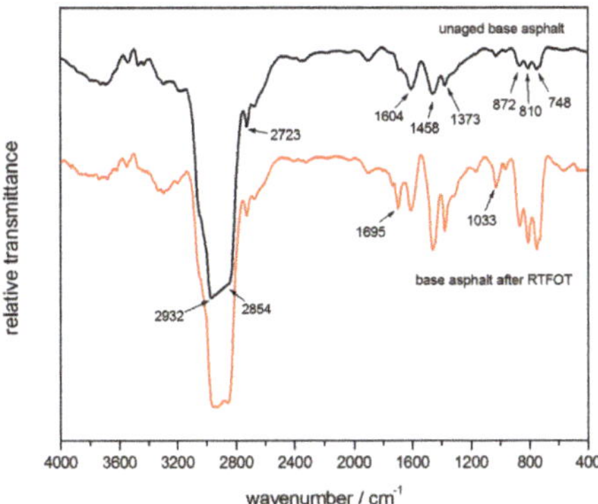

Figure 9. Infrared spectra of the base asphalt, before and after the RTFOT aging.

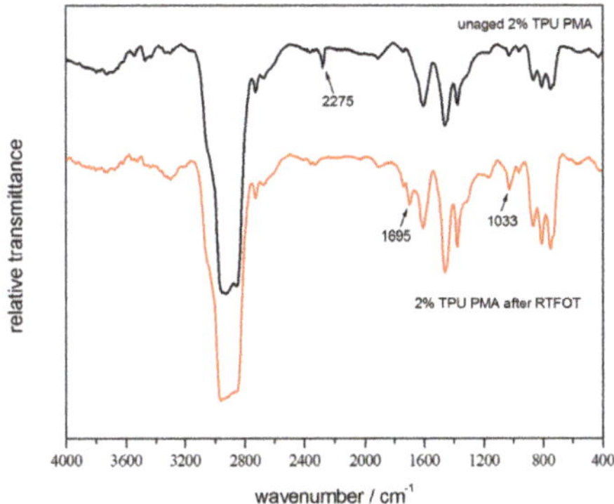

Figure 10. Infrared spectra of the modified-asphalt, before and after the RTFOT aging.

4. Conclusions

Compared with the base asphalt, the penetration ratio and the softening point increment of the TPU-modified asphalt was increased and decreased respectively, and continued to increase and decrease with the raised-TPU content, which confirmed that the aging resistance of the TPU-modified asphalt was enhanced. With respect to the combustion of the modified-asphalt, the degradation of the TPU absorbed the heat and delayed the decomposition of asphalt, thus, improving the asphalt thermal stability. TPU increased the thermal activity of the modified-asphalt and reduced its flame resistance. The weakened characteristic peak of the asphalt aging in the infrared spectrum also reflected the improvement in the aging-resistance of the TPU-modified asphalt. In general, crosslinking was formed and the asphalt-rich phase and the polymer-rich phase had interacted together, due to the reaction of the NCO reactive group in the TPU and the active hydrogen in the asphalt. Thus, the modified asphalt system could be stably stored and used. The fracture and degradation of the TPU chains, after the aging, further improved the compatibility of the polymer-modifying agent and the asphalt. TPU's excellent properties could be reflected in the improvement of the modified-asphalt's properties. The asphalt modification not only included the physical process but also included the chemical process, and further increased the comprehensive properties of the material.

Author Contributions: Conceptualization, R.Y. and C.F.; Methodology, R.Y. and X.Z.; Investigation, R.Y. and M.Z.; Writing-Original Draft Preparation, R.Y.; Writing-Review & Editing, R.Y.; Visualization, M.Z.; Funding Acquisition, R.Y. and C.F.

Funding: This research was funded by the Project in Scientific Innovation of Colleges and Universities of Shanxi Province of China (Grant No. 2017155) and the National Natural Science Foundation of China (Grant No. 51372200).

Conflicts of Interest: The authors declare no conflict of interest.

References

1. McNally, T. Introduction to Polymer Modified Bitumen. In *Polymer Modified Bitumen: Properties and Characterization*; McNally, T., Ed.; Woodhead Publishing Limited: Cambridge, UK, 2011; pp. 1–21, ISBN 978-0-85709-048-5.
2. Chen, H.X.; Xu, Q.W. Experimental Study of Fibers in Stabilizing and Reinforcing Asphalt Binder. *Fuel* **2010**, *89*, 1616–1622. [CrossRef]
3. Yu, R.E.; Zhu, X.J.; Zhou, X.; Kou, Y.F.; Zhang, M.R.; Fang, C.Q. Rheological properties and storage stability of asphalt modified with nanoscale polyurethane emulsion. *Petrol. Sci. Technol.* **2018**, *36*, 85–90. [CrossRef]
4. Jeong, K.D.; Lee, S.J.; Amirkhanian, S.N.; Kim, K.W. Interaction Effects of Crumb Rubber Modified Asphalt Binders. *Constr. Build. Mater.* **2010**, *24*, 824–831. [CrossRef]
5. Pang, L.; Liu, K.Y.; Wu, S.P.; Lei, M.; Chen, Z.W. Effect of LDHs on the Aging Resistance of Crumb Rubber Modified Asphalt. *Constr. Build. Mater.* **2014**, *67*, 239–243. [CrossRef]
6. Modarres, A. Investigating the Toughness and Fatigue Behavior of Conventional and SBS Modified Asphalt Mixes. *Constr. Build. Mater.* **2013**, *47*, 218–222. [CrossRef]
7. Bai, M. Investigation of Low-temperature Properties of Recycling of Aged SBS Modified Asphalt Binder. *Constr. Build. Mater.* **2017**, *150*, 766–773. [CrossRef]
8. Özen, H. Rutting Evaluation of Hydrated Lime and SBS Modified Asphalt Mixtures for Laboratory and Field Compacted Samples. *Constr. Build. Mater.* **2011**, *25*, 756–765. [CrossRef]
9. Behnood, A.; Olek, J. Rheological Properties of Asphalt Binders Modified with Styrene-butadiene-styrene (SBS), Ground Tire Rubber (GTR), or Polyphosphoric Acid (PPA). *Constr. Build. Mater.* **2017**, *151*, 464–478. [CrossRef]
10. Yu, R.E.; Liu, X.L.; Zhang, M.R.; Zhu, X.J.; Fang, C.Q. Dynamic Stability of Ethylene-vinyl Acetate Copolymer/Crumb Rubber Modified Asphalt. *Constr. Build. Mater.* **2017**, *156*, 284–292. [CrossRef]
11. Liang, M.; Xin, X.; Fan, W.Y.; Ren, S.S.; Shi, J.T.; Luo, H. Thermo-stability and Aging Performance of Modified Asphalt with Crumb Rubber Activated by Microwave and TOR. *Mater. Des.* **2017**, *127*, 84–96. [CrossRef]
12. Rodríguez-Alloza, A.M.; Gallego, J.; Pérez, I.; Bonati, A.; Giuliani, F. High and Low Temperature Properties of Crumb Rubber Modified Binders Containing Warm Mix Asphalt Additives. *Constr. Build. Mater.* **2014**, *53*, 460–466. [CrossRef]
13. Polacco, G.; Filippi, S.; Merusi, F.; Stastna, G. A Review of the Fundamentals of Polymer-modified Asphalts: Asphalt/polymer Interactions and Principles of Compatibility. *Adv. Colloid Interface Sci.* **2015**, *224*, 72–112. [CrossRef] [PubMed]
14. Zhang, H.L.; Yu, J.Y.; Wang, H.C.; Xue, L.H. Investigation of Microstructures and Ultraviolet Aging Properties of Organo-Montmorillonite/SBS Modified Bitumen. *Mater. Chem. Phys.* **2011**, *129*, 769–776. [CrossRef]
15. Zhang, H.L.; Yu, J.Y.; Wu, S.P. Effect of Montmorillonite Organic Modification on Ultraviolet Aging Properties of SBS Modified Bitumen. *Constr. Build. Mater.* **2012**, *27*, 553–559. [CrossRef]
16. Drobny, J.G. *Handbook of Thermoplastic Elastomers*; William Andrew Publishing: New York, NY, USA, 2007; p. 179, ISBN 978-0-323-22136-8.
17. Cuadri, A.A.; García-Morales, M.; Navarro, F.J.; Partal, P. Processing of Bitumens Modified by a Bio-Oil-Derived Polyurethane. *Fuel* **2014**, *118*, 83–90. [CrossRef]
18. Martín-Alfonso, M.J.; Partal, P.; Navarro, F.J.; García-Morales, M.; Gallegos, C. Use of a MDI-Functionalized Reactive Polymer for the Manufacture of Modified Bitumen with Enhanced Properties for Roofing Applications. *Eur. Polym. J.* **2008**, *44*, 1451–1461. [CrossRef]
19. Carrera, V.; Partal, P.; García-Morales, M.; Gallegos, C.; Páez, A. Influence of Bitumen Colloidal Nature on the Design of Isocyanate-Based Bituminous Products with Enhanced Rheological Properties. *Ind. Eng. Chem. Res.* **2009**, *48*, 8464–8470. [CrossRef]
20. Carrera, V.; Garcia-Morales, M.; Partal, P.; Gallegos, C. Novel Bitumen/Isocyanate-Based Reactive Polymer Formulations for the Paving Industry. *Rheol. Acta* **2010**, *49*, 563–572. [CrossRef]
21. Xia, L.; Cao, D.W.; Zhang, H.Y.; Guo, Y.S. Study on the Classical and Rheological Properties of Castor Oil-Polyurethane Prepolymer (C-PU) Modified Asphalt. *Constr. Build. Mater.* **2016**, *112*, 949–955. [CrossRef]

Polymers **2018**, *10*, 1189

22. Cong, P.L.; Chen, S.F.; Yu, J.Y.; Wu, S.P. Effects of Aging on the Properties of Modified Asphalt Binder with Flame Retardants. *Constr. Build. Mater.* **2010**, *24*, 2554–2558. [CrossRef]

23. Dai, Z.; Shen, J.N.; Shi, P.C.; Zhu, H.; Li, X.S. Nano-sized Morphology of Asphalt Components Separated from Weathered Asphalt Binders. *Constr. Build. Mater.* **2018**, *182*, 588–596. [CrossRef]

24. Bonati, A.; Merusi, F.; Polacco, G.; Filippi, S.; Giuliani, F. Ignitability and Thermal Stability of Asphalt Binders and Mastics for Flexible Pavements in Highway Tunnels. *Constr. Build. Mater.* **2012**, *37*, 660–668. [CrossRef]

25. Singh, B.; Tarannum, H.; Gupta, M. Use of Isocyanate Production Waste in the Preparation of Improved Waterproofing Bitumen. *J. Appl. Polym. Sci.* **2003**, *90*, 1365–1377. [CrossRef]

26. Navarro, F.J.; Partal, P.; García-Morales, M.; Martínez-Boza, F.J.; Gallegos, C. Bitumen Modification with a Low-Molecular-Weight Reactive Isocyanate-Terminated Polymer. *Fuel* **2007**, *86*, 2291–2299. [CrossRef]

27. Fang, C.Q.; Liu, P.; Yu, R.E.; Liu, X.L. Preparation Process to Affect Stability in Waste Polyethylene-Modified Bitumen. *Constr. Build. Mater.* **2014**, *54*, 320–325. [CrossRef]

28. Sengoz, B.; Isikyakar, G. Analysis of Styrene-Butadiene-Styrene Polymer Modified Bitumen Using Fluorescent Microscopy and Conventional Test Methods. *J. Hazard. Mater.* **2008**, *150*, 424–432. [CrossRef] [PubMed]

29. Zhang, F.; Yu, J.Y.; Han, J. Effects of Thermal Oxidative Ageing on Dynamic Viscosity, TG/DTG, DTA and FTIR of SBS- and SBS/Sulfur-Modified Asphalts. *Constr. Build. Mater.* **2011**, *25*, 129–137. [CrossRef]

30. Fang, C.Q.; Yu, R.E.; Li, Y.; Zhang, M.Y.; Hu, J.B.; Zhang, M. Preparation and Characterization of an Asphalt-Modifying Agent with Waste Packaging Polyethylene and Organic Montmorillonite. *Polym. Test.* **2013**, *32*, 953–960. [CrossRef]

Article

Synthesis and Characterization of Isosorbide-Based Polyurethanes Exhibiting Low Cytotoxicity Towards HaCaT Human Skin Cells

Barbara S. Gregorí Valdés [1,2], Clara S. B. Gomes [3], Pedro T. Gomes [3,*], José R. Ascenso [3,*], Hermínio P. Diogo [3], Lídia M. Gonçalves [2], Rui Galhano dos Santos [1], Helena M. Ribeiro [2] and João C. Bordado [1,*]

[1] CERENA, Departamento de Engenharia Química, Instituto Superior Técnico, Universidade de Lisboa, Av. Rovisco Pais, 1049-001 Lisboa, Portugal; barbara.valdes@tecnico.ulisboa.pt (B.S.G.V.); rui.galhano@tecnico.ulisboa.pt (R.G.d.S.)

[2] Research Institute for Medicine and Pharmaceutical Science (iMed.ULisboa), Faculty of Pharmacy, Universidade de Lisboa, Av. Prof. Gama Pinto, 1649-003 Lisboa, Portugal; lgoncalves@ff.ulisboa.pt (L.M.G.); hribeiro@campus.ul.pt (H.M.R.)

[3] Centro de Química Estrutural, Departamento de Engenharia Química, Instituto Superior Técnico, Universidade de Lisboa, Av. Rovisco Pais, 1049-001 Lisboa, Portugal; clara.gomes@tecnico.ulisboa.pt (C.S.B.G.); hdiogo@tecnico.ulisboa.pt (H.P.D.)

* Correspondence: pedro.t.gomes@tecnico.ulisboa.pt (P.T.G.); jose.ascenso@tecnico.ulisboa.pt (J.R.A.); jcbordado@tecnico.ulisboa.pt (J.C.B.); Tel.: +35-121-841-9612 (P.T.G.); +35-121-841-9417 (J.R.A.); +35-121-841-9182 (J.C.B.)

Received: 21 September 2018; Accepted: 17 October 2018; Published: 20 October 2018

Abstract: The synthesis of four samples of new polyurethanes was evaluated by changing the ratio of the diol monomers used, poly(propylene glycol) (PPG) and D-isosorbide, in the presence of aliphatic isocyanates such as the isophorone diisocyanate (IPDI) and 4,4′-methylenebis(cyclohexyl isocyanate) (HMDI). The thermal properties of the four polymers obtained were determined by DSC, exhibiting T_g values in the range 55–70 °C, and their molecular structure characterized by FTIR, ^{1}H, and ^{13}C NMR spectroscopies. The diffusion coefficients of these polymers in solution were measured by the Pulse Gradient Spin Echo (PGSE) NMR method, enabling the calculation of the corresponding hydrodynamic radii in diluted solution (1.62–2.65 nm). The molecular weights were determined by GPC/SEC and compared with the values determined by a quantitative ^{13}C NMR analysis. Finally, the biocompatibility of the polyurethanes was assessed using the HaCaT keratinocyte cell line by the MTT reduction assay method showing values superior to 70% cell viability.

Keywords: biocompatibility; GPC/SEC; keratinocyte cells; NMR; polyurethane; renewable sources

1. Introduction

Polyurethanes (PUs) are extremely versatile polymers due to their easy structural tunability and to their elastomeric and thermoplastic behavior. Their syntheses from the combination of new monomers are being extensively studied by several groups [1–6].

PU materials are usually composed of two types of phases, in which *hard* glassy phase-enriched domains are dispersed in a matrix of *soft* rubbery segments. These *hard* segments contribute to the modulus, mechanical strength and elevated temperature properties owing to the strong intermolecular interactions such as hydrogen bonding among the urethane groups, whereas the *soft* segments afford elasticity and low-temperature mechanical properties [2]. Moreover, the properties of PUs are remarkably affected by the content, type, and molecular weight of the soft segments. Therefore, the choice of monomers used to synthesize PUs is dependent on the final applications of the materials [7]. For example, monomers offering excellent durability are selected for polymers to be used for prostheses [8], those increasing the compatibility of PU with tissues are applied for pacemakers [9], and monomers offering thermal stability are carefully chosen for catheters that require sterilization processes [7,10].

We have been interested in the synthesis of linear polyurethanes from carbohydrates suitably functionalized for applications in the pharmaceutical industry, namely, for the use in nail lacquer formulations, as they are being considered as a solution to deliver relevant drugs to infected nails [11]. However, the currently available polymer excipients present some disadvantages for the formulation of vehicles for an efficient drug delivery using therapeutic nail lacquers. Isosorbide is a monomer already employed in the synthesis of biocompatible polymers [3,7,12,13]. This glycol (Figure 1) is a chiral and quite thermostable diol, which raises the T_g of the corresponding polymers, and can be obtained by the reduction of glucose followed by dehydration [14]. On the other hand, the poly(propylene glycol) (PPG) is a biocompatible polyether that has been extensively investigated for application as a biomedical material conferring elastomeric properties to polyurethanes [15–17]. This monomer was then chosen as a second glycol in the outlined synthesis. To avoid the use of monomers that probably yield toxic compounds, the isophorone diisocyanate (IPDI) and 4,4′-methylenebis(cyclohexyl isocyanate) (HMDI) were envisaged as options to access nontoxic formulations since those isocyanates have already been used in the synthesis of several biocompatible polyurethanes [16,18,19].

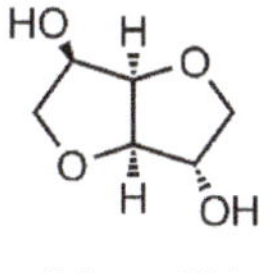

D-Isosorbide

Figure 1. The molecular structure of D-isosorbide.

The present work describes the synthesis of new polyurethanes from D-isosorbide by step-growth polymerization involving diisocyanates, such as IPDI or HDMI, and PPG as soft segments, and the characterization of their chemical structures, molecular weights, and thermal properties by Fourier-Transform Infrared (FTIR) and ^{1}H and ^{13}C Nuclear Magnetic Resonance spectroscopies (NMR), Gel Permeation Chromatography/Size-Exclusion Chromatography (GPC/SEC), and Differential Scanning Calorimetry (DSC). The novelty of this work consists in the synthesis and characterization of new polyurethanes derived from unprecedented combinations of aliphatic monomers (PPG-HMDI-Isosorbide and PPG-IPDI-Isosorbide), which are biocompatible with keratinocyte cells and can participate as drug delivery supports in the formulation of nail lacquers used in the treatment of onychomycosis. In fact, to the best of our knowledge, there are no reports on the application of D-isosorbide-based polyurethanes in therapeutic nail lacquers.

2. Experimental Section

2.1. Materials

The D-isosorbide was provided by Acrös (Geel, Belgium), HMDI by Bayer Material Science (Leverkusen, Germany), and IPDI and 2,2′-dimorpholinyldiethylether (DMDEE) by Sigma-Aldrich (Ludwigshafen, Germany). PPG (VORANOL 1010L, average molecular weight 1000 g mol^{-1}, OH number range (Phthalic Anhydride-Pyridine Solution) 106–114 mg KOH g^{-1}) was supplied by Dow Chemical (Midland, MI, USA). Anhydrous ethanol was purchased from Carlo Erba (Barcelona, Spain). DMSO-d_6, CDCl$_3$, and tetramethylsilane (TMS) (99.9% purity) used in the NMR determinations were acquired from Sigma-Aldrich. The HPLC grade tetrahydrofuran (THF) and the polystyrene (PS) standards used in GPC/SEC were obtained from Aldrich and TSK Tosoh Co. (Tokyo, Japan), respectively.

2.2. Polymer Synthesis

The polyurethanes reported in the present work were synthesized in two steps. The first one involved the quasi-prepolymerization method by reacting the IPDI or HMDI monomers and PPG for ca. 3 h, at 80 °C. Subsequently, the second step comprised the reaction of the previously afforded quasi-prepolymers with D-isosorbide diol monomer for 1 h, at 80 °C.

2.2.1. Synthesis of Quasi-Prepolymers

Three polyurethane quasi-prepolymers were prepared by reacting IPDI or HMDI and PPG in a 250 mL glass flask, under a dry nitrogen atmosphere, to prevent and avoid the presence of moisture and the subsequent formation of urea bonds during the synthesis. The reaction took place with the dropwise addition of PPG to the quasi-prepolymer, under mechanical stirring (400 rpm), at 80 °C and in the presence of 0.5 mL of the catalyst DMDEE. The reactions were practically completed in 3 h, as determined by the isocyanate (-NCO) groups content existing in the reaction mixtures (determined by back titration with an excess of *N*,*N*-dibutylamine with standard HCl [20]). Viscometry measurements of the reaction mixtures by cone-plate rheometry were also used to evaluate the course of the reaction [21] (using a shear stress-controlled ICI Cone and Plate rheometer—London, LTD—at 25 °C and 20 Hz, using parallel plates, with an upper plate diameter of 20 mm and a gap of 0.4 mm). The molar ratios of the monomers used in each synthesized pre-polymer are presented in Table 1.

Table 1. The molar ratios of the monomers employed in the synthesis of the quasi-prepolymer.

Monomers	Pre-1	Pre-2	Pre-3
IPDI	6	6	-
HMDI	-	-	6
PPG	1	2	1

2.2.2. Syntheses of Polyurethanes PU1–PU4

The syntheses of the four polyurethanes **PU1–PU4** were carried out in a 250 mL reactor by the addition of an appropriate amount of D-isosorbide to the IPDI/PPG or HMDI/PPG polyurethane quasi-prepolymers obtained previously (see Section 2.2.1), according to the predetermined molar and mass ratios of monomers and prepolymer, which are presented in Table 2. For comparison, the hard segment content, as obtained by the NMR for each **PU**, is also presented (see below in Table 5). The reactions took place in 1 h at 80 °C under a dry nitrogen atmosphere, in quantitative yields.

Table 2. The molar and mass ratios of the monomers and prepolymer employed in the synthesis of the polyurethanes **PU1–PU4**, and the hard segment content (wt %).

Monomers	Polymers			
	PU1 n (mol)/m (g)	PU2 n (mol)/m (g)	PU3 n (mol)/m (g)	PU4 n (mol)/m (g)
IPDI	0.06/13.33	0.06/13.33	0.06/13.33	-
HMDI	-	-	-	0.06/15.74
PPG	0.01/10.00	0.02/20.00	0.01/10.00	0.01/10.00
D-isosorbide	0.05/7.30	0.05/7.30	0.06/8.76	0.05/7.30
Hard segment content (wt %) [a]	67.4	50.8	68.8	69.7

[a] Hard segment content (wt %) = $(m_{isocyanate} + m_{D\text{-}isosorbide})/(m_{isocyanate} + m_{D\text{-}isosorbide} + m_{PPG})$.

2.3. FTIR Characterization

The quasi-prepolymers' FTIR spectra were acquired with a Nexus Thermo Nicolet spectrometer equipped with attenuated total reflectance (ATR) device for the quasi-prepolymers. The solid samples of polyurethane (approximately 1 mg) were finely powdered and dispersed in a KBr matrix (200 mg). For each polymer, a pellet was then formed by compressing the sample at 784 MPa. The FTIR spectra were obtained at room temperature, in the range of 4000 to 600 cm^{-1}, after 128 scans with a resolution of 4 cm^{-1}.

2.4. NMR Spectroscopy Characterization

^{1}H and ^{13}C NMR spectra were obtained using a Bruker AVANCE III 500 MHz spectrometer (Bruker Corporation, Billerica, MA, USA) equipped with a 5 mm BBO probe. To prepare each of the polyurethanes samples, 20–53 mg of the polymers were dissolved in 0.6 mL of DMSO-d_6. Quantitative ^{13}C NMR spectra were obtained under inverse gated decoupling with a delay of 9 s and an 80-degree pulse. Two-dimensional COSY, HSQC, and HMBC spectra were obtained with Bruker standard sequences. All the spectra were run at room temperature employing TMS as an internal standard.

2.5. Measurement of the Diffusion Coefficients by NMR Spectroscopy

The Pulse Gradient Spin Echo (PGSE) method is a relatively simple technique of measuring diffusion coefficients in solutions. It combines NMR spin echoes with pulsed-field gradients of variable strength. To determine the diffusion coefficient of a molecule in a solution, a series of ^{1}H spectra is obtained in which the strength of the gradients is increased and the attenuation of the intensity of the proton peaks due to diffusion is monitored [22]. In our study, the diffusion coefficient (D) of each polymer was determined at five different concentrations. For this purpose, a series of five solutions was prepared with concentrations 0.03%, 0.07%, 0.15%, 0.3%, and 1% (w/v) in DMSO-d_6. The diffusion coefficients of the polymers at infinite diffusion (D_0) were obtained from plots of D versus concentration (Tables S1–S4 and Figure S12 in the Supplementary Material). The D values were measured using the PGSE method in an NMR Bruker AVANCE III 500 MHz spectrometer with a 3-mm BBO probe and a z-gradient shielded coil. This combination gives a maximum possible gradient of 0.56 T m^{-1}. A bipolar stimulated echo sequence (STE) with smoothed square gradients and WATERGATE solvent suppression were used [23]. The signal intensity was monitored as a function of the square of the gradient amplitude and the resulting self-diffusion coefficients D were calculated according to the echo attenuation equation for STE:

$$I = I_0 \exp[-D\,(\gamma\delta g)^2\,(\Delta - \delta/3)] \tag{1}$$

where I_0 is the intensity in the absence of gradient pulses, δ is the duration of the applied gradient, γ is the gyromagnetic ratio of the nucleus, and Δ is the diffusion time. For each polymer, the areas of three or four single proton peaks were used in the fittings and the average D value was taken. For more details see the Supplementary Materials.

The solutions of the polymers were transferred to 3 mm NMR tubes to a total volume of 0.3 mL. To guarantee reproducibility of the results this volume was kept constant in all the samples. The temperature was controlled at 30 °C by a BCU05 Bruker unit with an air flow of 521 Lh^{-1}.

2.6. Gel Permeation Chromatography/Size-Exclusion Chromatography (GPC/SEC) Characterization

The analyses were made in an HPLC Waters chromatograph containing a Waters 515 isocratic pump and a Waters 2414 refractive index detector. In this apparatus, the oven was stabilized at 35 °C and the elution of samples was carried out through two PolyPore columns (Agilent, Santa Clara, CA, USA), protected by a PolyPore Guard column (Agilent). The software Empower® performed the acquisition and data processing.

THF was used as the eluent, at a flow rate of 1.0 mL min^{-1}. Before use, the solvent was filtered through 0.45 μm of PTFE membranes Fluoropore (Merck Millipore, Burlington, MA, USA) and degassed in an ultrasound bath for 45 min. The oligomer/polymer samples were also filtered across 0.20 μm PTFE filters Durapore (Merck Millipore). The molecular weights were calibrated relative to polystyrene standards (TSK Tosoh Co.). As a result, it should be taken into account that small deviations could have occurred when the present polymer samples were analyzed.

2.7. Thermal Behavior Characterization by Differential Scanning Calorimetry (DSC)

The characterization of the thermal behavior of the polymer samples was performed with a 2920 MDSC system from TA Instruments Inc. (New Castle, DE, USA), equipped with a refrigerated cooling accessory (LNCA), which provided automatic and continuous programmed sample cooling down to 123 K.

The temperature scale of the instrument was calibrated with five standard compounds and the heat flow scale was calibrated with indium and tin. Further details regarding the calibration process can be found in the literature [24]. The samples were accurately weighed (± 0.1 µg) in aluminum pans on a Mettler UMT2 ultra-micro balance in the air. All the measurements were performed under dry high purity helium gas (Air Liquide N55 – Alphagaz 1, Paris, France), at a flow rate of 30 mL min^{-1}.

The protocol used in the DSC measurements consisted of two consecutive identical thermal cycles: samples were cooled to $-110\,^\circ$C and then heated until 230 $^\circ$C, at 10 K min^{-1}.

2.8. In Vitro Cytotoxicity Assay

The polyurethanes were completely dissolved in ethanol under stirring (200 rpm), at a concentration of 250 mg mL^{-1}. Later, this solution was spread on one side of a 10 mm diameter glass coverslip and dried for 30 min.

The biocompatibility of the polymer was evaluated in vitro by direct contact with cells, following the ISO 10993-5:2009 recommendation guidelines [25]. The procedure used was previously published in Reference [11]. Briefly, to each polyurethane glass coverslip was added 1.25×10^5 cells HaCaT (Cultured Human Keratinocyte) cell suspension in the culture medium, in sterile 24-well plates. The plates were incubated for 72 h in a humidified atmosphere of 5% CO_2 at 37 $^\circ$C without refreshing the culture medium. For cell proliferation quantification, the general cell viability endpoint MTT reduction was used (MTT = 3-(4,5-dimethylthiazol-2-yl)-2,5-diphenyltetrazolium bromide).

The data are expressed as the mean and the respective standard deviation (mean $\pm$ SD) of 6 experiments. The statistical evaluation of data was performed using one-way analysis of variance (ANOVA). The Tukey–Kramer multiple comparison tests (GraphPad PRISM 5 software, GraphPad Software Inc., La Jolla, CA, USA) were used to compare the significance of the difference between the groups; a $p < 0.05$ was accepted as statistically significant.

3. Results and Discussion

3.1. Syntheses of the PUs

A family of aliphatic biocompatible PUs, based on IPDI or HDMI, D-isosorbide and PPG, were synthesized in order to convey the active principle. The PUs were designed to be composed of various ratios of soft segments (PPG based blocks) and hard segments (D-isosorbide based blocks).

A two-stage step-growth polymerization method was employed. In an early step, the pre-polymerization of an excess of a diisocyanate monomer (IPDI or HMDI) with the PPG diol, catalyzed by DMDEE, was carried out by stirring the reaction mixture for 3 h, at 80 $^\circ$C. Then, a complementary step consisted in the addition of a pre-defined amount of the D-isosorbide diol monomer to the previous pre-polymer reaction mixtures (catalyst included), at the abovementioned temperature, for an additional hour. The polyurethanes **PU1–PU3** were obtained from the reaction of IPDI, PPG, and D-isosorbide, according to the ratios listed in Table 2, whereas **PU4** was obtained from the reaction of HMDI, PPG, and D-isosorbide (Figure 2). The four PU products obtained in these reactions were characterized by FTIR, NMR, and GPC/SEC without further purification.

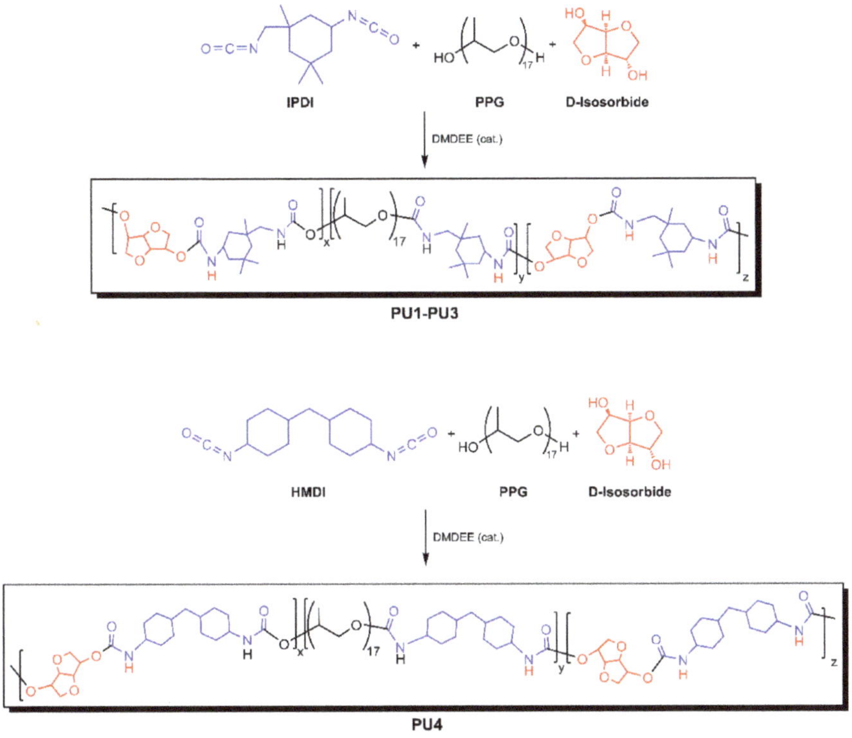

Figure 2. The synthesis of polyurethanes **PU1–PU4**.

3.2. FTIR of the PUs

The FTIR spectra of the three quasi-prepolymers, which were synthesized according to the monomer molar ratios defined in Table 1, are presented in Figure 3. In all the quasi-prepolymers, a band at 2252 cm^{-1} is observed, indicating the presence of isocyanate groups (–NCO) corresponding to the monomers IPDI and HMDI, which is in agreement with the reports of the literature (2270 to 2250 cm^{-1} for the N=C=O stretching vibration) [26–28]. The spectra of the quasi-prepolymers also showed bands at 2927 cm^{-1}, corresponding to –CH$_2$- groups, and the peak of the ether group at 1527 cm^{-1}.

These bands correspond to the PPG segments (pure PPG presents a –CH$_2$– band between 2970–2850 cm^{-1} and another one corresponding to the ether group at 1526 cm^{-1} [29]). The band corresponding to the carbonyl groups is observed between 1677–1708 cm^{-1}, which is indicative of the condensation reaction between the –OH group of PPG and the –NCO groups of HMDI and IPDI [30–32].

The FTIR spectra of the resulting IPDI based polyurethanes **PU1–PU3** and the corresponding monomers are shown in Figure 4. The characteristic band of the isosorbide monomer –OH groups is observed at 3366 cm^{-1}, whereas the band at 2973 cm^{-1} corresponds to the isosorbide methylene group (–CH$_2$–) [33]. In the PUs spectra, a broad signal between 3700 and 3200 cm^{-1} remains. This signal likely corresponds to the superimposition of –NH– stretching vibrations of the urethane groups of the PUs hard segments as well as to terminal isosorbide –OH groups and terminal urea –NH$_2$ groups (the latter resulting from the hydrolysis of the PUs' former –NCO urethane terminal groups upon exposure to air after the reaction). The –CH$_2$– bands, identified in isosorbide and PPG, are detected in the spectra of the PUs between 3050 and 2800 cm^{-1}. The hard segments carbonyl bands can be observed between 1705–1691 cm^{-1} [30,34]. The disappearance of –N=C=O stretching vibration, around 2250 cm^{-1}, in the spectra of the synthesized polyurethanes suggests that there are no unreacted isocyanate groups [19], i.e., all the isocyanate groups reacted with the monomers –OH groups (and the remaining terminal ones with H_2O from moisture) [35].

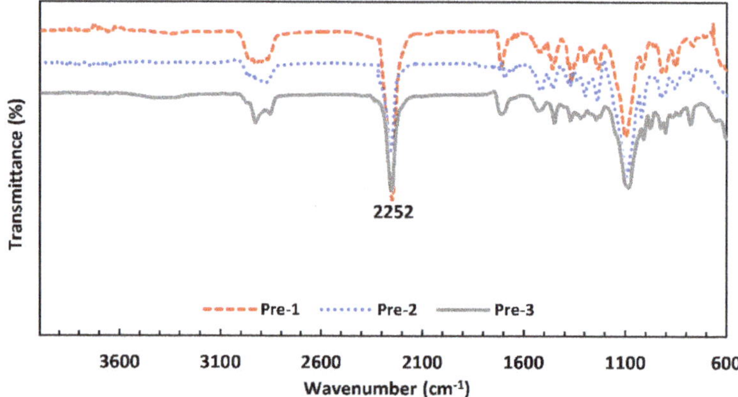

Figure 3. The FTIR spectra of the three polyurethane quasi-prepolymers.

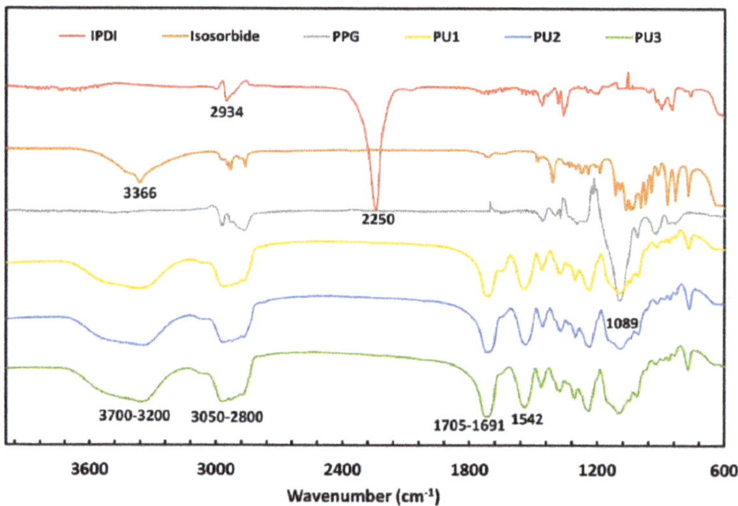

Figure 4. The FTIR spectra of the monomers Isosorbide, PPG, IPDI, and of the polyurethanes **PU1-PU3**.

On the other hand, the –NH– bending bands of the synthesized PUs were identified at 1542 cm^{-1} (Figure 4) [27]. These bands were observed in the reaction between polycaprolactone and 2-isocyanate ethylmethacrylate [36]. The –NH– bending of the urethane group was also detected by da Silva et al. at the same wavenumber [37]. Furthermore, C–O–C stretching was observed at 1089 cm^{-1} [34].

The FTIR spectrum of the HMDI based polyurethane **PU4** is depicted in Figure 5. In this spectrum, the –CH$_2$– stretching bands (from PPG) are observed between 2927 and 2854 cm^{-1}, the band of the urethane carbonyl being detected at 1704 cm^{-1}. The C–N stretching bands, combined with those of the bending of N–H in the plane, are typically observed at 1523 cm^{-1} and 1446 cm^{-1}, demonstrating the occurrence of the reaction between the hydroxyl group and the isocyanate [38]. The C–O–C characteristic band was identified at 1079 cm^{-1}.

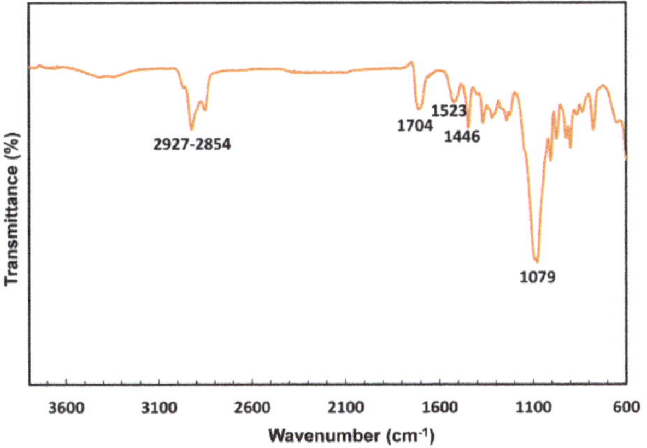

Figure 5. The FTIR spectrum of the polyurethane **PU4**.

3.3. NMR of the PUs

The ^{1}H NMR spectra of polyurethanes **PU1**, **PU2**, and **PU3** are shown in Figure 6.

The presence of amide protons near 7.0 ppm of the urethane groups –OCONH– indicates that the reaction between the isosorbide hydroxyl groups and the pre-polymer isocyanate groups has occurred. In the research published by Besse et al., the characteristic –NH– of the urethane group was identified at 7.1 ppm as a result of the synthesis of polyhydroxyurethanes by the reaction of isosorbide dicyclocarbonate with four commercial diamines [3]. Other authors mentioned that the proton from the –OCONH– group, obtained from the reaction between IPDI and a polyester, can also be observed at 6.8, 6.82, and 7.2 ppm [39]. The literature data shows data in the polyurethanes resulting from the reaction between isosorbide, hexamethylene diiosocyanate (HDI), and poly(tetramethylene glycol) (PTMG); the –NH– peak is described at 8.07 ppm [7]. The ^{1}H NMR spectra of the in-chain isosorbide units, peaks from 4.0 to 5.2 ppm; of the IPDI units, peaks from 2.6–4.0 ppm; and of the IPDI methyl groups, ca. 1.0 ppm, were assigned on the basis of the existing literature of pre-polymers made from the reaction of PPG with IPDI monomers [17]. The assignment of the isosorbide and IPDI resonances in the spectra of **PU1**–**PU3** were further confirmed by means of 2D NMR experiments (COSY, HSQC, and HMBC). Some of these spectra are shown in Figures S1 and S3–S7 of the Supplementary Materials. The peaks of the ^{1}H NMR spectra of the PPG soft-segment are observed as multiple resonances between 3.2 and 4.0 ppm for both the CH and CH$_2$ protons, and in the range 1.0–1.2 ppm for the methyl groups. The ^{1}H NMR spectrum of **PU4** is presented in Figure 7. The peaks ca. 7.2 ppm in Figure 7 corresponds to the NH of urethane groups. The signals at 4.0–5.2 ppm are a piece of evidence for the presence of isosorbide groups in the structure of polyurethane [18]. The strong peaks centered at 1.2 and 3.4 ppm belong to the CH$_3$ and CH+CH$_2$ protons of the PPG segment, respectively. The peaks characteristic of HMDI units are identified at 0.75–3.6 ppm [31,40]. The assignment of the spectra of the isosorbide and HMDI units were further confirmed by means of 2D NMR experiments (HSQC and HMBC). They are shown in Figures S8–S11 of the Supplementary Materials. The unreacted hydroxyl in *exo* position of the isosorbide end group appears as a minor resonance at ca. 5.1 ppm and is characteristic of polyurethanes with isosorbide as a chain extender [19].

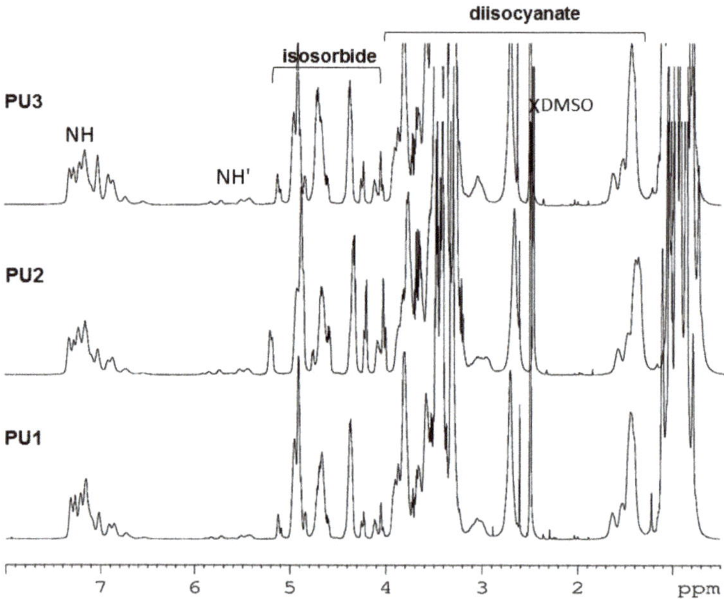

Figure 6. The ^{1}H NMR spectra of **PU1**, **PU2**, and **PU3** in DMSO-d_6. Protons of the terminal groups are primed.

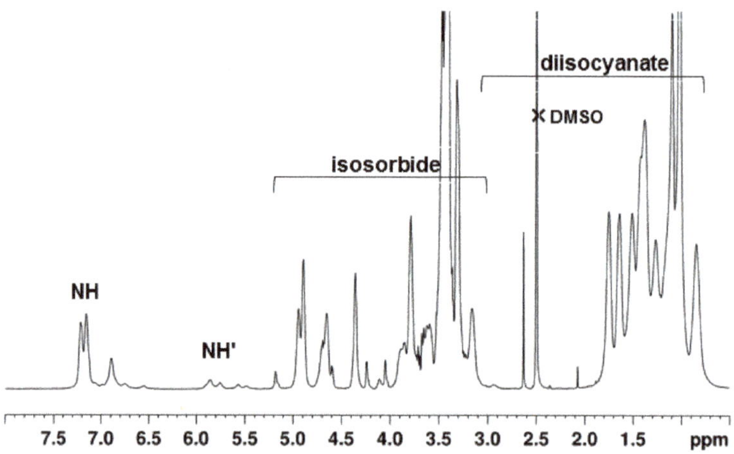

Figure 7. The ^{1}H NMR spectrum of **PU4** in DMSO-d_6. Protons of the terminal groups are primed.

The assignment of the ^{13}C NMR spectra of polyurethanes **PU1**, **PU2**, and **PU3** presented in Figure 8 was complicated owing to the fact that monomer IPDI is available as a mixture of the *cis* and *trans* isomers (3:1), and we had to use HSQC and HMBC spectra to fully assign the in-chain carbons (see Figures S2 and S4–S7 of the Supplementary Materials). The peaks from the isosorbide units are observed between 65 and 90 ppm, while those of IPDI appear at a higher field from 20 to 60 ppm. The carbonyl resonances appear between 153–157 ppm while the strong peaks centered at 72.4 and 74.6 ppm correspond to the methine and methylenic carbons of the PPG segments, respectively. The methyl peaks appear at around 17.2 ppm. The experimental ^{13}C chemical shifts agree with those referenced in the literature [17] for pre-polymers made of PPG and IPDI monomers.

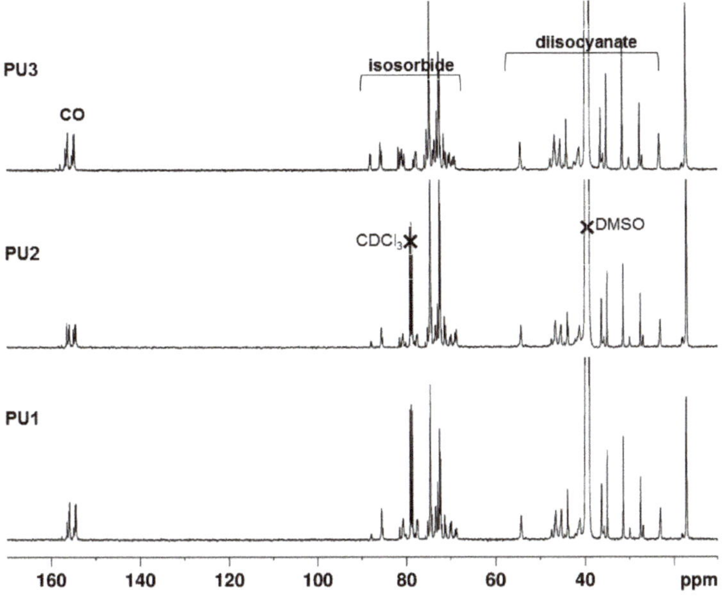

Figure 8. The ^{13}C{^{1}H} NMR spectra of **PU1**, **PU2**, and **PU3** in DMSO-d_6. Traces of CDCl$_3$ are seen in the spectra of **PU1** and **PU2**.

No peaks of the primary and secondary isocyanate groups of the IPDI terminal groups or monomers appear at 121.7 and 122.6 ppm in the spectra [41], which indicates that all the IPDI monomers/units have reacted with the -OH groups of the diol monomers, in accordance with the FTIR results.

The ^{13}C{^{1}H} NMR spectrum of **PU4** presented in Figure 9 is similar to those of the previous PUs, except that the higher field region is occupied by the peaks of the HDMI monomers. HSQC and HMBC spectra were used to fully assign the in-chain carbons of **PU4** (see Figures S9–S11 of the Supplementary Materials). This similarity is found in the characterization of polyether urethane urea obtained by the reaction between HMDI and polyethylene glycol and ethylenediamine [42].

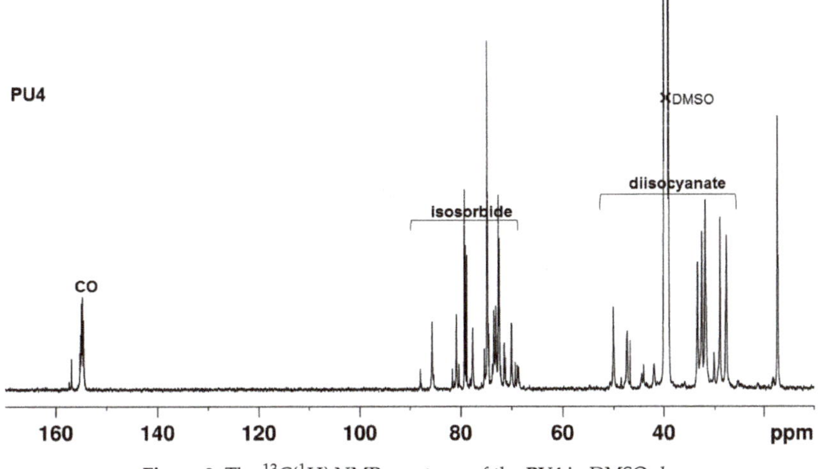

Figure 9. The ^{13}C{^{1}H} NMR spectrum of the **PU4** in DMSO-d_6.

The disappearance of the ^{13}C NMR peak at 124 ppm, characteristic of the isocyanate group, indicates that all the diisocyanate groups have fully reacted with the diol monomers [31]. This result was also confirmed by the FTIR analysis.

3.4. Determination of Hydrodynamic Radii of the PUs

The determination of the diffusion coefficients of the PUs of this work by the Pulse Gradient Spin Echo (PGSE) NMR method, enabled the calculation of the corresponding hydrodynamic radii (r_h) using the Stokes-Einstein equation for diluted solutions of large molecules [43]:

$$D_0 = \frac{k_B T}{6\pi\eta r_h} \tag{2}$$

where k_B is the Boltzmann's constant, T is the temperature, η is the viscosity of the solution (N s m^{-2} = kg s^{-1} m^{-1}) and r_h is the hydrodynamic radius (m). The viscosity of DMSO-d_6 at 30 °C, η = 1.951 $\times$ 10^{-3} Pa.s, was estimated from that of the non-deuterated solvent taking into account the isotopic effects of the viscosity [44].

The self-diffusion coefficient D_0 represents the translational movement of the solute at infinite dilution and, as shown by Equation (2), is dependent on the size and shape of the solute, temperature, and viscosity of the solvent. The dependence of the diffusion coefficient D versus solute concentration was plotted for all the PUs (Figure S12 in Supplementary Materials), being approximately linear for all polymers because the solutions are diluted. The self-diffusion coefficients D_0 at infinite dilution (c = 0) are summarized in Table 3. The corresponding values of r_h obtained from Equation 2 are also presented in Table 3.

Table 3. The measured average values of D_0 for the PUs determined by PGSE NMR and the corresponding hydrodynamic radii (r_h) in DMSO-d_6, at 30 °C.

Polyurethane	$D_0 \times 10^{11}$ (m^2/s^{-1})	r_h (nm)
PU1	7.04	1.62 ± 0.04
PU2	4.49	2.65 ± 0.15
PU3	6.74	1.69 ± 0.04
PU4	4.51	2.52 ± 0.19

In the family of polyurethanes based on IPDI diisocyanate (**PU1**, **PU2**, and **PU3**), the hydrodynamic radius increases with M_n (see below in Section 3.5.2). When the IPDI is replaced by HMDI diisocyanate, as in the case of **PU4**, a substantial increase in the r_h is observed in comparison with polymers with similar molecular weights, such as **PU1** and **PU3**. The fragment derived from HMDI behaves like a larger (and more flexible) unit compared to IPDI, resulting in polymers of larger sizes for a similar molar ratio of the monomers used in the synthesis of the linear polyurethanes.

3.5. Determination of the Molecular Weights of the PUs

3.5.1. Determination by GPC/SEC

The highest number-average molecular weight was obtained for **PU4**, the lowest one corresponds to **PU3** (Table 4), which are in agreement with the hydrodynamic radii of these polyurethanes. All the synthesized polyurethanes presented similar relatively narrow dispersities ($Đ = M_w/M_n$).

Since the GPC/SEC measurements (see chromatograms in Figure S13 in the Supplementary Materials) were calibrated with polystyrene (PS) standards, the values of M_n and M_w are relative to PS and may be quite different from the absolute values because the hydrodynamic volumes of the present PUs are considerably different from those of polystyrenes with equivalent molecular weights.

Table 4. The average molecular weights and dispersities determined by GPC/SEC [a].

Polyurethane No.	M_w [a] (Dalton)	M_n [a] (Dalton)	$Đ$ (M_w/M_n)
PU1	15,600	10,300	1.50
PU2	19,200	12,800	1.48
PU3	9200	7200	1.27
PU4	23,500	15,100	1.55

[a] In polystyrene units.

3.5.2. Determination by ^{13}C NMR Spectroscopy

Quantitative ^{1}H and ^{13}C NMR are powerful tools to count monomer groups in homopolymers and copolymers of very well defined composition and narrow molecular weight distributions. The segmented linear polyurethanes herein described satisfy the above-mentioned criteria as they present well defined linear chains, low molecular weights, and dispersity indices below 1.5.

To improve the determination of M_n values of the present polyurethanes and attempt to obtain absolute values, the area of methyl Ca of PPG segments, centered at 17.2 ppm, which corresponds to an average of 17 carbons per PPG segment, was used as the reference instead of those of the copolymer end groups (see Tables S5–S8 of the Supplementary Materials for composition calculations). The polyurethanes average compositions obtained in this way are shown in Table 5 together with the calculated M_n values. These compositions agree with the monomer ratios used in the synthesis of polyurethanes **PU1–PU4** (see Table 2).

Table 5. The average compositions and M_n of **PU1–PU4** determined by NMR spectroscopy.

Polyurethane No.	Number of Isosorbide Units [a]	Number of IPDI Units [a]	Number of HDMI Units [a]	Number of PPG Segments	M_n [b] (Dalton)
PU1	5 (t)	6 (t)	-	1	3040
PU2	5 (t)	7 (t,b)	-	2	4270
PU3	6 (t)	6	-	1	3210
PU4	5 (t)	-	6 (t)	1	3380

[a] t = at least one monomer is terminal; b = one monomer is in between PPG segments. [b] M_n values with uncertainties up to 15%.

The values obtained for M_n calculated by NMR are substantially lower than those determined by GPC/SEC (see Table 4), meaning that the hydrodynamic volumes of these PUs are considerably higher than those of the polystyrene standards with equivalent molar masses.

3.6. Thermal Properties of the PUs

The evaluation of the thermal behavior of the samples (polymers) is crucial for biomedical applications. For instance, if the value of the glass transition temperature (T_g) of the polymer is higher than the human body temperature, the polymer is rigid (glass region). In contrast, if the T_g value is below the body temperature, the sample is in a rubber or elastomeric state. Figure 10 displays the DSC thermograms of the four materials under study in the range −100 to +230 °C. The DSC results presented refer to the second thermal cycle and the T_g value was determined as the onset point of the thermal event detected. All the samples are in the amorphous state and do not present phase separation. Thus, a unique glass transition above the human body temperature is detectable, at 54.6, 57.7, 69.0, and 68.0 °C, for the samples **PU1** to **PU4**, respectively, which indicate that these polyurethanes exhibit a lacquer/varnish behavior. No other thermal events were detected in addition to these glass transition temperatures, except in the case of **PU3**, which exhibits two additional small T_g events at lower temperatures (−7.2 and 14.4 °C), with the ΔC_p values ca. being three times smaller.

The **PU3** sample has a higher content of low oligomers than the remaining samples, as can be observed in the corresponding GPC/SEC chromatograms (Figure S13 of the Supplementary Materials), which can possibly give rise to very small domains characterized by lower T_g values (likely corresponding to isolated short hard segments).

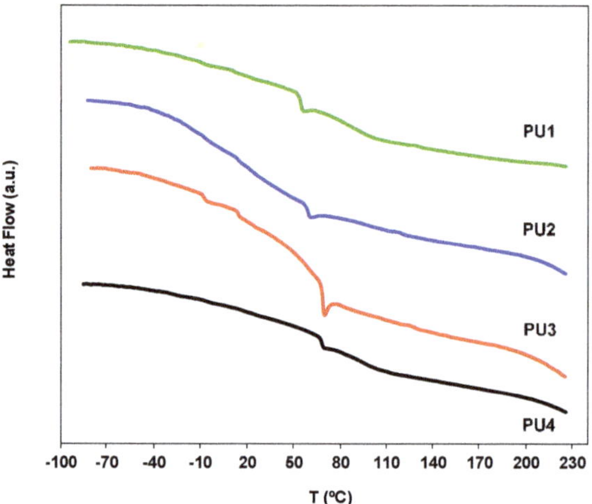

Figure 10. The DSC thermograms of the polyurethanes samples **PU1–PU4** (exo up).

3.7. Cytotoxicity Assay

The HaCaT keratinocyte cell line was selected to evaluate the cytotoxicity of the PUs synthesized in this work since these cells are the predominant cell type in the adult human epidermis, the outermost layer of the skin, thus, presenting an adequate in vitro cell model that better mimics in vivo conditions. The HaCaT cell viability, with 72 h of exposure to polyurethanes **PU1–PU4**, was evaluated by the MTT reduction assay [45] (Figure 11). The tests corresponding to **PU1–PU3** do not differ significantly from the control experiment (glass coverslip), whereas the cell viability for **PU4** is 80% $\pm$ 5% (i.e., significantly different from control for $p < 0.05$).

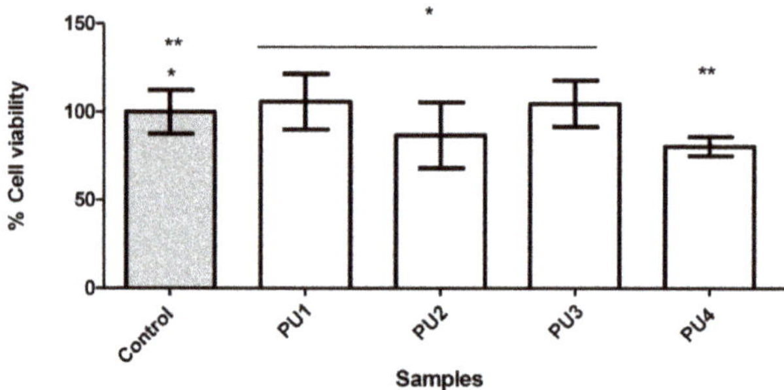

Figure 11. The HaCaT cell viability by MTT after proliferation under polyurethanes **PU1–PU4**. Control (glass slide). (mean $\pm$ SD) ($n = 6$). (* Not significantly different ($p > 0.05$), ** significantly different ($p < 0.05$)).

The cell viability was higher than 70% for all the PUs tested. Thus, these materials can be considered biocompatible according to this in vitro assay. These results are in accordance with the previously published results, where the PU nail lacquers were evaluated and found to be material biocompatible, being proposed for nail applications as safe pharmaceutical excipients [11].

4. Conclusions

A convenient synthesis for polyurethanes from renewable sources is described. The structure of these polyurethanes was determined and confirmed by FTIR, ^{1}H, and ^{13}C NMR spectroscopies of all four samples. The structures found confirm the higher reactivity of the secondary NCO group of IPDI compared to the primary group.

The quantitative ^{13}C NMR spectroscopy analysis enabled the calculation of the polyurethanes compositions, number-average molecular weights, as well as of their diffusion coefficients (D) and hydrodynamic radii (r_h) from the PGSE NMR experiments. Both values are in accordance with the molar ratio employed in the synthesis of the polyurethanes.

The polyurethanes were also characterized by GPC/SEC leading to molecular weights, which are considerably higher than the previous values since they are determined relative to the polystyrene standards, and relatively narrow molecular weight distributions, with dispersity indices $M_w/M_n < 1.5$.

The DSC studies indicate that the four PU samples are in the amorphous state, being characterized by T_g values above the human body temperature.

The polyurethanes obtained were biocompatible with keratinocyte cells in the conditions used in the tests. This is a relevant property that must be taken into account when considering pharmaceutical excipients. For this reason, these newly synthesized polymers can be used in pharmaceutics, namely, as a component of nail lacquers for controlled-release drug delivery.

Supplementary Materials: The following are available online at http://www.mdpi.com/2073-4360/10/10/1170/s1.

Author Contributions: Conceptualization, P.T.G., H.M.R. and J.C.B.; Investigation, B.S.G.V., C.S.B.G., J.R.A., H.P.D., L.M.G. and R.G.d.S.; Methodology, P.T.G., J.R.A., H.M.R. and J.C.B.; Project administration, H.M.R. and J.C.B.; Supervision, P.T.G., H.M.R. and J.C.B.; Writing—original draft, B.S.G.V., C.S.B.G., P.T.G., J.R.A. and H.P.D.; Writing—review & editing, C.S.B.G., P.T.G. and H.M.R.

Funding: The authors gratefully acknowledge the support of CERENA and CQE strategic projects (FCT-UID/ ECI/04028/2013 and UID/QUI/00100/2013, respectively) funded by the Fundação para Ciência e a Tecnologia (FCT), Portugal. B.S.G.V., R.G.S. and C.S.B.G. also acknowledge FCT for fellowships (SFRH/BD/78962/2011, SFRH/BPD/105662/2015 and SFRH/BPD/107834/2015, respectively).

Acknowledgments: We also thank to PTNMR network for the NMR facility.

Conflicts of Interest: The authors declare no conflict of interest.

References

1. Sarkar, D.; Yang, J.C.; Sen Gupta, A.; Lopina, S.T. Synthesis and characterization of L-tyrosine based polyurethanes for biomaterial applications. *J. Biomed. Mater. Res. Part A* **2009**, *90*, 263–271. [CrossRef] [PubMed]

2. Han, J.; Chen, B.; Ye, L.; Zhang, A.Y.; Zhang, J.; Feng, Z.G. Synthesis and characterization of biodegradable polyurethane based on poly caprolactone and L-lysine ethyl ester diisocyanate. *Front. Mater. Sci. China* **2009**, *3*, 25–32. [CrossRef]

3. Besse, V.; Auvergne, R.; Carlotti, S.; Boutevin, G.; Otazaghine, B.; Caillol, S.; Pascault, J.P.; Boutevin, B. Synthesis of isosorbide based polyurethanes: An isocyanate free method. *React. Funct. Polym.* **2013**, *73*, 588–594. [CrossRef]

4. Chen, T.K.; Tien, Y.I.; Wei, K.H. Synthesis and characterization of novel segmented polyurethane clay nanocomposite via poly(epsilon-caprolactone)/clay. *J. Polym. Sci. Part A Polym. Chem.* **1999**, *37*, 2225–2233. [CrossRef]

5. Bachmann, F.; Reimer, J.; Ruppenstein, M.; Thiem, J. Synthesis of novel polyurethanes and polyureas by polyaddition reactions of dianhydrohexitol configurated diisocyanates. *Macromol. Chem. Phys.* **2001**, *202*, 3410–3419. [CrossRef]

6. Guelcher, S.A.; Gallagher, K.M.; Didier, J.E.; Klinedinst, D.B.; Doctor, J.S.; Goldstein, A.S.; Wilkes, G.L.; Beckman, E.J.; Hollinger, J.O. Synthesis of biocompatible segmented polyurethanes from aliphatic diisocyanates and diurea diol chain extenders. *Acta Biomater.* **2005**, *1*, 471–484. [CrossRef] [PubMed]

7. Kim, H.-J.; Kang, M.-S.; Knowles, J.C.; Gong, M.S. Synthesis of highly elastic biocompatible polyurethanes based on bio-based isosorbide and poly(tetramethylene glycol) and their properties. *J. Biomater. Appl.* **2014**, *29*, 454–464. [CrossRef] [PubMed]

8. Abraham, G.A.; Marcos-Fernández, A.; San Román, J. Bioresorbable poly(ester-ether urethane)s from L-lysine diisocyanate and triblock copolymers with different hydrophilic character. *J. Biomed. Mater. Res. Part A* **2006**, *76*, 729–736. [CrossRef] [PubMed]

9. Zhang, C.; Zhang, N.; Wen, X. Synthesis and characterization of biocompatible, degradable, light-curable, polyurethane-based elastic hydrogels. *J. Biomed. Mater. Res. Part A* **2007**, *82A*, 637–650. [CrossRef] [PubMed]

10. Spiridon, I.; Popa, V.I. Hemicelluloses: Major Sources, Properties and Applications. In *Monomers, Polymers and Composites from Renewable Resources*; Ch. 13, Belgacem, M.N., Gandini, A., Eds.; Elsevier: Amsterdam, The Netherlands, 2008; pp. 289–304.

11. Gregorí Valdes, B.S.; Serro, A.P.; Gordo, P.M.; Silva, A.; Gonçalves, L.; Salgado, A.; Marto, J.; Baltazar, D.; dos Santos, R.G.; Bordado, J.M.; et al. New Polyurethane Nail Lacquers for the Delivery of Terbinafine: Formulation and Antifungal Activity Evaluation. *J. Pharm. Sci.* **2017**, *106*, 1570–1577. [CrossRef] [PubMed]

12. Gogolewski, S.; Gorna, K.; Zaczynska, E.; Czarny, A. Structure-property relations and cytotoxicity of isosorbide-based biodegradable polyurethane scaffolds for tissue repair and regeneration. *J. Biomed. Mater. Res. Part A* **2008**, *85*, 456–465. [CrossRef] [PubMed]

13. Fenouillot, F.; Rousseau, A.; Colomines, G.; Saint-Loup, R.; Pascault, J.P. Polymers from renewable 1,4:3,6-dianhydrohexitols (isosorbide, isomannide and isoidide): A review. *Prog. Polym. Sci.* **2010**, *35*, 578–622. [CrossRef]

14. Stoss, P.; Hemmer, R. 1,4/3,6-Dianhydrohexitols. *Adv. Carbohydr. Chem. Biochem.* **1991**, *49*, 93–173. [CrossRef] [PubMed]

15. Marques, M.F.F.; Gordo, P.M.; De Lima, A.P.; Queiroz, D.P.; De Pinho, M.N.; Major, P.; Kajcsos, Z. Free-volume studies in polycaprolactone/poly(propylene oxide) urethane/urea membranes by positron lifetime spectroscopy. *Acta Phys. Pol. A* **2008**, *113*, 1359–1364. [CrossRef]

16. Chattopadhyay, D.K.; Raju, N.P.; Vairamani, M.; Raju, K.V.S.N. Structural investigations of polypropylene glycol (PPG) and isophorone diisocyanate (IPDI) based polyurethane prepolymer by matrix-assisted laser desorption/ionization time-of-flight (MALDI-TOF)-mass spectrometry. *Prog. Org. Coat.* **2008**, *62*, 117–122. [CrossRef]

17. Prabhakar, A.; Chattopadhyay, D.K.; Jagadeesh, B.; Raju, K.V.S.N. Structural investigations of polypropylene glycol (PPG) and isophorone diisocyanate (IPDI)-based polyurethane prepolymer by 1D and 2D NMR spectroscopy. *J. Polym. Sci. Part A Polym. Chem.* **2005**, *43*, 1196–1209. [CrossRef]

18. Lee, C.-H.; Takagi, H.; Okamoto, H.; Kato, M.; Usuki, A. Synthesis, Characterization, and Properties of Polyurethanes Containing 1,4:3,6-Dianhydro-D-sorbitol. *J. Polym. Sci. Part A Polym. Chem.* **2009**, *47*, 6025–6031. [CrossRef]

19. Marín, R.; Alla, A.; Ilarduya, A.M.; Munõz-Guerra, S. Carbohydrate-Based Polyurethanes: A Comparative Study of Polymers Made from Isosorbide and 1,4-Butanediol. *J. Appl. Polym. Sci.* **2012**, *123*, 986–994. [CrossRef]

20. Daniel-da-Silva, A.L.; Bordado, J.C.M.; Martín-Martínez, J.M. Moisture curing kinetics of isocyanate ended urethane quasi-prepolymers monitored by IR spectroscopy and DSC. *J. Appl. Polym. Sci.* **2008**, *107*, 700–709. [CrossRef]

21. Gregorí Valdés, B.S.; Bordado, J.M.; Ribeiro, H.; Bello, A.; Fernández, M.; Fernandes, S.; Vázquez, NA.R. Analysis and Characterization of Quasi-Prepolymers Obtained from Polyethylene Glycol 1500 and 4,4′-Diphenylmethane-DiIsocyanate. *Mater. Sci. Eng. Adv. Res.* **2015**, *1*, 10–15. [CrossRef]

22. Price, W.S. Pulsed-Field Gradient Nuclear Magnetic Resonance as a Tool for Studying Translational Diffusion. Part 1. Basic Theory. *Concepts Magn. Reson.* **1997**, *9*, 299–336. [CrossRef]

23. Geil, B. Measurement of translational molecular diffusion using ultrahigh magnetic field gradient NMR. *Concepts Magn. Reson.* **1998**, *10*, 299–321. [CrossRef]

24. Moura Ramos, J.J.; Taveira-Marques, R.; Diogo, H.P. Estimation of the fragility index of indomethacin by DSC using the heating and cooling rate dependency of the glass transition. *J. Pharm. Sci.* **2004**, *93*, 1503–1507. [CrossRef] [PubMed]

25. International Organization for Standardization. *Biological Evaluation of Medical Devices Part 5: Tests for In Vitro Cytotoxicity*; ISO 10993–5; ISO: Geneva, Switzerland, 2009; Volume 5, pp. 1–52.

26. Kuan, H.-C.; Chuang, W.-P.; Ma, C.-C.M.; Chiang, C.-L.; Wu, H.-L. Synthesis and characterization of a clay/waterborne polyurethane nanocomposite. *J. Mater. Sci.* **2005**, *40*, 179–185. [CrossRef]

27. Ferreira, P.; Pereira, R.; Coelho, J.F.J.; Silva, A.F.M.; Gil, M.H. Modification of the biopolymer castor oil with free isocyanate groups to be applied as bioadhesive. *Int. J. Biol. Macromol.* **2007**, *40*, 144–152. [CrossRef] [PubMed]

28. Ferreira, P.; Silva, A.F.M.; Pinto, M.I.; Gil, M.H. Development of a biodegradable bioadhesive containing urethane groups. *J. Mater. Sci. Mater. Med.* **2008**, *19*, 111–120. [CrossRef] [PubMed]

29. Li, G.; Li, P.; Qiu, H.; Li, D.; Su, M.; Xu, K. Synthesis, characterizations and biocompatibility of alternating block polyurethanes based on P3/4HB and PPG-PEG-PPG. *J. Biomed. Mater. Res. Part A.* **2011**, *98A*, 88–99. [CrossRef] [PubMed]

30. Alves, P.; Coelho, J.F.J.; Haack, J.; Rota, A.; Bruinink, A.; Gil, M.H. Surface modification and characterization of thermoplastic polyurethane. *Eur. Polym. J.* **2009**, *45*, 1412–1419. [CrossRef]

31. Rahman, M.M.; Hasneen, A.; Chung, I.; Kim, H.; Lee, W.-K.; Chun, J.H. Synthesis and properties of polyurethane coatings: The effect of different types of soft segments and their ratios. *Compos. Interfaces* **2013**, *20*, 15–26. [CrossRef]

32. Caracciolo, P.C.; Buffa, F.; Abraham, G.A. Effect of the hard segment chemistry and structure on the thermal and mechanical properties of novel biomedical segmented poly(esterurethanes). *J. Mater. Sci. Mater. Med.* **2009**, *20*, 145–155. [CrossRef] [PubMed]

33. Kricheldorf, H.R.; Chatti, S.; Schwarz, G.; Krüger, R.P. Macrocycles 27: Cyclic Aliphatic Polyesters of Isosorbide. *J. Polym. Sci. Part A Polym. Chem.* **2003**, *41*, 3414–3424. [CrossRef]

34. Ferreira, P.; Coelho, J.F.J.; Pereira, R.; Silva, A.F.M.; Gil, M.H. Synthesis and characterization of a poly(ethylene glycol) prepolymer to be applied as a bioadhesive. *J. Appl. Polym. Sci.* **2007**, *105*, 593–601. [CrossRef]

35. Vieira, A.P.; Ferreira, P.; Coelho, J.F.; Gil, M.H. Photocrosslinkable Polymers for Biomedical Applications. *Int. J. Biol. Macromol.* **2008**, *43*, 325–332. [CrossRef] [PubMed]

36. Ferreira, P.; Coelho, J.F.; Gil, M.H. Developmet of a new photocrosslinkable biodegradable bioadhesive. *Int. J. Pharm.* **2008**, *352*, 172–181. [CrossRef] [PubMed]

37. Daniel da Silva, A.L.; Martín-Martinez, J.M.; Bordado, J.C.M. Influence of the storage of reactive urethane quasi-prepolymers in their compositions and adhesion properties. *Int. J. Adhes. Adhes.* **2007**, *28*, 29–37. [CrossRef]

38. Zhu, R.; Wang, Y.; Zhang, Z.; Ma, D.; Wang, X. Synthesis of polycarbonate urethane elastomers and effects of the chemical structures on their thermal, mechanical and biocompatibility properties. *Heliyon* **2016**, *2*, e00125. [CrossRef] [PubMed]

39. Zhang, S.; Cheng, L.; Hu, J. NMR studies of water-borne polyurethanes. *J. Appl. Polym. Sci.* **2003**, *90*, 257–260. [CrossRef]

40. Rahman, M.M.; Hasneen, A.; Jo, N.J.; Kim, H.I.; Lee, W.K. Properties of Waterborne Polyurethane Adhesives with Aliphatic and Aromatic Diisocyanates. *J. Adhes. Sci. Technol.* **2011**, *25*, 2051–2062. [CrossRef]

41. Götz, H.; Beginn, U.; Bartelink, C.F.; Grünbauer, H.J.M.; Möller, M. Preparation of isophorone diisocyanate terminated star polyethers. *Macromol. Mater. Eng.* **2002**, *287*, 223–230. [CrossRef]

42. Wang, H.; Kao, H.; Digar, M.; Wen, T. FTIR and Solid State 13 C NMR Studies on the Interaction of Lithium Cations with Polyether Poly(urethane urea). *Macromolecules* **2011**, *34*, 529–537. [CrossRef]

43. Mondiot, F.; Loudet, J.C.; Mondain-Monval, O.; Snabre, P.; Vilquin, A.; Würger, A. Stokes-Einstein diffusion of colloids in nematics. *Phys. Rev. E Stat. Nonlinear Soft Matter Phys.* **2012**, *86*, 010401. [CrossRef] [PubMed]

44. Holz, M.; Mao, X.A.; Seiferling, D.; Sacco, A. Experimental study of dynamic isotope effects in molecular liquids: Detection of translation-rotation coupling. *J. Chem. Phys.* **1996**, *104*, 669–679. [CrossRef]

45. Mosmann, T. Rapid colorimetric assay for cellular growth and survival: Application to proliferation and cytotoxicity assays. *J. Immunol. Methods* **1983**, *65*, 55–63. [CrossRef]

Article

A Mild Method for Surface-Grafting PEG Onto Segmented Poly(Ester-Urethane) Film with High Grafting Density for Biomedical Purpose

Lulu Liu [†], Yuanyuan Gao [†], Juan Zhao, Litong Yuan, Chenglin Li, Zhaojun Liu and Zhaosheng Hou *

College of Chemistry, Chemical Engineering and Materials Science, Shandong Normal University, Jinan 250014, Shandong, China; Lulu0214@163.com (L.L.); mgaoyuanyuan@163.com (Y.G.); zhaojuan1016@yeah.net (J.Z.); ylt199710@163.com (L.Y.); chenglinli01@163.com (C.L.); zhaojun0403 @126.com (Z.L)
* Correspondence: houzs@sdnu.edu.cn
† These authors contributed equally to this work.

Received: 16 September 2018; Accepted: 8 October 2018; Published: 10 October 2018

Abstract: In the paper, poly(ethylene glycol) (PEG) was grafted on the surface of poly(ester-urethane) (SPEU) film with high grafting density for biomedical purposes. The PEG-surface-grafted SPEU (SPEU-PEG) was prepared by a three-step chemical treatment under mild-reaction conditions. Firstly, the SPEU film surface was treated with 1,6-hexanediisocyanate to introduce -NCO groups on the surface with high density (5.28×10^{-7} mol/cm^2) by allophanate reaction; subsequently, the -NCO groups attached to SPEU surface were coupled with one of -NH$_2$ groups of tris(2-aminoethyl)amine via condensation reaction to immobilize -NH$_2$ on the surface; finally, PEG with different molecular weight was grafted on the SPEU surface through Michael addition between terminal C = C bond of monoallyloxy PEG and -NH$_2$ group on the film surface. The chemical structure and modified surface were characterized by FT-IR, ^{1}H NMR, X-ray photoelectron spectroscopy (XPS), and water contact angle. The SPEU-PEGs displaying much lower water contact angles (23.9–21.8°) than SPEU (80.5°) indicated that the hydrophilic PEG chains improved the surface hydrophilicity significantly. The SPEU-PEG films possessed outstanding mechanical properties with strain at break of 866–884% and ultimate stress of 35.5–36.4 MPa, which were slightly lower than those of parent film, verifying that the chemical treatments had minimum deterioration on the mechanical properties of the substrate. The bovine serum albumin adsorption and platelet adhesion tests revealed that SPEU-PEGs had improved resistance to protein adsorption (3.02–2.78 µg/cm^2) and possessed good resistance to platelet adhesion (781–697 per mm^2), indicating good surface hemocompatibility. In addition, due to the high grafting density, the molecular weight of surface-grafted PEG had marginal effect on the surface hydrophilicity and hemocompatibility.

Keywords: segmented poly(ester-urethane); poly(ethylene glycol); surface grafting; chemical treatment; high grafting density; hemocompatibility

---

1. Introduction

Segmented polyurethane (SPU) is widely used in biomedical fields, such as cardiovascular devices, artificial organs, and tissue engineering scaffolds, due to its long-term bio-stability, excellent mechanical properties, and relatively superior biocompatibility [1–6]. However, when SPU is used as long-term blood-contacting materials, the surface of SPU films will result in significant adsorption of proteins, and induce platelet adhesion by activating the coagulation pathway, eventually leading to the formation of microscopic thrombi [7,8]. In addition, the biofouling on SPU surfaces can reduce its mechanical properties [9,10]. To further improve their hemocompatibility, much attention has been paid to producing a nonspecific protein repelling surface by surface modification and creating highly effective non-thrombogenic devices. A preferred strategy is to immobilize natural or synthetic materials onto the hydrophobic surfaces that shield the surface, thus introducing a high activation barrier to repel proteins [11–13]. Among them, grafting poly(ethylene glycol) (PEG) onto the SPU surface has attracted considerable interest because PEG can effectively prevent protein adsorption and platelet adhesion mostly due to its low interfacial free energy with water, unique solution properties, hydrophilicity, high chain mobility, and steric stabilization effect [14]. Additionally, much theoretical work is generated to explain the early discovery that grafted PEG chains resist protein adsorption to a high degree [15,16].

Many kinds of surface modification approaches, including chemical treatment, strong oxidation, plasma treatment, UV irradiation, and laser treatment have been used to modify the surface of medical SPU [17–20]. Among the techniques, chemical treatment possesses some advantageous uniqueness, such as has clearer mechanism and predictable products, and the reaction rate can accurately be controlled by adjusting the reaction parameters. Moreover, the chemical treatment methods including click chemistry [21] and NHS-amine reaction [22] are adopted to modify poly(ε-caprolactone) dendrimer and poly(ethylene-co-acrylic acid) films, respectively. The most commonly used surface chemical treatment to modify the SPU film is allophanate reaction, which is to attack on N-H bonds of urethane groups in the backbone by small molecular diisocyanate to immobilize free -NCO groups on the surface, and subsequently PEG (or PEG derivatives) is grafted on the surface using the chemical reaction [13,23,24]. However, in order to obtain high -NCO density on the surface, the allophanate reaction should be carried out at high reaction temperature of 50–70 °C, which inevitably deteriorates the bulk properties of the substrates. On the other hand, even if using the high reaction temperature, the -NCO density on the surface was unsatisfactory due to the low -NH- content in SPU.

In our previous report [25], a new biodegradable segmented poly(ester-urethane) (SPEU) with uniform-size hard segments was prepared using aliphatic diurethane diisocyanate (1,6-hexanediisocyanate-1,4-butanediol-1,6-hexanediisocyanate, HBH) as a chain extender. The SPEU, which exhibited excellent mechanical properties comparable to MDI-based PU, could meet the requirement of long-term implant biomaterial. The PEG was grafted on the surface of SPEU films via aminolysis. However, the grafting density was only 2.74×10^{-7} mol/cm^2, and the results of protein adsorption and platelet adhesion tests showed that the surface hemocompatibility needed to be further improved. In consideration of there being four urethane groups in each hard segment, the SPEU possesses higher −NH- content than MDI-based PU, and thus, the high grafting density on the film surface could be obtained by chemical treatment with diisocyanate.

In this study, we graft PEG on the surface of SPEU film to obtain a high grafting density without deteriorating the intrinsic mechanical properties of the substrates. The surface grafting of PEG on SPEU films was prepared by a three-step chemical treatment (Figure 1), which were all carried out under moderate conditions. The synthetic scheme involved coupling 1,6-hexanediisocyanate (HDI) to the surfaces of SPEU through allophanate reaction (Figure 1a); the -NCO groups attached to SPEU surface (SPEU-NCO) were then coupled -NH$_2$ groups of tris(2-aminoethyl)amine (TAEA) via condensation reaction (Figure 1b) to immobilize -NH$_2$ on the SPEU surface (SPEU-NH$_2$); finally, the PEG was grafted on the SPEU (SPEU-PEG) surface through Michael addition (Figure 1c) between double bond of monoallyloxy poly(ethylene glycol) (APEG) and a primary amino group on the film surface. The chemical structure and modified surface were characterized by Fourier transform infrared spectroscopy (FT-IR), H Nuclear magnetic resonance (^{1}H NMR), X-ray photoelectron spectroscopy (XPS), and water contact angle measurements. The influence of chemical treatments on the mechanical properties of the films was researched, and surface hemocompatibility of the PEG-grafted SPEU films was evaluated by protein adsorption and platelet adhesion tests. Furthermore, the effect of molecular weight of grafted PEG on the surface hydrophilicity and hemocompatibility was preliminarily studied.

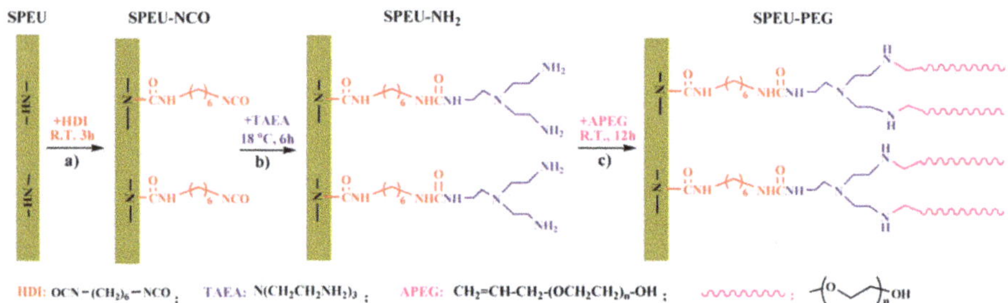

Figure 1. Schematic diagram of grafting PEG onto SPEU surface via (**a**) allophanate reaction; (**b**) condensation reaction; and (**c**) Michael addition reaction.

2. Materials and Methods

2.1. Materials

Poly(ε-caprolactone) (PCL, M_n = 2000 g/mol) was supplied by Shenzhen Polymtek Biomaterial Co., Ltd. (Shenzhen, China) and dried for 4 h at 100 °C under vacuum prior to use. HBH was synthesized in our lab according our published paper [4] and the chemical structure was confirmed by ^{1}H NMR, ^{13}C NMR, and HR-MS. HDI and dibutyltin dilaurate (DBTDL) were purchased from Sigma-Aldrich Chemical Co. (St Louis, MO, USA) and used without further purification. TAEA, (> 97%, Shanghai Macklin Biochemical Co., Ltd. Shanghai, China) was dried over 4-Å molecular sieves and redistilled before use. APEG (M_n = 1200, 2400, 4000 g/mol) was obtained from Shanghai Aladdin Reagent Co. (Shanghai, China) and was used as received. *N,N*-dimethylformamide (DMF, AR grade, Beijing Chemical Reagent Co., Ltd, Beijing, China) was dried with phosphorus pentoxide and distilled under reduced pressure before use. Other reagents were AR grade and purified by standard methods.

2.2. Preparation of SPEU and SPEU Films

SPEU was synthesized according to our previous publication [25]. Briefly, the DMF solution of HBH (25 wt %) was added dropwise into the PCL containing DBTDL (0.3 wt % of PCL) with vigorous mechanical stirring under dried nitrogen atmosphere at 80°C. The molar ratio of -NCO/-OH was controlled at 1.02. After that, the reaction mixture was allowed to proceed at the same temperature until the -NCO peak (~2270 cm^{-1}) in the FT-IR spectrum disappeared (~3.5 h), and subsequently diluted with DMF to approximately 4.5 g/100 mL. The diluted solution was poured into a Teflon mold. The solvent was removed by natural volatilizing at 50 °C for 4 days, and the semitransparent films with 0.20 ± 0.02 mm thickness were subsequently vacuum dried at 40 °C for 1 day to remove the last traces of solvent. GPC (THF): M_w = 111000, M_n = 83000, M_w/M_n = 1.33.

2.3. Grafting of PEG on the SPEU Film Surface

Before chemical treatments, the surfaces of the SPEU film discs with ~10 mm diameter were ultrasonically cleaned in 50% ethanol for 30 min and then dried in a vacuum. First, HDI (1.0 g) and DBTDL (0.02 g) were dissolved in anhydrous toluene (10 mL) to get a homogeneous solution, SPEU disc was immersed in the solution and the reaction was allowed to shake for 3 h at room temperature. The film disc was taken out and rinsed with anhydrous toluene to remove unreacted HDI from the surface completely. The SPEU-NCO film achieved was then soaked in 10 mL anhydrous toluene containing 0.13 g TAEA. After gentle shaking at 18 °C for 6 h, the film disc was removed from the solution and flushed with anhydrous toluene to obtain the PEU-NH$_2$ film. Finally, the PEU-NH$_2$ film was immersed in 10 mL absolute ethanol containing APEG with different molecular weight (0.3 mmol/mL). After gentle shaking for 12 h at room temperature, the films were rinsed with absolute ethanol thoroughly and dried under vacuum at room temperature to constant weight. Three

SPEU-PEGs based on APEG with M_n = 1200, 2400 and 4000 g/mol were named as SPEU-PEG-a, SPEU-PEG-b, and SPEU-PEG-c, respectively.

2.4. Instruments and Characterization

FT-IR: FT-IR spectra were recorder on an Alpha infrared spectrometer (Bruker, Germany) equipped with a Bruker platinum ATR accessory in the range of 4000-400 cm^{-1} with the resolution of 4 cm^{-1}.

^{1}H NMR: NMR spectra were recorded with a 400 MHz Avance II spectrometer (Bruker, BioSpin GmbH, Rheinstetten, Germany) with CDCl$_3$ as the solvent.

GPC: The number average molecular weight (M_n) and molecular weight distribution (M_w/M_n) were measured by gel permeation chromatography (GPC, Waters Alliance GPC 2000). The continuous phase was tetrahydrofuran, and monodisperse polystyrene was used as the calibration standards.

Mechanical properties: Tensile strength properties were determined with a single-column tensile test machine (Model HY939C, Dongguan Hengyu Instruments, Ltd., Dongguan, China) at room temperature. Dumbbell-shaped specimens were punched from the films with a punching die of 12 mm width and 75 mm length, the neck width and length were 4.0 and 30 mm, respectively. The crosshead speed was controlled at 50 mm/min. At least five specimens were measured and the averaged results were reported.

XPS: Chemical composition of the film surface was characterized by XPS (ESCALAB250Xi, Thermo Scientific, Waltham, UK) with a mono-chromated Al-Kα radiation source (energy 1486.68 eV). The base pressure for the measurement was ~3 $\times$ 10^{-7} Pa. The binding energies were referenced to the C1s line at 284.8 eV from adventitious carbon. The measurement was carried out at room temperature with a 90° take-off angle of the photoelectron. The atomic concentrations of the elements were determined by their corresponding peak areas.

Water contact angle: The static water-contact angles of blank and modified SPEU films were measured on a drop shape analysis system (CAM 200, KSV Instruments, Helsinki, Finland) at room temperature. Ultrapure water was used as test fluid, and the contact angles were read within 3 s after each drop by using a microscopy. At least six replicate measurements were performed on each specimen and the average was calculated

Protein adsorption: Bradford's protein determining method was used to assay the albumin adsorbed on the film surface with bovine serum albumin (BSA) as the model protein [26,27]. Before measurement, all the film discs (~10 mm in diameter) were aged in phosphate buffer saline (PBS, pH = 7.4) for one day to remove the physically adsorbed impurities and to achieve complete hydration. The BSA concentration (initial concentration: 1.0 mg/mL) was diluted to 45 µg/mL with PBS. In a typical adsorption experiment, a film disc was immersed in 10 mL of BSA solution at 37°C for 1 h. At the end of the predetermined equilibrium period (1 h), the disc was taken out and rinsed with fresh PBS for several times remove the unbound BSA. The adsorbed protein on the surface was detached in 1 wt % sodium dodecylsulfonate aqueous solution by the stirring method (100 rpm for 1 h). The concentration of the adsorbed BSA was determined with micro-Bradford protein assay method [25,28], and the adsorbed amount was calculated according to the standard curve. The reported values were the average of three independent measurements.

Platelet adhesion: The interaction between the film surface and blood was evaluated by platelet adhesion tests. The platelet-rich plasma (PRP) was prepared from fresh rabbit blood containing sodium citrate as an anticoagulant (Shandong Success Biological Technology Co., Ltd., Jinan, China) according the reported literature [29]. The film discs with ~10 mm diameter were firstly soaked with PBS (pH = 7.4) for 24 h to achieve complete hydrated surface, and then immerged in 1.0 mL PRP and incubated for 60 min at 37 °C. After that, the discs were taken out and rinsed with PBS to remove the non-adherent platelet. Subsequently, the platelets adhering on the surface were fixed with a 2.5% glutaraldehyde for 30 min at 37 °C, and the discs were thoroughly rinsed with PBS and dehydrated by treating with gradual ethanol/water solution from 50% to 100% ethanol (*v:v*) with a step of 10% for 30 min in each step. Finally, the platelet-attached surfaces were allowed to dry at room temperature and coated with gold prior to being observed with a Cold Field Emission Scanning Electron Microscope (FE-SEM, Hitachi SU8010, Hitachi, Tokyo, Japan). Different fields were randomly observed. Quantification of adhered platelets was calculated from eight different areas of one specimen using FE-SEM images.

3. Results and Discussion

3.1. Preparation and Characterization

3.1.1. Activating SPEU Surface with HDI (SPEU-NCO)

The free -NCO groups were grafted onto SPEU film surface by the allophanate reaction (Figure 1a) between -NH- proton of urethane group and -NCO group of HDI in the presence of DBTDL catalyst. The reaction temperature was controlled at room temperature for the purpose of minimum deterioration on the SPEU substrate. The density of free -NCO group on the surface measured by di-n-butylamine back titration method [30] was 5.28×10^{-7} mol/cm^2, which was much higher than that of MDI-based PU (2.5×10^{-8} mol/cm^2, Pellethane$^®$, Dow Chemical Co., Midland, MI, USA) in the reported paper in Reference [31]. It was obviously attributed to the high content of urethane (-NH-) groups in SPEU, in which each hard segment contained four urethane groups. The calculated -NH- content in SPEU was 1.67×10^{-3} mol/g, while the -NH- content in Pellethane$^®$ PU was only 0.9×10^{-3} mol/g (the result is based on the same soft segment content in SPEU with Pellethane$^®$ PU). After introducing the -NCO groups onto the SPEU surface via allophanate reaction with HDI, a sharp peak at 2270 cm^{-1}, the characteristic symmetric stretching vibration peak of -NCO, was observed in the FT-IR spectrum of SPEU-NCO film (Figure 2b), which strongly supported that -NCO groups had been immobilized onto the SPEU film surface.

3.1.2. Introducing -NH$_2$ Group onto SPEU-NCO Film (SPEU-NH$_2$)

The free -NCO groups on the surface were reacted with -NH$_2$ groups of TAEA by condensation reaction (Figure 1b) to introduce -NH$_2$ on the film surface. Obviously, one TAEA molecule contains three -NH$_2$ groups, two -NH$_2$ groups can remain on the surface after one -NH$_2$ reacts with the -NCO group, thus producing more -NH$_2$ groups on the SPEU surface. However, because of the high reactive activity of -NCO group, the neighboring -NCO groups on surface may react with the -NH$_2$ groups in one TAEA molecule, which can reduce the density of -NH$_2$ groups on film surface [20]. With the purpose of decrease side reaction, the condensation reaction is allowed at low temperature of 18 °C for 12 h without catalyst. The density of -NH$_2$ groups on the film surface, which was determined by an Acid Orange II assay according to reported method [32], was 9.98×10^{-7} mol/cm^2. The result was almost twice as much as that of -NCO groups on the SPEU-NCO film surface, indicating a minimum side reaction. In the FT-IR spectrum of SPEU-NH$_2$ (Figure 2c), the significant change is the disappearance of the peak at 2265 cm^{-1}, due to -NCO groups completely reacted with -NH$_2$ groups of TAEA. Comparing with the FT-IR spectrum of SPEU-NCO (Figure 2b), the absorption intensity of the peak at 3324 cm^{-1}, 1678 cm^{-1}, and 1529 cm^{-1} increased obviously, which was attributed to the N-H stretching vibration, amide I and amide II of the newly formed ureido. In addition, the characteristic peak at 1611 cm^{-1}, a symmetric bending vibration peak of -NH$_2$, confirmed the introduction of -NH$_2$ on the surface [33].

3.1.3. Grafting of PEG onto SPEU-NH$_2$ Film (SPEU-PEG)

The terminal double bond of APEG reacted with primary amino group on the surface of PEU-NH$_2$ film by Michael addition (Figure 1c) to realize grafting PEG onto the PEU film. As we know, Michael addition not only possesses excellent features including mild reaction conditions, high functional group tolerance and high conversions, but also can avoid the use of cytotoxic free-radicals [34]. The SPEU-PEG films were characterized by FT-IR and ^{1}H NMR (SPEU-PEGs had the similar spectra), the spectra of SPEU-PEG-a are shown in Figures 2d and 3c, respectively. From the FT-IR spectrum, it could be found that the -NH$_2$ absorption peak at 1611 cm^{-1} disappeared thoroughly, and the peak intensity at 3329 cm^{-1} increased slightly which was due to the reaction between -NH$_2$ group and APEG. In addition, the characteristic absorption bands at 1166 cm^{-1} and 1095 cm^{-1} belonged to the stretching vibration of ester bonds C-O-C of SPEU and ether bonds C-O-C of grafted PEG. In the ^{1}H NMR spectrum, the proton signals of double bond at δ 5.25 and δ 5.92 ppm (Figure 3a) disappeared completely, alongside the proton signals at δ 3.65 ppm belonging to the protons of repeat units of grafted PEG, demonstrating the complete Michael addition. This was another key feature indicating that the PEG was grafted on the film surface.

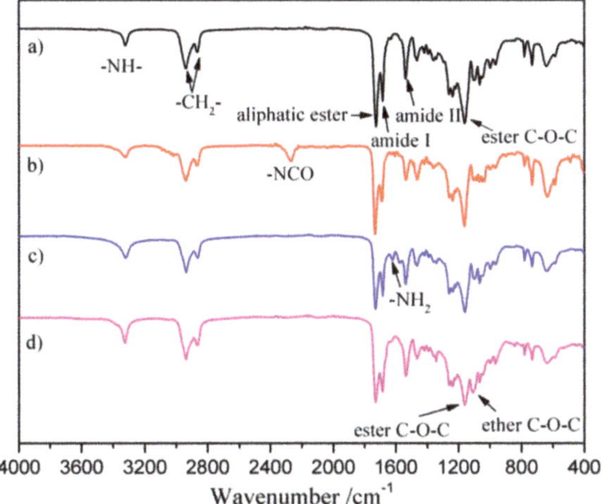

Figure 2. FT-IR spectra of (**a**) SPEU; (**b**) SPEU-NCO; (**c**) SPEU-NH$_2$; and (**d**) SPEU-PEG-a films.

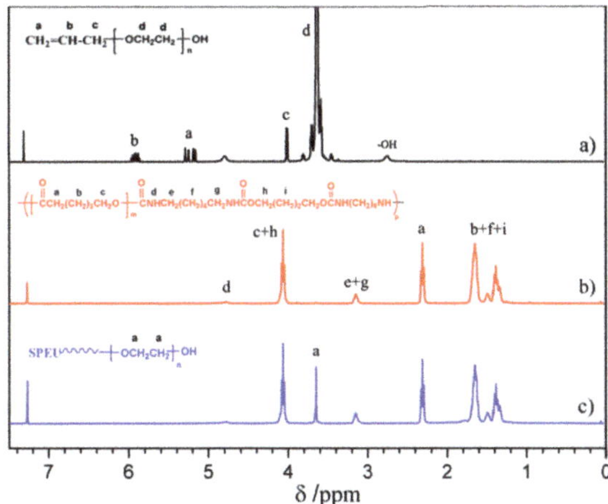

Figure 3. ^{1}H NMR spectra of (**a**) APEG; (**b**) SPEU; and (**c**) SPEU-PEG-a.

3.2. XPS Analysis

The presented XPS investigations were performed in order to extract quantitative results about the surface composition. The overview XPS spectra of the SPEU, SPEU-NH$_2$ and SPEU-PEG-a films (SPEU-PEGs had the similar XPS spectra) are illustrated in Figure 4, and the relevant compositions of the surface elements (O1s, C1s and N1s) calculated from XPS results are summarized in Table 1.

The blank SPEU film exhibited N1s signal at 398.8 eV with the content of 3.1%, while the N1s signal of SPEU-NH$_2$ film increased obviously with the content increasing to 10.3%, which indicated that -NH$_2$ groups had been immobilized onto the surface. Compared to SPEU-NH$_2$, the nitrogen content of SPEU-PEG decreased significantly, and with the increasing molecular weight of surface-grafted PEG, the nitrogen content decreased gradually. After the PEG is grafted on the surface with high grafting density, the PEG chains can form a thin layer coating on the surface which hinders the inner nitrogen element to be detected, leading to the low surface nitrogen content. The results confirmed that PEG was successfully immobilized onto SPEU surface.

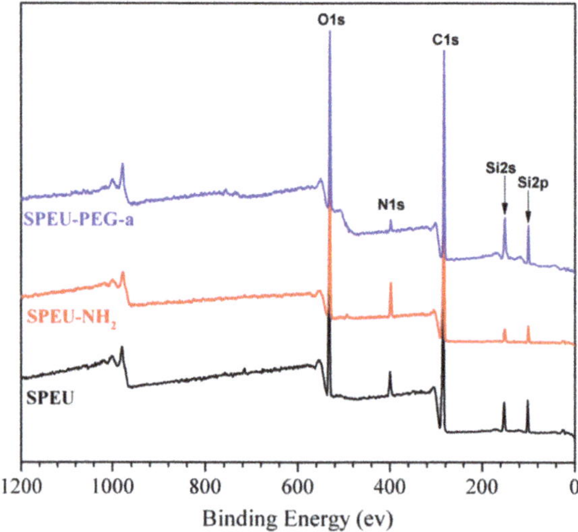

Figure 4. Overview XPS spectra of the SPEU, SPEU-NH$_2$, and SPEU-PEG-a films.

Table 1. Surface elemental composition of the SPEU, SPEU-NH$_2$, and SPEU-PEG films by XPS.

Films	Atomic Content/%		
	C1s	O1s	N1s
SPEU	78.2	18.7	3.1
SPEU-NH$_2$	74.8	14.9	10.3
SPEU-PEG-a	72.1	25.7	2.2
SPEU-PEG-b	71.3	27.1	1.6
SPEU-PEG-c	70.2	28.4	1.4

3.3. Mechanical Properties

Mechanical property was one of the most important properties for long-term implant biomaterials. The stress-strain behaviors of the blank and modified films are displayed in Figure 5 and the corresponding characteristic values derived from these curves are shown in Table 2. Two different regions being visible clearly from the curves manifested that all the films behaved as a soft elastic material, showing a smooth transition in stress-strain behaviors from elastic to plastic deformation regions [35]. The blank SPEU film exhibited outstanding mechanical properties with a strain at break of 896% and an ultimate stress of 38.1 MPa. The excellent mechanical properties should be attributed to the compact physical-linking network structure formed by multiple H-bonds existing among urethane groups and between urethane and ester groups, as per the description in our previous report [36]. The modified films, including SPEU-NH$_2$, SPEU-PEG-a, SPEU-PEG-b, and SPEU-PEG-c films displayed analogous stress-strain behaviors with the strain at break of 866–884% and ultimate stress of 35.5–36.4 MPa (Table 2), which were only slightly lower than that of parent

SPEU film. The results indicated that the chemical treatments had minimum deterioration on the intrinsic properties of the substrate and the molecular weight of surface-grafted PEG had a marginal influence on the mechanical properties except for initial modulus. It seemed that the initial modulus decreased with the increment of molecular weight of PEG, which needs further studies. Obviously, the mechanical properties of surface-modified films could also meet the clinical requirements of long-term implant biomaterials.

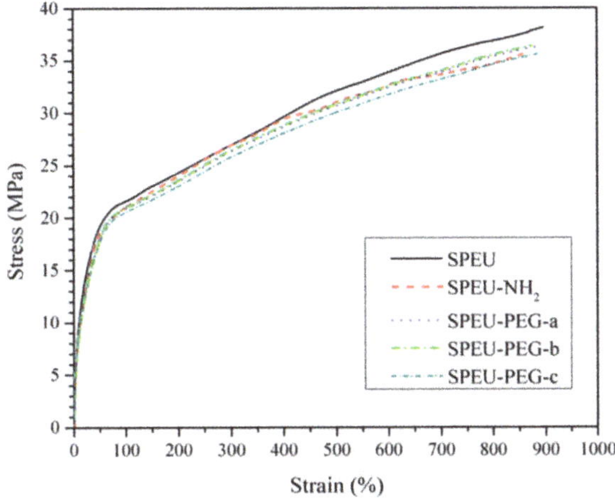

Figure 5. Stress-strain behaviors of blank and modified SPEU films.

Table 2. Mechanical properties of CPU films.

Films	Strain at Break (%)	Ultimate Stress (MPa)	Yield Strain (%)	Yield Stress (MPa)	Initial Modulus (MPa)
SPEU	896 ± 25	38.1 ± 2.1	46.7 ± 2.6	20.2 ± 1.2	43.3
SPEU-NH$_2$	866 ± 16	35.7 ± 1.8	45.8 ± 2.1	19.3 ± 1.1	42.1
SPEU-PEG-a	873 ± 18	36.1 ± 1.4	52.7 ± 2.2	19.8 ± 0.9	37.6
SPEU-PEG-b	879 ± 21	36.4 ± 1.9	56.8 ± 2.0	20.1 ± 1.1	35.4
SPEU-PEG-c	884 ± 20	35.5 ± 1.7	58.7 ± 1.9	19.9 ± 1.3	34.0

3.4. Surface Hydrophilicity and Swellability

The surface hydrophilicity of the blank and modified SPEU films were characterized by sessile contact angle measurement, and the results are presented in Figure 6. The blank SPEU film exhibited a characteristic hydrophobic surface with a high-water contact angle of 80.5°, whereas the SPEU-NH$_2$ film had low water contact angle of 47.1°, which indicated that the surface hydrophilicity had been improved after introducing -NH$_2$ groups on surface. After the PEG was grafted onto the surface, the water contact angle decreased dramatically. The water contact angle of SPEU-PEG-a, SPEU-PEG-b, and SPEU-PEG-c was 23.9°, 22.2°, and 21.8°, respectively. It needs to be emphasized that the measure should be carried out immediately (within one s) after the ultrapure water is dripped onto the film surface, otherwise the water drop will spread out on the surface. The PEG grafted on the surface can interact with water molecules by the role of hydrogen bonding to form hydration layer on the film surface, resulting in low water contact angle. In addition, SPEU-PEGs exhibiting similar water contact angle indicated that the molecular weight of grafted PEG slightly affected the surface hydrophilicity, which should be due to the high grafting density of PEG on surface. These results supported that PEG had been grafted onto the SPEU surface and the hydrophilicity of PEG-grafted SPEU was improved significantly.

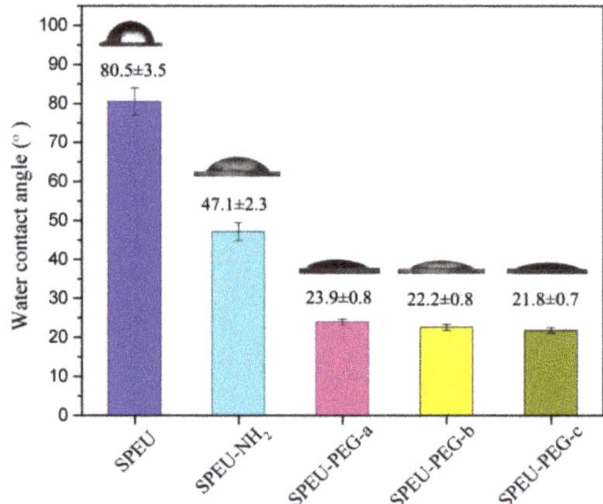

Figure 6. Water contact angle and images of the blank and modified SPEU films.

3.5. Protein Adsorption

Protein adsorption on the surface of a material is always considered as the first step to evaluate the blood compatibility of the implanted or blood-contact biomaterials [37]. Figure 7 shows the adsorption behaviors of BSA on the surface of blank and SPEU-PEG films. The amount of adsorbed protein on the PEG-grafted surface (SPEU-PEG-a: 3.02 $\mu g/cm^2$; SPEU-PEG-b: 2.88 $\mu g/cm^2$; SPEU-PEG-c: 2.78 $\mu g/cm^2$) was significantly lower than that of blank surface (SPEU: 11.87 $\mu g/cm^2$), which was ascribed to the surface-grafted PEG. PEG chain not only has excellent hydrophilicity, but also possesses high flexibility. The highly flexible and well-hydrated chains can affect the fluidity of blood and hinder the protein adsorption on the surface [14]. The SPEU-PEGs exhibited similar protein adsorption quantity, indicating that the molecular weight of PEG grafted on the surface had marginal effect on the protein adsorption capacity. The result is inconsistent with that of PEG-grafted surface of commercial polyurethane (TT-1095A, Thermedics, Wilmington, MA, USA) in previous report [13], which should be due to the high grafting density of PEG on the surface. The lower protein adsorption capacity of SPEU-PEG surface means better surface hemocompatibility.

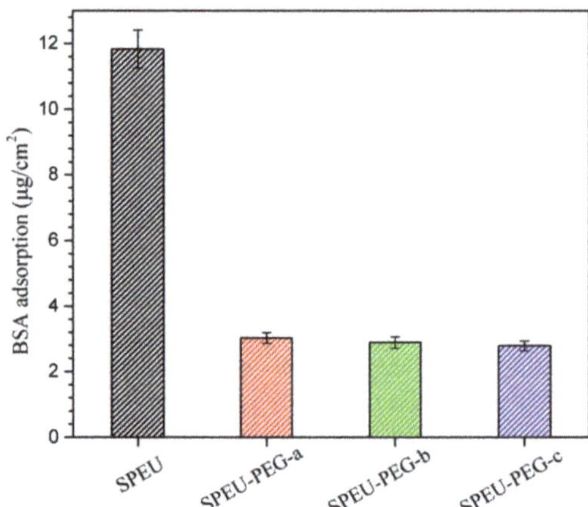

Figure 7. Adsorption behaviors of BSA on the surface of blank and PEG-grafted SPEU films at 37 ± 0.5 °C.

3.6. Platelet Adhesion

When blood is exposed to a foreign surface, initial protein adsorption on the surface is followed by platelet adhesion and release, which brings about cellular thrombogenesis [38]. Thus, platelet adhesion is rather important for the hemocompatibility of blood-contacting implantable materials. The morphologies of the platelets adherent on the surfaces of the blank and PEG-grafted SPEU films were assessed by FE-SEM observation, and the typical micrographs and quantification of adhered platelets are given in Figure 8 and Table 3, respectively. Massive platelets adhering on the surface of blank SPEU film (Figure 8a). Some of the platelets aggregated to some extent, some presented shape variation and some were spread on the surface, which indicated a highly activated state. After PEG was grafted on the surface, as shown in the Figure 8b–d, the amount of adhered platelets was reduced dramatically, and no obvious aggregation and deformation of platelets appeared, which proved a better anti-platelet adhesion surface. The possible explanation for this excellent anti-platelet adhesion is that the hydrophilic surface decreases the blood and plasma proteins interfacial energy on the PEG-grafted surface, which suppresses platelet adhesion [39]. The SPEU-PEG films having similar quantity of adhered platelets (Table 3) indicated that the molecular weight of PEG grafted on the surface marginally affected on the anti-platelet adhesion capacity. Due to the grafting density of PEG on the surface being relatively high, the denser short chain of PEG can form a layer coating the entire surface. Analogous effects have been reported for PEG-grafted silica surfaces by Alstine [40].

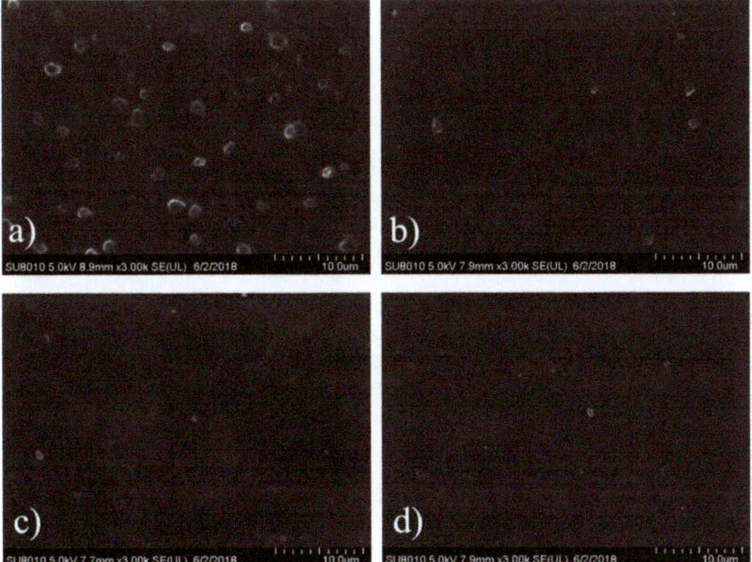

Figure 8. Representative SEM micrographs of platelet adhesion on the film surface of (**a**) SPEU; (**b**) SPEU-PEG-a; (**c**) SPEU-PEG-b; and (**d**) SPEU-PEG-c.

Table 3. Quantification of platelets adhering to the blank and MPC-grafted PEU surface.

Films	SPEU	SPEU-PEG-a	SPEU-PEG-b	SPEU-PEG-c
Quantity of adhered platelets (per mm^2)	20,702 ± 880	781 ± 56	731 ± 57	697 ± 52

4. Conclusions

In the paper, PEG was grafted on the surface of SPEU film with high grafting density to improve surface hemocompatibility. The PEG-grafted SPEU (SPEU-PEG) was prepared by three-step chemical treatments (allophanate reaction, condensation reaction, and Michael addition reaction) under mild reaction conditions. The surfaces were characterized by FT-IR, ^{1}H NMR, XPS, and water contact angle. The SPEU-PEGs displaying a much lower water contact angle than SPEU indicated that the hydrophilic PEG chains improved the surface hydrophilicity significantly. The mechanical properties of the SPEU-PEG films were slightly lower than that of parent film, verifying that the chemical treatments had minimum deterioration on the mechanical properties of the substrate. The low BSA adsorption quantity (2.78–3.02 $\mu g/cm^2$) and good anti-platelet adhesion capacity (781–697 per mm^2) revealed that SPEU-PEGs had improved surface hemocompatibility. Moreover, due to the high grafting density, the molecular weight of grafted PEG had marginal effect on the surface hydrophilicity and hemocompatibility. The PEG-grafted SPEU films possessed outstanding mechanical properties (strain at break: 866–884%; ultimate stress: 35.5–36.4 MPa) and good surface hemocompatibility, implying its high potential to be applied as long-term implants and blood-contacting biomaterials.

Author Contributions: Z.H. and L.L. conceived and designed the experiments; L.L., Y.G., L.Y., J.Z., C.L. and Z.L. performed the experiments; Z.H., Y.G. and L.L. analyzed the data and wrote the paper.

Funding: This research was funded by Shandong Provincial Natural Science Foundation, China (Project No. ZR2018MEM024), National Undergraduate Training Programs for Innovation and Entrepreneurship, China (Project No. 201810445115) and Scientific Research Fund Project of Undergraduates, Shandong Normal University, China (Project No.2017BKSKY50, 2018BKSKYJJ53).

Conflicts of Interest: The authors declare no conflict of interest.

References

1. Ghanbari, H.; Viatge, H.; Kidane, A.G.; Burriesci, G.; Tavakoli, M.; Seifalian, A.M. Polymeric heart valves: New materials, emerging hopes. *Trends Biotechnol.* **2009**, *27*, 359–367. [CrossRef] [PubMed]
2. He, W.; Hu, Z.; Xu, A.; Liu, R.; Yin, H.; Wang, J.; Wang, S. The preparation and performance of a new polyurethane vascular prosthesis. *Cell Biochem. Biophys.* **2013**, *66*, 855–866. [CrossRef] [PubMed]
3. Jozwiak, A.B.; Kielty, C.M.; Black, R.A. Surface functionalization of polyurethane for the immobilization of bioactive moieties on tissue scaffolds. *J. Mater. Chem.* **2018**, *18*, 2240–2248. [CrossRef]
4. Qu, W.Q.; Xia, Y.R.; Jiang, L.J.; Zhang, L.W.; Hou, Z.S. Synthesis and characterization of a new biodegradable polyurethanes with good mechanical properties. *Chin. Chem. Lett.* **2016**, *27*, 135–138. [CrossRef]
5. Tenorio-Alfonso, A.; Sánchez, M.C.; Franco, J.M. Preparation, characterization and mechanical properties of bio-based polyurethane adhesives from isocyanate-functionalized cellulose acetate and castor oil for bonding wood. *Polymers* **2017**, *9*, 132. [CrossRef]
6. Hsu, S.; Chen, C.W.; Hung, K.C.; Tsai, Y.C. Li, S. Thermo-responsive polyurethane hydrogels based on poly(ε-caprolactone) diol and amphiphilic polylactide-poly(ethylene glycol) block copolymers. *Polymers* **2016**, *8*, 252. [CrossRef]
7. Puskas, J.E.; Chen, Y. Biomedical application of commercial polymers and novel polyisobutylene-based thermoplastic elastomers for soft tissue replacement. *Biomacromolecule* **2004**, *5*, 1141–1154. [CrossRef] [PubMed]
8. Bochynska, A.I.; Hammink, G.; Grijpma, D.W.; Buma, P. Tissue adhesives for meniscus tear repair: An overview of current advances and prospects for future clinical solutions. *J. Mater. Sci. Mater. Med.* **2016**, *27*, 85–102. [CrossRef] [PubMed]
9. Zhao, Q.; Topham, N.; Anderson, J.M.; Hiltner, A.; Lodoen, G.; Payer, C.R. Foreign-body giant cells and polyurethane biostability: In vivo correlation of cell adhesion and surface cracking. *J. Biomed. Mater. Res.* **1991**, *25*, 177–183. [CrossRef] [PubMed]
10. Zhao, Q.H.; McNally, A.K.; Rubin, K.R.; Renier, M.; Wu, Y.; Rose-Caprara, V.; Anderson, J.M.; Hiltner, A.; Urbanski, P.; Stokes, K. Human plasma a 2-macroglobulin promotes in vitro oxidation stress cracking of Pellethane 2363-80A: In vivo and in vitro correlations. *J. Biomed. Mater. Res.* **1993**, *27*, 379–389. [CrossRef] [PubMed]

11. Yang, J.; Lv, J.; Gao, B.; Zhang, L.; Yang, D.; Shi, C.; Guo, J. Modification of polycarbonateurethane surface with poly (ethylene glycol) monoacrylate and phosphorylcholine glyceraldehyde for anti-platelet adhesion. *Front. Chem. Sci. Eng.* **2014**, *8*, 188–196. [CrossRef]

12. Kara, F.; Aksoy, E.A.; Yuksekdag, Z.; Hasirci, N.; Aksoy, S. Synthesis and surface modification of polyurethanes with chitosan for antibacterial properties. *Carbohydr. Polym.* **2014**, *112*, 39–47. [CrossRef] [PubMed]

13. Li, D.; Chen, H.; Glenn, M.W.; Brash, J.L. Lysine-PEG-modified polyurethane as a fibrinolytic surface: Effect of PEG chain length on protein interactions, platelet interactions and clot lysis. *Acta Biomater.* **2009**, *5*, 1864–1871. [CrossRef] [PubMed]

14. Lee, J.H.; Lee, H.B.; Andrade, J.D. Blood compatibility of polyethylene oxide surfaces. *Prog. Polym. Sci.* **1995**, *20*, 1043–1079. [CrossRef]

15. Han, D.K.; Park, K.D.; Ryu, G.H.; Kim, U.K.; Min, B.G.; Kim, Y.H. Plasma protein adsorption to sulfonated poly(ethylene oxide)-grafted polyurethane surface. *J. Biomed. Mater. Res.* **1996**, *30*, 23–30. [CrossRef]

16. Vert, M.; Domurado, D. Poly(ethylene glycol): Protein repulsiveor albumin-compatible. *J. Biomater. Sci. Polym. Ed.* **2000**, *11*, 1307–1317. [CrossRef] [PubMed]

17. Najafabadi, S.A.A.; Keshvari, H.; Ganji, Y.; Tahriri, M.; Ashuri, M. Chitosan/heparin surface modified polyacrylic acid grafted polyurethane film by two step plasma treatment. *Surf. Eng.* **2012**, *28*, 710–714. [CrossRef]

18. Irving, M.; Murphy, M.F.; Lilley, F.; French, P.W.; Burton, D.R.; Dixon, S.; Sharp, M.C. The use of abrasive polishing and laser processing for developing polyurethane surfaces for controlling fibroblast cell behavior. *Mater. Sci. Eng. C* **2017**, *71*, 690–697. [CrossRef] [PubMed]

19. Li, X.; Deng, H.; Li, Z.; Xiu, H.; Qi, X.; Zhang, Q.; Wang, K.; Chen, F.; Fu, Q. Graphene/thermoplastic polyurethane nanocomposites: Surface modification of graphene through oxidation, polyvinyl pyrrolidone coating and reduction. *Compos. Part A Appl. Sci. Manuf.* **2015**, *68*, 264–275. [CrossRef]

20. Gao, B.; Feng, Y.; Lu, J.; Zhang, L.; Zhao, M.; Shi, C.; Khan, M.; Guo, J. Grafting of phosphorylcholine functional groups on polycarbonate urethane surface for resisting platelet adhesion. *Mater. Sci. Eng. C* **2013**, *33*, 2871–2878. [CrossRef] [PubMed]

21. Liu, X.; Miller, A.L., II; Fundora, K.A.; Yaszemski, M.J.; Lu, L. Poly(ε-caprolactone) dendrimer cross-Linked via metal-free click chemistry: Injectable hydrophobic platform for tissue engineering. *ACS Macro Lett.* **2016**, *5*, 1261–1265. [CrossRef]

22. Zhang, C.; Luo, N.; Hirt, D.E. Surface grafting polyethylene glycol (PEG) onto poly(ethylene-*co*-acrylic acid) films. *Langmuir* **2006**, *22*, 6851–6857. [CrossRef] [PubMed]

23. Chen, H.; Hu, X.; Zhang, Y.; Li, D.; Wu, Z.; Zhang, T. Effect of chain density and conformation on protein adsorption at PEG-grafted polyurethane surfaces. *Colloids Surf. B* **2008**, *61*, 237–243. [CrossRef] [PubMed]

24. Alves, P.; Coelho, J.F.J.; Haack, J.; Rota, A.; Bruinink, A.; Gil, M.H. Surface modification and characterization of thermoplastic polyurethane. *Eur. Polym. J.* **2009**, *45*, 1412–1419. [CrossRef]

25. Liu, X.; Xia, Y.; Liu, L.; Zhang, D.; Hou, Z. Synthesis of a novel biomedical poly(ester urethane) based on aliphatic uniform-size diisocyanate and the blood compatibility of PEG-grafted surfaces. *J. Biomater. Appl.* **2018**, *32*, 1329–1342. [CrossRef] [PubMed]

26. Sheikh, Z.; Khan, A.S.; Roohpour, N.; Glogauer, M.; Rehman, I.U. Protein adsorption capability on polyurethane and modified-polyurethane membrane for periodontal guided tissue regeneration applications. *Mater. Sci. Eng. C* **2016**, *68*, 267–275. [CrossRef] [PubMed]

27. Roohpour, N.; Wasikiewicz, J.M.; Moshaverinia, A.; Paul, D.; Grahn, M.F.; Rehman, I.U.; Vadgama, P. Polyurethane membranes modified with isopropyl myristate as a potential candidate for encapsulating electronic implants: A study of biocompatibility and water permeability. *Polymers* **2010**, *2*, 102–119. [CrossRef]

28. Xu, W.; Xiao, M.; Yuan, L.; Zhang, J.; Hou, Z. Preparation, physicochemical properties and hemocompatibility of biodegradable chitooligosaccharide-based polyurethane. *Polymers* **2018**, *10*, 580. [CrossRef]

29. Zhang, N.; Yin, S.; Hou, Z.; Xu, W.; Zhang, J.; Xiao, M.; Zhang, Q. Preparation, physicochemical properties and biocompatibility of biodegradable poly(ether-ester-urethane) and chitosan oligosaccharide composites. *J. Polym. Res.* **2018**, *25*, 212. [CrossRef]

30. Hou, Z.; Qu, W.; Kan, C. Synthesis and properties of triethoxysilane-terminated anionic polyurethane and its waterborne dispersions. *J. Polym. Res.* **2015**, *22*, 111. [CrossRef]

31. Park, K.D.; Kim, Y.S.; Han, D.K.; Kim, Y.H.; Lee, E.H.B.; Suh, H.; Choi, K.S. Bacterial adhesion on PEG modified polyurethane surfaces. *Biomaterials* **1998**, *19*, 851–859. [CrossRef]

32. Lee, K.W.; Kowalczyk, S.P.; Shaw, J.M. Surface modification of BPDA-PDA polyimide. *Langmuir* **1991**, *7*, 2450–2453. [CrossRef]

33. Gao, W.; Feng, Y.; Lu, J.; Khan, M.; Guo, J. Biomimetic surface modification of polycarbonateurethane film via phosphorylcholine-graft for resisting platelet adhesion. *Macromol. Res.* **2012**, *20*, 1063–1069. [CrossRef]

34. Zhou, Y.; Nie, W.; Zhao, J.; Yuan, X. Rapidly in situ forming adhesive hydrogel based on a PEG-maleimide modified polypeptide through Michael addition. *J. Mater. Sci. Mater. Med.* **2013**, *24*, 2277–2286. [CrossRef] [PubMed]

35. Caracciolo, P.C.; Queiroz, A.A.D.; Higa, Q.Z.; Buffa, F.; Abraham, G.A. Segmented poly(esterurethane urea)s from novel urea-diol chain extenders: Synthesis, characterization and in vitro biological properties. *Acta Biomater.* **2008**, *4*, 976–988. [CrossRef] [PubMed]

36. Yin, S.; Xia, Y.; Jia, Q.; Hou, Z.; Zhang, N. Preparation and properties of biomedical segmented polyurethanes based on poly(ether ester) and uniform-size diurethane diisocyanates. *J. Biomater. Sci. Polym. Ed.* **2017**, *28*, 119–138. [CrossRef] [PubMed]

37. Kwon, M.J.; Bae, J.H.; Kim, J.J.; Na, K.; Lee, E.S. Long acting porous microparticle for pulmonary protein delivery. *Int. J. Pharm.* **2007**, *333*, 5–9. [CrossRef] [PubMed]

38. Patel, R.; Jacobs, H.A.; Kim, S.W. Surface adsorption and fibrinogen interactions with hirudin-thrombin complex. *J. Biomed. Mater. Res.* **1996**, *32*, 11–18. [CrossRef]

39. Ren, Z.; Chen, G.; Wei, Z.; Sang, L.; Qi, M. Hemocompatibility evaluation of polyurethane film with surface-grafted poly(ethylene glycol) and carboxymethyl-chitosan. *J. Appl. Polym. Sci.* **2013**, *127*, 308–315. [CrossRef]

40. Malmsten, M.; Emoto, K.; Alstine, J.M.V. Effect of chain density on inhibition of protein adsorption by poly(ethylene glycol) based coatings. *J. Colloid Interface Sci.* **1998**, *202*, 507–517. [CrossRef]

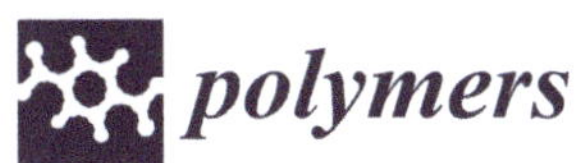 

Article

Branched Polyurethanes Based on Synthetic Polyhydroxybutyrate with Tunable Structure and Properties

Joanna Brzeska [1,*], Anna Maria Elert [2], Magda Morawska [1], Wanda Sikorska [3], Marek Kowalczuk [3,4,*] and Maria Rutkowska [1]

[1]   Department of Commodity Industrial Science and Chemistry, Gdynia Maritime University, 83Morska Street, 81-225 Gdynia, Poland; m.morawska@wpit.am.gdynia.pl (M.M.); m.rutkowska@wpit.am.gdynia.pl (M.R.)
[2]   Nanotribology and Nanostructuring of Surfaces, Federal Institute for Materials Research and Testing (BAM), Unter den Eichen 87, 12205 Berlin, Germany; anna-maria.elert@bam.de
[3]   Centre of Polymer and Carbon Materials, Polish Academy of Sciences, 34 M. Curie-Sklodowska Street, 41-819 Zabrze, Poland; wsikorska@cmpw-pan.edu.pl
[4]   School of Biology, Chemistry and Forensic Science, Faculty of Science and Engineering, University of Wolverhampton, Wolverhampton WV1 1SB, UK
*   Correspondence: j.brzeska@wpit.am.gdynia.pl (J.B.); marek.kowalczuk@cmpw-pan.edu.pl (M.K.)

Received: 20 June 2018; Accepted: 22 July 2018; Published: 26 July 2018

Abstract: Branched, aliphatic polyurethanes (PURs) were synthesized and compared to linear analogues. The influence of polycaprolactonetriol and synthetic poly([R,S]-3-hydroxybutyrate) (R,S-PHB) in soft segments on structure, thermal and sorptive properties of PURs was determined. Using FTIR and Raman spectroscopies it was found that increasing the R,S-PHB amount in the structure of branched PURs reduced a tendency of urethane groups to hydrogen bonding. Melting enthalpies (on DSC thermograms) of both soft and hard segments of linear PURs were higher than branched PURs, suggesting that linear PURs were more crystalline. Oil sorption by samples of linear and branched PURs, containing only polycaprolactone chains in soft segments, was higher than in the case of samples with R,S-PHB in their structure. Branched PUR without R,S-PHB absorbed the highest amount of oil. Introducing R,S-PHB into the PUR structure increased water sorption. Thus, by operating the number of branching and the amount of poly([R,S]-3-hydroxybutyrate) in soft segments thermal and sorptive properties of aliphatic PURs could be controlled.

Keywords: polyurethane structure; linear and branched polyurethanes; synthetic polyhydroxybutyrate; thermal properties; sorptive properties

---

1. Introduction

Typical linear polyurethane (PUR) is built with (i) a soft segment (based on oligomeric polyester- or polyetherdiol), which gives flexibility and softness of polymer and (ii) a hard segment (synthetized with diisocyanate and low molecular chain extender) the aim of which is to increase stiffness, hardness changes, etc. Immiscibility of soft and hard segments facilitates their reorganization into domains. Further ordering of chains in domains can lead to formation of crystallites. The polar nature of the urethane group influences their tendency to hydrogen bonding. Creation of hydrogen bonds between N–H group and C=O in urethanes facilitates their ordering and phase separation. Whereas hydrogen bonds of the urethane group with oxygen of the soft segment can reduce chains mobility and their ordering [1]. Generally, hydrogen bonds create physical reinforcement of the PUR network which causes increased strength and stiffness [2].

Thermodynamic incompatibility of hard and soft segments and connected with this, their insolubility, cause microphase separation in segmented PURs [3]. Ordering of chains in both separated domains of segments can lead to formation of crystallites.

Branched polymers are a kind of macromolecules, whose architecture is neither linear nor cross-linked. In addition to polymers with long or short side-chains, this group of polymers also includes densely branched structures with a large number of functional end groups (star like and hyperbranched polymers, and dendrimers).

The architecture of branched polymers causes an intermediate structure between cross-linked and linear polymers. The presence of side chains in their structure significantly influences the properties of polymers.

Branched PURs are generally characterized by good solubility in many organic solvents and compatibility with different materials, low viscosity (both in the molten state and in the solution) and they have free spaces inside the network. In consequence, they can be used as drug carriers, catalysts, chemical sensors, coatings, binders, elastomers, sorption mats, etc. [2,4].

In branched polymers linear side-chains diverge (uniformly or randomly) from branching points of the linear flexible chain (backbone). In this case, introducing trifunctional compound into the PUR structure reduces a possibility of chains to order and form soft and hard domains. However, local aggregation of soft or hard segments and their microphase separation can often still occur even in cross-linked PURs [1]. The presence of branching points generally introduces irregularities in the polymer structure and consequently leads to lowered crystallinity and creation of smaller crystals with the lower melting point in comparison to linear polymers [5]. Tendency of urethane groups to create hydrogen bonds is strongly affected by side-chains. Both short and long side-chains can influence a possibility of interaction between polymer chains.

Polyhydroxybutyrate (PHB) is a polymer that naturally exists in prokaryote and eukaryote cells. Biosynthesized PHB is high crystalline, stiff and brittle material, which makes it difficult to use. Therefore, it is valuable to us its synthetic analogue, which can be obtained as an amorphous, elastic polymer.

Synthetic PHB can be obtained via ring-opening polymerization (ROP) of β-butyrolactone to isotactic, atactic and syndiotactic poly(3-hydroxybutyrate) [6]. Atactic (R,S)-PHB is an amorphous polymer with low glass transition temperature and it maintains elastomer properties at room temperature. Because of the secondary hydroxyl group in the PHB_{diol} structure methyl side-groups are present in its chains (Figure 1A).

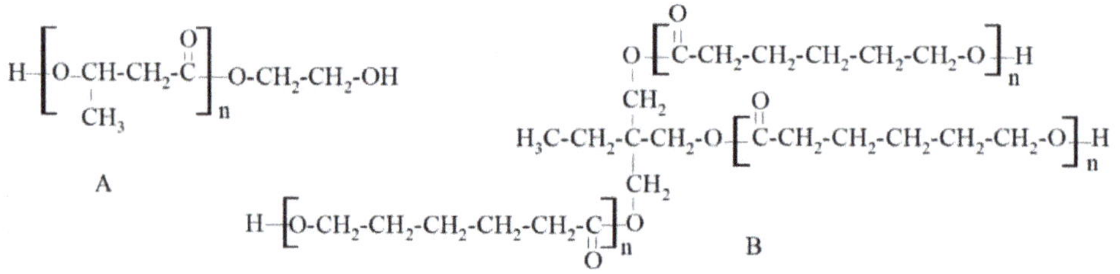

A

B

Figure 1. Schematic structure of polyhydroxybutyratediol (**A**) and polycaprolactonetriol (**B**).

Through the right combination of polycaprolactonetriol (PCL_{triol}) (Figure 1B) with polycaprolactonediol (PCL_{diol}), branched PURs with the highest possible molecular weight without reaching gelation and formed into foils could be obtained [5]. Knowing that introduction of amorphous R,S-PHB influences thermal, sorptive and surface [7] properties of linear and cross-linking PUR materials and their degradability profile [8], such effects are also expected in branched PURs.

In this paper branched PURs, as potential sorptive materials and their linear analogues were comparatively studied. The influence of PCL_{triol} and R,S-PHB presence in the soft segments structure on thermal and sorptive properties of obtained PURs was estimated.

2. Experimental Section

2.1. Materials

β-butyrolactone (Aldrich, Steinheim, Germany) was purified [9], 18-crown-6 complex (Fluka, Bucharest, Romania), 3-hydroxybutyric acid sodium salt and 2-bromoethanol (Aldrich) were used as received. Oligomerols of R,S-PHB (M_n 1700), PCL_{triol} (M_n 900, Aldrich, St. Louis, MO, USA) and PCL_{diol} (M_n 1870, Aldrich, St. Louis, MO, USA) were dried by heating at 60–65 °C under reduced pressure (1.4 hPa). 4,4'-methylene dicyclohexyl diisocyanate (H_{12}MDI) (Aldrich, St. Louis, MO, USA) was vacuum distilled; 1,4-butanediol (1,4-BD) (Aldrich, Steinheim, Germany) was distilled azeotropically with benzene; *N,N*-dimethylformamide (DMF) (Chempur, Piekary Śląskie, Poland) was dehydrated over diphosphorous pentoxide (P_2O_5) and distilled under low pressure. Catalyst tin(II) octanoate (OSn) (Alfa Aesar, Karlsruhe, Germany) was used as received.

2.2. Synthesis of Linear PURs

Oligomerole of telechelic (with OH groups on both sides of chains) poly([R,S]-3-hydroxybutyrate) (R,S-PHB) was obtained by anionic ring opening polymerization of ß-butyrolactone initiated by 3-hydroxybutyric acid sodium salt/18-crown-6 complex at room temperature and terminated with 2-bromoethanol [10]. Obtained R,S-PHB was generally atactic and almost completely amorphous polymer (with very small melting enthalpy about 4 J/g).

The synthesis of PURs was carried out in a two-step reaction with the molar ratio of NCO:OH = 2:1 in the prepolymer step [7]. Prepolymer of PURs was synthesized for 3 h at 70–75 °C under vacuum with oligomer (PCL_{diol} or blend of PCL_{diol} with R,S-PHB) and H_{12}MDI in the presence of OSn. NCO-terminated prepolymer was dissolved in DMF, and next its molecular weight was increased by a reaction with the chain extender (1,4-BD) for 2 h at 60 °C. The final molar ratio of NCO:OH in PURs was 1:1. PUR foil was formed by pouring the polymer solution on Teflon plates and heating it at 80–105 °C in the vacuum heater for 6 h.

2.3. Synthesis of Branched PURs

Branched PURs were synthesized in the same way as linear PURs. Soft segments were built with PCL_{triol}, but PCL_{diol} and R,S-PHB were additionally added to vary a distance between branch points. This branched PUR turned out to be easily soluble in organic solvents (dimethylformamide, toluene, acetone, ethanol and chloroform).

Amounts of reactants in the synthesis of PUR are shown in Table 1.

Table 1. The code of linear and branched polyurethanes (PURs) and amounts of reactants [g] in their synthesis (calculated for 100 g of final polymer).

	PUR	PCL_{triol}	PCL_{diol}	R,S-PHB	H_{12}MDI	1,4-BD
linear	l-PUR A	-	74.62	-	21.66	3.72
	l-PUR B	-	59.75	14.96	21.56	3.72
branched	b-PUR C	5.50	66.13	-	24.10	4.27
	b-PUR D	5.51	44.69	21.85	23.90	4.05
	b-PUR E	2.20	35.24	35.20	23.28	4.08

Substrates used for PURs determined how they were built. All obtained PURs were aliphatic polyesterurethanes, differing in architecture of the structure (Figure 2). Linear PURs differed in side-chain methyl groups from R,S-PHB. Whereas, using PCL$_{triol}$ to build soft segment, introduced branch nodes into the structure of branched PURs. So, they differed in the amount of short (methyl groups from R,S-PHB) and long (polycaprolactone chains from PCL$_{triol}$) side-chains. These lateral branches should affect interactions between chains (e.g., hydrogen bonds creation), which consequently should influence sorption and thermal properties.

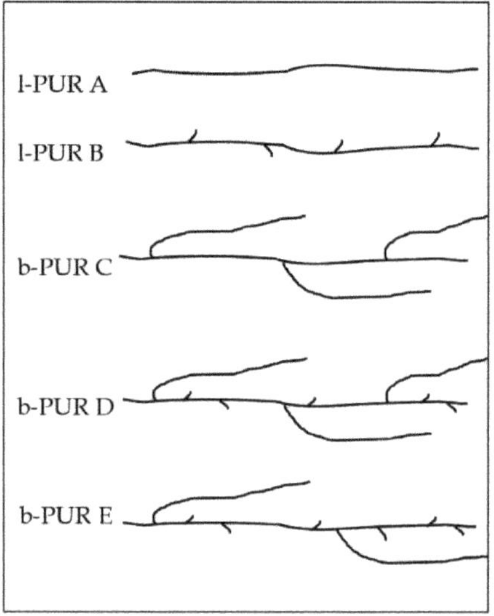

Figure 2. Illustration of the schematic structure of linear and branched PURs.

2.4. Methods

2.4.1. Differential Scanning Calorimetry (DSC)

Thermal properties of PURs and their blends were determined using the Setaram thermal analyzer (Setaram, Caluire, France). Indium and lead were used for calibration. Specimens (with mass about 20 mg) were sealed in aluminum pans and scanned from 20 to 200 °C with the heating rate of 10 °C/min. All experiments were conducted in a flow of dry N$_2$.

2.4.2. Raman Spectroscopy

All Raman spectra were recorded in backscattering geometry by a WiTec 300R Alpha device (WiTec, Ulm, Germany). The spectra are shown without background correction. A ruled 600 grooves/mm grating was chosen in the optical spectrometer (WiTEC, UHTS 300), which was equipped with a Peltier-cooled CCD camera (ANDOR, iDus DV401A) operating at a temperature of 210 K. The resulting wave number resolution was <2 cm^{-1}. Investigated solids were excited at a power level of 2.5 mW using the 532 nm emission line of a continuous wave laser (Spectra Physics, Excelsior). The Raman-laser radiation was focused on the sample surface with a microscope (Zeiss EC Epiplan 20', NA 0.4) probing a circular spot of ~4 μm in the diameter. At the given experimental conditions, the Raman-laser induced oxidation processes could be excluded. The spectra were recorded with the integration time of 10 s and 10 accumulations each. After measurement, spectra were background subtracted using WITec Project 2.08 software.

2.4.3. ATR IR Spectroscopy

FTIR investigations of solids were performed using attenuated total reflection (ATR, Smart Orbit Accessory) in a Nicolet 6700 FTIR spectrometer (Bruker Optics, Ettlingen, Germany) with a DTGS KBr detector. To obtain a spectrum, 32 scans were taken at optical resolution of 4 cm^{-1}. For ATR-FTIR investigations the materials were pressed on the diamond cell to achieve surface sensitive test results.

2.4.4. Density

Density of polymer samples was determined according to ISO 1183 standard and using analytical balance equipped with a density determination kit (Radwag, Radom, Poland).

2.4.5. Oil and Water Sorption

PURs and blends samples were immersed in sunflower oil at 37 °C for 24 h and next they were weighed after wiping off the oil with filter paper [11]. For water sorption estimation, investigated polymers were immersed in deionized water for 14 days at 37 °C. Next the swollen samples were gently blotted with filter paper and weighed. Sorption was calculated (using the gravimetric method) from the weight after incubation (w_i) and the initial weight (w_0) by:

Sorption% = $(w_i - w_0)/w_0 \times 100\%$. The results were the average of three measurements.

3. Results and Discussion

The chemical structure of synthetized PURs was confirmed with FTIR and Raman spectroscopy. Figure 3 shows FTIR spectra of linear and branched PURs.

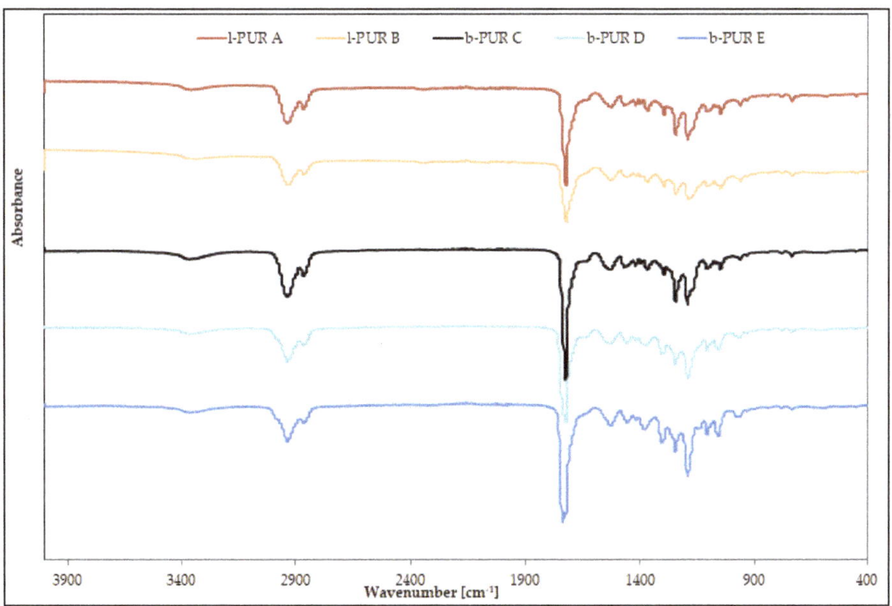

Figure 3. FTIR spectra of linear and branched PURs.

Infrared spectra of linear and branched PURs were very similar and showed characteristic bands typical for PURs. The absence of bands ascribing functional groups of starting substrates on FTIR spectra indicated completion of the polyaddition reaction.

The region between 3410 and 3200 cm^{-1} was ascribed to N–H stretching vibration, whereas the region between 2990 and 2840 cm^{-1} to C–H stretching vibration (of both soft and hard segments). Infrared absorptions between 1780 and 1620 cm^{-1} indicated the presence of the carbonyl group (amide I), whereas bands 1550–1510 cm^{-1} ascribed deformation vibrations of N–H groups (amide II) and 1242–1240 cm^{-1} C–O stretching of the carbonate group. Bands corresponding to asymmetric and symmetric C–H stretching vibrations of aliphatic –CH_2 groups were observed at 2905 and 2860 cm^{-1} respectively (Figure 3).

Spectra of carbonyl moiety in PURs can be highly complex due to numerous origins of carbonyl and hydrogen bond interactions within the polymer network. The observed on FTIR spectrum of PCL_{diol} (Figure 4) carbonyl peak at 1720 cm^{-1} was narrow and slender, which suggested homogeneous nature of this moiety. The presence of branch nodes in PCL_{triol} and the lateral methyl group in chains of R,S-PHB (only with CH_2 and CH groups between ester moiety in comparison to five CH_2 in PCL) changed molecular interaction, causing in these cases the carbonyl peak to be wider and less uniform than for PCL_{diol}. Moreover, the peak appeared at a higher wavenumber (1730 cm^{-1}).

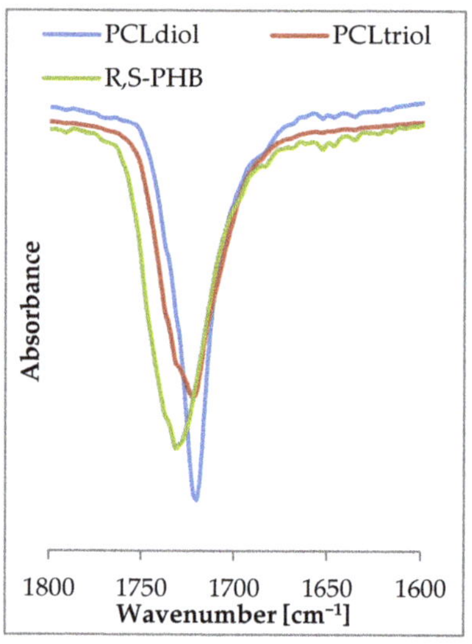

Figure 4. FTIR spectrum of C=O stretching region of oligomeroles used for the PURs synthesis.

On FTIR spectra observed carbonyl bands were a result of overlapping of the peaks of carbonyl moieties from oligomeroles and urethane groups. In the case of linear l-PUR A with soft segments built with only PCL_{diol} frequency (1720 cm^{-1}) and shape of the carbonyl stretching peak were identical as in pure PCL_{diol} (Figure 5 compared to Figure 4). Only a small shoulder at right sight of the peak was observed. It was the overlapping band of carbonyl groups from urethane (with wavenumber 1690 cm^{-1}). After introducing the R,S-PHB into soft segments the band in the stretching carbonyl region became wider and less sharp. The overlapped urethane C=O peak was also observed on l-PUR B FTIR spectra as a shoulder at lower frequencies (at 1690 cm^{-1}) in relation to that of the ester group. This band was almost no discernible at 1720 cm^{-1}.

Type of PUR   N–H Region   C=O Region

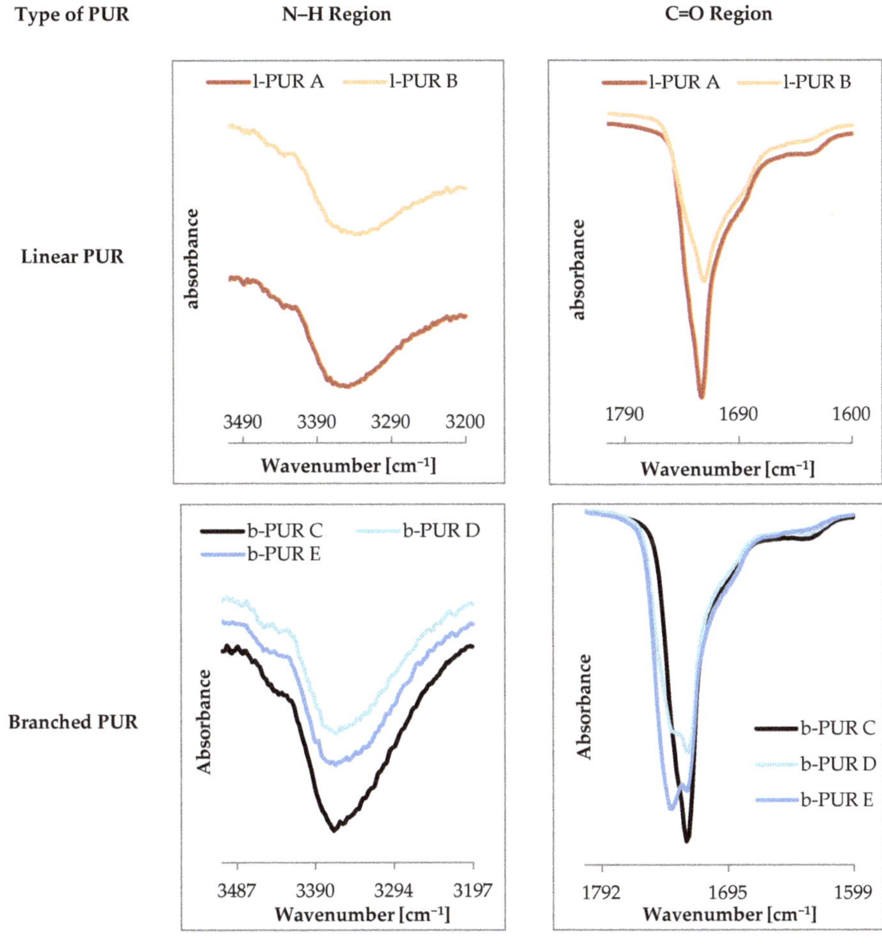

Figure 5. FTIR spectrum of N–H and C=O stretching regions of linear and branched PURs.

Using PCL_{triol} to build soft segments caused that shape of the carbonyl band to be more complicated (Figure 5). Besides two mentioned peaks at 1740 cm^{-1} another one appeared, which could be attributed to free C=O groups. On FTIR spectra of linear PURs no bands were detected above 3500 cm^{-1} (Figure 5), which suggested that all N–H groups were engaged in formation of hydrogen bonds. However, at 3430–3460 cm^{-1} on FTIR spectra of branched PURs (Figure 5), a little intensity shoulder was observed. It was often described as the overtone of the band at 1720 cm^{-1} [12]. Taking into consideration that small intensity bands indicating the presence of non-bonded C=O stretching of the urethane and carbonate groups, were visible at 1740 cm^{-1} (Figure 5), it was concluded that this peak (at 3430–3460 cm^{-1}) ascribed non-associated N-H groups. Moreover, increasing R,S-PHB in the structure of branched PURs inhibited a tendency of urethane groups to hydrogen bonding. In b-PUR E intensity of bands ascribed free C=O ester groups were larger than those bound by hydrogen bonding. Probably, the presence of the side-chain methyl group in the R,S-PHB chain hindered formation of hydrogen bonds. As it was said, the band attributed to stretching vibrations of C=O in the urethane group strongly overlapped with these of the ester group, so no quantitative information about relations between free and bonded ester and urethane groups could be obtained.

The peak observed at 1381 cm^{-1} on FTIR spectra of R,S-PHB (Figure S1) was ascribed as banding deformation vibration of CH$_3$ according to Eldessouki et al. [13]. Its presence was noted on FTIR spectra of branched PURs based on R,S-PHB at 1380 cm^{-1}. Intensity of the peak was the highest in the case of b-PUR E, because of the high amount of R,S-PHB (Figure 3). On spectra of l-PUR B (obtained with the smaller amount R,S-PHB in comparison to branched PURs) this peak overlapped with C–H bending vibration (Figure 3).

Raman spectroscopy, which is a complementary technique to infrared spectroscopy, also measures vibrational energy levels [14].

The very strong FTIR spectra band at about 1740–1720 cm^{-1}, ascribed the free C=O ester groups, was quite weak in Raman spectra of both types of PURs (Figure 6). In addition. because of polarity of urethane, the N–H presence on Raman spectra was invisible. The largest bands were asymmetric at 2930–2917 cm^{-1} and symmetric at 2890–2863 cm^{-1}, with CH$_2$ stretching vibration and CH$_2$ bending vibration at $\approx$1448 cm^{-1}.

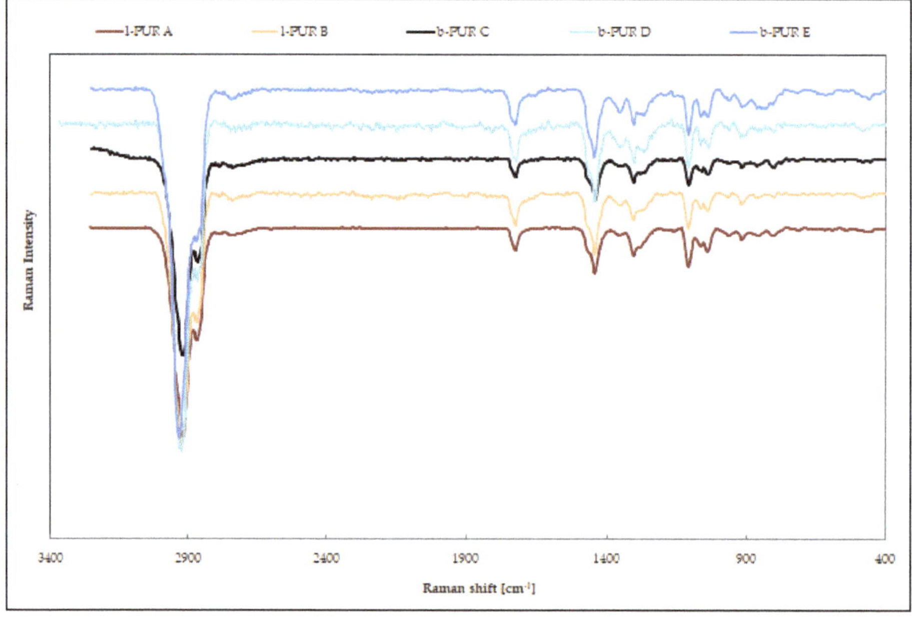

Figure 6. Raman spectra of linear and branched PURs.

DSC thermograms of investigated samples were typical for segmented PURs. Two endotherms, ascribed to the melting crystalline phase of soft (with melting temperature T_{m1}) and hard (with melting temperature T_{m2}) segments were visible on DSC diagrams of linear and branched PURs (Figure 7). The data read from thermograms are shown in Table 2.

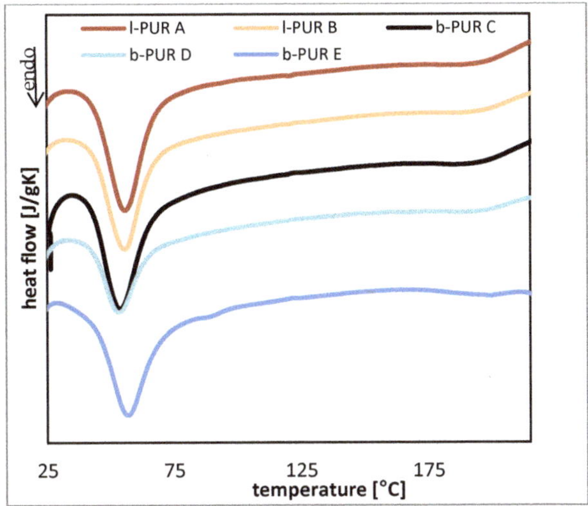

Figure 7. Differential scanning calorimetry (DSC) thermograms of linear and branched PURs.

Table 2. Melting temperature, enthalpy of soft [T_{m1}, ΔH_1] and hard [T_{m2}, ΔH_2] segments and density (with standard deviation SD) of PURs.

PUR	Density ± SD [g/cm^3]	T_{m1} [°C]	ΔH_1 [J/g]	T_{m2} [°C]	ΔH_2 [J/g]
l-PUR A	1.16 ± 0.06	55.6	53.2	195.7	17.1
l-PUR B	1.13 ± 0.02	55.3	54.4	194.1	10.4
b-PUR C	1.02 ± 0.08	53.6	46.2	194.3	11.4
b-PUR D	1.15 ± 0.02	53.2	30.4	196.2	5.5
b-PUR E	1.17 ± 0.03	56.6	46.7	198.8	1.7

Using semicrystalline PCL to build soft segments caused obtained PURs to be characterized by partial crystallinity of soft segment domains.

Total melting enthalpies (ΔH) of both soft and hard segments of linear PURs were higher than branched PURs, which suggested that linear PURs were more crystalline than branched ones, as it was expected.

Melting temperatures of soft and hard segments depended on segments separation. In case of linear PURs (l-PUR A and l-PUR B) and branched b-PUR E (with low amount of PCL$_{triol}$) T_{m1} was closer to melting temperature of PCL$_{diol}$ (T_m = 60.1 °C) than for b-PUR C and b-PUR D, suggesting the good separation of soft segment chains. However, the endothermic peak ascribing melting of crystalline phase of soft segments on DSC thermograms of b-PUR E was very wide and was integrated to baseline, including a small peak about 88 °C. It was concluded that this broad peak represents melting of all kinds of spherulites with different volume that formed in range of soft segments. This was a reason of high ΔH_1 b-PUR E.

Mobility of long aliphatic chains of PUR in its structure facilitated re-organization of PCL and, in consequence, formation of crystals. The presence of amorphous R,S-PHB did not restrict a tendency of PCL to crystallize. A different situation was observed for branched PURs. In the case of reduction of chain mobility via branch nodes, partial replacement of PCL$_{diol}$ chains by amorphous R,S-PHB reduced crystallinity of soft segments. In b-PUR E with a very small amount of trifunctional PCL$_{triol}$ the tendency was the same as in linear samples. This dependence of crystallinity with the number of branch was also observed by Mahapatra et al. in hyperbranched PURs [15].

What was interesting was the presence of R,S-PHB in the PURs structure reduced melting enthalpy of hard segments. It was supposed that the presence of the lateral methyl group in R,S-PHB hindered ordering of the PUR chain. As mentioned before the presence of non-bonded C=O in PURs containing R,S-PHB was confirmed by FTIR and contribution of C=O urethane groups in formation of hydrogen bonds could not be determined. However, reduced crystallinity of hard segments (lowered ΔH_2) suggested the absence of hydrogen bonds between urethane groups and connected with it chains ordering.

Architecture of PURs structure and their hydrophilicity also influenced tendency to oil sorption. The data presented in Figure 8 indicated that branched PUR without R,S-PHB absorbed more vegetable oil than linear l-PUR A. The presence of free spaces in the polymer network of branched PURs, which were a result of the presence of branching points, ought to facilitate migration of oil molecules. However, the ability of investigated PURs to sorb oil was also connected with their soft segment building. Soft segments, both linear and branched PURs, were mainly built with chains of hydrophobic polycaprolactone. Oil sorptions of l-PUR A and b-PUR C were higher than appropriate PURs containing R,S-PHB in their structure. Increasing PURs hydrophilicity after using R,S-PHB to build soft segments was observed earlier [7]. In consequence, branched b-PUR C with soft segments built only with polycaprolactone chains and characterized by the lowest density absorbed highest amount of lipid among all investigated PURs.

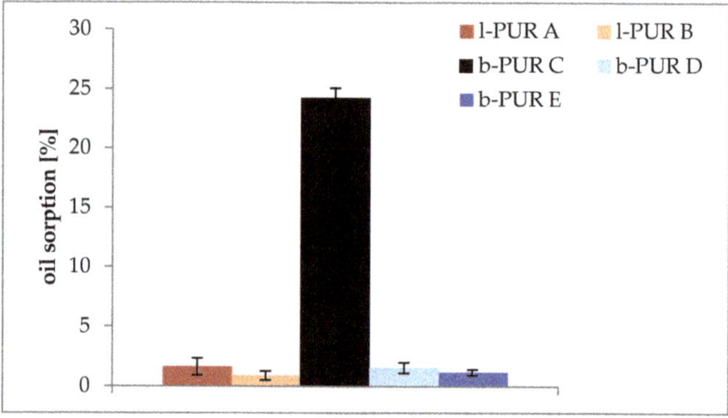

Figure 8. Oil sorption by linear and branched PURs.

The water sorption is a synergistic effect of crystallinity and the chemical structure of chains in PURs. Introducing R,S-PHB into the PURs (both linear and branched) structure, instead of hydrophobic polycaprolactone chains, increased their hydrophilicity (Figure 9), as it was observed earlier [16]. Crystallinity of samples and the presence of free spaces in polymer network were very important for water sorption. Comparing the structure of l-PUR A and b-PUR C, it could be said that their chemical building was very similar to each other. They differed only in architecture, i.e., the presence of long, side chains in the branched PUR structure resulting from the introduction of tri-functional PCL$_{triol}$ (Figure 1). In consequence, the free spaces in the b-PUR C structure were formed, which facilitated migration of water molecules into polymer bulk. However, as it was said before, more important was using amorphous, hydrophilic R,S-PHB as a part of soft segments. Its presence significantly increased the ability of investigated PURs to absorb water. Branched b-PUR E with the high amount of R,S-PHB and with very low crystallinity of hard segments absorbed the most water (Figure 9).

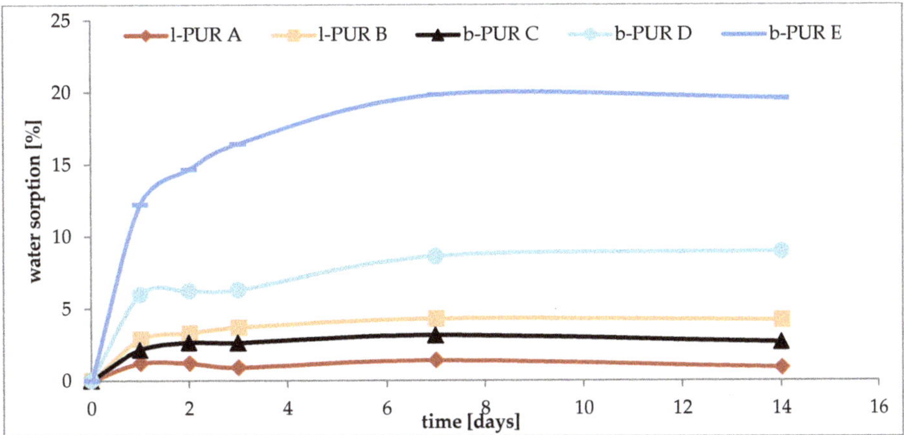

Figure 9. Water sorption by linear and branched PURs.

4. Conclusions

Aliphatic PURs with synthetic poly([R,S]-3-hydroxybutyrate) (R,S-PHB) in soft segments were synthesized with the prepolymer method. Branched and linear PURs differed in architecture of structure even where they were based on the same chemical structure of starting substrates.

The spectroscopies analysis (FTIR and Raman spectra), used for structure estimation, indicated that urethane linkage appeared and isocyanate peaks related 4,4'-methylene dicyclohexyl diisocyanate (H_{12}MDI) disappeared in all the samples after the synthesis. All N–H groups were engaged in formation of hydrogen bonds in linear PURs, whereas in branched PURs the presence of non-associated N–H groups was confirmed. Introducing R,S-PHB into the structure of branched PURs inhibited a tendency of urethane groups to hydrogen bonding. Introducing a small amount of R,S-PHB did not change crystallinity of linear PURs, but reduced melting enthalpy of branched PUR. Significant changes of sorptive properties (increased water sorption and decreased oil sorption) were observed for both linear and branched PURs after incorporation of R,S-PHB as a soft segments building block into the macromolecular structure. Polymer materials with the desired sorption characteristic can be obtained by manipulating architecture of the PUR structure and the amount of added synthetic R,S-PHB.

Supplementary Materials: The supplementary materials are available online at http://www.mdpi.com/2073-4360/10/8/826/s1.

Author Contributions: Conceptualization, J.B.; Investigation, A.M.E., M.M. and W.S.; Validation, M.K. and M.R.

Funding: This research was funded by the statutory activity No. DS/413 of the Department of Commodity Industrial Science and Chemistry AMG. Partial financial support from the European Regional Development Fund Project EnTRESS No 01R16P00718 and UM0-2016/22/Z/STS/00692 PELARGODONT Project financed under the M-ERA.NET 2 Programme of Horizon 2020 as well as Polish National Science Centre Project contract DEC-2013/11/B/ST5/02222 are gratefully acknowledged.

Conflicts of Interest: The authors declare no conflict of interest.

References

1. Yilgör, I.; Yilgör, E.; Wilkes, G.L. Critical parameters in designing segmented polyurethanes and their effect on morphology and properties: A comprehensive review. *Polymer* **2015**, *58*, A1–A36. [CrossRef]

2. Chattopadhyay, D.K.; Raju, K.V.S.N. Structural engineering of polyurethane coatings for high performance applications. *Prog. Polym. Sci.* **2007**, *32*, 352–418. [CrossRef]

3.   Klinedinst, D.B.; Yilgör, I.; Yilgör, E.; Zhang, M.; Wilkes, G.L. The effect of varying soft and hard segment length on the structure-property relationships of segmented polyurethanes based on a linear symmetric diisocyanate, 1,4-butanediol and PTMO soft segments. *Polymer* **2012**, *53*, 5358–5366. [CrossRef]

4.   Mao, H.; Qiang, S.; Yang, F.; Zhao, C.; Wang, C.; Yin, Y. Synthesis of blocked and branched waterborne polyurethanes for pigment printing applications. *J. Appl. Polym. Sci.* **2015**, 1–9. [CrossRef]

5.   Rivero, R.S.; Solaguren, P.B.; Zubieta, K.G.; Peponi, L.; Marcos-Fernández, A. Synthesis, kinetics of photo-dimerization/photo-cleavage and physical properties of coumarin-containing branched polyurethanes based on polycaprolactones. *Express Polym. Lett.* **2016**, *10*, 84–95. [CrossRef]

6.   Rydz, J.; Sikorska, W.; Kyulavska, M.; Christova, D. Polyester-Based (Bio)degradable Polymers as Environmentally Friendly Materials for Sustainable Development. *Int. J. Mol. Sci.* **2015**, *16*, 564–596. [CrossRef] [PubMed]

7.   Brzeska, J.; Morawska, M.; Heimowska, A.; Sikorska, W.; Wałach, W.; Hercog, A.; Kowalczuk, M.; Rutkowska, M. The influence of chemical structure on thermal properties and surface morphology of polyurethane materials. *Chem. Pap.* **2018**, *72*, 1249–1256. [CrossRef] [PubMed]

8.   Brzeska, J.; Morawska, M.; Sikorska, W.; Tercjak, A.; Kowalczuk, M.; Rutkowska, M. Degradability of cross-linked polyurethanes based on synthetic polyhydroxybutyrate and modified with polylactide. *Chem. Pap.* **2017**, *71*, 2243–2251. [CrossRef] [PubMed]

9.   Kurcok, P.; Smiga, M.; Jedliński, Z. β-Butyrolactone polymerization initiated with tetrabutylammonium carboxylates: A novel approach to biomimetic polyester synthesis. *J. Polym. Sci. Part A* **2002**, *40*, 2184–2189. [CrossRef]

10.   Arslan, H.; Adamus, G.; Hazer, B.; Kowalczuk, M. Electrospray ionisation tandem mass spectrometry of poly[(R,S)-3-hydroxybutanoic acid] telechelics containing primary hydroxyl end groups. *Rapid Commun. Mass Spectrom.* **1999**, *13*, 2433–2438. [CrossRef]

11.   Szelest-Lewandowska, A.; Skupień, A.; Masiulanis, B. Syntezy i właściwości nowych poliuretanów dla medycyny. *Elastomery* **2002**, *6*, 3–14.

12.   Heller, N.W.M.; Clayton, C.R.; Giles, S.L.; Wynne, J.H.; Walker, M.E.; Mark, J.; Wytiaz, M.J. Delineating crosslink density gradients via in-situ solvation of immiscibly blended polyurethane thermosets. *Colloid Polym. Sci.* **2017**, *295*, 2019–2030. [CrossRef]

13.   Eldessoukia, M.; Buschle-Dillerc, G.; Gowayedd, Y. Solution-based synthesis of a four-arm star-shaped poly(L-lactide). *Des. Monomers Polym.* **2016**, *19*, 180–192. [CrossRef]

14.   Ghobashy, M.M.; Abdeen, Z.I. Radiation Crosslinking of Polyurethanes: Characterization by FTIR, TGA, SEM, XRD, and Raman Spectroscopy. *J. Polym.* **2016**. [CrossRef]

15.   Mahapatra, S.S.; Yadav, S.K.; Yoo, H.J.; Cho, J.W.; Park, J.-S. Highly branched polyurethane: Synthesis, characterization and effects of branching on dispersion of carbon nanotubes. *Compos. Part B* **2013**, *45*, 165–171. [CrossRef]

16.   Brzeska, J.; Dacko, P.; Janeczek, H.; Kowalczuk, M.; Janik, H.; Rutkowska, M. Wpływ syntetycznego polihydroksymaślanu na wybrane właściwości nowych, otrzymanych z jego udziałem, poliuretanów do zastosowań medycznych. Cz.II. Poliuretany z cykloalifatycznym diizocyjanianem w segmencie sztywnym. *Polimery* **2011**, *1*, 27–34.

MDPI
St. Alban-Anlage 66
4052 Basel
Switzerland
Tel. +41 61 683 77 34
Fax +41 61 302 89 18
www.mdpi.com

Polymers Editorial Office
E-mail: polymers@mdpi.com
www.mdpi.com/journal/polymers